Ecological Developmental Biology

Integrating Epigenetics, Medicine, and Evolution

Ecological Developmental Biology

Integrating Epigenetics, Medicine, and Evolution

Scott F. Gilbert
Swarthmore College

David Epel
Hopkins Marine Station, Stanford University

Sinauer Associates, Inc. • Publishers
Sunderland, Massachusetts U.S.A.

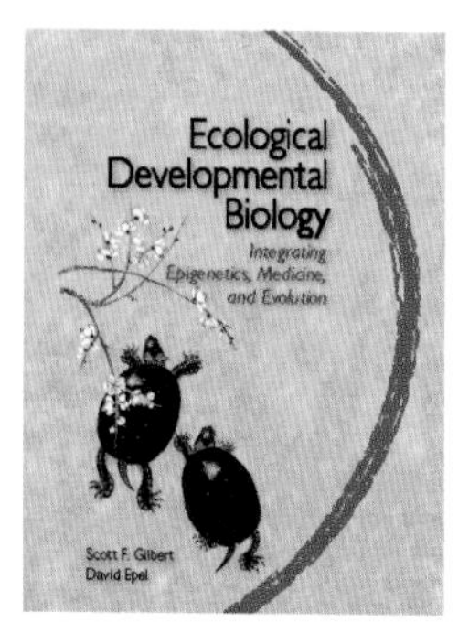

The Cover

"Turtles and Wild Plum" by David Carroll. The development of the spotted turtle *Clemmys guttata* is strongly influenced by environmental cues. The sex of the hatchlings is determined by the nest temperature in the middle of their incubation period, with higher temperatures inducing females. The development of the plum (its growth and flowering) is likewise regulated by environmental conditions, including temperature and photoperiod. (Original watercolor © David Carroll.)

Ecological Developmental Biology: Integrating Epigenetics, Medicine, and Evolution

Sinauer Associates Inc.
23 Plumtree Road
Sunderland, MA 01375 USA
FAX 413-549-1118
publish@sinauer.com, orders@sinauer.com

Visit our website at www.sinauer.com

Library of Congress Cataloging-in-Publication Data
Gilbert, Scott F., 1949-
Ecological developmental biology : integrating epigenetics, medicine, and evolution / Scott F. Gilbert, David Epel.
p. cm.
Includes bibliographical references and index.
ISBN 978-0-87893-299-3 (paperbound)
1. Phenotypic plasticity. 2. Developmental biology. 3. Epigenesis. 4. Evolution (Biology) I. Epel, David. II. Title.

QH438.5G55 2009
572.8′65—dc22 2008039921

Printed in China

6 5 4 3 2

To Anne

and

To Lois

Brief Contents

Contents

PART I Environmental Signals and Normal Development

CHAPTER 3 Developmental Symbiosis: Co-Development as a Strategy for Life 79

CHAPTER 4 Embryonic Defenses: Survival in a Hostile World 119

PART 2 Ecological Developmental Biology and Disease States

PART 3 Toward a Developmental Evolutionary Synthesis

Preface

There are more things in heaven and earth, Horatio, than are dreamt of in your philosophy.

HAMLET, ACT 1, SCENE V

One doesn't write a new book if one is satisfied with what exists. This book arises largely from our belief that there is a new field of biology that should have its own synthetic treatment in a book that can be read by students as well as by scientists. This field is ecological developmental biology, a realm that studies the developing organism in relation to its environment and that integrates the molecular and anatomical aspects of development with ecology, evolution, and medicine.

Ecological developmental biology studies those interactions where "embryology meets the real world" (Gilbert 2001). The realization that developmental biology was missing this critical ecological component was captured in the title of one of David Epel's papers, "Beakers versus breakers: How fertilisation in the laboratory differs from fertilisation in nature" (Mead and Epel 1995). The idea that development in nature can be importantly different from development in the laboratory is intrinsic to this book.

Ecological developmental biology is the intersection set of numerous disciplines. Indeed, when the first symposium on this subject was held in 2001, it included representatives from endocrine disruptor research, vertebrate behavior, insect physiology, plant ecology, developmental biology, biomechanics, conservation biology, evolutionary biology, and phenotypic plasticity (see Dusheck 2002; Gilbert and Bolker 2003). The symposium continued past the time allotted, and its success was cemented when the various scientists exchanged business cards. Since that time, ecological developmental biology has been seen as a harbinger of a new "integrative biology," synthesizing development, ecology, and evolution (Wake 2004). With this book, we hope to give a new voice to this discipline.

Ecological developmental biology provides a framework where the molecular biology of epigenetic development meets and interacts with the ecological and evolutionary aspects of phenotypic plasticity in an entirely new fashion. Through these interactions, ecological developmental biology provides the missing piece to many disciplines, not just to developmental biology. We have tried to show that ecological developmental biology provides new and important perspectives on ecology, genetics, evolution, conservation biology, cancer biology, aging research and medicine.

Indeed, in several of these disciplines, there is a growing dissatisfaction with the dominant genetic paradigm that had displaced and marginalized any other approach. In developmental biology, embryogenesis came to be seen solely as the processes of gene expression whereby the genes in a zygote's nucleus created the phenotype. Environmental factors influencing phenotype production and specificity had been ignored until very recently. In evolutionary biology, the population genetic approach marginalized notions of development, viewing evolution exclusively in terms of the changing frequency of different gene alleles within an interbreeding population. The same explanatory triad of gene mutations, translocations, and genetic drift was seen in the clinical microcosm to cause aging syndromes and cancers, and modern medicine became organized around the Human Genome Project and other genetic programs in an effort to determine the allelic bases for human disease.

Recently these disciplines have had their scientific critics, each calling for explanations broader than that of genetic programming. It takes more than a series of accumulated mutations to produce macroevolutionary diversity, and there is much more to the generation of cancer and the aging syndrome than the accumulation of mutations in somatic cells. One might be born with a predisposition to diabetes or hypertension not only from one's genes, but from one's mothers' diet. And can development be explained solely by the interactions of materials present in the fertilized egg? Fred Nijhout (1999), Scott Gilbert (2001), Lenny Moss (2003) and others have shown that this hypothesis is no longer tenable.

One cannot displace an existing paradigm without having something "better" (that is, more predictive, or explaining more observations) to replace it. There are currently several "revolutions" taking place in biology, each claiming to offer a better explanatory principle. One of these is the "epigenetics revolution," wherein inheritance is shown to involve the faithful transmission not only of gene nucleotide sequences, but also of gene expression patterns. Another revolutionary trend is the expansion of developmental biology into medicine, promoting developmental explanations for human diseases. A third revolution has been sparked by the synthesis of developmental and evolutionary biology ("evo-devo"), in which developmental biology and phenotypic plasticity are seen as driving forces in the creation of biodiversity. Indeed, numerous news and media reports have heralded a "modernization" or "paradigm shift" in evolutionary biology (e.g., Erwin 2007; Pennisi 2008). A fourth revolution concerns a changing mindset. Rather than analyzing independent "things," a new focus of developmental biology concerns "relationships." Nothing, it seems, exists except as part of a network of interactions. This new focus has resulted in the emergence of "systems biology," "integrative biology," and even new buildings designed to facilitate collaborative research and communication (Lawler 2008).

How much more of a revolution would it be if we could unite these four trends—epigenetics, the developmental origins of adult disease, evolutionary developmental biology, and systems biology—into a common cause without sacrificing the uniqueness of each revolutionary concept? We see each of these revolutions as different fronts of the same scientific enterprise, and the common front, the common shared feature of each of them is—developmental biology. Thus we emphasize developmental biology in this volume, and we try to show its importance to each of these ongoing revolutions. Development weaves genotype and environment into phenotype. As such, the causes for the origins of biodiversity, cancer, and the prevalent phenotypes of organisms and communities must have developmental explanations. So one of the major reasons for writing this book has been to put forth a fresh and challenging way of looking at biology, a perspective that uses development to integrate disciplines as disparate as ecology and chromatin dynamics and which should cause one to ask a different set of questions than before.

Another reason to write this book is that we think ecological developmental biology is critically needed to diagnose and perhaps help cure the problems of our planet. When we sent the chapters out for review, we found that we were not alone in this feeling. One reviewer summarized this need passionately, saying:

> The field of eco-devo, I think, is *the* most important field of science at the moment. I don't think I'm overstating the case. Ecology isn't prepared to analyze at the molecular level the ills of our present world, genetics doesn't contain the background in tissue interactions, and developmental biology has the tools but is only just now turning its attention to an environment outside of cells and the individual organism. Eco-devo is the synthesis that combines all of the above, and we and our students desperately need to have a basic understanding of this new field to become proper stewards of our planet.

The relationships between developing organisms and their environments is crucial for our world's future. We need to understand the former if we are to protect the latter.

This book is intended both for students and for our scientific colleagues. While it would help students to have had courses in developmental biology, cell biology, ecology, or evolution, a good first-year biology course should be adequate. It is also a book written for specialists who would like to learn something about how their particular subdiscipline might interact with other biological sciences. We hope that the examples presented here will reinforce the sense of wonder that biologists find in the world, and at the same time be a jumping-off point for discussions about the integration of different areas of biology, as well as for discussions of the increasingly critical question of biology's relation to public policy. While we have tried to be integrative, we realize that we are still bound by our past history and training. So we hope that college students, still relatively undifferentiated, will come up with their own connections and syntheses and that they will see patterns that we haven't yet imagined.

This book is a "first approximation," a "version 1.0." It is certainly not a definitive text, but rather a "snapshot" of a revolution in progress. Moreover, in order to keep this book focused, we have had to leave out two major areas of research that are related to and which interact with the material in this text. One of them concerns the plasticity of plant development. This is such an enormous topic (and so critical to plant development and physiology) that we could not do it justice here, although we have called attention to some examples of plant developmental plasticity when they are relevant to other studies discussed in the book. Second, we have not attempted to cover the immense and important body of studies being done on environmental input to mammalian postnatal brain development. Both of these fields have their own books; here we will focus on embryonic and larval development in animals.

We hope that this book will be used as a text upon which to base seminars wherein students can update the material with fresh studies from the current literature. We also hope that this integration of different areas of biology stimulates new research in all of them, and that it engenders new ways of thinking about these respective areas. Finally, we hope the ideas in this book evoke a way of approaching nature, an approach exemplified in the banner that hangs over the library of the Woods Hole Marine Biology Laboratory, reminding us that we should "Study Nature, Not Books." Good advice.

Acknowledgments

Both authors have written and tested these chapters with members of their respective undergraduate seminars and laboratories. Scott Gilbert would like to thank J. Braslow, M. Fisher, L. Gyi, F. Hussein, N. Oberfoell, J. Robertson, J. Sachs, I. Yarett B. Zee, and especially Lucy McNamara, who wrote the earliest draft of the chapter on developmental symbiosis. David Epel would especially like to thank his colleague Amro Hamdoun and his students G. Del Rey, T. Hoang, D. Hu, C. O'Neill, M. Stegner, R. Wable, and A. Wicklund.

These chapters have been read by several experts in their respective fields, including John Beatty, Greg Davis, John Gerhart, Mark Hanson, Eva Jablonka, Margaret McFall-Ngai, Sahotra Sarkar, Carlos Sonnenschein, Ana Soto, Richard Strathmann, and Mary Jane West-Eberhard. H. Fred Nijhout and Mary Stott Tyler reviewed the entire manuscript from the perspectives of expert investigator and expert teacher. In addition, the coda was gone over by philosophers of biology L. van Speybroeck, W. M. Johnson, L. Moss, and G. Auletta. We hope each of the reviewers agrees with our changes and that they enjoy the finished chapters as much (if not more) than they did when they first read them. Any errors, of course, are ours. In addition, there have been hidden environmental influences, including those of teachers and colleagues such as Donald P. Abbott, Bob Auerbach, N. J. Berrill, John Gerhart, Donna Haraway, Donal Manahan, Barbara Migeon, V. Nanjundiah, Tokindo Okada, Steve Palumbi, George Somero, Michael Somers, Fred Tauber, and Cor van der Weele.

Publishing a new type of textbook takes one into uncharted waters, and one could not be on a better ship than Sinauer Associates. We thank Andy Sinauer, David McIntyre, Chris Small, Lou Doucette, Elizabeth Morales, Janice Holabird, Marie Scavotto, Dean Scudder, and especially Carol Wigg, the "secret weapon" of Sinauer Associates, who over the years has done so much to make biology books that students can simultaneously understand and enjoy.

Patient and bemused spouses are part of the scientific tradition. Mrs. Darwin and Mrs. Haeckel had to endure their respective husbands' obsessions with nature and writing, and Maria Merian's husband had to deal with his wife's compulsion to study and paint plants and the caterpillars that ravaged them. Our spouses are in fields that look at the human side of biology, so we hope that they will understand and forgive our spending so much time on this project. It's done!

References

Dusheck, J. 2002. Evolutionary developmental biology: It's the ecology, stupid. *Nature* 481: 578–579.

Erwin, D. H. 2008. Darwin still rules, but some biologists dream of a paradigm shift. *New York Times*, June 26.

Gilbert, S. F. 2001. Ecological developmental biology: Developmental biology meets the real world. *Dev. Biol.* 233: 1–12.

Gilbert, S. F. and J. A. Bolker. 2003. Ecological developmental biology: Preface to the symposium. *Evol. Dev.* 5: 3–8.

Lawler, A. 2008. Steering Harvard toward collaborative science. *Science* 321: 190–192.

Mead, K. S. and D. Epel. 1995. Beakers versus breakers: How fertilisation in the laboratory differs from fertilisation in nature. *Zygote* 3: 95–99.

Moss, L. 2003. *What Genes Can't Do.* MIT Press, Cambridge, MA.

Nijhout, H. F. 1999. Control mechanisms of polyphenic development in insects. *BioScience* 49: 181–192.

Pennisi, E. 2008. Modernizing the modern synthesis. *Science* 321: 196–197.

Wake, M. H. 2004. Integrative biology: The nexus of development, ecology, and evolution. *Biol. Internatl.* 46: 1–18 (25 February 2008; www.iubs.org/test/bioint/46/bi46.htm).

PART I

Environmental Signals and Normal Development

Chapter 1

The Environment as a Normal Agent in Producing Phenotypes

A single genotype can produce many phenotypes, depending on many contingencies encountered during development. That is, phenotype is an outcome of a complex series of developmental processes that are influenced by environmental factors as well as genes.

H. F. Nijhout, 1999

Imagine a young aquatic organism developing in a particular pond. This organism has the ability to sense soluble biochemicals in the water—chemicals given off in the saliva or urine of its major predators. In the presence of these signals, the organism's pattern of development changes, resulting in a phenotype that is less likely to be eaten by its predators. In the presence of the dragonfly larvae that feed on them, tadpoles of the gray tree frogs *Hyla chrysoscelis* and *H. versicolor* develop bright red tails that deflect the predators' attention, and a set of trunk muscles that enables them to make "ice hockey turns" to escape being eaten (Figure 1.1A; McCollum and Van Buskirk 1996; Relyea 2003a).

Imagine an organism that develops different phenotypes depending on the season. *Nemoria arizonaria* larvae hatching on oak trees in the spring have a form that blends remarkably with young oak flowers ("catkins"). But caterpillars that hatch in the summer would be very conspicuous if they looked like the long-fallen oak flowers; thus the summer caterpillars resemble newly formed twigs (Figure 1.1B). Here, it is the larva's diet that deter-

(A)

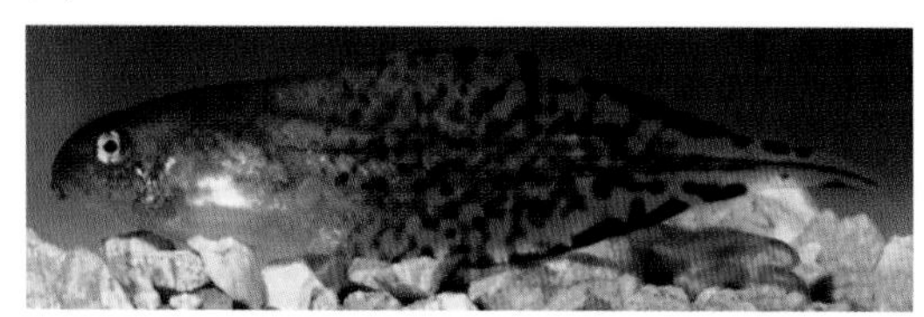

Predator present

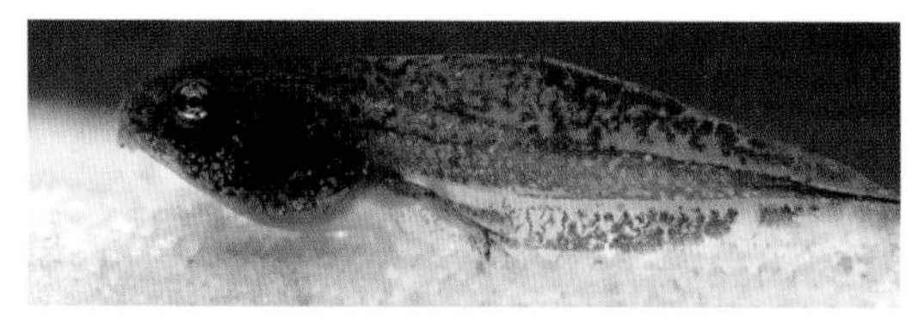

Predator absent

(B)

Spring morph among catkins

Summer morph on twig

(C)

◀ **FIGURE 1.1** Environmental cues can result in the development of completely different phenotypes in individuals of the same species. (A) Tadpoles of the tree frog *Hyla chrysoscelis* developing in the presence of cues from a predator's larvae (left) develop strong trunk muscles, and a red "warning" coloration. When predator cues are absent (right), the tadpoles grow longer and sleeker. (B) *Nemoria arizonaria* caterpillars that hatch in the spring (left) eat young oak leaves and develop a cuticle that resembles the oak's flowers (catkins). Caterpillars that hatch in the summer (right), after the catkins are gone, eat mature oak leaves and develop a cuticle that resembles young twigs. (C) A single male blue-headed wrasse (*Thalassoma bifasciatum*) swims with a cohort of the less colorful females. Should the male die, one of the females will grow testes, changing phenotype completely to become a male. (A courtesy of T. Johnson/USGS; B courtesy of E. Greene; C © Frederick R. McConnaughey/Photo Researchers, Inc.)

mines its phenotype. Larvae who feed on young oak leaves will look like the catkins, while larvae eating older leaves (which have a different chemical composition) will develop to resemble twigs (Greene 1989).

Next imagine an organism whose sex is determined not by its chromosomes, but by the environment the embryo experiences during a particular time during its development. In many species of fish, turtles, and alligators, sex is determined by the temperature of incubation. The same egg placed at one temperature will be male, but at another temperature will be female. The blue-headed wrasse (*Thalassoma bifasciatum*), a Caribbean reef fish, is one of several fish species whose sex depends on the other fish it encounters (Figure 1.1C). When an immature wrasse reaches a reef where a single male lives and defends a territory with many females, the newcomer develops into a female. If the same immature wrasse had reached a reef that was undefended by a male, it would have developed into a male (Warner 1984). If the territorial male dies, one of the females (usually the largest) becomes a male; within a day, its ovaries shrink and testes grow (Godwin et al. 2000, 2003).

Consider now an organism with a set of cells that can recognize and attack invading viruses and bacterial cells. It has billions of such immune cells, each of which will divide and produce antibodies only when it binds to a particular virus or bacterium. While its genetic repertoire allows this organism to form billions of different types of immune cells, the actual number of different antibody-producing cell types in a given individual is only a fraction of this potential and will depend on which bacteria and viruses infect that individual. This same organism has the ability to regulate its muscular phenotype such that continued physical stress on a particular muscle will cause that muscle to grow. Furthermore, the brain development of this organism can be altered by experience, making learning possible. Moreover, parts of its digestive system develop in response to the many different bacteria residing symbiotically in its gut. This species, in which so much of the phenotype is due to environmental circumstances, is *Homo sapiens*.

Plasticity Is a Normal Part of Development

In each of the above instances, the environment has profound effects on the animal's phenotype. In other words, everything one needs for phenotype production is *not* packaged in the fertilized egg. This ability of a single individual to develop into more than one phenotype has been called **phenotypic plasticity** (Nilsson-Ehle 1914) and was well known to late nineteenth-century embryologists, who showed that different environmental conditions produced different phenotypes during normal development (Nyhart 1995). Today we define phenotypic plasticity as the abil-

Bonellia viridis: When the Environment Determines Sex

Some early studies of the effect of the environment on development included the fascinating case of sex determination in echiuroid worms, specifically *Bonellia viridis* (the green spoonworm). Females of this marine species have a deep green, round body that burrows into rock crevasses and gravel on the seafloor, from which they extend a projection (proboscis) that can grow up to a meter long. Males of *B. viridis* have an amazingly different phenotype; indeed, the males are rarely seen, being colorless, only about 3 mm long, and living parasitically inside the female's genital sac, where their sole function is to produce sperm to fertilize the female's eggs.

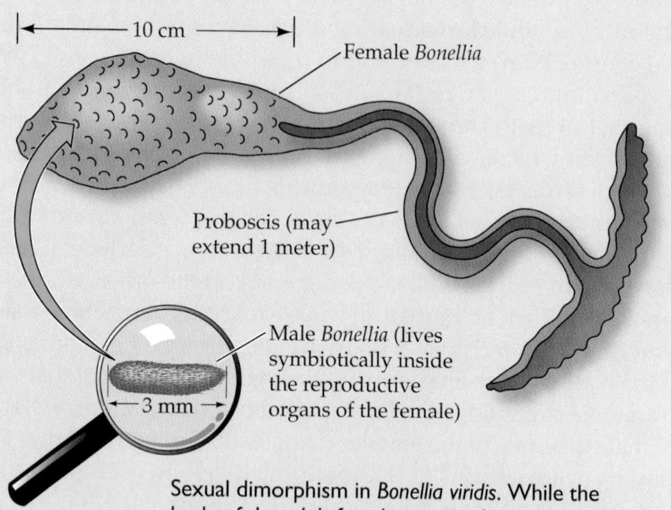

Sexual dimorphism in *Bonellia viridis*. While the body of the adult female remains buried in rocks or ocean sediments, her proboscis extends widely across the seafloor, where it is used for feeding. The proboscis also produces chemical signals that attract other *B. viridis* larvae, which, upon landing on the female, develop into males.

It has been known since the nineteenth century that the sex of a *B. viridis* individual depends solely on the environment in which the larva develops (Baltzer 1914; Leutert 1974). Fertilized *B. viridis* eggs are expelled into the seawater. Larvae that settle and develop on the seafloor become female, maturing over several years as the proboscis extends. The female's cells, especially those of its proboscis, generate a vivid green pigment called bonellin that is toxic to most other animals but is, along with other chemical signals emitted by the female, a powerful attractant to *B. viridis* larvae. Larvae passing within range of these signals will land on the proboscis of the sessile female and then crawl up into her mantle and/or be sucked into her gut, where they develop into the miniscule males.

ity of an organism to react to an environmental input with a change in form, state, movement, or rate of activity (West-Eberhard 2003). This plasticity is the property of the trait, not the individual; indeed, most individuals have several plastic traits. When seen in embryonic or larval stages of animals or plants, phenotypic plasticity is often referred to as **developmental plasticity**.

A century of studies

In his 1894 volume *The Biological Problem of Today: Preformation or Epigenesis?*, Oscar Hertwig summarized the studies demonstrating that development involved not only the interactions between embryonic cells, but also important interactions between developing organisms and their environments. He cited numerous cases of developmental plasticity, especially instances in which the sex of an organism was determined by the environment. These included the well-known case of *Bonellia* (box, left), as well as temperature-dependent sex determination in rotifers, nutrition-dependent production of workers and queens in ant colonies (see Figure 1.8), and temperature-dependent pigmentation patterns of butterfly wings (see Figure 1.5). Said Hertwig (1894, p. 122), "These seem to me to show how very different final results may grow from identical rudiments, if these, in their early stages of development, be subjected to different external influences."

In 1909, two publications brought the concept of phenotypic plasticity to the awareness of many biologists. The Danish biologist Wilhelm Johannsen's *Elemente der Exakten Erblichkeitslehre* made clear the distinction between the genotype and phenotype (Figure 1.2). Rejecting August Weismann's 1893 proposal that all the causes of an embryo's development were compressed into the nucleus of the egg, Johannsen stated specifically that

FIGURE 1.2 One hundred years ago, Wilhelm Johannsen noted that the phenotype is the product of both the genome and environmental circumstances. Here he writes on the board that *Anlaegspraeg* (genotype) + *Kaar* (Danish for "conditions" or "circumstances") gives *Fremtonigspraeg* (phenotype). (Photograph from a movie of Professor Johannsen at http://www.wjc.ku.dk/library/video/original.avi.)

phenotype (what the organism looks like and how it behaves) is not merely the expression or actualization of the genotype (the set of inherited genes), but rather depends on the interactions of inherited genes with components of the environment. Like Hertwig, Johannsen felt that early development was genetically controlled but that the environment could effect changes in the later developmental stages (Moss 2003; Roll-Hansen 2007). Johannsen also believed that Weismann's refutation of the inheritance of acquired characteristics had not been complete.

Another important paper published in 1909 was from the German biologist Richard Woltereck, who reported that pure (i.e., genetically identical) lines of *Daphnia* (a water flea that reproduces asexually) could produce different phenotypes during different times of the year (see Figure 1.15). Moreover, different pure lines responded differently. Woltereck argued that what actually was inherited was the *potential* to generate an almost infinite number of small variations in phenotype. He called this potential the ***Reaktionsnorm*** (**reaction norm** or **norm of reaction**).*

A contextually integrated view of life

The view expressed in the preceding examples is that the environment is not merely a filter that selects existing variations. Rather, it is a *source of variation*. The environment contains signals that can enable a developing organism to produce a phenotype that will increase its fitness in that particular environment. This isn't the view of life usually presented in today's textbooks or popular presentations of biology.

Since World War II, the dominant paradigm for explaining biodiversity has been genetics. Indeed, there has been marked antipathy against the notion of phenotypic plasticity and the inheritance of nonallelic phenotypic variation (Sarkar 2006; see Appendices A and C). Ernst Mayr, François Jacob, and numerous other influential biologists have emphasized the gene as being the core of animal identities and the "master molecule" of life. James Watson (1989) claimed that "We used to think our fate was in the stars. Now we know in large measure, our fate is in our genes." In his popular book *The Selfish Gene*, Richard Dawkins (1976) wrote of the genome as "the book of life" and proposed that our bodies are merely transient vehicles for the survival and propagation of our immortal DNA. In 1995, Nelkin and Lindee reviewed the scientific and popular accounts of DNA and con-

*Sarkar (1999) has pointed out that Woltereck actually concluded the *Reaktionsnorm* is what is inherited and that hereditary change consists of an alteration of the *Reaktionsnorm*. Woltereck identified the *Reaktionsnorm* with the genotype: "*Der 'Genotypus' … eines Quantitativmerkmals ist die vereberte Reaktionsnorm*" (Woltereck 1909, p. 136). Johannsen agreed, saying that Woltereck's *Reaktionsnorm* was "nearly synonymous" with his own conception of genotype (Johannsen 1911, p. 133).

cluded that DNA is being perceived as the secular equivalent of the soul. It is the essence of our being; it determines our behaviors, and it is that from which we can be resurrected after death (*vide* Jurassic Park). Richard Lewontin (1993) has also documented the dominance acquired by genetic determinism as an explanation of behavioral phenotypes.

But as we have seen (and will see much more of), genes are not the only explanation for animal diversity. During the past decade, interest in the environmental mechanisms of variation proposed early in the twentieth century has been renewed, fueled by new findings in conservation biology, developmental biology, public health, and evolutionary biology. The breakthroughs in molecular biology that have led to our exponentially expanded genomic knowledge also led to our current understanding of molecular signaling, casting a brilliant new light on the work of a century earlier. Just as in the 1600s the microscope revolutionized our view of life by revealing a previously invisible world, so in the twenty-first century technologies such as PCR (the polymerase chain reaction) and high-flowthrough RNA analysis have allowed us glimpses of a hitherto unsuspected world of interactions and interrelationships between genes and the environment. The result is a new perspective on life, its origins, and its interconnections.

"Eco-Devo" and Developmental Plasticity

Ecological developmental biology, casually known as **eco-devo**, is an approach to embryonic development that studies the interactions between a developing organism and its environment (Gilbert 2001; Sultan 2007). It focuses on how animals have evolved to integrate signals from the environment into their normal developmental trajectories. In many ways, ecological developmental biology is the extension of embryology to levels above that of the individual.

In standard embryology, the focus has always been on the internal dynamics through which the genes of an individual's cell nuclei produce the phenotype of the organism. Within the past century, we have discovered that cell-cell communication is key to this phenomenon. By itself, the genetic information in a cell's nucleus cannot directly produce the many differentiated cell types in a multicellular organism; cells must interact, reciprocally instructing each other as they differentiate. Molecular signals called **paracrine factors** are released by one set of cells and induce gene expression changes in the cells adjacent to them. These neighboring cells, with their newly acquired characteristics, then produce their own paracrine factors that can change the gene expression of *their* neighbors—sometimes including the cell that originally induced them! By such cooperative signaling between cells, organs are formed.

But as Paul Weiss (1970) and others hypothesized, such molecular signals are not limited to the internally generated paracrine factors, but can also come from sources outside the organism. Oscar Hertwig (1894), Curt Herbst (1901), and others catalogued these environmental agents and discussed them as normal components in determining the phenotype of the embryo. Thus the same genotype can generate different phenotypes depending on what cues are present in the environment, allowing the embryo to change its developmental trajectory in response to environmental input. Sonia Sultan (2007, p. 575) summarized the modern status of ecological developmental biology in a recent review:

> [E]cological developmental biology ("eco-devo") examines how organisms develop in "real-life" environments ... [and] aims to provide an integrated framework for investigating development in its ecological context. ... Eco-devo is not simply a repackaging of plasticity studies under a new name. ... Whereas plasticity studies draw on quantitative genetic and phenotypic selection analyses to examine developmental outcomes and their evolution as adaptive traits, eco-devo adds an explicit focus on the molecular and cellular mechanisms of environmental perception and gene regulation underlying these responses, and how these signaling pathways operate in genetically and/or ecologically distinct individuals, populations, communities, and taxa.

Developmental plasticity is usually adaptive—that is, it makes the organism more fit for its environment. However, as we will see in later chapters, there are times when plasticity is maladaptive—either when environmental cues alter development in a pathological manner, or when there is a mismatch between the phenotype induced by the embryonic environment and the environment experienced by the organism later in life. In both instances, developmental plasticity can give rise to disease, as will be discussed in Chapters 6 and 7.

In most developmental interactions, the genome provides specific instructions, while the environment is permissive. That is to say, the genes determine what structures get made, and the only requirement of the environment is that it support and not disturb the developmental processes. Dogs will generate dogs and cats will beget cats, even if the animals live in the same house. However, in most species, there are instances in development when the *environment* plays the instructive role and the genome is merely permissive. In these instances, the environment determines what type of structure is made—but the genetic repertoire has to be capable of building that structure. The genetic ability to respond to environmental factors has to be inherited, of course, but in these cases it is the environment that directs the formation of the specific phenotype (Sarkar 1998; Gilbert 2001; Jablonka and Lamb 2005).

Reaction norms and polyphenisms

Two main types of phenotypic plasticity are currently recognized: reaction norms and polyphenisms (Woltereck 1909; Schmalhausen 1949; Stearns et al. 1991; West-Eberhard 2003). As mentioned earlier, in a reaction norm, the genome encodes a *continuous range* of potential phenotypes, and the environment the individual encounters determines the phenotype. For instance, the length of the male's horn in some dung beetle species is determined by the quantity and quality of food (i.e., dung) the larva eats prior to metamorphosis. The upper and lower limits of a reaction norm (e.g., the size threshold a male larva must reach in order to produce horns and the longest and shortest possible horns in larvae that reach this threshold; see Figure 1.9B) are also a property of the genome that can be selected. Different species of beetle are expected to differ in the direction and amount of plasticity they are able to express (Gotthard and Nylin 1995; Via et al. 1995).

The second type of phenotypic plasticity, **polyphenism**, refers to *discontinuous* (either/or) phenotypes elicited by the environment. One obvious example is sex determination in turtles, where one range of temperatures induces female development in the embryo, while a different set of temperatures elicits male development. Between these two sets of temperatures is a small band of temperatures that will produce different proportions of males and females—but these intermediate temperatures do not induce intersexual animals.

An important example of polyphenism is seen in the migratory locust *Schistocerca gregaria*. These plant-eating grasshoppers exist in two mutually exclusive forms (Figure 1.3): they are either short-winged, uniformly colored, and solitary; or they are long-winged, brightly colored, and gregarious (Pener 1991; Rogers et al. 2003, 2004). The phenotypic differences between these two morphs are so striking that only in 1921 did the Russian biologist Boris Usarov finally realize they were the same species. Cues in the environment determine which morphology a young locust will develop. The major stimulus appears to be population density, as measured by the rubbing of legs. When locust nymphs get crowded enough that a certain neuron in the hind femur is stimulated by other nymphs, their development pattern changes, and the next time they molt they emerge with long, brightly colored wings and migratory behaviors.

The different phenotypes induced by the environment are sometimes called **morphs** or **ecomorphs**. Genetically identical animals can have different morphs depending on the season, their larval diet, or other signals present in the environment. Confusingly, the phenotypes produced by different genetic alleles are called either *mutants* (if rare) or *polymorphisms* (if common—arbitrarily defined as found in more than 5% of the population). Polymorphism is therefore a condition where variation is the product of genetic differences, while morphs are the result of environmental signals.

FIGURE 1.3 Density-induced polyphenism in the desert (or "plague") locust *Schistocerca gregaria.* (A) The low-density morph has green pigmentation and miniature wings. (B) Triggered by crowding, the high-density morph develops with deep pigmentation and wing and leg development suitable for migration. (From Tawfik et al. 1999, courtesy of S. Tanaka.)

Epigenetics

In 1968, Waddington coined the term *epigenetics* to describe a way of integrating the series of ordered interactions in development (epigenesis) with genetics (see Van Speybroeck 2002). Since then, epigenetics has been redefined as the set of mechanisms involved in regulating gene activity during development. **Epigenetics** is defined here as those genetic mechanisms that create phenotypic variation without altering the base-pair nucleotide sequence of the genes. Specifically, we use this term to refer to those mechanisms that cause variation by altering the *expression* of genes rather than their sequence.

As we will see in the next chapter, epigenetic mechanisms can integrate genomic and environmental inputs to generate the instructions for producing a particular phenotype. Epigenetic investigations have become focused on the mechanisms of phenotypic plasticity and on how changes in gene expression patterns mediated by the environment can cause diseases such as cancers and hypertension.

The term **epigenetic inheritance** has been used to denote heritable phenotypes that are not encoded in the genome (Jablonka and Lamb 2005; Jablonka and Raz 2008). Epigenetic inheritance includes:

- Variations inherited over *cell* generations, such as changes in chromatin that stabilize a particular cell type during normal development. For instance, during the course of their differentiation, mammalian liver precursor cells obtain a chromatin configuration that instructs their liver-specific genes to function, and henceforth all liver cells "remember" this chromatin configuration and maintain it over progressive cell divisions.

- Variations inherited from one *organismal* generation to the next. For instance, certain drugs can induce changes in the chromatin structure in the nuclei of mouse cells, including the mouse's germ cells. The progeny of such a mouse can inherit the drug-induced chromatin change from its parents, even if the drug is no longer present (see Chapter 7).

Among humans (and possibly among other animals; see Avital and Jablonka 2000), variations in cultural inheritance represent another inheritance pattern that is not mediated by changes in DNA.

Agents of developmental plasticity

Nearly every embryo probably has environmentally determined components in its phenotype. Therefore, a complete list of organisms with phenotypic plasticity would resemble a survey of all the eukaryotes on the Tree of Life. Examples of how the environment acts in normal development are given in Table 1.1, which shows some of the numerous environmental agents that contribute to producing normal phenotype, including:

- Temperature
- Nutrition
- Pressure and gravity
- Light
- The presence of dangerous conditions (predators or stress)
- The presence or absence of conspecifics (other members of the same species)

The remainder of this chapter describes how environmental cues affect the course of normal development in a variety of species. In subsequent chapters, more specific details about the mechanisms of developmental plasticity will be discussed.

Temperature-Dependent Phenotypes

Enzyme activity as a function of temperature

Nearly all enzyme activity is temperature-dependent. This concept is often expressed as the enzyme's Q_{10}, or the ratio of its activities at two temperatures, one 10°C higher than the other. Temperature can cause changes in the way a protein folds and thereby determine the shape of an enzyme's active site and the sites of interaction with other proteins. One example of such a protein is the tyrosinase enzyme variant found in Siamese cats and

TABLE 1.1 Some environmental contributors to phenotype development

CONTEXT-DEPENDENT NORMAL DEVELOPMENT (CHS. 1, 2)

A. Morphological polyphenisms
1. Nutrition-dependent (*Nemoria*, hymenoptera castes)
2. Temperature-dependent (*Arachnia*, *Bicyclus*)
3. Density-dependent (locusts)
4. Stress-dependent (*Scaphiopus*)

B. Sex determination polyphenisms
1. Location-dependent (*Bonellia*)
2. Temperature-dependent (*Menidia*, turtles)
3. Social-dependent (wrasses, gobys)

C. Predator-induced polyphenisms
1. Adaptive predator-avoidance morphologies (*Daphnia*, *Hyla*)
2. Adaptive immunological responses (mammals)
3. Adaptive reproductive allocations (ant colonies)

D. Stress-induced bone formation
1. Prenatal (fibular crest in birds)
2. Postnatal (patella in mammals; lower jaw in humans?)

CONTEXT-DEPENDENT LIFE CYCLE PROGRESSION (CHS. 2, 3)

A. Larval settlement
1. Substrate-induced metamorphosis (bivalves, gastropods)
2. Prey-induced metamorphosis (gastropods, chitons)
3. Temperature/photoperiod-dependent metamorphosis

B. Diapause
1. Overwintering in insects
2. Delayed implantation in mammals

C. Sexual/asexual progression
1. Temperature/photoperiod-induced (aphids, *Megoura*)
2. Temperature/colony-induced (*Volvox*)

D. Symbioses/parasitism
1. Blood meals (*Rhodnius*, *Aedes*)
2. Commensalism (*Euprymna*/*Vibrio*; eggs/algae; Mammalian gut microbiota)
3. Parasites (*Wollbachia*)

E. Developmental plant-insect interactions

ADAPTATIONS OF EMBRYOS AND LARVAE TO ENVIRONMENTS (CH. 4)

A. Egg protection
1. Sunscreens against radiation (*Rana*, sea urchins)
2. Plant-derived protection (*Utetheisa*)

B. Larval protection
1. Plant-derived protection (*Danaus*; tortoise beetles)

Source: After Gilbert 2001.

Note: This list should not be thought to be inclusive. For example, the list is limited to animals; plant developmental plasticity and many plant-animal interactions have not been included here.

Himalayan rabbits (Figure 1.4). Tyrosinase is critical for making melanin, the dark pigment of vertebrate skin. (Indeed, mutations that block melanin production result in albinism, the lack of dark pigment throughout the body.) The mutation that creates the phenotype of Siamese cats and Himalayan rabbits transforms tyrosinase from an enzyme that is not temperature-dependent (in the physiological ranges expected in an organism) into a temperature-dependent enzyme. In these animals, tyrosinase folds properly at relatively cold temperatures but does not fold properly—and thus does not work—at warmer temperatures. Cooler temperatures are normally found at the extremities (the tips of the ears, the paws and tail, and part of the snout), with warmer temperatures throughout the major

FIGURE 1.4 The dark pigment melanin is synthesized only in the colder areas of the vertebrate skin in Siamese cats and Himalayan rabbits. This is due to a mutation in the gene for tyrosinase—the rate-limiting enzyme of melanin synthesis—that renders the protein heat-sensitive. In the colder extremities of the body, tyrosinase folds properly and melanin (dark pigment) is produced. In warmer regions of the body, however, the enzyme folds improperly and cannot function, thus limiting the production of melanin. (Left, courtesy of Leslie and Ben Cook; right © Jim Strawser/Alamy.)

parts of the body.* Thus, tyrosinase functions (and melanin pigment is made) only in the extremities of Siamese cats and Himalayan rabbits, demonstrating that enzymes are affected by temperature and that their subsequent responses can have large impacts on phenotype.†

There are analogous conditions in humans in which only the hair at the extremities is pigmented (Berson et al. 2000). These conditions result from a single G → A mutation that replaces the positively charged amino acid arginine at position 402 of tyrosinase with the uncharged glutamine.

*Yes, this means that if you shave the back of a Himalayan rabbit and continually replace an ice pack on the shaven area, the hair will grow in black (see Schmalhausen 1949). In conditions such as hair pigmentation, where cells are continuously made and sloughed off during the animal's lifetime, the inducing conditions must be continuously present for the induced effect to occur. In other cases, such as turtle sex determination, the inducer need only be present for one period of time, and the decision is set from that time forth.

†Both Siamese and Burmese cats possess mutations of the tyrosinase gene, but they occur at slightly different sites within the gene. This apparently produces different thresholds for gene activity, allowing the Burmese breeds to have darker body color (Schmidt-Küntzel et al. 2005).

The induction of melanin pigment by the environment is part of our plastic response to the environment and is the basis for suntanning (D'Orazio et al. 2006; April and Barsh 2007). Plants also have an inducible system for melanin production. When certain fruits are cut, melanin is induced, creating a dark, protective meshwork that prevents bacterial and fungal penetration. This is why apples, potatoes, and bananas turn brown when sliced. (See Szent-Györgyi 1966 and Bachem et al. 2004 for discussion of the importance of this reaction.)

Seasonal polyphenism in butterflies

Since enzymes (and presumably other proteins, such as transcription factors) can be influenced by temperature, it is not surprising that animals have evolved such that thermal cues can cause different phenotypes at different seasons. Ecologists have long known that in North America, the pigmentation of many butterfly species follows a seasonal pattern. Throughout much of the Northern Hemisphere, one can see such a polyphenism in butterflies of the family Pieridae (the cabbage whites), with phenotypes that differ between individuals that eclose during the long days of summer and those that eclose at the beginning of the season, in the shorter, cooler days of spring. The hindwing pigments of the spring forms are darker than those of the summer butterflies (Figure 1.5). Pigmentation has a functional advantage during the cooler months: darker pigments absorb sunlight more effi-

FIGURE 1.5 Polyphenic variation in *Pontia* (Pieridae) butterflies. The top row shows summer morphs: *P. protodice* female (left) and male (center), *P. occidentalis* male (right). The bottom row shows spring morphs, which have a more highly pigmented ventral hindwing: *P. protodice* female (left) and male (center), *P. occidentalis* male (right). (Photograph courtesy of T. Valente.)

ciently than lighter ones, raising the body temperature more rapidly (Shapiro 1968; Watt 1968; see also Nijhout 1991). As we will see later, temperature may effect these changes in color by affecting the production of hormones needed for growth and differentiation.

Temperature and sex

Aristotle—a noteworthy naturalist and history's first embryologist—made few major errors in his embryological descriptions. One of these, however, was to attribute human sex determination to temperature (Aristotle 355 BCE). He felt that maleness was generated through the heat of the semen, and he encouraged elderly men to mate in the summertime if they desired male heirs.

Although Aristotle was wrong about temperature having a role in human sex determination, in many species, temperature *does* control whether an embryo develops testes or ovaries. Indeed, among certain reptile groups (turtles and crocodilians), there are many species in which the temperature at which an embryo develops determines whether an individual is male or female (Figure 1.6). This type of environmental sex determination, which also is found in certain fishes, has advantages and disadvantages.

One probable advantage is that it gives the species the benefits of sexual reproduction without tying the species to a 1:1 sex ratio. In crocodiles, in which temperature extremes produce females while moderate temperatures produce males, the sex ratio may be as great as 10 females to each male (Woodward and Murray 1993). In instances where the population size

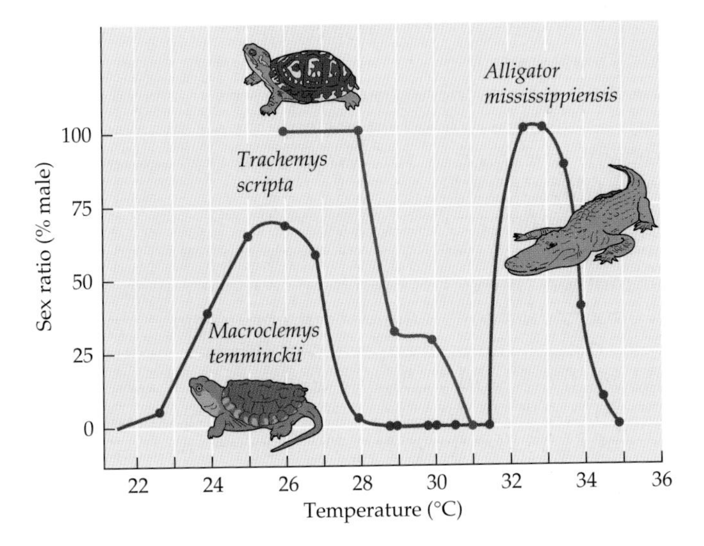

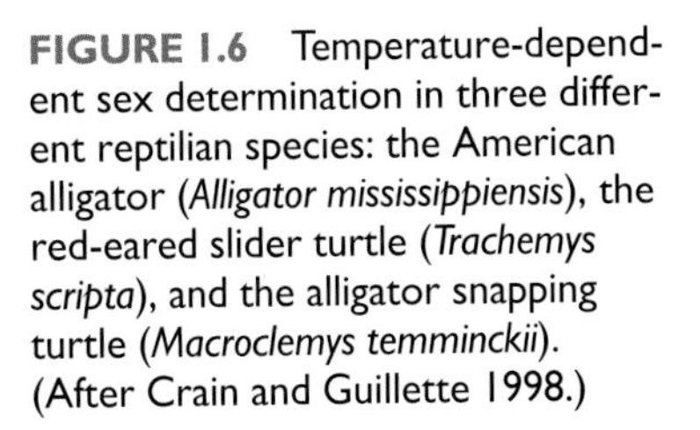
FIGURE 1.6 Temperature-dependent sex determination in three different reptilian species: the American alligator (*Alligator mississippiensis*), the red-eared slider turtle (*Trachemys scripta*), and the alligator snapping turtle (*Macroclemys temminckii*). (After Crain and Guillette 1998.)

is limited by the number of females, such a ratio is more advantageous than the 1:1 ratio usually resulting from genotypic sex determination.

The major disadvantage of temperature-dependent sex determination may involve its narrowing of the temperature range within which a species can persist. Thus, thermal pollution (either local or due to global warming) could conceivably eliminate a species from a given area (Janzen and Paukstis 1991). Researchers have speculated that dinosaurs may have had temperature-dependent sex determination, and that their sudden demise may have been the result of a slight change in temperature that created conditions wherein only males or only females hatched (Ferguson and Joanen 1982; Miller et al. 2004). Unlike turtles, which have long reproductive lives, can hibernate for years, and whose females can store sperm, dinosaurs may have had a relatively narrow window of time in which to reproduce and lacked the ability to hibernate through long stretches of bad times.

Charnov and Bull (1977) argued that environmental sex determination would be adaptive in those habitats characterized by patchiness—that is, habitats having some regions where it is more advantageous to be male and other regions where it is more advantageous to be female. Conover and Heins (1987) provided evidence for this hypothesis. In certain fish species, females benefit from being larger, since larger size translates into higher fecundity. If you are a female Atlantic silverside (*Menidia menidia*), it is advantageous to be born early in the breeding season, because you have a longer feeding season and thus can grow larger. (The size of males in this species doesn't influence mating success or outcomes.) In the southern range of *M. menidia*, females are indeed born early in the breeding season, and temperature appears to play a major role in this pattern. However, in the northern reaches of its range, the species shows no environmental sex determination. Rather, a 1:1 sex ratio is generated at all temperatures (Figure 1.7). Conover and Heins speculated that the more northern populations have such a short feeding season, there is no reproductive advantage for females in being born earlier. Thus, this fish has environmental sex determination in those regions where it is adaptive and genotypic sex determination in those regions where it is not.

In mammals, primary sex determination is controlled by chromosomes and not by hormones. This is important because we develop inside the hormonal milieu of our mothers. If the determination of mammalian gonads were accomplished through hormones, there would be no males. In mammals, two stages of sex determination have evolved. Primary sex determination is controlled by the X and Y chromosomes, which determine whether the gonads differentiate as ovaries or as testes. Secondary sex determination is accomplished by the hormones (testosterone, estrogen, and others) made by the gonads. This second stage is responsible for the male- and female-specific external genitalia, as well as for the differentia-

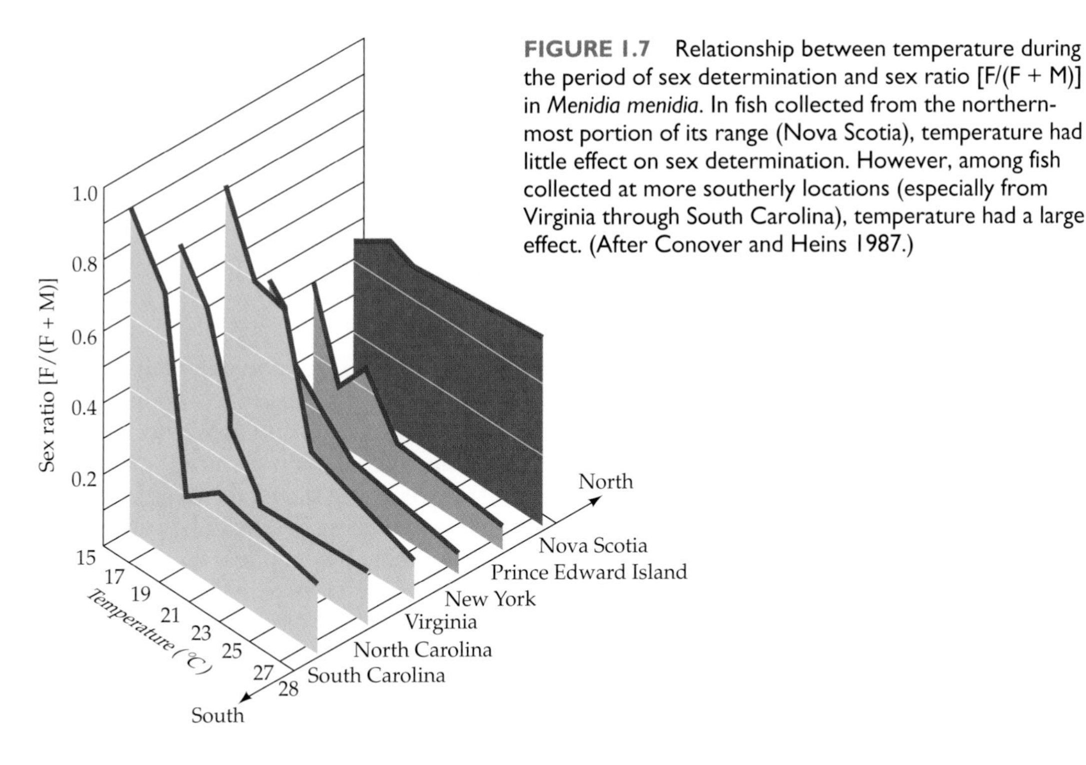

FIGURE 1.7 Relationship between temperature during the period of sex determination and sex ratio [F/(F + M)] in *Menidia menidia*. In fish collected from the northernmost portion of its range (Nova Scotia), temperature had little effect on sex determination. However, among fish collected at more southerly locations (especially from Virginia through South Carolina), temperature had a large effect. (After Conover and Heins 1987.)

tion of the uterus and oviducts in females and the development of the spermatic ducts in males.

In other vertebrates (including fishes, amphibians, and birds), the hormones estrogen and testosterone appear to be responsible for making ovaries or testes, respectively. The enzyme responsible for controlling the ratio of these hormones is **aromatase**, which converts testosterone into estrogen. Aromatase has been found to be temperature-regulated in several vertebrate species (Kroon et al. 2005). As we will see in later chapters, the enzyme is a target for environmental mutagens that can seriously alter the sexual development of a number of vertebrate species.*

*This sex-altering property of aromatase was useful, however, in an experiment that demonstrated the adaptive value of temperature-dependent sex determination. In the jacky dragon lizard (*Amphibolourus muricatus*), males are produced at intermediate temperatures (around 27°C), whereas both higher and lower temperatures produce females. By using aromatase inhibitors to block the conversion of testosterone to estrogen, Warner and Shine (2008) were able to produce males throughout the temperature range 23–33°C. There were no morphological differences among the males produced at any of these temperatures, but males produced at the intermediate temperatures had significantly better fitness (i.e., they sired more progeny) than the males produced at the extreme (normally female-producing) temperatures. The reason for this increased fitness as yet remains unknown.

Nutritional Polyphenism: What You Eat Becomes You

The food an organism eats may contain powerful chemical signals that induce phenotypic changes. We saw at the start of the chapter that the larval phenotype of the moth *Nemoria arizonaria* depends on its diet (see Figure 1.1B). Such effects are not uncommon among insects.

Royal jelly and egg-laying queens

In hymenopteran insects (bees, wasps, and ants), the determination of queen and worker castes can be effected by several factors, including genes, nutrition, temperature, and even volatile chemicals secreted by other members of the hive. In the honeybee, new queens are generated within 2 weeks after the death of the preceding queen (or in anticipation of the colony's splitting and a second queen being needed); they are almost never produced otherwise. Queen formation is dependent almost entirely on diet. A larva fed "royal jelly" (a protein-rich food that contains secretions from the workers' salivary glands) for most of its larval life will be a queen (with functional ovaries), while a larva fed a poorer diet (and given royal jelly for only a brief time late in larval development) will become a sterile worker.

Larvae can become queens if they reach a certain size before metamorphosis. A larva continually fed royal jelly from a relatively early stage retains the activity of a structure called the corpora allata throughout its larval stages. The corpora allata secretes juvenile hormone (JH), which delays metamorphosis, allowing the larva to grow larger and to have functional ovaries (Brian 1974, 1980; Plowright and Pendrel 1977). The rate of JH synthesis in the "queen larvae" is 25 times greater than the rate of synthesis in larvae not fed royal jelly. Applying large amounts of JH to worker larvae late in life can transform them into queens (Wirtz 1973; Rachinsky and Hartfelder 1990). Thus, the queen does not achieve her large and fertile status due to a genetic predisposition, but from nutritional supplementation.

Similarly, ant colonies are predominantly female, and the females can be very different in size and function. The much larger reproductive females ("gynes" or "queens") have functional ovaries; the workers do not (Figure 1.8). These striking differences in anatomy and physiology are also regulated through juvenile hormone (Wheeler 1991). The influence of the environment on hormone levels and gene expression in ants was analyzed by Abouheif and Wray (2002), who found that nutrition-induced JH levels regulated wing formation. In the queen, both the forewing and the hindwing disc undergo normal development, expressing the same genes as *Drosophila* wing discs. However, in the wing imaginal discs of workers, some of these genes remain unexpressed, and the wings fail to form.

FIGURE 1.8 Gyne (reproductive queen) and worker of the ant *Pheidologeton*. This picture shows the remarkable dimorphism between the large queen and the small worker (seen near the queen's antennae). The difference between these two sisters involves larval feeding and juvenile hormone synthesis. (Photograph © Mark W. Moffett/Minden Pictures.)

Horn length in the male dung beetle

The structural and behavioral male phenotypes of some male dung beetles depend on the quality and quantity of the nutrition—in the form of maternally provided dung—that they have access to during development (Emlen 1997; Moczek and Emlen 2000). In dung beetle species such as *Onthophagus taurus* and *O. acuminatus*, males have the ability to grow horns while females do not. The hornless female beetle gathers manure, digs tunnels, and places balls of dung in brood chambers that she constructs at the ends of the tunnels. She then lays a single egg on each cluster of dung, and when the larvae hatch, they eat the dung. Metamorphosis occurs when the dung cluster is consumed.

The amount of food affects the titer of juvenile hormone present during the developing beetle's last larval molt. In the males, the last organs to form are the horns. The size of the larva at metamorphosis determines the titer of JH, and the titer of JH affects the growth of the ectodermal regions that make the horns (Figure 1.9A; Emlen and Nijhout 1999; Moczek 2005). If juvenile hormone is added to a *Onthophagus* male larva during the sensitive period of his last molt, the cuticle in its head expands to produce a horn. The male horn does not grow unless the beetle larva reaches a certain size. Above this threshold body size, horn size is proportional to body size. Thus, although body size has a normal distribution, there is a bimodal distribution of horn sizes. About half the males (the small-bodied ones) have no horns, while the other half have horns of considerable length (Figure 1.9B).

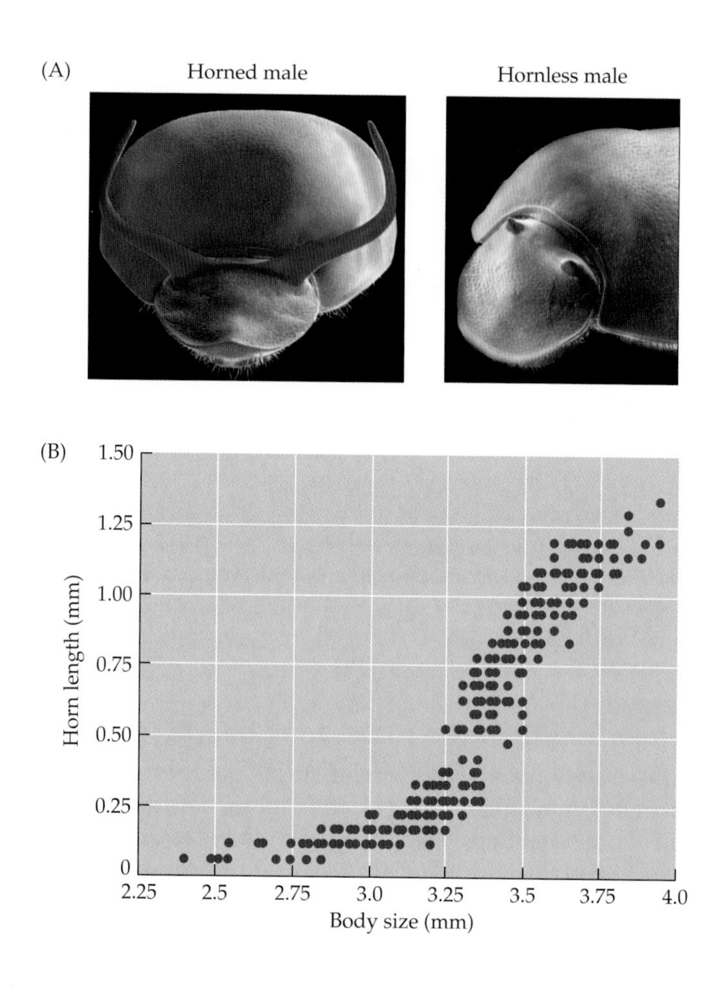

FIGURE 1.9 (A) Horned and hornless males of the dung beetle *Onthophagus acuminatus* (horns have been artificially colored). (B) Whether a male of this species is horned or hornless is determined by the titer of juvenile hormone at the last molt. This hormone titer depends in turn on the size of the larva. There is a sharp threshold of body size, before which horns fail to form and after which horn growth is linear with the size of the beetle. This threshold effect produces populations with no horns and with large horns, but very few with horns of intermediate size. (After Emlen 2000; photographs courtesy of D. Emlen.)

Horned males guard the females' tunnels and use their horns to prevent other males from mating with the females; the male with the biggest horns wins such contests. But what about the males with no horns? Hornless males do not fight with the horned males for mates. Since they, like the females, lack horns, they are able to dig their own tunnels. These "sneaker" males dig tunnels that intersect those of the females and mate while the horned males stand guard at the tunnel entrances (Figure 1.10; Emlen 2000; Moczek and Emlen 2000). Both strategies appear to be highly successful.

The heritability of horn length is zero; it is a phenotype that is environmentally determined by the response of the endocrine system to food intake. However, large horns also appear to correlate with reduced penis and testis size (Simmons and Emlen 2006; Parzer and Moczek 2008). This **trade-off** is probably due to altered allocation of resources during development, since experimentally ablating the male genital disc (from which the penis originates) results in a male with larger horns (Moczek and Nijhout 2004).

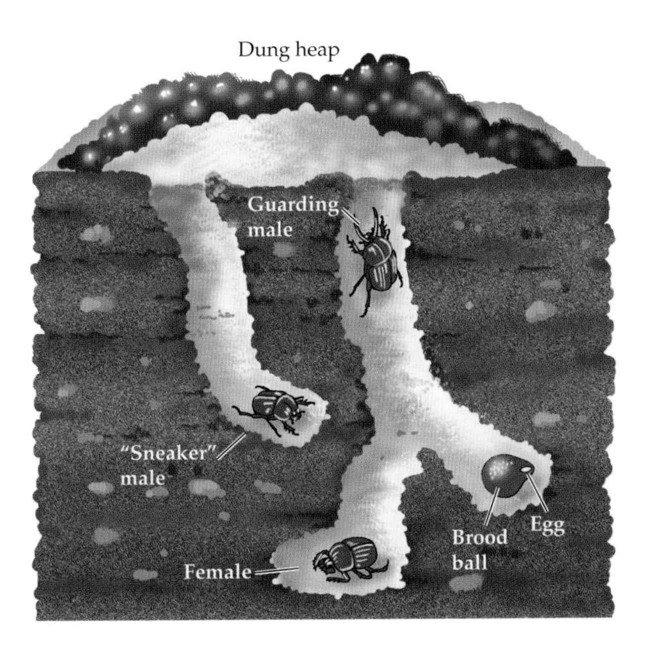

FIGURE 1.10 The presence or absence of horns determines the male reproductive strategy in some dung beetle species. Females dig tunnels in the soil beneath a pile of dung and bring dung fragments into the tunnels. These will be the food supply of the larvae. Horned males guard the entrances to the tunnels and mate repeatedly with the females. They fight to prevent other males from entering the tunnels, and those males with long horns usually win such contests. Smaller, hornless males do not guard tunnels. Rather, they dig their own tunnels to connect with those of females, mate, and exit. (After Emlen 2000.)

Gravity and Pressure

Embryologists have long appreciated the critical role that gravity plays in frog and chick body axis formation. For instance, if a frog egg is rotated during the first cell division cycle, the axes will not form correctly, or several axes will form—in which case the embryo will have more than one head. Bird eggs use the force of gravity to form their anterior-posterior (head-to-tail) axis (see Gilbert 2006).

More recent experiments have shown that the human body expects a 1-G gravitational field in order for bones and muscles to develop properly. Astronauts experiencing weightless conditions undergo severe muscle atrophy. As Figure 1.11 shows, weightlessness results in dramatic structural changes in the muscles, leading to tears and loss of strength and coordination. Spending 11 days in microgravity (without exercising) can cause a 30% shrinkage in the mass of certain muscles (NASA 2003). Several genes necessary for muscle differentiation and maintenance—including those genes encoding the transcription factors MyoD and myogenin—are not expressed in microgravity conditions (Inobe et al. 2002). Moreover, the genes encoding proteins that support mitochondria also fail to function without normal gravity (Nikawa et al. 2004).

In addition to muscle, the formation of several vertebrate bones is dependent on gravity (or on pressure from the environment). Such stresses are known to be responsible for the formation of the human patella

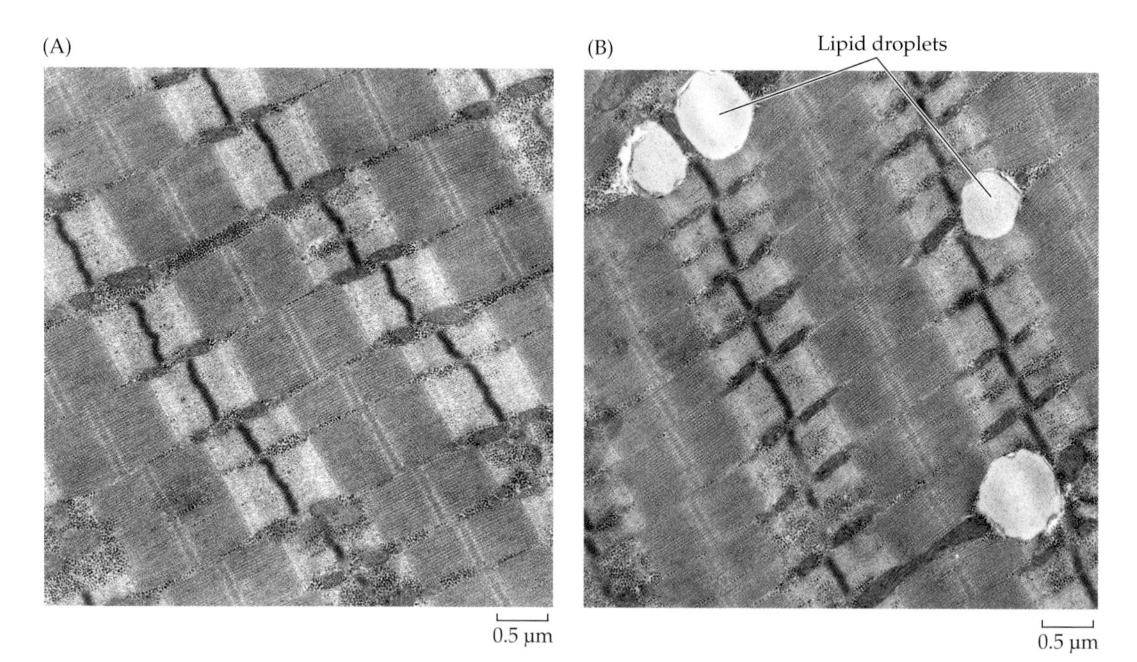

FIGURE 1.11 Human soleus muscle tissues, showing the effects of exercise in weightless conditions. (A) Exercised tissue, where gravitational load stimulates the production of proteins that keep muscle fibers strong. (B) After 17 days in microgravity, muscle protein synthesis has slowed down. The muscle cells have grown more irregular and show signs of atrophy. The prevalence of lipid droplets indicates that in microgravity, the muscle cells store fat instead of using it for energy. (From Widrick et al. 1999.)

(kneecap) after birth and have also been found to be critical for jaw growth in humans and fish. The jaws of cichlid fish differ enormously, depending on the food they eat (Figure 1.12; Meyer 1987). Similarly, normal human jaw development may be predicated on expected tension due to grinding food: mechanical tension appears to stimulate the expression of the *indian hedgehog* gene in mammalian mandibular cartilage, and this paracrine factor stimulates cartilage growth (Tang et al. 2004). If an infant monkey is given soft food, its lower jaw is smaller than normal. Corruccini and Beecher (1982, 1984) and Varela (1992) have shown that people in cultures where infants are fed hard food have jaws that "fit" better, and they speculate that soft infant food explains why so many children in Western societies need braces on their teeth. Indeed, the notion that mechanical tension can change jaw size and shape is the basis of the functional hypothesis of modern orthodontics (Moss 1962, 1997).

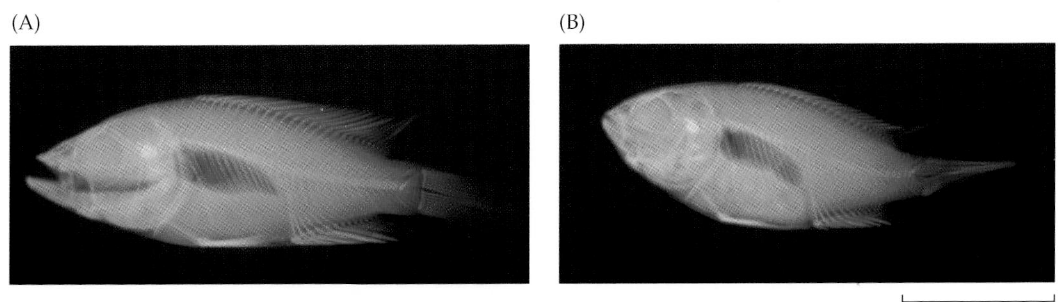

FIGURE 1.12 X-rays of cichlid fish fed different diets for 8 months. (A) Fish fed shrimp larvae developed an acute-angled jaw. (B) Fish fed commercial flaked food and nematodes had an obtuse jaw. (From Meyer 1987; photographs courtesy of Axel Meyer.)

In the chick, several bones do not form if the embryo's movement inside the egg is suppressed. One of these bones is the fibular crest, which connects the tibia directly to the fibula. This direct connection is believed to be important in the evolution of birds, and the fibular crest is a universal feature of the bird hindlimb (Müller and Steicher 1989). When the chick is prevented from moving within its egg, the fibular crest fails to develop (Figure 1.13; Wu et al. 2001; Müller 2003).

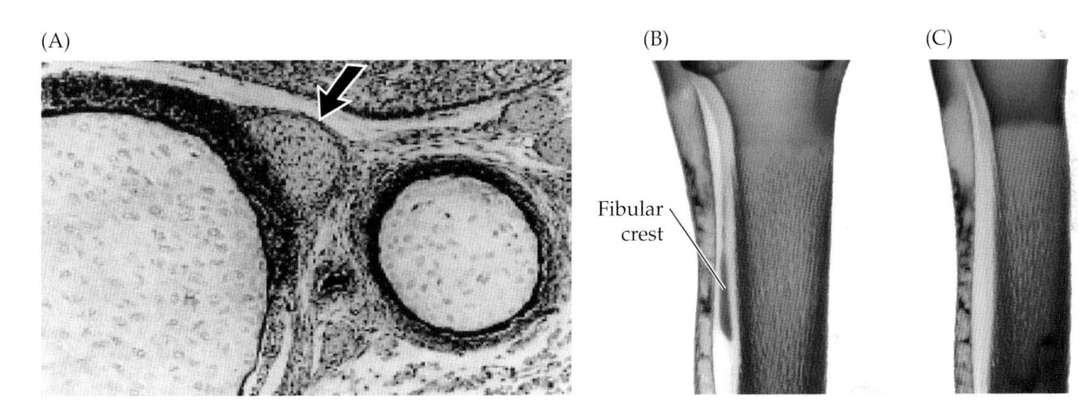

FIGURE 1.13 Activity-induced formation of the fibular crest. The fibular crest (syndesmosis tibiofibularis) is formed when the movement of the embryo in the egg puts stress on the tibia. (A) Transverse section through the 10-day embryonic chick limb, showing the condensation (arrow) that will become the fibular crest. (B) A 13-day chick embryo, showing fibular crest forming between the tibia and fibula. (C) Absence of fibular crest in the connective tissue of a 13-day embryo whose movement was inhibited. The blue dye stains cartilage; the red dye stains the bone elements. (From Müller 2003; photographs courtesy of G. Müller.)

Predator-Induced Polyphenisms

At the beginning of this chapter, we asked you to imagine an animal that is frequently confronted by a particular predator. One could then imagine an individual that could recognize soluble molecules secreted by that predator and that could use those molecules to activate the development of structures that would make this individual less likely to be eaten. This ability to modulate development in the presence of predators is called predator-induced defense, or **predator-induced polyphenism**.

To demonstrate predator-induced polyphenism, one has to show that the phenotypic modification is caused by the presence of the predator. In addition, many investigators also say that the modification should increase the fitness of its bearers when the predator is present (see Adler and Harvell 1990; Tollrian and Harvell 1999). Figure 1.14 shows both the typical and predator-induced morphs for several species. In each case, the induced morph is more successful at surviving the predator, and soluble filtrate

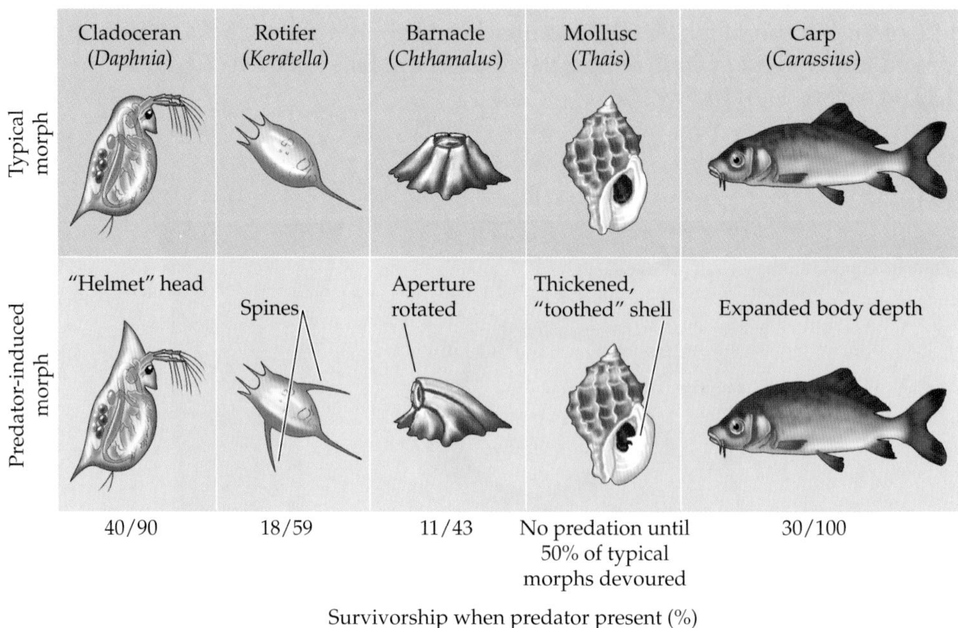

FIGURE 1.14 Predator-induced defenses. Typical (upper row) and predator-induced (lower row) morphs of various organisms are shown. The numbers beneath each column represent the percentages of organisms surviving predation when both induced and uninduced individuals were presented with predators (in various assays). (Data from Adler and Harvell 1990 and references cited therein.)

from water surrounding the predator is able to induce the changes. The chemicals that are released by a predator and that induce defenses in its prey are called **kairomones**.

One important concept to remember is that, as with the larger horns of male dung beetles mentioned earlier, there is usually a trade-off. That is, the energy and material used to produce the adaptation to the predator often come at the expense of making other organs or cells. Thus, what is adaptive in one environment is less adaptive in another. *Daphnia*, for instance, make a spiked "helmet" in the presence of kairomones (see Figure 1.15). However, helmeted *Daphnia* individuals produce fewer eggs than their smaller counterparts. Similarly, tadpoles that develop quickly in order to escape predators are usually less robust than those that take the full time developing (see Figure 1.16).

Predator-induced polyphenism in invertebrates

Several rotifer species will alter their morphology when they develop in pond water in which their predators were cultured (Dodson 1989; Adler and Harvell 1990). The predatory rotifer *Asplanchna* releases a soluble compound that induces the eggs of a prey rotifer species, *Keratella slacki*, to develop into individuals with slightly larger bodies and anterior spines 130% longer than they otherwise would be (see Figure 1.14), making the prey more difficult to eat. Also shown in Figure 1.14, the snail *Thais lamellosa* develops a thickened shell and a "tooth" in its aperture when exposed to water that once contained the crab species that preys on it. In a mixed snail population, crabs will not attack the thicker snails until more than half of the typical-morph snails are devoured (Palmer 1985).

The predator-induced polyphenism of the parthenogenetic water flea *Daphnia* is beneficial not only to itself, but also to its offspring. When *Daphnia cucullata* encounter the predatory larvae of the fly *Chaeoborus*, their heads grow to twice the normal size, becoming long and helmet-shaped (Figure 1.15). This increase in size lessens the chances that *Daphnia* will be eaten by the fly larvae. This same helmet induction occurs if the *Daphnia* are exposed to extracts of water in which the fly larvae had been swimming. Agrawal and colleagues (1999) have shown that the offspring of such induced *Daphnia* are born with this same altered head morphology. It is possible that the *Chaeoborus* kairomone regulates gene expression both in the adult and in the developing embryo.

Predator-induced polyphenism in vertebrates

Predator-induced polyphenism is not limited to invertebrates. Indeed, predator-induced polyphenisms are abundant among amphibians. Tadpoles found in ponds or reared in the presence of other species may differ

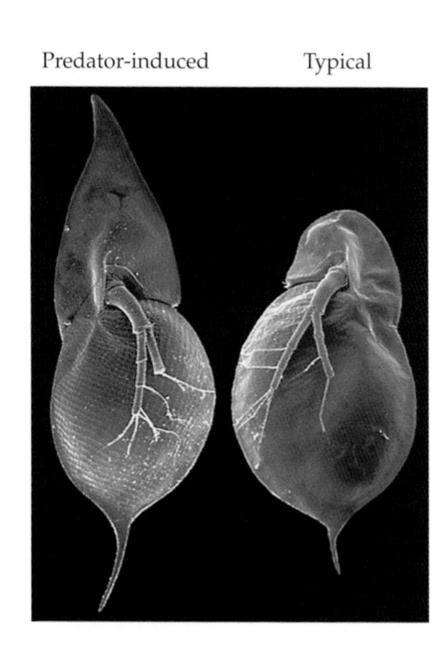

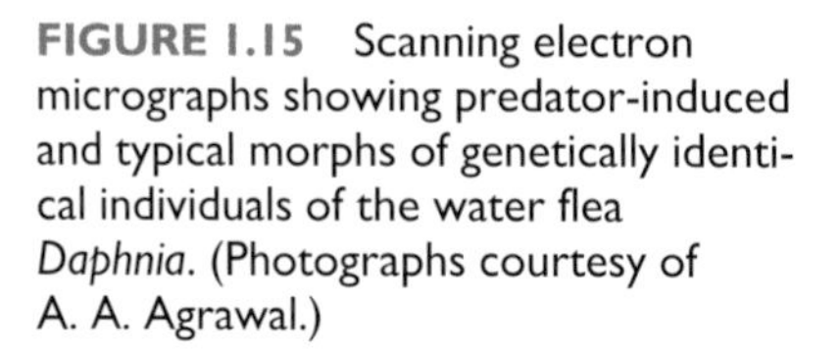
FIGURE 1.15 Scanning electron micrographs showing predator-induced and typical morphs of genetically identical individuals of the water flea *Daphnia*. (Photographs courtesy of A. A. Agrawal.)

significantly from tadpoles reared by themselves in aquaria. For instance, newly hatched wood frog tadpoles (*Rana sylvatica*) reared in tanks containing the predatory larval dragonfly *Anax* (confined in mesh cages so that they cannot eat the tadpoles) grow smaller than those reared in similar tanks without predators. Moreover, as with the *Hyla* species shown in Figure 1.1A, the wood frog tadpoles' tail musculature deepens, allowing faster turning and swimming speeds (Van Buskirk and Relyea 1998). The addition of more predators to the tank causes a continuously deeper tail fin and tail musculature, and in fact what initially appeared to be a polyphenism may be a reaction norm that responds to the number (and type) of predators. In some species, phenotypic plasticity is even reversible, and removing the predators can restore the original phenotype (Relyea 2003a).

Predator-induced defensive reactions in some other frogs involve responding to specific vibrational cues produced by predators. The embryos of the Costa Rican red-eyed tree frog (*Agalychnis callidryas*) use vibrations transmitted through the egg mass to escape egg-eating snakes. These egg masses are laid on leaves that overhang ponds. The embryos usually develop into tadpoles within 7 days, at which time the tadpoles wiggle out of the eggs and fall into the pond. However, when snakes attempt to feed on the frog eggs (Figure 1.16A), the vibrations from the snakes' movements cue any embryos remaining inside the eggs to begin the twitching movements that initiate their hatching (within seconds!), and they drop into the pond, escaping the snakes. Embryos are competent to begin these hatching movements as soon as day 5 (Figure 1.16B). Interest-

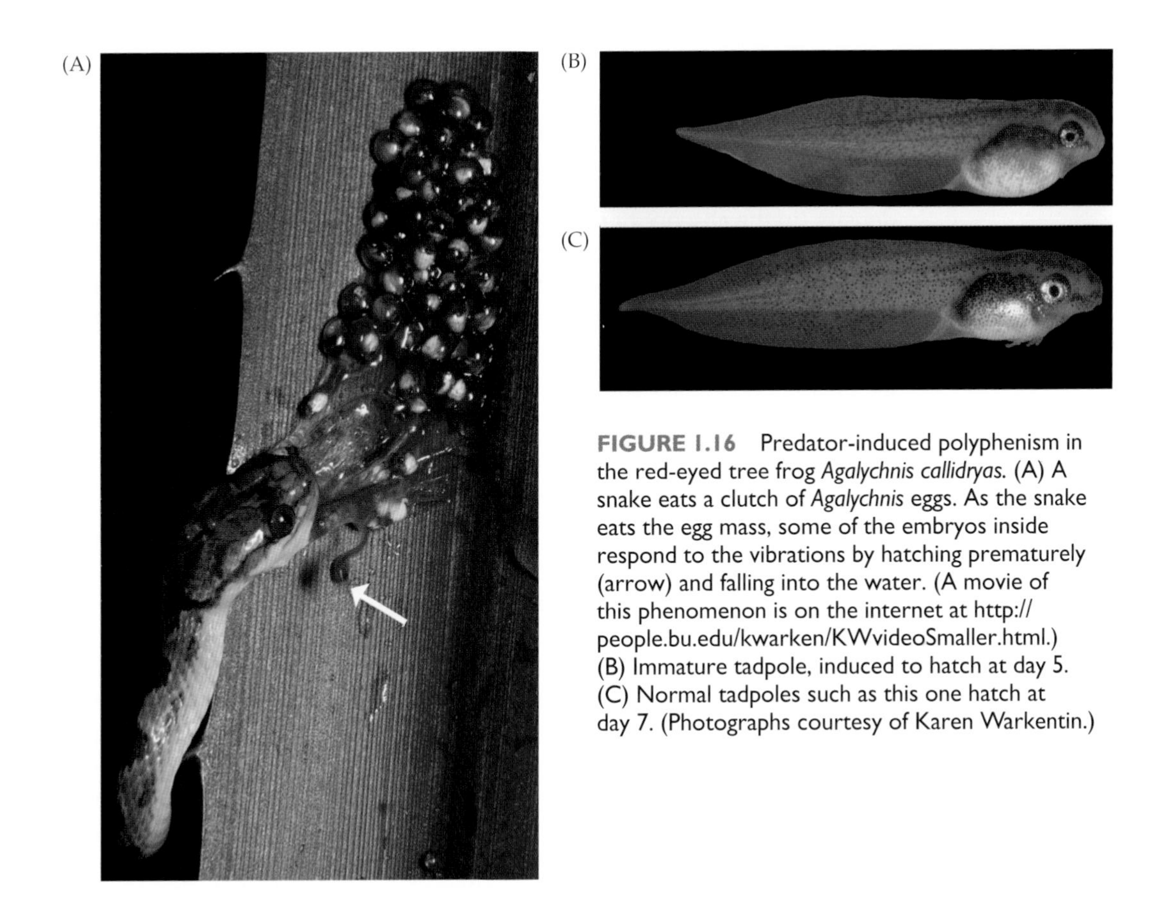

FIGURE 1.16 Predator-induced polyphenism in the red-eyed tree frog *Agalychnis callidryas.* (A) A snake eats a clutch of *Agalychnis* eggs. As the snake eats the egg mass, some of the embryos inside respond to the vibrations by hatching prematurely (arrow) and falling into the water. (A movie of this phenomenon is on the internet at http://people.bu.edu/kwarken/KWvideoSmaller.html.) (B) Immature tadpole, induced to hatch at day 5. (C) Normal tadpoles such as this one hatch at day 7. (Photographs courtesy of Karen Warkentin.)

ingly, the frog embryos respond this way only to vibrations given at a certain frequency and interval, and research has shown that these vibrations alone (and not sight or smell) cue the hatching movements (Warkentin 2005; Warkentin et al. 2006). Up to 80% of the remaining embryos can escape snake predation in this way. However, though these embryos escape their snake predators, they are at greater risk from water-borne predators than are "full-term" embryos, since the musculature of the early hatchers has not fully developed (Figure 1.16C).

The Presence of Conspecifics: It's Who You Know

Cues to change phenotype can come not only from predators but also from conspecifics (organisms of the same species); an individual in a large population can have a markedly different phenotype from that of an individual of the same species who is solitary. As in the gray tree frogs described at the beginning of this chapter, the presence of predators induces wood frog (*Rana sylvatica*) tadpoles to develop thicker trunk muscles and shorter bodies. In contrast, when the tadpoles are raised together in high population density (in the absence of predators), development is slowed down, resulting in shallower tails and longer bodies relative to those raised in isolation. Thus, predation results in the development of short, muscular tadpoles, while competition from conspecifics results in long, sleek tadpoles (Relyea 2004). A developing tadpole apparently integrates competing signals from predators and conspecifics to produce a body shape that will optimize its performance.

A swarm of locusts: Polyphenism through touch

Crowding among conspecifics can produce remarkably different phenotypes; this phenomenon is especially obvious in migratory desert locusts, *Schistocerca gregaria*. In this species, mechanoreceptors are responsible for the induction of the crowding phenotype. Locusts in the low-density "solitary" phase are usually green, have short wings and large abdomens, and avoid each other. However, when forced to crowd together (as in small areas of patchy food), they actively aggregate to form a high-density "gregarious" or "migratory" phase. Individuals change their behavior (from avoidance to attraction), change their color (from green to brown, black, and orange), and molt into adults with longer wings and more slender abdomens (see Figure 1.3). These profound changes in color and behavior can be accomplished by subjecting solitary locust nymphs to buffeting with small papier-mâchè balls (Roessingh et al. 1998).

The behavioral phase of the phenotypic change is mediated by direct physical contact among locusts, and the major sites of this mechanosensory input are the femurs of the hind legs. Repeatedly touching a minute region of the outer surface area of a hind femur with a fine paintbrush produces full behavioral gregarization within 4 hours (Simpson et al. 2001; Rogers et al. 2003). The colorization of the gregarious phenotype, however, may come from different cues. The smell of other locusts appears able to induce dark coloration in solitary nymphs by inducing the secretion of the neuropeptide hormone corazonin (also known as dark pigmentropin) (Lester et al. 2005).

The green coloration of the solitary stage blends in with the background, making the grasshopper harder for predators to see. In contrast, the black-and-orange pattern on the gregarious locusts functions as a warning, telling potential predators that these nymphs have been feeding on toxicity-conferring plants. Moreover, gregariousness is thought to enhance the efficiency of this protective coloration. Thus the cryptic (hiding) coloration of the solitary morph and the aposematic (warning) coloration of the gregarious morph both serve as predator-avoidance strategies (Sword et al. 2000).

Polyphenisms and Conservation Biology

Phenotypic plasticity means that animals in the wild may develop differently than those in the laboratory. Embryos and larvae in the wild develop in the presence of particular plants, predators, and conspecifics, and they experience variations of temperature and day length. In contrast, animals developing in a laboratory are usually grown in a monoculture of conspecific organisms, under a single particular temperature regime. This has important consequences when we apply knowledge gained in the lab to a field science such as conservation biology.

For instance, the metabolism of predator-induced morphs may differ significantly from that of the uninduced morphs, and this phenomenon has important consequences. Relyea (2003b, 2004) has found that in the presence of the chemical cues emitted by predators, pesticides such as carbaryl (Sevin®) can become up to 46 times more lethal than they are without the predator cues. Bullfrog and green frog tadpoles were especially sensitive to carbaryl when they were exposed simultaneously to predator chemicals. Relyea (2003b) has related these findings to the global decline of amphibian populations, saying that governments should test the toxicity of the chemicals under natural conditions, including that of predator stress. He concluded that "ignoring the relevant ecology can cause incorrect estimates of a pesticide's lethality in nature, yet it is the lethality of pesticides under natural conditions that is of utmost interest."

Temperature-dependent polyphenisms are also important for conservation biology and are likely to become more so with global warming. The significance of thermal polyphenisms was highlighted by Morreale and colleagues (1982) in a paper documenting temperature-dependent sex determination in a range of sea turtle species. Prior to that time, conservation biologists interested in restoring sea turtle populations had been growing the eggs in laboratory incubators set at a certain temperature, or culturing them in a single area of a beach. But these practices result in turtles of only one sex. Thus, Morreale and colleagues concluded that "current practices threaten conservation of sea turtles" rather than enhance them. They suggested protecting existing nests from predators, thereby maintaining the normal sex ratio.

One of the most interesting types of polyphenisms involves larval cues for metamorphosis. Environmental cues are critical to metamorphosis in many species, and some of the best-studied examples are the settlement cues used by marine larvae. A free-swimming marine larva often needs to settle near a source of food or on a firm substrate on which it can metamorphose. If prey, conspecifics, or substrates give off soluble molecules, these molecules can be used by the larvae as cues to settle and begin metamorphosis. Among the molluscs, there are often very specific cues for settlement (Hadfield 1977). In some cases, the prey supply the cues, while in other cases the substrate itself gives off molecules used by the larvae to initiate settlement. These cues may not be constant, but they need to be part of the environment if normal development is to occur (Pechenik et al. 1998).

The importance of substrates for larval settlement and metamorphosis was demonstrated in 1880 when William Keith Brooks, an embryologist at Johns Hopkins University, was asked to help the ailing oyster industry of Chesapeake Bay. For decades, oysters had been dredged from the bay, and there had always been a new crop to take their place. But suddenly, each year brought fewer oysters. What was responsible for the decline? Experimenting with larval oysters, Brooks discovered that the American oyster (*Crassostrea virginica*) requires a hard substrate on which to metamorphose. For years, oystermen had simply thrown the molluscs' shells back into the water, but with the advent of suburban sidewalks, they began selling the shells to cement factories. Brooks's solution: go back to returning the oyster shells to the bay. The oyster population responded, and Baltimore wharves still sell their descendants.*

Knowledge of phenotypic plasticity is critical in conservation biology and is necessary for making informed decisions that will benefit the environment. We will revisit this theme several times in this book.

* Sadly, Brooks's victory was not complete. Although his discovery restored the Chesapeake Bay oyster population, others sought to exploit it commercially. As an outcome of ill-tempered negotiations among scientists, state legislators, private dredging firms, and oystermen, the bay's oysters were dramatically overharvested. Brooks was politically naive and thought that science would easily sway the private interests. His is one of the first cases of scientific data coming into conflict with economic interests and losing (see Keiner 1998).

Sexual polyphenism by the community environment

Fish of many species change their sex based on social interactions; the blue-headed wrasse described at the start of the chapter is one good example. Marine gobys are among the few fishes that can change their sex more than once—and in either direction. A female goby can become male if the male of the group dies. However, if a larger male enters the group, such males revert to being female (Black et al. 2005). Grober and Sunobe (1996) induced females to become males, males to become females, and females to become males and then females again, merely by changing their companions. A goby can change its sex in about 4 days. In both gobys and the blue-headed wrasse, the shift of sex is mediated by hormonal changes caused by the environmental conditions. These endocrine changes include alterations in the glucocorticoid stress hormones that activate neuropeptides in the fishes' hypothalamus (Godwin et al. 2000, 2003; Perry and Grober 2003) as well as alterations in the ratio of testosterone to estrogen (Kroon et al 2005).

Convergence on Favorable Phenotypes

One principle of environmentally induced polyphenisms worth stressing is the fact that different environmental cues can produce the same favorable phenotype. For instance, the helmet-and-spiked-tailed *Daphnia* morph can be induced by different predators, and the chemical signals eliciting this phenotype are probably different. The water conditioned by dragonfly larvae can induce this phenotype, but so can chemicals released by dead *Daphnia* individuals being digested inside a fish's gut (Stabell et al. 2003). Similarly, the hatch-early-into-the-pond behavior of the red-eyed tree frog can be induced not only by snake vibrations, but also by wasp predation and by fungal infection (Warkentin 2000; Warkentin et al. 2001). The gregarious phase change of desert locusts can be induced by mechanostimulation of the hindlimb neurons described on page 30, but can also take place in the presence of a combination of visual and olfactory stimuli. In these developing locusts, either cause will induce a rise in the levels of serotonin in the thoracic ganglia of solitary individuals and subsequent components of the phase change. Drugs that block the action or synthesis of serotonin will prevent the phase change in both cases (S. J. Simpson, pers. comm.).

Summary

Phenotype is not just the expression of one's inherited genome. Rather, there are interactions between an organism's genotype and its environment that elicit a particular phenotype from a genetic repertoire of possible phenotypes. Environmental factors such as temperature, diet, physical stress, the presence of predators, and crowding can generate a phenotype that is suited for that particular environment. Environment is therefore considered to play a role in the generation of phenotypes, in addition to its well-established role in the natural selection of which phenotypes will survive and reproduce. Thus, in addition to helping decide the survival of the fittest, the environment is also important in formulating the arrival of the fittest.

References

Abouheif, E. and G. A. Wray. 2002. Evolution of the gene network underlying wing polyphenism in ants. *Science* 297: 249–252.

Adler, F. R. and C. D. Harvell. 1990. Inducible defenses, phenotypic variability, and biotic environments. *Trends Ecol. Evol.* 5: 407–410.

Agrawal, A. A., C. Laforsch and R. Tollrian. 1999. Transgenerational induction of defenses in animals and plants. *Nature* 401: 60–63.

April, C. S. and G. S. Barsh. 2007. Distinct pigmentary and melanocortin-1 receptor-induced components of cutaneous defense against ultraviolet radiation. *PLoS Genetics* 3(1): e9.

Aristotle. 355 BCE. *Generation of Animals.* 737a 27-29; 766B 29-767A 2.

Avital, E. and E. Jablonka. 2000. *Animal Traditions: Behavioural Inheritance in Evolution.* Cambridge University Press, Cambridge.

Bachem, C. W. B, G.-J. Speckmann, P. C. G. van der Linde, F. T. M. Verheggen, M. D. Hunt, J. C. Steffens and M. Zabeau. 2004. Antisense expression of polyphenol oxidase genes inhibits enzymatic browning in potato tubers. *Biotechnology* 12: 1101–1105.

Baltzer, F. 1914. Die Bestimmung und der Dimorphismus des Geschlechtes bei *Bonellia. Sber. Phys.-Med. Ges. Würzb.* 43: 1–4.

Berson, J. F., D. W. Frank, P. A. Calvo, B. M. Bieler and M. S. Marks. 2000. A common temperature-sensitive allelic form of human tyrosinase is retained in the endoplasmic reticulum at the nonpermissive temperature. *J. Biol. Chem.* 275: 12281–12289.

Black, J. E., K. R. Issacs, B. J. Anderson, A. A. Alcantara and W. T. Greenough. 1990. Learning causes synaptogenesis, whereas motor activity causes angiogenesis, in cerebellar cortex of adult rats. *Proc. Natl. Acad. Sci. USA* 87: 5568–5572.

Black, M.P., T.B Moore, A.V.M. Canario, D. Ford, R.H. Reavis and M.S. Grober. 2005. Reproduction in context: Field-testing a lab model of socially controlled sex change in *Lythrypnus dalli. J. Exp. Marine Biol. Ecol.* 318: 127–143.

Bradshaw, A. D. 1965. Evolutionary significance of phenotypic plasticity in plants. *Adv. Genet.* 13: 115–155.

Brian, M. V. 1974. Caste differentiation in *Myrmica rubra:* The role of hormones. *J. Insect Physiol.* 20: 1351–1365.

Brian, M. V. 1980. Social control over sex and caste in bees, wasps, and ants. *Biol. Rev.* 55: 379–415.

Brooks, W. K. 1880. *Development of the American Oyster: Report of the Commission of Fisheries of Maryland.* Inglehart, Annapolis, MD.

Charnov, E. L. and J. J. Bull. 1977. When is sex environmentally determined? *Nature* 266: 828–830.

Conover, D. O. and S. W. Heins. 1987. Adaptive variation in environmental and genetic sex determination in a fish. *Nature* 326: 496–498.

Corruccini, R. S. and C. L. Beecher. 1982. Occlusal variation related to soft diet in a nonhuman primate. *Science* 218: 74–76.

Corruccini, R. S. and C. L. Beecher. 1984. Occlusofacial morphological integration lowered in baboons raised on soft diet. *J. Craniofacial Genet. Dev. Biol.* 4: 135–142.

Crain, D. A. and L. L. Guillette, Jr. 1998. Reptiles as models of contaminant-induced endocrine disruption. *Anim. Reprod. Sci.* 53: 77–86.

Dawkins, R. 1976. *The Selfish Gene.* Oxford University Press, New York.

Denno, R. F., L. W. Douglass and D. Jacobs. 1985. Crowding and host plant nutrition: Environmental determinants of wing form in *Prokelisia marginata. Ecology* 66: 1588–1596.

Dodson, S. 1989. Predator-induced reaction norms. *BioScience* 39: 447–452.

D'Orazio J. A. and 11 others. 2006. Topical drug rescue strategy and skin protection based on the role of Mc1r in UV-induced tanning. *Nature* 443: 340–344.

Emlen, D. J. 1997. Alternative reproductive tactics and male dimorphism in the horned beetle *Onthophagus acuminatus* (Coleoptera: Scarabaeidae). *Behav. Ecol. Sociobiol.* 141: 335–341.

Emlen, D. J. and H. F. Nijhout. 1999. Hormonal control of male horn length dimorphism in the horned beetle *Onthophagus taurus. J. Insect Physiol.* 45: 45–53.

Emlen, D. J. 2000. Integrating development with evolution: A case study with beetle horns. *BioScience* 50: 403–418.

Ferguson, M. W. J. and T. Joanen. 1982. Temperature of egg incubation determines sex in *Alligator mississippiensis. Nature* 296: 850–853.

Gilbert, S. F. 2001. Ecological developmental biology: Developmental biology meets the real world. *Dev. Biol.* 233: 1–12.

Gilbert, S. F. 2004. Mechanisms for the environmental regulation of gene expression. *Birth Defects Research: Embryo* 72: 291–299.

Gilbert, S. F. 2006. *Developmental Biology*, 8th Ed. Sinauer Associates, Sunderland, MA.

Godwin, J., J. A. Luckenbach and R. J. Borski. 2003. Ecology meets endocrinology: Environmental sex determination in fishes. *Evol. Dev.* 5: 40–49.

Godwin, J., R. Sawby, R. R. Warner, D. Crews and M. S. Grober. 2000. Hypothalamic arginine vasotocin mRNA abundance variation across sexes and with sex change in a coral reef fish. *Brain Behav. Evol.* 55: 77–84.

Gotthard, K. and S. Nylin. 1995. Adaptive plasticity and plasticity as an adaptation: A selective review of plasticity in animal morphology and life history. *Oikos* 74: 3–17.

Greene, E. 1989. A diet-induced developmental polymorphism in a caterpillar. *Science* 243: 643–646.

Grober, M. S. and T. Sunobe. 1996. Serial adult sex change involves rapid and reversible changes in forebrain neurochemistry. *Neuroreport* 7: 2945–2949.

Hadfield, M. G. 1977. Metamorphosis in marine molluscan larvae: An analysis of stimulus and response. In R.-S. Chia and M. E. Rice (eds.), *Settlement and Metamorphosis of Marine Invertebrate Larvae*. Elsevier, New York, pp. 165–175.

Herbst, C. 1901. Formative Reize in der tierischen Ontogenese, Ein Beitrag zum Verständnis der tierischen Embryonalentwicklung. Arthur Georgi, Leipzig. Quoted in J. M. Oppenheimer. 1991. Curt Herbst's contribution to the concept of embryonic induction. In S. F. Gilbert (ed.), *A Conceptual History of Modern Embryology*. Plenum, New York, pp. 63–89.

Hertwig, O. 1894. *Zeit- und Streitfragen der Biologie* I. *Präformation oder Epigenese?* Grundzüge einer Entwicklungstheorie der Organismen. Gustav Fischer, Jena. Translated as *The Biological Problem of Today: Preformation or Epigenesis?* P. C. Mitchell (trans.). Macmillan, New York.

Inobe, M., I. Inobe, G. R. Adams, K. M. Baldwin and S. Takeda. 2002. Effects of microgravity on myogenic factor expressions during postnatal development of rat skeletal muscle. *J Appl. Physiol.* 92: 1936–1942.

Jablonka, E. and M. J. Lamb. 2005. *Evolution in Four Dimensions: Genetic, Epigenetic, Behavioral, and Symbolic Variation in the History of Life*. MIT Press, Cambridge, MA.

Jablonka, E. and G. Raz. 2008. Transgenerational epigenetic inheritance: Prevalence, mechanisms, and implications for the study of heredity. *Q. Rev. Biol.* In press.

Jacob, F. 1976. *The Logic of Life: A History of Heredity*. B. E. Spillman (trans.). Vintage Books, New York, p. 224.

Janzen, F. J. and G. L. Paukstis. 1991. Environmental sex determination in reptiles: Ecology, evolution, and experimental design. *Q. Rev. Biol.* 66: 149–179.

Johannsen, W. 1909. *Elemente der Exakten Erblichkeitslehre*. Gustav Fischer, Jena.

Johannsen, W. 1911. The genotype conception of heredity. *Amer. Nat.* 45: 129–159.

Keiner, C. 1998. W. K. Brooks and the oyster question: Science, politics, and resource management in Maryland, 1880–1930. *J. Hist. Biol.* 31: 383–424.

Kroon, F. J., P. L. Munday, D. A. Westcott, J. P. Hobbs and N. R. Liley. 2005. Aromatase pathway mediates sex change in each direction. *Proc. Biol. Sci.* 272: 1399–1405.

Lester, R. L., C. Grach, M. P. Pener and S. J. Simpson. 2005. Stimuli inducing gregarious colouration and behaviour in nymphs of *Schistocerca gregaria*. *J. Insect Physiol.* 51: 737–747.

Leutert, T. R. 1974. Zur Geschlechtsbestimmung und Gametogenese von *Bonellia viridis* Rolando. *J. Embryol. Exp. Morphol.* 32: 169–193.

Lewontin, R. C. 1993. *Biology as Ideology: The Doctrine of DNA*. Harper Collins, New York.

McCaffery, A. R., S. J. Simpson, M. S. Islam and P. Roessingh. 1998. A gregarizing factor present in egg pod foam of the desert locust *Schistocerca gregaria*. *J. Exp. Biol.* 201: 347–363.

McCollum, S. A. and J. Van Buskirk. 1996. Costs and benefits of a predator-induced polyphenism in the gray treefrog *Hyla chrysoscelis*. *Evolution* 50: 583–593.

Meyer, A. 1987. Phenotypic plasticity and heterochrony in *Cichlasoma managues* (Pisces, Cichlidae) and their implications for speciation in cichlid fishes. *Evolution* 41: 1357–1369.

Miller, D., J. Summers and S. Silber. 2004. Environmental versus genetic sex determination: A possible factor in dinosaur extinction? *Fertil. Steril.* 81: 954–964.

Moczek, A. P. 2005. The evolution of development of novel traits, or how beetles got their horns. *BioScience* 55: 937–951.

Moczek, A. P. and D. J. Emlen. 2000. Male horn dimorphism in the scarab beetle *Onthophagus taurus:* Do alternative tactics favor alternative phenotypes? *Anim. Behav.* 59: 459–466.

Moczek, A. P. and H. F. Nijhout. 2004. Trade-offs during development of primary and secondary sexual traits in a horned beetle. *Amer. Nat.* 163: 184–191.

Morreale, S. J., G. J. Ruiz, J. R. Spotila and E. A. Standora. 1982. Temperature-dependent sex determination: Current practices threaten conservation of sea turtles. *Science* 216:1245–1247.

Moss, L. 2003. *What Genes Can't Do*. MIT Press, Cambridge, MA.

Moss, M. L. 1962. The functional matrix. In B. Kraus and R. Reidel (eds.), *Vistas in Orthodontics*. Lea & Febiger, Philadelphia, pp. 85–98.

Moss, M. L. 1997. The functional matrix hypothesis revisited. IV. The epigenetic antithesis and the resolving synthesis. *Amer. J. Orthod. Dentofac. Orthop.* 112: 410–417.

Müller, G. B. 2003. Embryonic motility: Environmental influences on evolutionary innovation. *Evo. Dev.* 5: 56–60.

Müller, G. B. and J. Steicher. 1989. Ontogeny of the syndesmosis tibiofibularis and the evolution of the bird hindlimb: A caenogenetic feature triggers phenotypic novelty. *Anat. Embryol.* 179: 327–339.

National Aeronautics and Space Administration (NASA). 2003. Pumping iron in microgravity. *Space Research Office Biol. Physical Res. Newsltr.* http://spaceresearch.nasa.gov/general_info/pumpingiron.html.

Nelkin, D. and M. S. Lindee. 1996. *The DNA Mystique: The Gene as a Cultural Icon.* W.H. Freeman, New York.

Nijhout, H. F. 1991. *The Development and Evolution of Butterfly Wing Patterns.* Smithsonian Institution Press, Washington, DC.

Nikawa, T. and 13 others. 2004. Skeletal muscle gene expression in space-flown rats. *FASEB J.* 18: 522–524.

Nilsson-Ehle, H. 1914. Vilka erfarenheter hava hittills vunnits roerande moejligheten av vaexters acklimatisering? *Kungl. Landtbruks-Akademiens Handlinger och Tidskrift* 53: 537–572.

Nyhart, L. 1995. *Biology Takes Form.* University of Chicago Press, Chicago.

Palmer, A. R. 1985. Adaptive value of shell variation in *Thais lamellosa:* Effect of thick shells on vulnerability to and preference by crabs. *Veliger* 27: 349–356.

Parzer, H. F. and A. P. Moczek. 2008. Rapid antagonistic co-evolution between primary and secondary sexual charactrs in horned beetles. *BioEssays.* In press.

Pechenik, J. A., D. E. Wendt and J. N. Jarrett. 1998. Metamorphosis is not a new beginning. *BioScience* 48: 901–910.

Pener, M. P. 1991. Locust phase polymorphism and its endocrine relations. *Adv. Insect Physiol.* 3: 1–79.

Perry, A. N. and M. S. Grober. 2003. A model for social control of sex change: Interactions of behavior, neuropeptides, glucocorticoids, and sex steroids. *Horm. Behav.* 43: 31–38.

Plowright, R. C. and B. A. Pendrel. 1977. Larval growth in bumble-bees. *Can. Entomol.* 109: 967–973.

Rachinsky, A. and K. Hartfelder. 1990. Corpora allata activity, a prime regulating element for caste-specific juvenile hormone titre in honey bee larvae (*Apis mellifera carnica*). *J. Insect Physiol.* 36: 329–349.

Relyea R. A. 2003a. Predators come and predators go: The reversibility of predator-induced traits. *Ecology* 84: 1840–1848.

Relyea, R. A. 2003b. Predator cues and pesticides: A double dose of danger for amphibians. *Ecol. Applic.* 13: 1515–1521.

Relyea, R. 2004. Fine-tuned phenotypes: Tadpole plasticity under 16 combinations of predators and competitors. *Ecology* 85: 172–179.

Roessingh, P., A. Bouaïchi and S. J. Simpson. 1998. Effects of sensory stimuli on the behavioural phase state of the desert locust, *Schistocerca gregaria. J. Insect Physiol.* 44: 883–893.

Rogers, S. M., T. Matheson, E. Despland, T. Dodgson, M. Burrows and S. J. Simpson. 2003. Mechanosensory-induced behavioural gregarization in the desert locust *Schistocerca gregaria. J. Exp. Biol.* 206: 3991–4002.

Rogers, S. M., T. Matheson, K. Sasaki, K. Kendsrick, S. J. Simpson and M. Burrows. 2004. Substantial changes in central nervous system neurotransmitters and neuromodulators accompany phase change in the locust. *J. Exp. Biol.* 207: 3603–3617.

Roll-Hansen, N. 2007. *Sources of Johannsen's genotype theory.* Meeting of the International Society for the History, Philosophy, and Social Studies of Biology, Exeter, England.

Sarkar, S. 1998. *Genetics and Reductionism.* Cambridge University Press, Cambridge.

Sarkar, S. 1999. From the *Reaktionsnorm* to the adaptive norm: The norm of reaction, 1906–1960. *Biol. Philos.* 14: 235–252.

Sarkar, S. 2006. From genes as determinants to DNA as resource: Historical notes on development and genetics. In E. Neumann-Held and C. Rehmann-Sutter (eds.), *Genes in Development: Re-Reading the Molecular Paradigm.* Duke University Press, Durham, NC, pp. 77–95.

Schmalhausen, I. I. 1949. *Factors of Evolution: The Theory of Stabilizing Selection.* University of Chicago Press, Chicago.

Schmidt-Küntzel, A., E. Eizirik, S. J. O'Brien and M. Menotti-Raymond. 2005. Tyrosinase and tyrosinase-related protein 1 genetic variants specify domestic cat coat color alleles of the albino and brown loci. *J. Hered.* 96: 289–301.

Shapiro, A. M. 1968. Photoperiodic induction of vernal phenotype in *Pieris protodice* Boisduval and Le Conta (Lepidoptera: Pieridae). *Wasmann J. Biol.* 26: 137–149.

Simmons, L. W. and D. J. Emlen. 2006. Evolutionary trade-off between weapons and testes. *Proc. Natl. Acad. Sci. USA* 103: 16346–16351.

Simpson, S. J. and G. A. Sword. 2008. Phase polyphenism in locusts: Mechanisms, population consequences, adaptive significance and evolution. In T. Ananthakrishnan and D. Whitman (eds.), *Phenotypic Plasticity of Insects: Mechanisms and Consequences.* Science Publishers, Plymouth, UK. In press.

Simpson, S. J., E. Despland, B. F. Hägele and T. Dodgson. 2001. Gregarious behavior in desert locusts is evoked by touching their back legs. *Proc. Natl. Acad. Sci. USA* 98: 3895–3897.

Stabell, O. B., F. Ogbeto and R. Primicerio. 2003. Inducible defenses in *Daphnia* depend on latent alarm signals from conspecific prey activated in predators. *Chem. Senses* 28: 141–153.

Stearns, S. C., G. de Jong and R. A. Newman. 1991. The effects of phenotypic plasticity on genetic correlations. *Trends Ecol. Evol.* 6: 122–126.

Sultan, S. E. 2007. Development in context: The timely emergence of eco-devo. *Trends Ecol. Evol.* 22: 575–582.

Sword, G. A., S. J. Simpson, O. T. M El Hadi and H. Wilps. 2000. Density-dependent aposematism in the desert locust. *Proc. R. Soc. London B* 267: 63–68.

The environmental agents involved in altering phenotypes usually do so by altering gene expression during development. This can be accomplished by at least three major routes (Gilbert 2005):

- **Direct transcriptional regulation**. Environmental factors can alter gene expression patterns by chemically modifying particular regions of DNA. These modifications regulate whether particular genes are activated (transcribed) or turned off (repressed).
- **The neuroendocrine system**. In most of these cases, the nervous system receives signals from the environment, and chemical signals from the nervous system cause changes in the hormone (i.e., endocrine) milieu within the organism. Hormones produced in response to such neural signals can alter gene expression patterns and regulate phenotype production.
- **Direct induction**. The environmental factor may interact directly with the cell to activate or repress the signal transduction cascades that activate gene expression and alter cell behaviors.

Regulation of Gene Transcription

Because every cell in an individual's body is derived by cell division from the same fertilized egg, every cell (with a few interesting exceptions such as the mammalian lymphocytes) has the same set of genes; this set of genes comprises one's **genome**. Thus, the genes in pancreas cells are the same as the genes in the lens precursor cells of the eye. What makes the pancreatic cells different from the lens cells are the genes that are activated, or *expressed*. In the lens precursor cell, a cascade of tissue-specific events activates genes producing crystalline proteins that allow the lens tissue to be transparent, whereas in certain cells in the pancreas, the gene for insulin is activated (and the lens-specific genes are turned off). To understand how such **differential gene expression** is accomplished, we must briefly review some molecular biology.

Differential gene expression

In addition to the structural portion of the gene—the DNA that encodes the amino acid sequence of the protein—there are regions of the gene that regulate where, when, and how efficiently the gene will be active. These regulatory regions (sometimes called *cis*-regulatory elements) include the promoter and the enhancers (Figure 2.1):

- Every gene has a **promoter**—a region of DNA containing a specific nucleotide sequence (usually TATA) that binds the enzyme **RNA polymerase**. This enzyme unwinds the double helix of DNA and initiates the synthesis of messenger RNA (mRNA), which in turn specifies a sequence of amino acids that fold to form a functional protein.

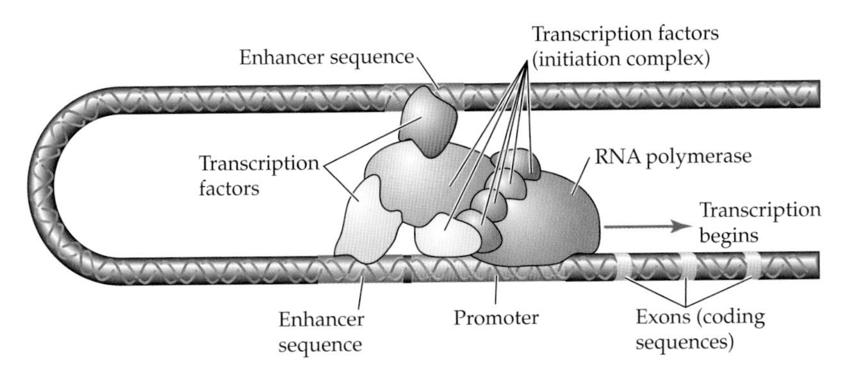

FIGURE 2.1 RNA polymerase is stabilized on the promoter site of the DNA by transcription factors recruited by the enhancers. The TATA sequence at the promoter binds a protein that serves as a "saddle" for RNA polymerase. However, RNA polymerase would not remain bound long enough to initiate transcription were it not for the stabilization provided by the transcription factors.

- **Enhancers** are DNA sequences that bind a set of proteins called **transcription factors**. The resulting complex of transcription factors interacts with the promoter to tell the gene when to be active, where to be active, and how much messenger RNA it should synthesize.

RNA polymerase makes mRNA that is complementary to the DNA strand, or **template**, it is reading from. The process whereby RNA polymerase synthesizes this complementary mRNA is called **transcription**. The promoter sequence orients RNA polymerase so that it reads in the correct direction and binds at the beginning of the gene. This makes sense, since if RNA polymerase could bind anywhere on the DNA, mRNAs would have no way of defining the starting point of a protein. The promoter can thus be thought of as a punctuation mark that tells RNA polymerase (the "reader") where to start its transcription of the structural gene. Once RNA polymerase binds to the promoter and transcription begins, the gene is said to be activated.

If every cell contains the same genes, how does each cell type get its own unique constellation of messenger RNAs? This differential gene transcription is accomplished by the enhancers. The enhancer sequences of the DNA (see Figure 2.1) are the same in every cell type; what differs is the combination of transcription factor proteins present. Enhancers can bind several transcription factors, and it is the specific combination of transcription factors that allows a gene to be active in a particular cell type. That is, the same transcription factor, in conjunction with different other transcription factors, will activate different promoters in different cells. Moreover, the same gene can have several enhancers, with each enhancer binding transcription factors that enable that same gene to be expressed in different cell types. Figure 2.2 illustrates this phenomenon for expression of the mammalian *Pax6* gene. First, there are several enhancer elements that can activate the *Pax6*

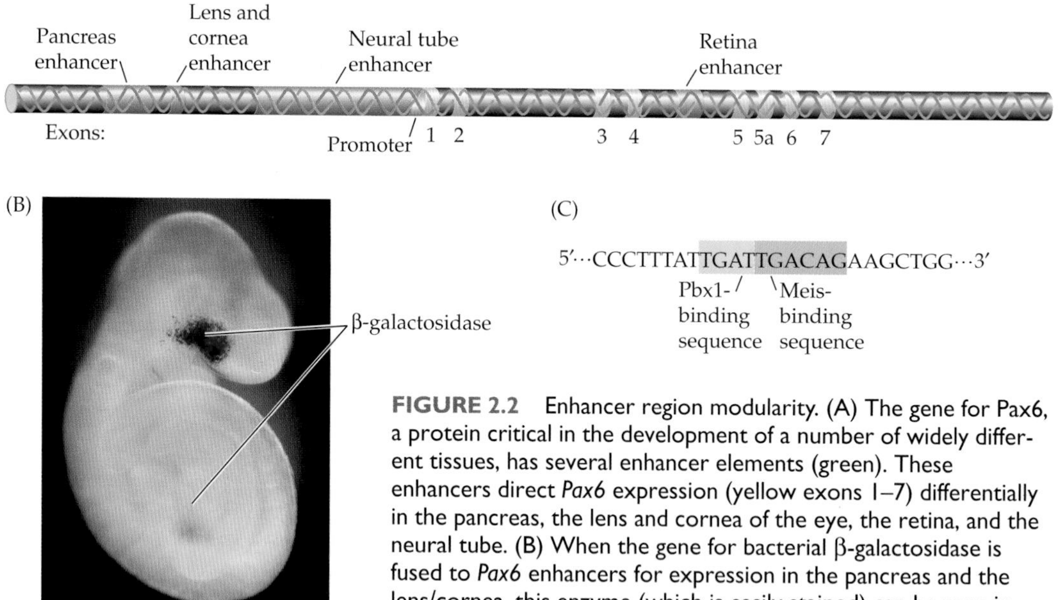

FIGURE 2.2 Enhancer region modularity. (A) The gene for Pax6, a protein critical in the development of a number of widely different tissues, has several enhancer elements (green). These enhancers direct *Pax6* expression (yellow exons 1–7) differentially in the pancreas, the lens and cornea of the eye, the retina, and the neural tube. (B) When the gene for bacterial β-galactosidase is fused to *Pax6* enhancers for expression in the pancreas and the lens/cornea, this enzyme (which is easily stained) can be seen in those tissues. (C) The pancreas-specific enhancer element has binding sites for the Pbx1 and Meis transcription factors; both of these factors must be present in order to activate the *Pax6* gene in the pancreas. (Photograph from Williams et al. 1998.)

gene differentially in several different cell types (Figure 2.2A,B). Second, within each of these enhancer regions there are sites that bind specific transcription factors (Figure 2.2C).

Transcription factor complexes generally activate gene expression in one of two ways. First, and most simply, the transcription factors bound on an enhancer can stabilize the RNA polymerase so that it will stay on the promoter and begin transcription. Second, the complex can clear a space in the chromosome, enabling RNA polymerase to find the promoter.

Finding a promoter is not easy, because the DNA is usually so wound up that the promoter sites are not accessible. Indeed, over 6 *feet* of DNA is packaged into chromosomes of each human cell nucleus (Schones and Zhao 2008). The protein particles that wrap the DNA are called **nucleosomes**. In most cases, the nucleosomes wrap the DNA so tightly they prevent any other protein (including RNA polymerase) from binding to the DNA. However, some enhancers bind transcription factors that modify the nucleosomes such that they fall apart and the promoter site becomes accessible.

Each nucleosome is made of a core of eight **histone proteins** (two molecules each of histones H2A, H2B, H3, and H4; Figure 2.3A). Adjacent nucle-

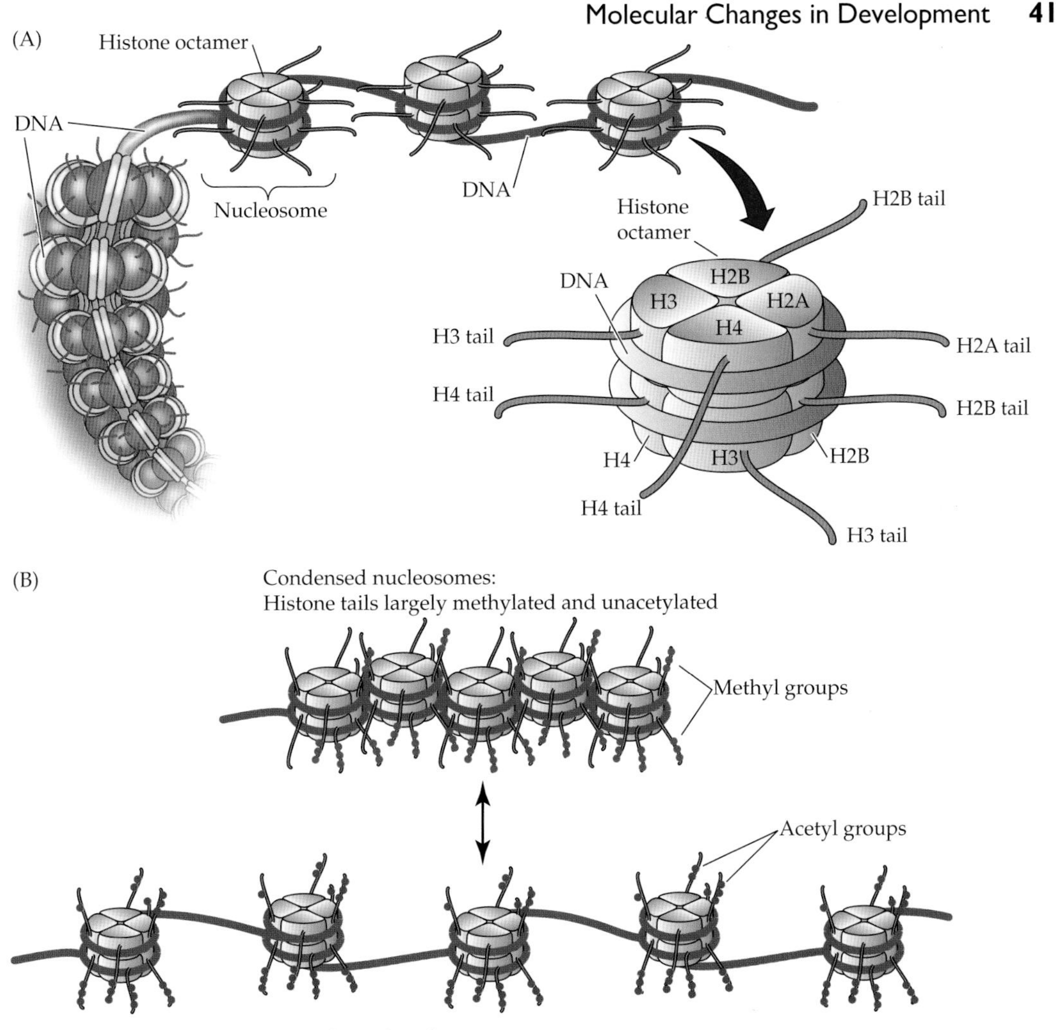

FIGURE 2.3 Nucleosomes condense the DNA into tight configurations on the chromosomes. (A) Nucleosomes are composed of histone subunits wound tightly by DNA. Histone "tails" protruding from the subunits allow for the attachment of chemical groups. (B) Methyl groups condense nucleosomes more tightly, preventing access to promotor sites and thus preventing gene transcription. Acetylation loosens the nucleosome packing, exposing the DNA to RNA polymerase and other transcription factors that will activate the genes.

osomes are packed together into tight arrays that prevent transcription factors and RNA polymerases from gaining access to the genes (Thoma et al. 1979; Schlissel and Brown 1984). It is generally thought that the "default" condition of chromatin is a repressed state, and that cell- and tissue-specific genes become activated by local interruption of this repression (Weintraub 1985). Repression can be locally strengthened (so that it is very difficult to transcribe those genes within the nucleosomes) or relieved (so that it becomes relatively easy to transcribe them) by modifying the histones.

Even in condensed nucleosomes, "tails" projecting from the histones are accessible to important enzymes that can place specific chemical groups on the DNA (Figure 2.3B). In general, the addition of acetyl groups (**acetylation**) to histone tails will loosen the nucleosome and allow transcription, while the addition of methyl groups (**methylation**) to histone tails tends to recruit proteins that will aggregate the nucleosomes more tightly and repress transcription. Enzymes (histone acetyltransferases) that place acetyl groups on histones (especially on lysines in histones H3 and H4) destabilize the nucleosome so that it comes apart and the DNA is exposed (Figure 2.3C). Enzymes that remove acetyl groups (histone deacetylases) stabilize nucleosomes and prevent transcription. The addition of methyl groups to histones by histone methyltransferases generally represses transcription even further, whereas removing methyl groups brings the nucleosomes back to their baseline stability (see Strahl and Allis 2000; Cosgrove et al. 2004).

DNA methylation

Methylation is one of the most important and best studied mechanisms of keeping genes inactive. Certain cytosine (C) bases on DNA are especially susceptible to methylation. (These are CpG sequences in animals and CpNpGp triplets in plants.) Many transcription factor proteins recognize a specific DNA sequence, but cannot find that sequence if the DNA is methylated. Therefore, one way of keeping a gene inactive is to methylate the promoter and enhancer regions of that gene. This inactivation can be transmitted at each cell division by enzymes called DNA methyltransferases that recognize a methylated C on a CpG sequence of the template strand and hence methylate the C on the newly synthesized complementary strand.

Methylated cytosines in DNA can bind particular proteins such as MeCP2. Once connected to a methylated cytosine, MeCP2 binds to histone deacetylases and histone methyltransferases, which remove acetyl groups from (Figure 2.4A) or add methyl groups to (Figure 2.4B) the histones. As a result, the nucleosomes form tight complexes with the DNA and don't allow other transcription factors and RNA polymerases to find the genes. Other proteins, such as HP1, also bind and aggregate methylated histones (Fuks 2005).

One of the enzymes recruited to the DNA by MeCP2 is DNA methyltransferase-3 (Dnmt3). This enzyme methylates previously unmethylated cytosines on the DNA. The newly established methylation pattern is then promulgated by DNA methyltransferase-1 (Dnmt1) which, as mentioned above, recognizes methyl groups on one strand of DNA and places methyl

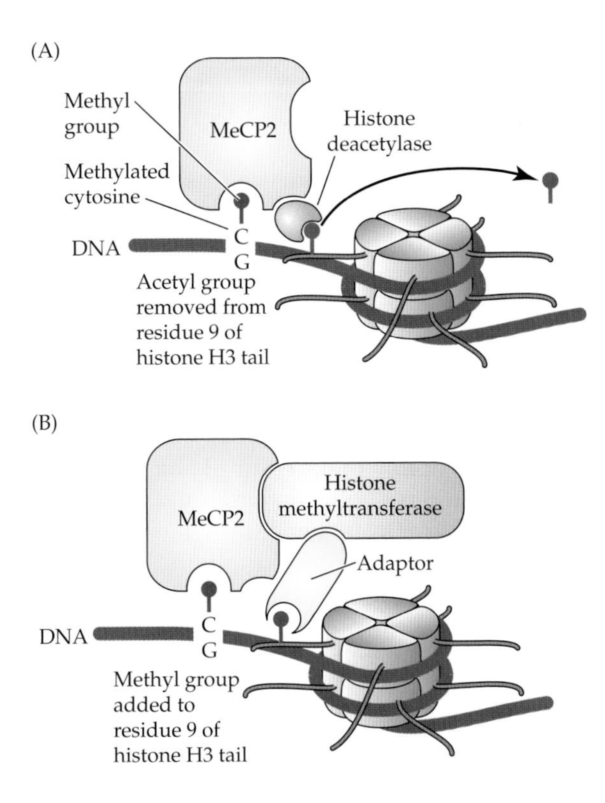

FIGURE 2.4 Modifying nucleosomes through methylated DNA. The MeCP2 protein recognizes methylated cytosines on DNA. It binds to the DNA and is thereby able to recruit (A) histone deacetylases (which remove acetyl groups) or (B) histone methyltransferases (which add more methyl groups). Both of these modifications promote the stability of the nucleosome and the tight packing of DNA, thereby repressing gene expression. (After Fuks 2005.)

groups on the newly synthesized strand (see Bird 2002; Burdge et al. 2007a). In this way, once the DNA methylation pattern is established in a cell, it can be stably inherited by all the progeny of that cell. Such heritable changes in methylation state are **epigenetic,** referring to the fact that these alterations are not mutations but are instructions that affect the expression, not the transmission, of the gene (see Chapter 1).

Thus, during normal development, differences in cytosine methylation are critical in telling a nucleus which genes can be expressed, and which genes are expressed determines what type of cell it will become. Moreover, thanks to Dnmt1, this pattern of gene expression is stably transmitted to that cell's progeny. Appendix B provides further details on how such chromatin alterations are passed on epigenetically from one cell generation to another.

Environmental agents and direct DNA methylation

Environmental agents can significantly alter gene expression and the resulting phenotype by changing DNA methylation patterns. One study by

Waterland and Jirtle (2003) used a dominant mutation of the *agouti* gene in mice to show that changes in maternal diet can produce changes in DNA methylation, and that these methylation changes can affect offspring phenotype.

The normal (wild-type) allele of the mouse *agouti* gene results in brown-pigmented fur and normal body size. The variant *Agouti* allele, however, is dominant over *agouti* and produces proteins that lead to obese mice with yellowish fur (Figure 2.5A). Waterland and Jirtle's study made use of a mutant allele of *Agouti* called *viable yellow* that contains an inserted sequence that can be methylated, thus blocking transcription of *Agouti*.

Because the mutant *Agouti* allele is dominant, we would predict that all the offspring of *Agouti* mothers would be yellow-furred and tend to obesity. However, when Waterland and Jirtle fed pregnant *viable yellow Agouti* mice different amounts of methyl-donor dietary supplements (folate, choline, and betaine), the offspring did not show the predicted phenotype (Figure 2.5B). Even though the offspring were genetically identical, they had strikingly different phenotypes due to their mothers' diets. The researchers found that the more methyl donors in the mother's diet, the greater the methylation of the *viable yellow* insertion in the offspring, and the darker (i.e., more normal) their pigmentation. Additionally, the obesity characteristic of the *Agouti* phenotype was reduced, and other methylation differences were seen throughout the body. In other words, the dietary methyl groups fed to the mother turned off the *Agouti* gene in her offspring.

There are several studies showing that DNA methylation is affected by diet and other environmental agents experienced during early development. As we will discuss in Chapter 7, feeding a pregnant rat a low-protein diet causes her offspring to develop phenotypes that predispose them to diabetes, hypertension, and obesity later in life. It has been proposed that this susceptibility to disease is caused by the inappropriate methylation of the genes encoding enzymes used in liver metabolism and kidney development. For example, rats starved for protein in utero produced much more *PPARα* mRNA in their livers than did rats whose mothers were fed a normal diet (Lillycrop et al. 2007); PPARα is a transcription factor that activates genes used in fat production. Further experiments suggest that a low-protein diet interferes with the activation of *Dnmt1*, leading to lower amounts of promoter methylation and thus higher levels of gene expression for both of these genes (Burdge et al. 2007b; Lillycrop et al. 2007).

The effects of maternal behavior on gene methylation

It may seem like science fiction, but scientists can tell whether a rat received good maternal care by looking at the methylation pattern of its DNA. In rats, behavioral differences in the response to stressful situations have been correlated with the number of glucocorticoid receptors (GRs) in the brain's hippocampus. The more GRs, the better the adult rat is able to downregulate adrenal hormones and deal with stress. The number of GRs appears to depend on the quality of grooming and licking the rat pup experiences during the first week after its birth.

How is the adult phenotype regulated by these perinatal (i.e., near the time of birth) experiences? Weaver and colleagues (2004, 2007) have shown that the difference involves the methylation of the binding site for a particular transcription factor (Egr1) in an enhancer region of the glucocorticoid receptor (*GR*) gene. Before birth, there is no methylation at this site; one day after birth, the site is methylated in all rat pups. However, in those pups that experience intensive grooming and licking during the first week after birth, this site *loses* its methylation, while methylation is retained in those rats that do not receive extensive care. Moreover, this methylation differ-

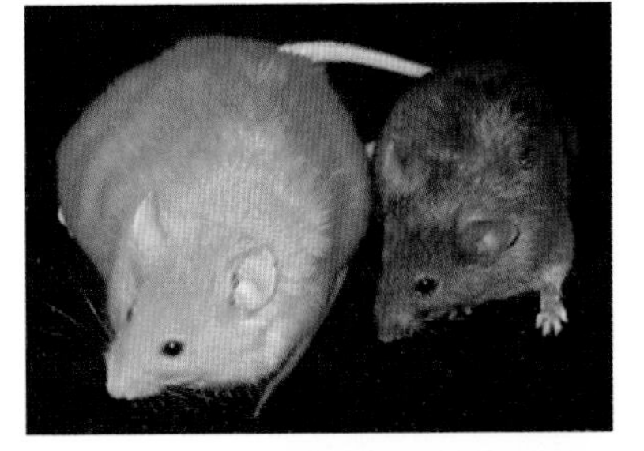

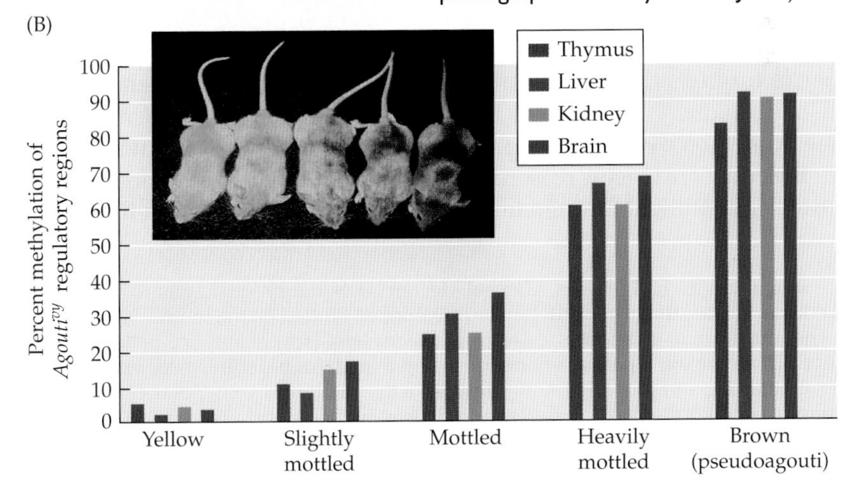

FIGURE 2.5 Maternal diet can affect phenotype. (A) These two mice are genetically identical; both are heterozygous for the *viable yellow* allele of the *Agouti* gene, whose protein product converts brown pigment to yellow and accelerates fat storage. The obese yellow mouse is the offspring of a mother whose diet was not supplemented with methyl donors (e.g., folic acid) during her pregnancy. The embryo's *Agouti* gene was not methylated, and Agouti protein was made. The sleek brown mouse was born of a mother whose prenatal diet was supplemented with methyl donors. The *Agouti* gene was turned off, and no Agouti protein was made. (B) Norm of reaction for the *viable yellow Agouti* gene. As methyl donor supplementation increased, genetically identical mice became progressively more brown. (The mice in B are younger than the pair in A.) The histogram shows coat color correlated with the amount of methylation of this gene in all tissues studied. (After Waterland and Jirtle 2003; photographs courtesy of R. L. Jirtle.)

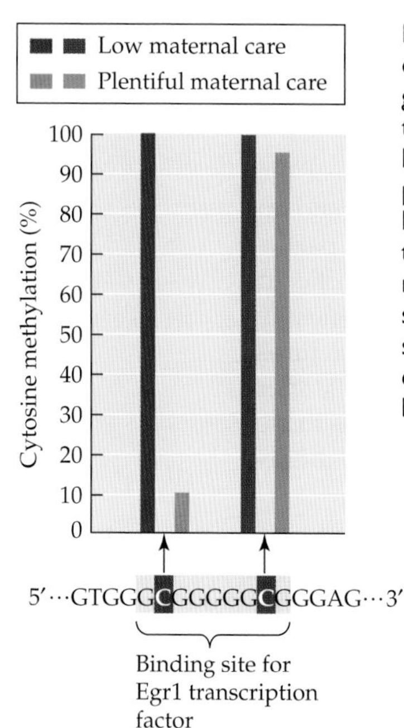

FIGURE 2.6 Differential DNA methylation due to behavioral differences in parental care. A portion of an enhancer sequence of the rat glucocorticoid receptor gene is shown, indicating the binding site for the Egr1 transcription factor. Two cytosine residues within this site have the potential to be methylated. The cytosine at the 5′ end is completely methylated in the brains of pups that did not receive extensive licking and grooming from their mothers (red bar). The Egr1 transcription factor did not bind these methylated sites, and thus the *GR* gene remained inactive. If the pups received proper maternal care, this same site was largely unmethylated (orange bar), and the gene was transcribed in the brain. As a control, the cytosine at the 3′ end of the enhancer (blue bars) was always methylated and had no effect on Egr1 binding. (After Weaver et al. 2004.)

ence is not seen at other sites in or near the gene (Figure 2.6). When unmethylated, the hippocampal enhancer DNA binds the Egr1 transcription factor and is associated with active (i.e., acetylated) nucleosomes. Egr1 does not bind when this DNA has been methylated, and in such cases the surrounding chromatin (containing the *GR* gene) is not activated. These chromatin differences, established during the first week after birth, appear to be retained throughout the life of the rat. Thus, adult rats that received extensive perinatal grooming have more glucocorticoid receptors in their brains and are able to deal with stress better than rats that received less care.*

Signal Transduction from Environment to Genome via the Neuroendocrine System

In studying the metamorphosis of insects, H. Fred Nijhout (1999, 2003) came to formulate one of the most important principles of developmental plasticity: namely, that actions of the neuroendocrine system can change gene expression by transducing sensory information from the environment into the body. This seems reasonable, since the nervous system monitors the outside environment through our sensory neurons. Signals from these neu-

*By switching pups and parents, Michael Meaney's laboratory (see Laplante et al. 2002) also demonstrated that the methylation difference depended on the mother's care and was not the result of differences in the pups themselves. How a mother's behavior can alter DNA methylation patterns in her offspring remains to be established, but it is possible that licking and grooming may stimulate a thyroid hormone-dependent increase in serotonin activity, with increased serotonin leading to a signaling cascade that alters methylation of the Egr1 binding site (Laplante et al. 2002; Weaver et al. 2007).

Vernalization: Temperature-Dependent Chromatin Changes

One reason why Western scientists have ignored the roles of the environment in normal development is that the idea of environmentally induced phenotypes formed the center of an ideologically biased version of heredity. From the 1930s through the mid 1960s, the Soviet biologist Trofim Lysenko fought for an environmentally based, anti-Mendelian science of heredity, and in 1948 he succeeded in making the study of genetics illegal in the Soviet Union (see Appendix A).

The paradigm for Lysenko's approach was vernalization, a phenomenon whereby the seeds of certain plants need to be exposed to cold temperatures in order to flower. The vernalization requirement of winter strains of cereals such as wheat means that their seeds are planted in the fall and experience the winter cold, becoming competent to flower the following spring. Lysenko's fervent claims that the vernalized state could be inherited by seeds that had never been exposed to cold temperatures failed miserably as a model of heredity (and probably led to crop failures in the Ukraine and China). But vernalization is a very real natural phenomenon in which the plant's development is altered by the environment. Moreover, the phenomenon has a genetic basis.

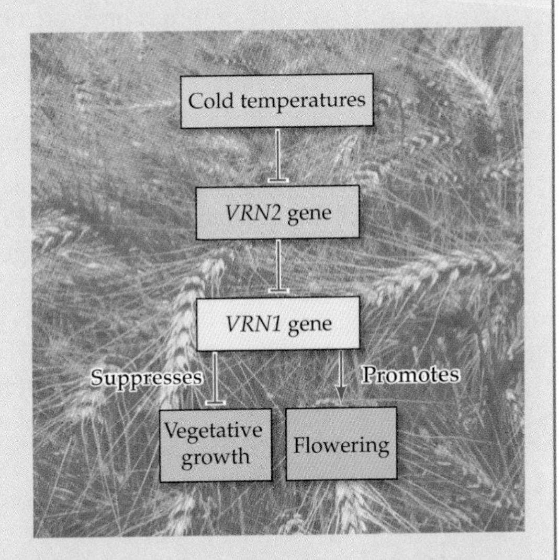

General pathway for vernalization in wheat. Cold temperatures inhibit the expression of the *VRN2* gene. *VRN2* encodes a transcription factor that actively represses the *VRN1* gene, the products of which promote flowering and suppress vegetative growth. Thus, cold temperatures inhibit the inhibitor of flowering.

The genetic basis of vernalization appears to be straightforward: Cold temperatures repress a repressor of flowering. In wheat, for instance, expression of the *VNR1* (*vernalization-1*) gene promotes flowering. *VNR1* encodes a transcription factor that *activates* genes that promote flowering while at the same time repressing genes that maintain vegetative (nonreproductive) growth. The activity of *VNR1*, however, is repressed by the transcription factor encoded by the *VNR2* gene. Cold conditions repress *VNR2*, thus releasing the inhibition of *VNR1* and allowing flowering to occur (see figure). This repression continues through the spring, allowing the wheat to become competent to flower as early as possible.

In the wild mustard *Arabidopsis* (the model system widely used in plant genetics because of its rapid life cycle and the ability to obtain mutants of flowering in the laboratory), the *FLC* ("flowering control") gene is equivalent to wheat's *VRN2* gene. Although the details of vernalization differ between cereals and *Arabidopsis*, the mechanisms are similar (see Sung and Amasino 2005). Using genetic screens of *Arabidopsis* to find mutations that interfere with vernalization has led to the discovery of genes whose protein products are critical in retaining the repression of the *FLC* gene. This repression is accomplished by methylating and deacetylating the gene's regulatory region. Thus, while it is true that Lysenko was an ideologue and a poor scientist, his concept that the environment could give plants new properties—and that cold treatment might induce a longer and safer growing season in wheat— is something that is being actively revisited today.

rons can then inform the endocrine glands to secrete (or to not secrete) hormones—and the functions of hormones include regulating gene expression. In this manner, an environmental stimulus can cause changes in gene expression. It is important to recognize that such responses to the environment must already be part of the genetic repertoire of the organism. Expression of such genes, however, requires an environmental signal; otherwise the genes will remain inactive.

Insect hormones work largely by activating transcription factors. Hormones such as the metamorphic molting agent ecdysone (from *ecdysis*, "to shed") induce many changes in the developing insect. Ecdysone functions by binding to an inactive transcription factor, the ecdysone receptor. Once ecdysone binds, it converts the receptor to an active form that can enter the nucleus, interact with other transcription factors, and initiate gene expression on particular promoters (Figure 2.7). Juvenile hormone (which prevents metamorphosis into an adult, hence its name) probably works in a similar manner.

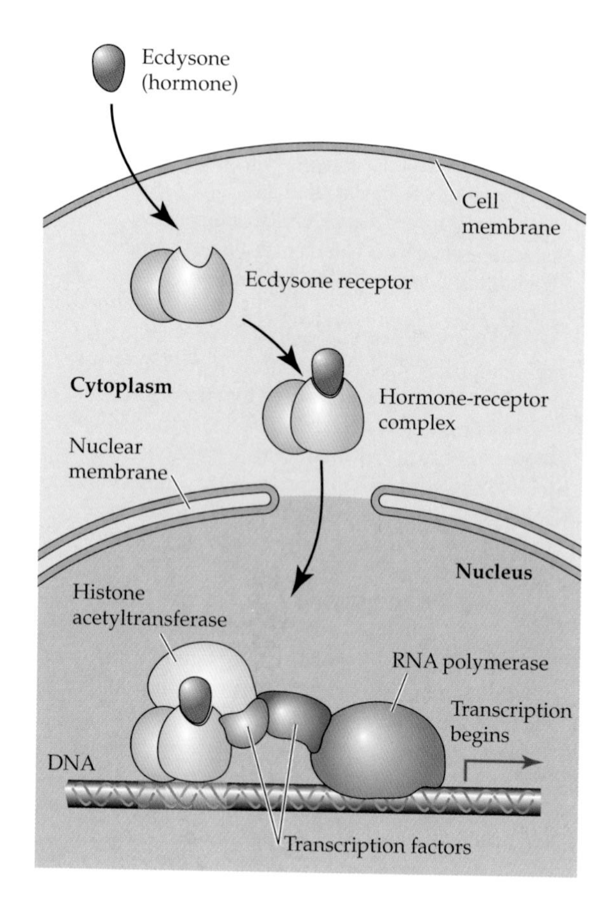

FIGURE 2.7 Simplified representation of the proposed mechanism of ecdysone action. Like other steroid hormones, ecdysone enters the cell readily (it is a lipid). Once inside, it binds to its receptor, a protein that is always present in the cytoplasm but is nonfunctional until it binds ecdysone. Once the receptor binds the steroid, this complex can enter the nucleus and activate gene expression by recruiting histone acetyltransferase enzymes and other transcription factors.

Neuroendocrine regulation of temperature-dependent polyphenism in insects

One of the earliest documented cases of environmental determination of phenotype came from studies on the European map butterfly (*Araschnia levana*). The butterflies eclosing from their pupal cases in the spring are so different from the butterflies eclosing in the summer that Linnaeus' *Systema Naturae* (1758) classified them as two different species. The spring morph is bright orange with black spots, while the summer form is mostly black with a white band. In 1875, August Weismann demonstrated that these butterflies were the same species but that temperature conditions during their larval development resulted in different phenotypes. By incubating caterpillars in different conditions, Weismann could produce the spring form in the summer.

It is now known that the normal change from spring to summer morph is controlled by changes in both day length and temperature during the larval period (Koch and Buckmann 1987; Nijhout 1991). Moreover, we now know that the temperature and daylight signals produce the phenotype by regulating the amount of the hormone ecdysone during the larval stage of development. Larvae that develop in the spring do not experience this pulse of ecdysone; larvae developing in the longer, warmer days of summer do. When ecdysone is injected into larvae, the spring form can be converted to a summer form, and intermediate amounts of ecdysone produce intermediate phenotypes not found in any natural conditions (Figure 2.8; Nijhout 2003).

In some instances, hormones have been correlated to the expression of particular genes. Chapter 1 described how, in honeybee larvae, protein-rich diets ("royal jelly") increase juvenile hormone levels, resulting in the formation of fertile queens. Without this enriched diet, the larvae become sterile workers. Evans and Wheeler (2000) showed that the elevated juvenile hormone levels increased the transcription of some genes (notably those involving energy metabolism) and repressed others. Two of the genes whose transcription is upregulated by juvenile hormone encode an insulin-like protein and an insulin receptor (Wheeler et al. 2006). These proteins often regulate developmental decisions and may be critical in regulating the differential development between those larvae destined to become workers and those destined to become queens.

The royal jelly diet may also work, through juvenile hormone and insulin, to downregulate the Dnmt3-induced DNA methylation described earlier in this chapter. When Dnmt3 expression was inhibited in newly hatched bumblebee larvae, 72% of the resulting bees were fertile queens with fully developed ovaries (Kucharski et al. 2008). Thus, the environmental factor—diet—may act via the endocrine system to alter gene methylation, thereby causing the fertile phenotype to develop.

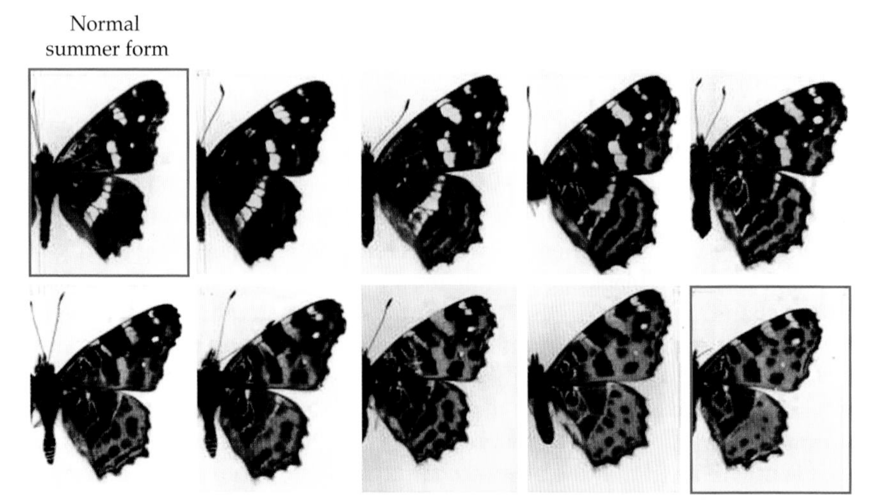

FIGURE 2.8 Hormonal regulation mediates the environmentally controlled pigmentation of *Araschnia*. In the wild, different generations experience significantly different photoperiods. In the short photoperiod (below the critical day length), there is no pulse of 20-hydroxyecdysone (20E) during early pupation, and the spring form of the butterfly is generated. When these spring butterflies mate, the larvae experience a long photoperiod and generate the summer pigmentation. In the laboratory, injections of 20E at different times during pupation can induce both phenotypes, as well as intermediate phenotypes not seen in the wild. (From Nijhout 2003; photographs courtesy of H. F. Nijhout.)

There is evidence that in some butterflies, the presence of eyespots is controlled by the hormonal regulation of the *Distal-less* gene. In the African country of Malawi, where there is a hot wet season and a cooler dry season, the butterfly *Bicyclus anynana* occurs in two environmentally induced phenotypes ("morphs"). The dry-season morph is somewhat sluggish, with cryptic coloration resembling the dead brown leaves of its habitat. The wet-season morph is an active flier and has ventral hindwing eyespots that deflect attacks from predatory birds and lizards. The determining factor in this seasonal polyphenism appears to be the temperature during pupation: low temperatures produce the dry-season morph, whereas high temperatures produce the wet-season morph (Figure 2.9; Brakefield and Reitsma 1991). The development of butterfly eyespots begins in the late larval stages, when the transcription of the *Distal-less* gene is restricted to a small focus that will become the center of each eyespot. During the early pupal stage, *Distal-less* expression is seen in a wider area, and this expression is thought to constitute the activating signal that determines the size of the spot. Last, the cells receiving the signal determine the color they will take.

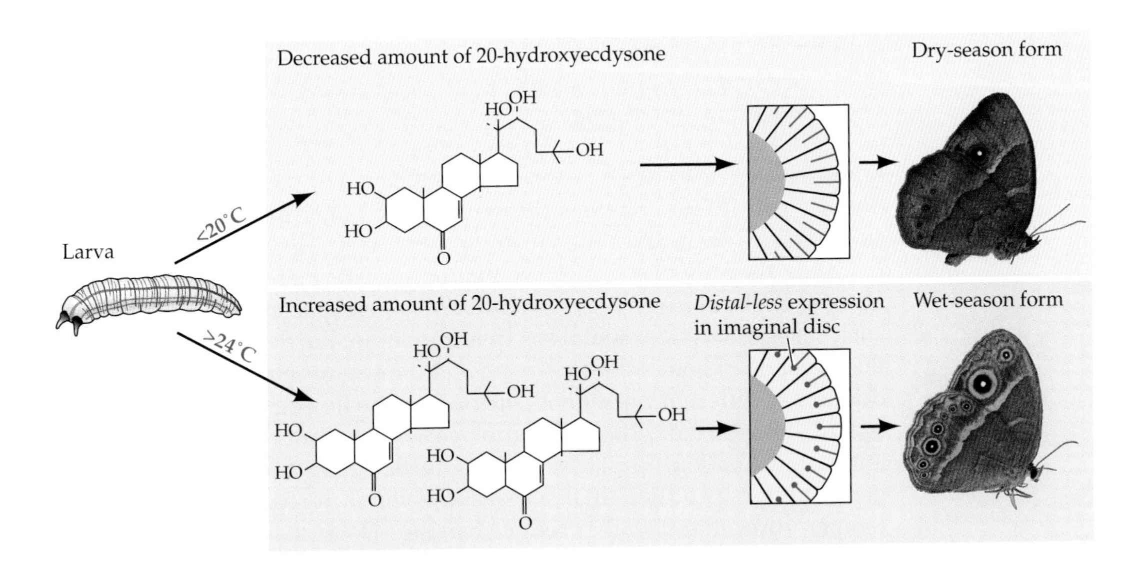

FIGURE 2.9 Phenotypic plasticity in *Bicyclus anynana* is regulated by temperature. High temperature (either in the wild or in controlled laboratory conditions) allows the accumulation of 20-hydroxyecdysone (20E), a hormone that is able to sustain *Distal-less* expression in the pupal imaginal disc. The region of *Distal-less* expression becomes the focus of each eyespot. In cooler weather, 20E is not formed, *Distal-less* expression in the imaginal disc begins but is not sustained, and eyespots fail to form. (Photographs courtesy of S. Carroll and P. Brakefield.)

The seasonal *Bicyclus* morphs appear to diverge at the later stages of signal activation and color differentiation, and this also appears to be regulated by the hormone ecdysone. When ecdysone is present early in pupal development, *Distal-less* expression is retained and the hot/wet season form of the wing, with its large eyespot, is formed (Brakefield et al. 1996; Koch and Buchmann 1987; Beldade et al. 2002). Conversely, in the dry season, ecdysone levels will not be sufficient to retain the expression of *Distal-less* in the wing primordia, and thus no eyespots will form. Thus, environmental temperature can be sensed and converted into an endocrine signal that regulates gene expression.

Neuroendocrine regulation of sex determination

Unlike the situation in mammals, where primary sex determination (i.e., whether gonads are ovaries or testes) is specified by the chromosomes, gonadal specification in many fishes, amphibians, and reptiles can be affected by hormones. While the detailed mechanisms are not known, the enzyme **aromatase** appears to play an important role in both temperature-dependent (turtles and crocodilians) and community-dependent (certain fishes) sex determination. Aromatase, as mentioned in Chapter 1, can convert testosterone into estrogens (Figure 2.10A).

RUNNING HOT AND COLD: SEX SPECIFICATION IN TURTLES Hormones may directly control the temperature-sensitive sex determination cascades of some species of turtles. When turtle embryos develop below a certain temperature, they are one sex; above that temperature, they develop into the other sex. The sex and set point differ between species (see Bull 1980). In such cases of environmental sex determination, hormones can determine the direction of gonad development, and the environment controls the production of different hormones. Injection of turtle eggs with estrogen can organize the gonadal rudiment into ovaries, no matter what their temperature of incubation. Similarly, injecting turtle eggs with inhibitors of estrogen synthesis produces male offspring, even if the eggs are incubated at temperatures that usually produce females (Dorizzi et al. 1994; Rhen and Lang 1994). The sensitive time for the effects of estrogens and their inhibitors coincides with the time when sex determination usually occurs (Bull et al. 1988; Gutzke and Chymiy 1988).

In some turtle species, aromatase is elevated at high temperatures, and estrogen is made at the expense of testosterone. In the laboratory, when aromatase inhibitors are used to inhibit estrogen synthesis in turtle eggs, male offspring are produced irrespective of temperature; this correlation is seen to hold under natural conditions as well. The aromatase activity of the pond turtle (*Emys*) is very low at the male-promoting temperature of 25°C (Figure 2.10B). At the female-promoting temperature of 30°C, aromatase activity increases dramatically during the critical period for sex determination (Desvages et al. 1993; Pieau et al. 1994; Ramsey et al. 2007). Temperature-dependent aromatase activity is also seen in diamondback terrapins, where its inhibition masculinizes their gonads (Jeyasuria et al. 1994). Aromatase also appears to be involved in the temperature-dependent differentiation of lizards and salamanders (Sakata et al. 2005).*

*What about mammals? Although we don't change gonadal sex when injected with hormones or hormone inhibitors, aromatase may still be important. It is known that the mammalian testes make aromatase and that this aromatase is used to synthesize estrogen from testosterone. (Estrogen is needed in male gonads for the proper functioning of the duct system.) It is also well known that if human or mouse testes don't descend, sperm production is abnormal. One explanation for the low sperm count in undescended testes is that the higher temperatures experienced within the abdomen increase aromatase function, thereby converting testosterone—required for efficient sperm production—into estrogen (Bilinska et al. 2003). Thus, the scrotum may have evolved because it holds the testes outside the main body wall, where these male gonads experience lower temperatures and the correct ratios of testosterone and estrogen can be maintained.

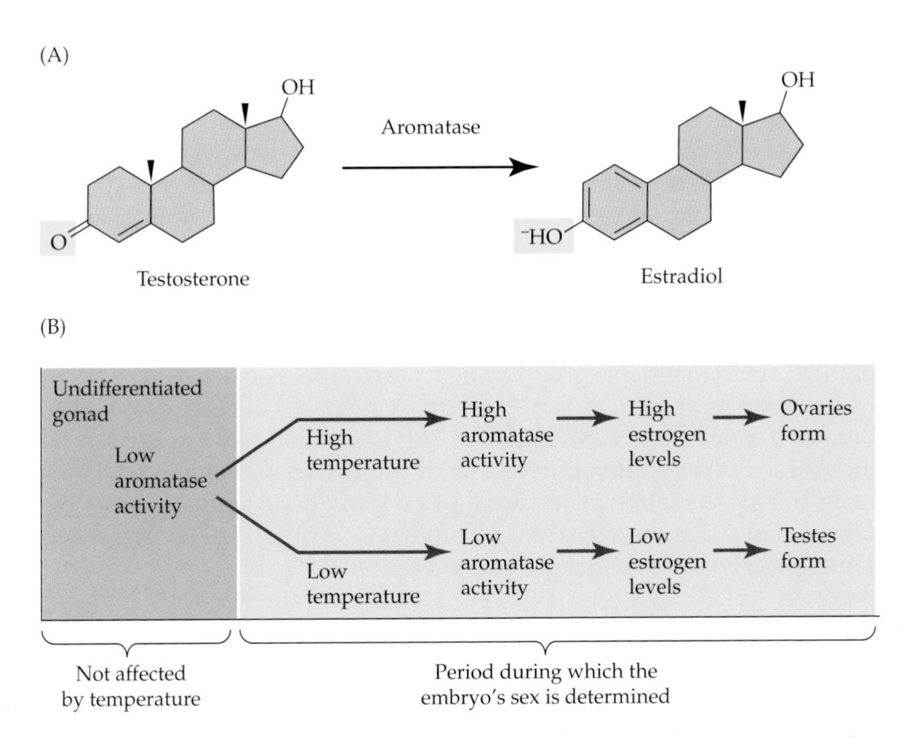

FIGURE 2.10 The enzyme aromatase converts androgens (such as testosterone) into estrogens (such as estradiol). (A) The name aromatase comes from its ability to aromatize the six-membered carbon ring by reducing the ring-stabilizing keto group (=O) to a hydroxyl group ($-OH^-$). This allows the hormones to bind to different receptors (dormant transcription factors) and activate different genes. (B) Scheme for sex determination in the turtle *Emys*. High temperatures during the sex-determining period cause high aromatase levels, leading to the production of estrogen and ovary formation. Aromatase is not promoted at low temperatures, thereby preventing ovaries from forming.

It is possible that the expression of aromatase is activated differently in different species. In some species, the aromatase *protein* itself may be temperature-sensitive. In other species, the expression of the aromatase *gene* may be differentially activated at high temperatures. In the red-eared slider turtle (*Trachemys scripta*), two of the same genes known to be involved in testes determination in mammals, *Sox9* and *Dmrt1*, are active very early during the period of sex determination and may act to turn on the aromatase gene (Murdock and Wibbels 2006; Shoemaker et al. 2007).

SEX AND SOCIAL INTERACTION: JOY TO THE FISHES IN THE DEEP BLUE SEA As mentioned in Chapter 1, many fishes can change sex based on social interactions, and these changes are thought to be mediated by the neuroendocrine system (Godwin et al. 2003). Interestingly, although the trigger of the sex change may be stress hormones (such as cortisol) that induce sex-specific neuropeptides, the effector of the change may once again be aromatase.

There are two aromatase genes in many animals; one is expressed in the brain, and one is expressed in the gonads. Black and colleagues (2005) have shown a striking correlation between changes in brain aromatase levels and changes in behavior during the sexual transitions of goby fish. The removal of the male from a stable group caused a rapid increase (more than 200%) in the aggressive behavior of the largest female, which is destined to become a male in about a week's time. This aggression could result from an increase in brain testosterone levels, since within hours upon removal of the male, these dominant females developed a lower brain aromatase than the other females (Figure 2.11). Gonadal aromatase levels, however, stayed the same, and gonadal sex change came later.

Sex, Aromatase, and Conservation Biology

Environmental sex determination has become a large issue in conservation biology because numerous man-made chemicals can alter hormone production. Atrazine, for instance, is the most widely used herbicide in the world; the United States alone uses 60 million pounds of it annually. But atrazine has effects beyond killing weeds. Atrazine can also induce aromatase expression and, thus, disrupt normal hormonal function (Crain and Guillette 1998; Fan et al. 2007). Such hormone-altering chemical compounds are known as **endocrine disruptors**. A variety of these environmental substances and their effects are discussed in detail in Chapter 6, but the effects of atrazine on amphibian gonadal development is of interest here.

One such case involves the development of hermaphroditic and demasculinized frogs after exposure to extremely low doses of atrazine. Hayes and colleagues (2002a) found that exposing tadpoles to atrazine concentrations as low as 0.1 part per billion (ppb) induced higher levels of aromatase and produced gonadal and other sexual anomalies in male frogs. Many male frogs developing at atrazine doses at 0.1 ppb and higher had ovaries in addition to testes. At concentrations of 1 ppb, the male vocal sacs (which the male frog must have in order to signal potential mates) failed to develop properly. The testosterone levels of adult male *Xenopus* were reduced nearly 90% (to the same level as control females) by exposure to 25 ppb atrazine for 46 days (Figure A).

These levels of atrazine are ecologically very relevant. The allowable amount of atrazine in our drinking water is 3 ppb, but levels are as high as 224 ppb in some streams in the agricultural heartland of the United States Midwest (Battaglin et al. 2000; Barbash et al. 2001). Given the amount of atrazine in the water and the sensitivity of frogs to this compound, the herbicide could be devastating to wild populations. Hayes and his colleagues collected leopard frogs (*Rana pipiens*) and water samples at eight sites across the central United States (Hayes

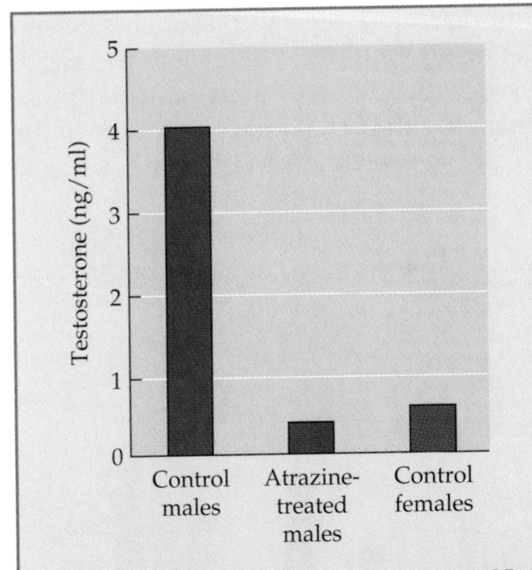

FIGURE A Effect of a 46-day exposure to 25 ppb atrazine on plasma testosterone levels in sexually mature male *Xenopus* frogs. Testosterone levels in control males were tenfold higher than in control females; levels in atrazine-treated males were at or below those of the control females. (After Hayes et al. 2002a.)

et al. 2002b, 2003). They sent the water samples to two separate laboratories for the determination of atrazine, and they coded the frog specimens so that the technicians dissecting the gonads did not know from which site an animal came. The results showed that all but one site contained atrazine—and that was the only site from which the frog specimens had no gonadal abnormalities. At concentrations as low as 0.1 ppb, leopard frogs displayed testicular dysgenesis (abnormally developed testes) or conversion of the testes into ovaries. In many examples, oocytes were found in the testes (Figure B).

These studies have obvious ramifications for humans. Indeed, in subsequent chapters we will see that mice and rats exposed to endocrine disruptors develop certain diseases (notably infertility syndromes and cancers) similar to those of humans. Many agricultural and social concerns mediate the amount of atrazine use, and the corporation that manufactures atrazine has lobbied against the work of some independent researchers whose results suggest a link between the chemical and reproductive malfunctions and cancers in wildlife and humans* (Blumenstyk 2003). However, concern over atrazine's apparent ability to disrupt sex hormones has resulted in bans on the use of this herbicide by France, Germany, Italy, Norway, Sweden, and Switzerland (Dalton 2002).

*Indeed, atrazine has been shown to lower the sperm count and testicular function in male rats, and some recent studies have shown that, in areas with higher amounts of pesticides, human sperm counts are low (Kniewald et al. 2000; Swan et al. 2003). In women, aromatase *inhibitors* are used clinically to lower the amount of estrogen in postmenopausal women who have estrogen-responsive breast cancer. The Environmental Protection Agency and the Centers for Disease Control are currently investigating whether agents that upregulate aromatase may put certain women at risk for breast cancer.

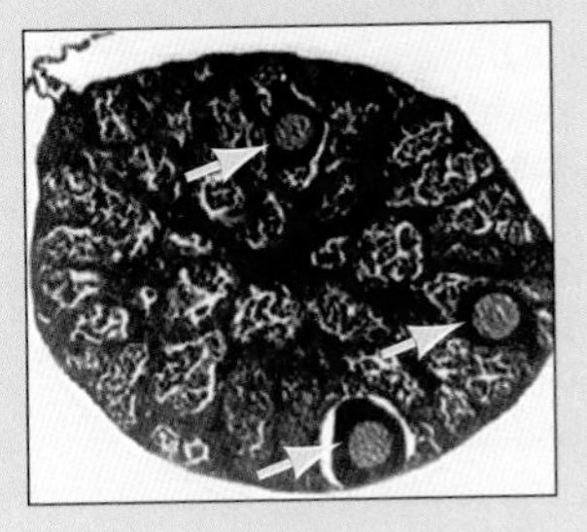
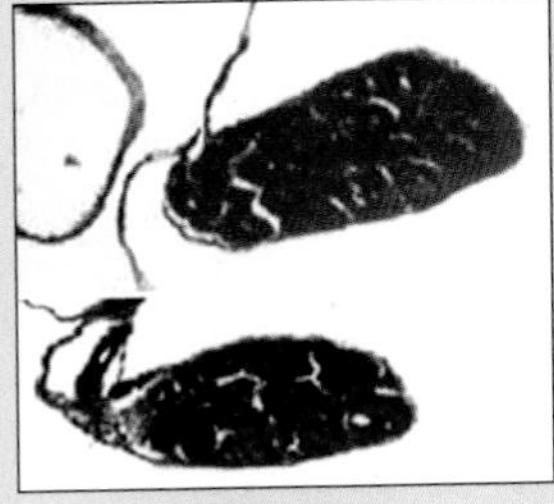

FIGURE B Demasculinization of frogs by low amounts of atrazine. (Left) Testis of a frog from a natural site having 0.5 ppb atrazine. The testis contains three lobules that are developing oocytes (arrows) as well as sperm. (Right) Two testes of a frog from a natural site containing 0.8 ppb atrazine. These organs show the severe testicular atrophy that characterized 28% of the frogs at that site. (From Hayes et al. 2003; photographs courtesy of T. Hayes.)

In porgy fish, aromatase inhibitors can block the natural sex change and induce male development (Lee et al. 2002). Thus, changes in the social group, perceived by the nervous system, became expressed by the hormonal system within hours, thereby changing the behavioral phenotype of the female fish. Interestingly, when it comes to behaviors in these fish, sex is in the brain before it is in the gonads.

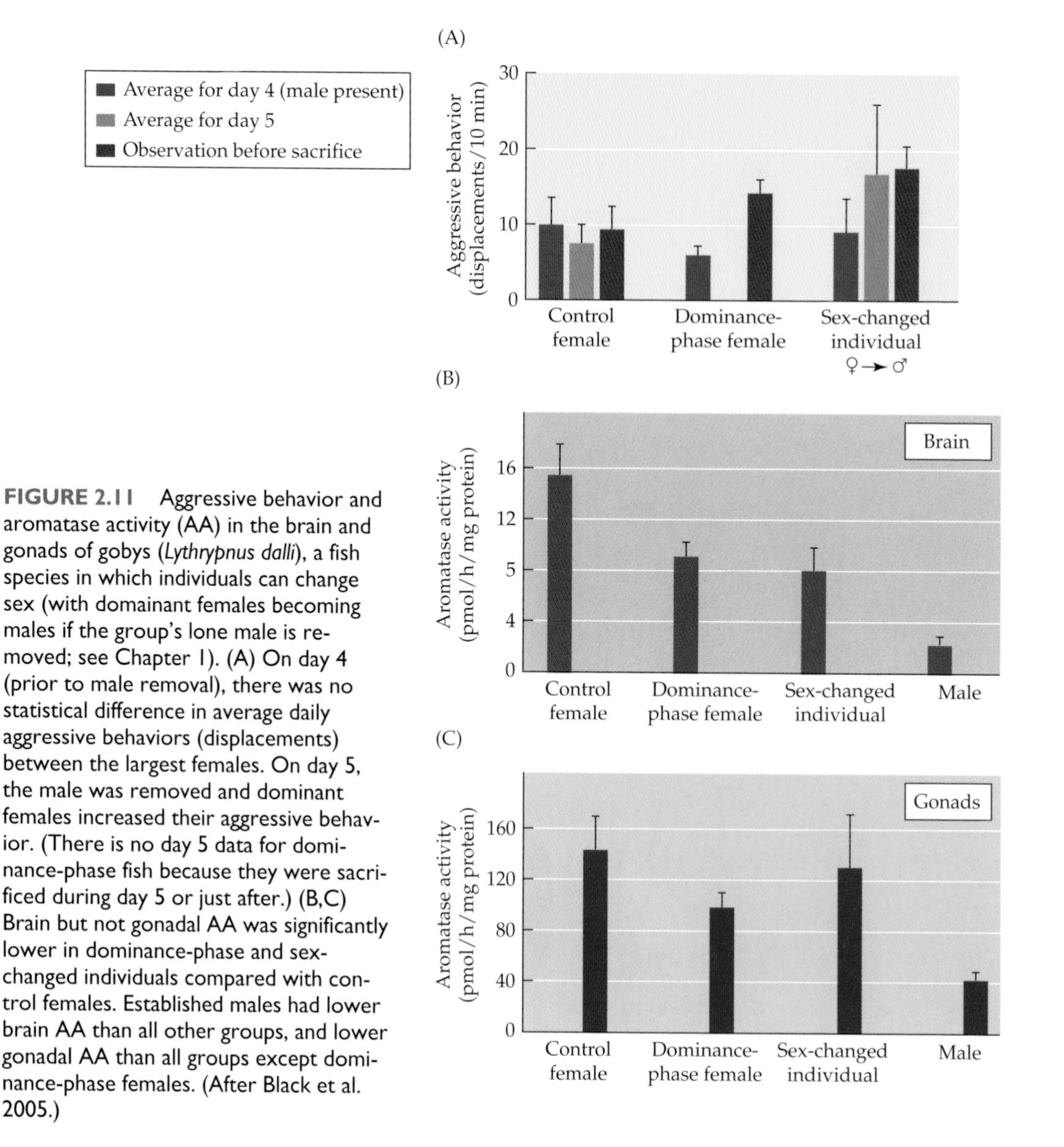

FIGURE 2.11 Aggressive behavior and aromatase activity (AA) in the brain and gonads of gobys (*Lythrypnus dalli*), a fish species in which individuals can change sex (with domainant females becoming males if the group's lone male is removed; see Chapter 1). (A) On day 4 (prior to male removal), there was no statistical difference in average daily aggressive behaviors (displacements) between the largest females. On day 5, the male was removed and dominant females increased their aggressive behavior. (There is no day 5 data for dominance-phase fish because they were sacrificed during day 5 or just after.) (B,C) Brain but not gonadal AA was significantly lower in dominance-phase and sex-changed individuals compared with control females. Established males had lower brain AA than all other groups, and lower gonadal AA than all groups except dominance-phase females. (After Black et al. 2005.)

An extreme phenotype for extreme times: Stress and cannibalism

Stress is a great inducer of polyphenisms, and it works through the production of adrenal corticosteroid hormones such as cortisol or corticosterone. Like insect juvenile hormone and ecdysone (and vertebrate sex hormones), corticosterone binds to an inactive transcription factor, converting the transcription factor to an active form that can enter the nucleus and regulate gene expression (see Figure 2.7). Corticosteroids are produced by the adrenal glands (or, in amphibians, by the interrenal glands) in response to the pituitary gland's secretion of adrenocorticotropic hormone (ACTH). ACTH is synthesized by corticotropin-releasing hormone (CRH), a small (41 amino acids) protein made in the hypothalamus. In amphibians and fish, CRH appears to induce the synthesis and secretion of both ACTH (for corticosteroids) and thyroid-stimulating hormone (for thyroxine) (Figure 2.12; see Denver 1999).

One of the most impressive of the stress-induced phenotypes is cannibalism in some spadefoot toads of the genus *Scaphiopus*. These amphibians have a remarkable strategy for coping with a very harsh environment. The toads are called out of hibernation by the thunder that accompanies the first spring storms in Arizona's Sonoran Desert. (Unfortunately, motorcycles produce much the same sound, causing the toads to come out of hibernation only to die in the scorching sun.) The toads breed in temporary ponds formed by the rain, and the embryos develop quickly into larvae. After the larvae metamorphose into toads, the young toads return to the desert, burrowing into the sand until the next year's storms bring them out.

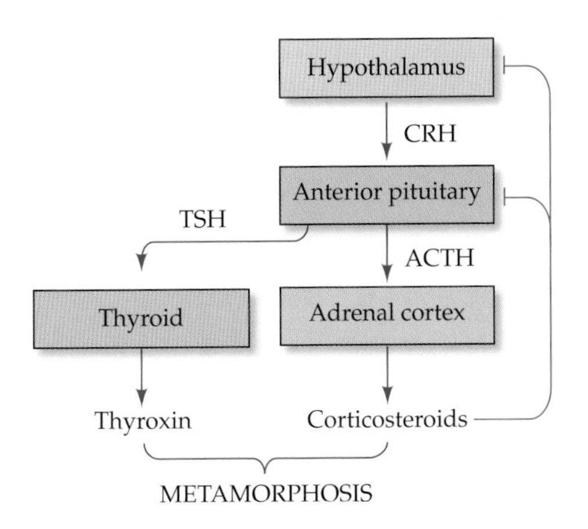

FIGURE 2.12 Hypothalmus-pituitary-adrenal-thyroid axis. The hypothalamus responds to internal or external stress signals by making corticotropin-releasing hormone (CRH), which tells the anterior portion of the pituitary gland to secrete adrenocorticotropic hormone (ACTH). ACTH circulates through the blood and tells the adrenal cortex to make corticosteroids. Corticosteroids circulate in the blood to activate various stress-response and polyphenism pathways; they also go to the pituitary and to the hypothalamus, where they bind to receptors that, in a negative feedback loop, downregulate CRH and ACTH production. The CRH stimulus from the hypothalamus to the pituitary is also responsible for the production of thyroid-stimulating hormone (TSH), resulting in the elevated levels of thyroid hormones that trigger metamorphosis.

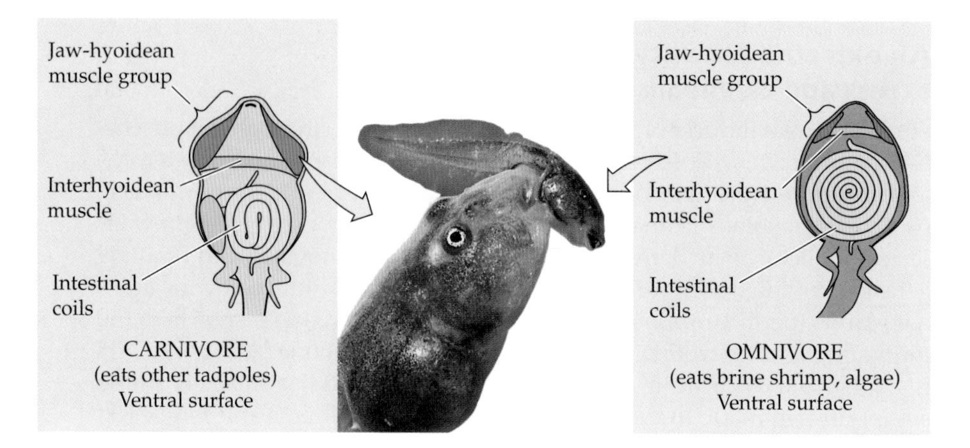

FIGURE 2.13 Polyphenism in tadpoles of a spadefoot toad (*Scaphiopus couchii*). The typical morph (right) is an omnivore, feeding on insects and algae. When ponds are drying out quickly, however, a carnivorous (cannibalistic) morph forms. It develops a wider mouth, larger jaw muscles, and an intestine modified for a carnivorous diet. The center photograph shows a cannibalistic tadpole eating a smaller pondmate. (Photograph © Thomas Wiewandt; drawings courtesy of R. Ruibel.)

Desert ponds are ephemeral pools that can either dry up quickly or persist, depending on the initial depth and the frequency of the rainfall. One might envision two alternative scenarios confronting a tadpole in such a pond: either (1) the pond persists until you have time to metamorphose, and you live; or (2) the pond dries up before your metamorphosis is complete, and you die. *Scaphiopus*, however, has evolved a third alternative. The timing of their metamorphosis is controlled by the pond. If the pond persists at a viable level, development continues at its normal rate, and the algae-eating tadpoles develop into juvenile toads. However, if the pond is drying out and getting smaller, some of the tadpoles embark on an alternative developmental pathway. They develop a wider mouth and powerful jaw muscles, which enables them to eat (among other things) other *Scaphiopus* tadpoles (Figure 2.13). These carnivorous tadpoles metamorphose quickly, albeit into a smaller version of the juvenile spadefoot toad. But they survive while other *Scaphiopus* tadpoles perish from desiccation (Newman 1989, 1992; Denver 1998).

The signal for accelerated metamorphosis appears to be the change in water volume. In the laboratory, *Scaphiopus* tadpoles are able to sense the removal of water from aquaria, and their acceleration of metamorphosis depends on the rate at which the water is removed. The stress-induced CRH signaling system appears to mediate this effect (Denver et al. 1998). This increase in brain CRH is thought to be responsible for the subsequent elevation of the thyroid hormones that initiate metamorphosis (see Figure 2.12) and the corticosteroids that accelerate it (Denver 1997; Boorse and Denver 2003). As in many other cases of polyphenism, the developmental changes in *Scaphiopus* are mediated through the endocrine system—sensory organs send a neural signal to regulate hormone release. The hormones then can alter gene expression in a coordinated and relatively rapid fashion. Moreover, hormonally mediated change means the environmental cue is able to affect several different target tissues, systematically altering the phenotype of the entire organism.

Corticosteroids may also mediate polyphenisms in response to intraspecies competition. When leopard frog (*Rana pipiens*) tadpoles grow at high densities, they tend to be smaller and develop more slowly. Their size and development are negatively correlated with the corticosteroid levels in their blood, suggesting that corticosteroid synthesis is responsible for slowing down development when population density is high (Glennemeier and Denver 2002).

"We will pump you up": Muscle hypertrophy

Physical stress affects gene expression in many places in the developing organism. Vertebrate cartilage cell differentiation and cartilage matrix production depend on mechanosensitive interactions among a number of genes and gene products. One of the most important of these genes is *Sox9*, which is upregulated by compressive force (Takahashi et al. 1998). The Sox9 protein is a transcription factor that activates numerous bone-forming genes. Tension forces also activate bone morphogenetic proteins (BMPs) and align chondrocytes (Bard 1990; Sato et al. 1999; Ikegame et al. 2001; Young and Badyaev 2007). Several studies implicate *indian hedgehog* as a key signaling molecule that is stimulated by stress and which activates the BMPs (Wu et al. 2001). Such examples of normal development's needing physical stress may explain certain birth defects. Abnormal muscle and joint forces on bones have been seen as causing the numerous bone deformities that afflict children with cerebral palsy after birth (Shefelbine and Carter 2004).

One of the most plastic traits of human beings is muscle mass. When muscles are strained, they get larger (Figure 2.14). Carpenters get enlarged biceps, weight lifters develop large pectoral muscles, and runners get big quadriceps. The basis of this hypertrophy (enlargement) is probably a modification of the injury repair system for muscle tissue. Muscle tissue represents fibers made of fused muscle cells. Next to the muscle tissue, usually within the muscular tissue matrix, are satellite cells—unfused muscle cells that have the ability to proliferate, grow, and fuse with muscle tissue upon stress. In other words, these satellites constitute a population of reserve cells that can be used to repair muscle tissue. When first stressed, muscles

FIGURE 2.14 Charles Atlas (born Angelo Siciliano in Calabria, Italy) won the title of "The Most Perfectly Developed Man" in 1922 and became an icon of male body building. Atlas pioneered resistance training for muscle hypertrophy. His "dynamic tension technique" was based on exercises that could be done within the confines of a small apartment, and did not rely on equipment or drugs. While biologists in the 1920s were saying that your genes were your destiny, Atlas' system, which pitted muscles against one another, delivered on his promise that "I can make you a new man!" (Photograph © AP Photo.)

respond by increased protein synthesis. Each nucleus in the muscle cell (a muscle cell has hundreds of nuclei because it was formed by a series of cell fusion events) has a region in which new protein synthesis occurs. However, if the physical stress continues, it appears that the muscle fibers cannot keep up with the amount of protein synthesis needed (Kadi et al. 2004). This may cause injury, or it may activate the same pathway that is used in injury. In either case, the muscle starts making a hormone-like protein called insulin-like growth factor-1 (IGF-1).

IGF-1 is made by both liver and muscle tissues. In physically stressed muscle, however, it appears that IGF-1 is produced in high amounts and remains near the muscle, where it can be used. Yang and colleagues (1996) found that in stressed (but not in unstressed) rabbit muscles, IGF-1 mRNA appeared. The protein products of this mRNA message could take many forms (due to alternative splicing), and one variant, seen chiefly in stressed muscles, gave rise to a protein called mechanogrowth factor (MGF). When complementary DNA (cDNA) derived from the mRNA for MGF was injected into mice, the animals showed a 25% increase in lean muscle within 3 weeks (Goldspink et al. 2004).

IGF-1/MGF appears to cause the satellite cells to proliferate and to fuse with the existing muscle fibers. This would be the basis of the hypertrophy. IGF-1 also acts on the muscle fibers through the pathway to further increase protein synthesis (Figure 2.15A). Lai and colleagues (2004) showed that when IGF-1 bound to its receptor, it initiated a cascade via a kinase called AKT. If AKT is mutated so it doesn't function, the mice have reduced musculature. Conversely, if AKT (without IGF-1) is artificially activated in mice, the mice develop three times the usual muscle mass, at the expense of the fat tissue (Figure 2.15B,C).

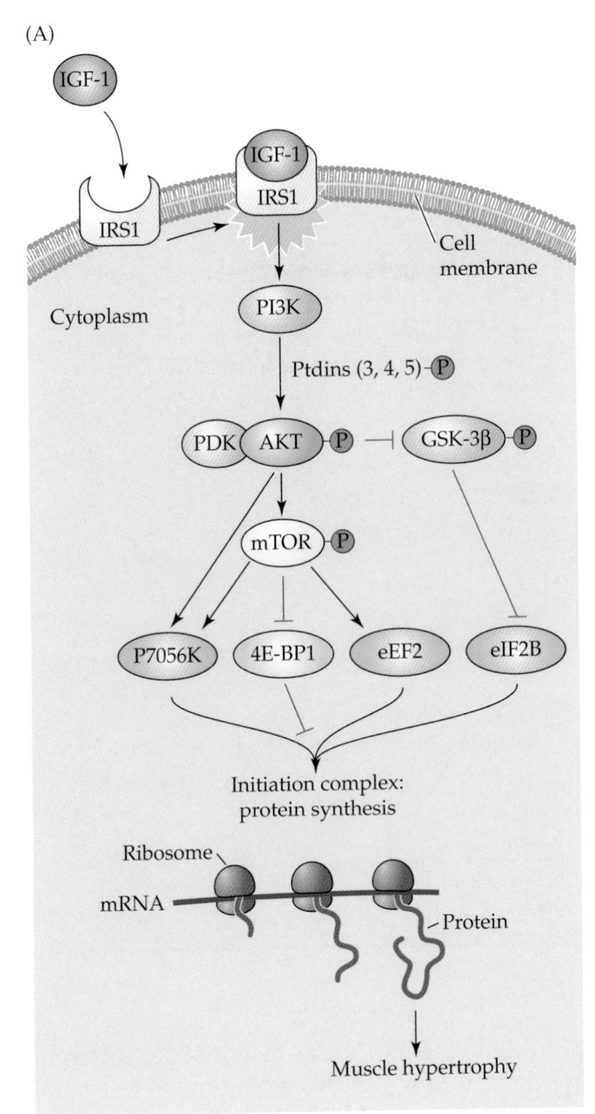

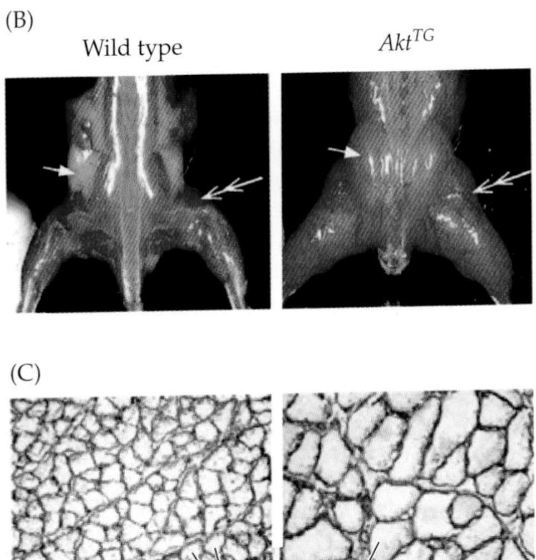

FIGURE 2.15 The actions of IGF-1 and AKT stimulate muscle development. (A) IGF-1 binds to its receptor (IRS1) on the cell membrane. After binding IGF-1, IRS1 becomes activated and it in turn activates compounds that phosphorylate AKT. Now AKT becomes active and can phosphorylate several target proteins, including GSK-3β, a protein that inhibits the protein-synthesis initiator eIF2B. When GSK-3β is phosphorylated, it cannot inhibit protein synthesis. Another protein phosphorylated by AKT is mTOR, which promotes the activity of the protein-synthesis initiators P7056K and eEF2; mTOR also blocks another protein-synthesis inhibitor, 4E-BP1. As a result, a protein synthesis initiation complex can be made on the ribosome and many more proteins can be produced, resulting in muscle growth. (B,C) Constitutive activation of the *Akt* gene in transgenic mice results in skeletal muscle hypertrophy. (B) The double-headed arrows point to a muscle that is significantly larger in the Akt^{TG} mouse than in its wild-type littermate. The single arrow points to a fat pad that is present in the wild type but absent in Akt^{TG} animals. (C) The micrographs show transverse sections of medial gastrocnemius muscle immunostained with an anti-laminin antibody to outline the perimeters of muscle fibers. When compared to its wild-type littermate, enlarged fibers are evident in the Akt^{TG} mouse. (B,C from Lai et al. 2004.)

Anabolic Steroids

Today's sports pages are full of stories about men (and some women) using anabolic steroids to increase their muscle mass and athletic performance. What are these substances, and do they work? Anabolic steroids are those steroid hormones whose functions include building cellular mass. Most commonly, these include testosterone and its derivatives. This makes them not only anabolic (muscle-building) but also androgenic (masculinizing) hormones.

It is well documented that injections of anabolic steroids increase muscle mass and strength in men whose testosterone levels are low due to disease or injury, and the rise in testosterone titers during male puberty is correlated with muscle growth at that time (Kopera 1985; Wilson 1996; Bross et al. 1999). The data supporting the value of testosterone in enhancing the performance of female athletes is uncontested, although derived largely from a study of the performance of East German athletes whose progress was charted by government officials (Figure A; Franke and Berendonk 1997).

However, as Cynthia Kuhn (2002) remarked in a review of anabolic steroids:

> There has been a tremendous disconnect between the conviction of athletes that these drugs are effective and the conviction of scientists that they aren't. In part, this disconnect results from the completely different dose regimens used by scientists to document the correction of deficiency states and by athletes striving to optimize athletic performance.

In other words, athletes in the gyms have been using much larger doses of steroids than the scientists in their laboratories.

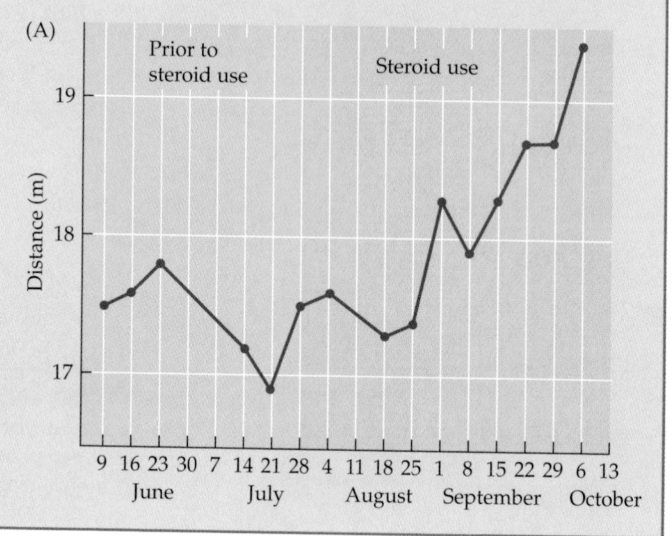

Interestingly for body builders, the molecular responses to different forms of exercise—isometric, lengthening-, or shortening-mode resistance exercise—seem to be the same (see Figure 2.14). Both exercise and passive stretching activate AKT (Sakamoto et al. 2003; Garma et al. 2007). Moreover, resistance training seems to be more effective for young men than for young women or the elderly (Petrella et al. 2006). This is reflected both in the muscle mass produced by training and in the amount of IGF-1 made by the cells. It is thought that older subjects lack satellite cells.

Thus we see that exercise activates the developmental mechanisms that promote the formation of muscles. Moreover, new studies show that without exercise, mitochondria fail to proliferate and produce fewer mRNA transcripts for the genes that promote energy metabolism. This leads to a vicious

Muscle biopsies from weight lifters show that anabolic steroid use increases both the number of fibers and the average fiber size in the trapezius muscle (Doumit et al. 1996; Kadi et al. 1999). Similar results were shown with animal models (Joubert and Tobin 1989). Both the number and size of muscle fibers depend on the activation and fusion of satellite cells within the muscle, and satellite cells contain testosterone receptors that would receive and utilize the anabolic steroids (Doumit et al. 1996). Furthermore, Inoue and colleagues (1994) showed that the steroid-induced hypertrophy does not happen if androgen antagonists are present during exercise. Therefore, one possible role of anabolic steroids is to stimulate the proliferation of satellite cells. This, in combination with stress-produced IGF-1, may be the cause of the increase in muscle mass.

Testosterone receptors abound in the body, so it is not surprising that extremely high amounts of anabolic/androgenic steroids have many other effects besides increasing muscle mass. These include the feedback inhibition of one's own testosterone production (leading to decreased sperm count), acne (sebaceous gland stimulation), and enlargement of the left part of the heart. In women, masculinization occurs (Figure B). Some of these masculinizing hormonal effects (acne, heightened libido, excess body hair) are reversible, while others (general muscle mass, lower voice due to larynx muscle enlargement, and clitoral hypertrophy) are not (see Kuhn 2002; Gruber and Pope 2000; Franke and Berendonk 1997).

(B)

(A) Effect of anabolic steroid use on the shot-put performance of a female athlete from East Germany in the year 1968. (B) East German shot-putter Heidi Krieger requested a sex-change operation due to the extent anabolic steroids (allegedly administered without her knowledge) had masculinized her voice and body. (A after Franke and Berendonk 1997; photograph © Tony Duffy/Getty Images.)

cycle in which lack of exercise diminishes mitochondrial function, which leads to lethargy and lack of exercise. Researchers at the University of Helsinki (Mustelin et al. 2008; Pietiläinen et al. 2008) have studied monozygotic twin pairs (who have identical genotypes) where one of the twins was obese and the other was not. The number of mitochondra in the obese twin was only about half that of the thinner twin. Moreover, the obese twins' mitochondria had significantly less expression of those genes responsible for oxidative phosphorylation. The obese twin's mitochondria also synthesized different enzymes, leading to their having lower energy metabolism, less ability to respond to insulin, and more fat accumulation. The researchers conclude that physical activity is important for establishing normal mitochondrial transcript and enzyme levels (Pietiläinen et al. 2008).

Signal Transmission from Environment to Genome through Direct Induction

When one tissue induces the differentiation of another during normal embryonic development (as when the presumptive neural retina tells the outer ectoderm to become the lens of the eye, or when the ureteric bud tells the mass of intermediate mesoderm next to it to become the kidney tubules), soluble factors are sent from the inducing tissue and are received by receptors in the tissue adjacent to it. In a similar fashion, microbes in the mammalian gut can be thought of as a tissue capable of inducing differentiation in the adjacent intestinal cells.

Microbial induction of gene expression in vertebrate intestines

In the next chapter, we will detail studies demonstrating that gut bacteria actually regulate some intestinal genes of vertebrates. Microarray analysis comparing gene expression in the digestive tracts of germ-free and conventionally raised zebrafish has revealed at least 200 microbially induced genes in the gut of the fish, including several that are involved in epithelial proliferation, nutrient metabolism, and the immune response (Rawls et al. 2004). One of the most important functions of the zebrafish microbial community is to complete the differentiation of the fish's gut (Figure 2.16). Without microbes, the fish intestine fails to form properly and even lacks the alkaline phosphatase-containing brush border that is characteristic of vertebrate intestinal cells. Such fish also lack the goblet cells and enteroendocrine cells normally found in the gut. (These cells help regulate the secretion of digestive enzymes and the contraction of the gut muscles.) Exposure of germ-free fish larvae to heat-killed preparations of the normal bacterial population will restore these cells and complete normal gut differentiation. One component of the bacteria, bacterial lipopolysaccharide, restores alkaline phosphatase activity and makes a brush border, but it does not restore the other functions of the bacteria (Bates et al. 2006; see Chapter 3 concerning a similar induction in squid). Thus, there appear to be several compounds made by the bacteria that induce gene expression and development in the vertebrate intestine.

In rabbits, a set of bacteria—including *Bacteroides fragilis* and *Bacillus subtilis*—appears to be required together in order to induce the formation of the **gut-associated lymphoid tissue**, or **GALT** (Perey and Good 1968; Tlaskalova-Hogenovava and Stepankova 1980; Cebra 1999). GALT mediates mucosal immunity and oral immune tolerance (so we don't have an allergic response to the food we eat), and GALT malfunction has been associated with allergies and with Crohn's disease in humans (see Rook and Stanford 1998; Steidler 2001). When introduced into germ-free rabbit appendices,

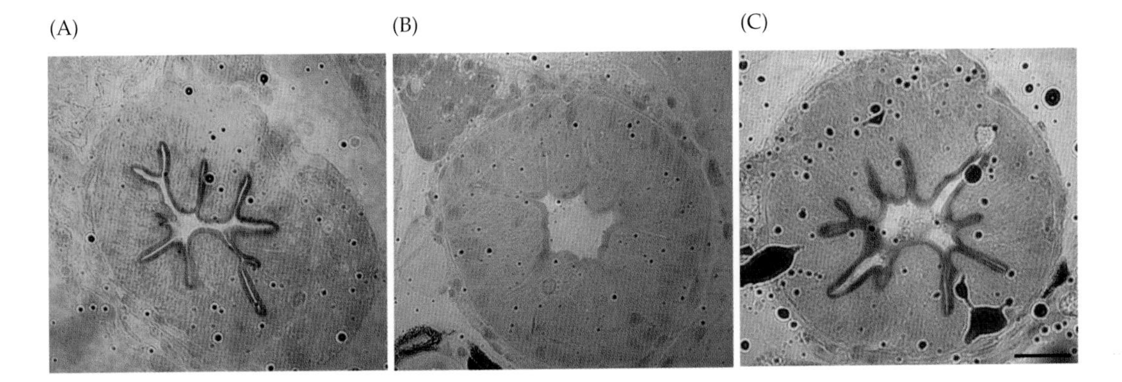

FIGURE 2.16 Zebrafish intestine fails to develop properly in the absence of microbial symbionts. Intestinal alkaline phosphatase activity (blue) was visualized on the brush border of transverse sections of the distal intestine at 8 days. (A) Alkaline phosphatase is present at the brush border in the intestine of conventionally raised zebrafish embryos at 8 days postfertilization. (B) No activity is seen in the 8-day germ-free zebrafish intestine. (C) Germ-free zebrafish given bacteria at day 5 synthesize alkaline phosphatase on their brush borders by day 8. (From Bates et al. 2006; photographs courtesy of Karen Guillemin.)

neither *B. fragilis* nor *B. subtilis* alone was capable of consistently inducing GALT or the mucosal B-cell repertoire. However, the combination of these bacteria consistently induced both (Rhee et al. 2004). Since *B. subtilis* induced some response, it is possible that it is the major inducer, receiving help from *B. fragilis*. Moreover, when *B. subtilis* is mutated such that it lacks the ability to produce spores or the YqxM protein (both regulated by the SpoOA stress system), there is no induction of GALT. As in zebrafish, it appears that chemical signals from the bacteria are being recognized by the rabbit intestinal cells and used as signals for differentiation.

Microbial induction of the vertebrate immune response

If predator-induced polyphenism is the alteration of development to counter potential predators, the mammalian immune system may be the acme of such polyphenic responses (see Frost 1999). The major predators of humans have never been lions, tigers, and bears, but rather bacteria, viruses, and protists. The mammalian immune system is an incredibly elaborate mechanism for sensing and destroying these microbes (unless they have remained compartmentalized in the gut or on the skin, where they may be essential partners).

Our immune system recognizes and destroys any material that is foreign to the body. When we are exposed to a foreign molecule (called an **antigen**),

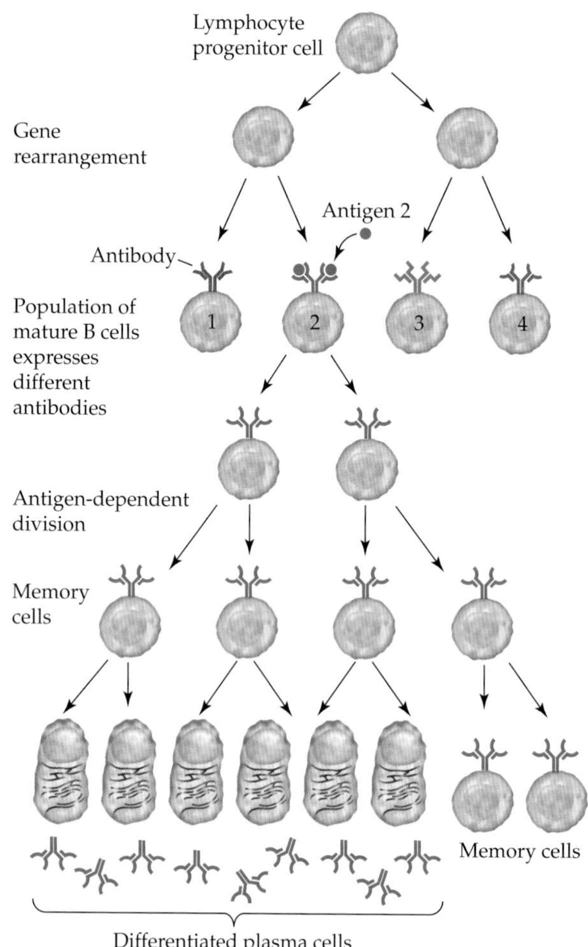

FIGURE 2.17 The clonal selection model. Each B cell produces only one type of the millions of potential antibody proteins. If and when the antigen to its specific antibody is presented, the B cell is stimulated to divide. (The steps of macrophage digestion and its presentation to the antibody are omitted for simplicity.) Most cells that result from this antigen-stimulated B-cell division differentiate, becoming plasma cells that secrete the antibody unique to that B-cell clone; some of the dividing cells become memory cells that will respond rapidly if the body is again exposed to the original antigen.

we manufacture **antibodies** and secrete them into our blood serum. These antibodies recognize and then inactivate or eliminate the antigen. When viruses or bacteria enter our bodies, they are seen as a collection of antigens, and an immune response is mounted against them. The basis for the antibody immune response is summarized in the five major postulates of the clonal selection hypothesis shown in Figure 2.17 (Burnett 1959):

1. Each B lymphocyte (B cell) can make one, and only one, type of antibody. It is specific for one shape of antigen only.
2. Each B cell places the antibodies it makes into its cell membrane with the specificity-bearing side pointing outward.

3. Bacteria and viruses are digested and internalized by macrophage cells, and the digested fragments (now considered antigens) are placed on their cell membranes. These antigens are presented by the macrophages to the antibodies on the B-cell membranes.
4. Only those B cells that bind to the antigen can complete their development into antibody-secreting plasma cells. These B cells divide repeatedly, produce an extensive rough endoplasmic reticulum, and synthesize enormous amounts of antibody molecules. These antibodies are secreted into the blood.
5. The specificity of the antibody made by the plasma cell is exactly the same as that which was on the cell surface of the B cells.

Out of the approximately 10 million types of antibody proteins a B cell can potentially synthesize, each B cell will make only one type. That is, one B cell may be making antibodies that bind to poliovirus, while a neighboring B cell might be making antibodies to diphtheria toxin. The type of antibody molecule on the B-cell surface is determined by chance.

B cells are continually being created and destroyed. However, when a specific antigen binds to a set of B cells, these cells are stimulated to divide and differentiate (see Figure 2.17). Most differentiate to become **plasma cells** that secrete the antibody; some divide to become **memory cells** that populate lymph nodes and can respond rapidly when exposed to the same antigen at a later time. The antigen activates the signal transduction cascade, which in turn activates transcription factors that activate the expression of those genes involved with lymphocyte differentiation and division. Thus, each person's constellation of plasma cells and memory cells differs depending on the antigens he or she has encountered. Even identical twins will have different populations of B-cell descendants in their spleens and lymph nodes.

This is, of course, a very simplified version of what actually happens. For our purposes here, the crucial point to understand is that the actual substances of the bacterial cell or virus—or anything else that is not seen as being a normal component of our bodies—can initiate the differentiation of a set of B cells into plasma cells. This not only shows that an environmental agent can induce differentiation directly, it also shows how sensitive our predator-induced polyphenism is.

Transgenerational Effects

That acquired characteristics cannot be inherited is one of the key concepts of genetics. And it is almost always the case. Children of weight lifters don't inherit their parents' physiques, and accident victims who have lost limbs can rest assured that their children will be born with normal arms and legs.

August Weismann cut off the tails of mice for several generations, "de-tailing" over 900 mice in all, but the following generations of mice were always born with normal tails. Darwin's colleague Thomas Huxley (see Richards 2004) remarked that these experiments demonstrated the truth of Hamlet's remark, "There's a divinity that shapes our ends/rough-hew them how we will." However, there have always been exceptions to this rule (see Appendix D), and some noteworthy exceptions involve the transgenerational continuity of polyphenisms.

Transgenerational polyphenism in locusts

The brown, gregarious, and long-winged migratory morph of the locust *Schistocerca gregaria* is retained for several generations after the crowding stimulus initiates the transformation from the solitary morph (see Chapter 1). This transgenerational effect is now known to be mediated during oviposition (egg laying) by a chemical agent introduced into the foam surrounding the eggs. Part of the gregarious female's phenotype involves the production of a particular chemical that she introduces into the foam surrounding her eggs. This chemical agent, which appears to be a modified form of the neurotransmitter L-dopa, is thought to be synthesized in the accessory glands of the female reproductive tract and to act during the time of egg laying (Hägele et al. 2000; Miller et al. 2008). If foam is transferred from egg masses laid by gregarious females to egg masses produced by solitary females, the solitary eggs turn into gregarious locusts. If the foam is washed off, the gregarious state reverts to the solitary phenotype after a few molts (Figure 2.18; McCaffery and Simpson 1998; Simpson and Sword, in press).

Transgenerational predator-induced polyphenisms

Ralph Tollrian's laboratory demonstrated that predator-induced polyphenism also can be stably transmitted from generation to generation (Agrawal et al. 1999). The investigators showed that, among plants, the wild radish has a very powerful predator-induced polyphenism. Compared to control plants, radish plants damaged by the caterpillar of the cabbage white butterfly (*Pieris rapae*) produce 10 times the concentration of mustard oil glycosides (which make the plant very distasteful), as well as 30% higher densities of trichomes (hooked, hairlike structures on the leaf, which stab the caterpillar's belly). Once these inducible defenses are generated, they are also found in the plant's offspring. Seedlings from induced plants were shown to germinate with more trichomes and higher levels of glycosides, even when the new generation was not exposed to herbivory by caterpillars.

Agrawal and colleagues also demonstrated that the predator-induced polyphenisms established in *Daphnia* were transmitted to their progeny.

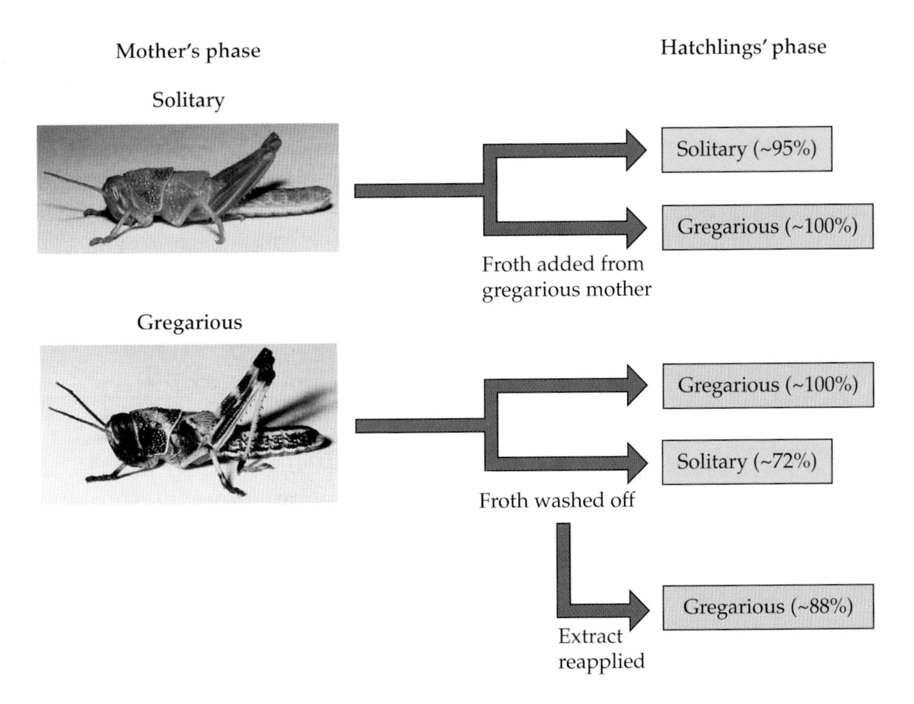

FIGURE 2.18 Summary of experiments showing that the source of the gregarizing agent in *Schistocerca* is an aqueous substance found in the froth surrounding the egg masses of gregarious females. (After McCaffery and Simpson 1998 and Simpson and Sword, in press.)

Daphnia cucullata is a parthenogenetic (all-female) species whose eggs develop within the parent's body and need not be fertilized. When the eggs of a *Daphnia* that underwent predator-induced polyphenism (exaggerated helmet and tail; see Figure 1.15) were grown in the absence of predators, they still developed the induced phenotype. The mechanisms for these polyphenisms and their inheritance are presently unknown.

Among the aphids, another group of parthenogenetic arthropods, winged morphs are induced either when the wingless form is at high density or when the larvae are exposed to the scent of a predator. When mature induced aphids are placed on clean plants to reproduce, they tend to give rise to winged offspring for several subsequent generations, even though the inducer is not present. In the pea aphid (*Acyrthosiphon pisum*), this transgenerational response is probably induced by the alarm pheremone β-farnesene (Mondor et al. 2005; Podjesak et al. 2005).

There is even evidence that predator-induced polyphenism can be transmitted in vertebrates. When Grindstaff and colleagues (2006) exposed birds to a bacterial surface product (bacterial lipopolysaccharide), the birds developed an immune response to it, as expected. Moreover, immunization of female birds before egg laying stimulated the production of antibodies against bacterial lipopolysaccharide in their chicks. Somehow, polyphenism appears to be transmitted across generations in the absence of the initiating cue. It is as if the organism evolves to expect the initiating condition: if the initiating cue is present for the parent (crowding, predators, bacteria), there is a good chance the situation will persist in the next generation. This certainly appears to be a Lamarckian situation, in which physiological changes induced in one organism are transmitted to its offspring. But how can this "evolutionary heresy" be happening? As it turns out, DNA methylation plays a role in many cases.

Methylation and transgenerational continuity: Toadflax

In 1999, studies of DNA methylation solved a long-standing mystery that had been vexing plant taxonomists since 1742—the origin of the symmetrical species of the toadflax plant, *Linaria vulgaris*. In the mid-eighteenth century, Linnaeus encountered some specimens of toadflax that were so different, they had to be classified as a new species (see Coen 1999). Normal *L. vulgaris* has flowers with definite upper and lower petals (Figure 2.19A). In the new form, the petals were symmetrically arranged, all on the same level—but every other part of the plant was obviously toadflax (Figure 2.19B). More-

(A)

(B)

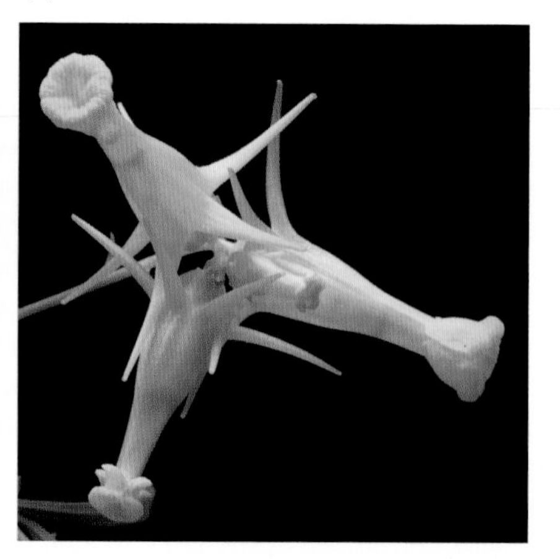

FIGURE 2.19 Epigenetic forms of toadflax. (A) *Linaria* with its relatively unmethylated *cycloidea* gene. (B) *Peloria* has a relatively heavily methylated *cycloidea* gene. Despite their different phenotypes, these two plants are the same species; *Peloria* is now considered to be a variant of *Linaria*. (Photographs courtesy of R. Grant-Downton.)

over, the symmetrical toadflax continued to generate such symmetrical forms, generation after generation. Since Linnaeus's binomial classification system for plants was based on petal and stamen parts, this caused a crisis: it meant that by Linnaean definitions, this was a newly formed species. For Linnaeus and his colleagues, species were invariant entities formed by God at the time of Creation. The concept of a new species was a dangerous thing to consider, and Linnaeus called the new form *Peloria* (from the Greek *pelor*, "monstrosity"), writing that "This is certainly no less remarkable than if a cow had given birth to a calf with a wolf's head."

In modern times, Coen's laboratory discovered that in fact the peloric form of toadflax is not a separate species (Coen 1999). Nor is there any mutation in the DNA between the linarian and peloric forms; rather, the DNA methylation pattern on their respective *cycloidea* genes is different. Compared with the same gene in *Linaria*, the *cycloidea* gene of *Peloria* is heavily methylated, and thereby inactivated. This hypermethylation is inherited from generation to generation (Cubas et al. 1999), thereby retaining the symmetrical phenotype. Other examples of inherited methylation differences causing stably inherited anatomical changes include flower genes in maize and *Arabidopsis* (Ronemus 1999).

Methylation and transgenerational continuity: Mice and rats

Earlier in this chapter, we described experiments that showed a female's diet during pregnancy controlled the coat color and obesity phenotypes of offspring carrying the *viable yellow* allele of the *Agouti* gene (see Figure 2.5). The expression of this gene was regulated by the differential DNA methylation in an inserted piece of DNA near the *Agouti* promotor site. What is just as amazing is that the epigenetic state of the *Agouti* gene (whether methylated or unmethylated) can be inherited transgenerationally: the coat colors of the grandoffspring resembled that of their mother, indicating that the *grandmother's* diet affected the coat color of her grandpups (Figure 2.20; Morgan et al. 1999). This same inheritance pattern has been noted in another mouse allele (*Axin*) that produces a polyphenism (kinky tail) due to DNA methylation differences (Rakyan et al. 2003). In this latter case, the transgenerational effect can arise through either the male or the female lineages, suggesting that the DNA methylation is on the germline cells, and that it is not erased during gamete production or during early development.* These transgenerational effects may have important considerations in medicine and in evolutionary biology.

*In mammals (at least in mice, where it has best been studied), there appear to be two waves of methylation erasure, the first of which takes place in the germ cell precursors and the second, in the early embryo prior to implantation. Thus, most DNA methylation has vanished by the time of implantation, except in a relatively small number of genes that are said to be *imprinted*. For more on this subject, see Appendix B.

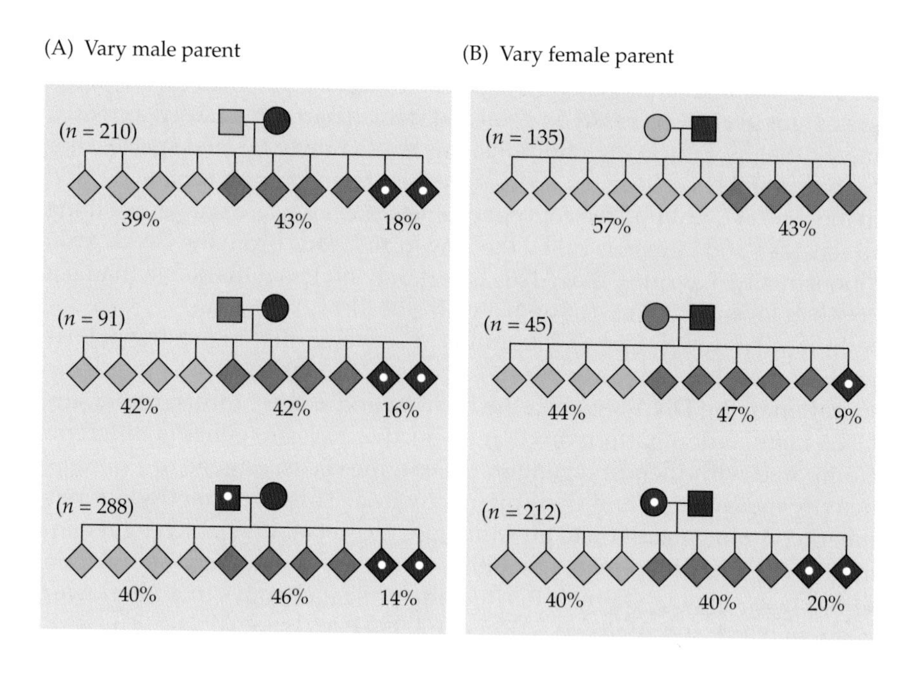

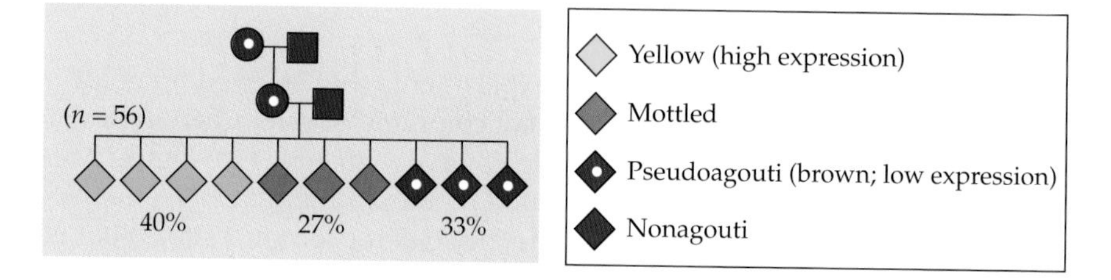

FIGURE 2.20 Transgenerational inheritance of an epigenetic phenotype. Heterozygous *viable yellow Agouti* (A^{vy}/a) mice of different phenotypes were mated with homozygous recessive nonagouti (*a*/*a*) mice. The expected *a*/*a* homozygotes were removed from this chart so, in all cases, the tested parent (of whatever color phenotype) has the same A^{vy}/a genotype. When the test parent is mated with a nonagouti partner (genotype *a*/*a*), the offspring are all A^{vy}/a heterozygotes. (A) There is no significant difference in the proportions of offspring phenotypes arising from yellow, mottled, and pseudoagouti (brown) fathers mated with nonagouti females. (B) The proportions of phenotypes arising from yellow, mottled, and pseudoagouti mothers differ significantly ($P < 0.0001$ for yellow versus mottled dams and mottled versus pseudoagouti dams). (C) A grandmaternal effect. Passage of the allele through two generations of pseudoagouti females produces significantly more pseudoagouti offspring than through only one generation of pseudoagouti dam (33% vs. 20%; $P = 0.003$). (After Morgan et al. 1999.)

As we will see in Chapter 7, mammalian embryos react to the maternal diet in utero, and levels of gene expression are set accordingly. If the maternal diet is poor in proteins, some of the genes involved in metabolism become more highly methylated, so the body will store food rather than utilize it. For instance, the offspring of pregnant rats fed a protein-restricted diet will have an unmethylated CG at a specific site in the promoter of the PPARα gene in the liver. This gene encodes a protein that regulates fatty acid metabolism, and the methylation pattern established in the embryo persists into adulthood (Lillycrop et al. 2008). Moreover, the methylation state of these genes appears to be transmitted transgenerationally.

Rats fed a low-protein diet for 12 generations demonstrated progressively slower fetal growth (Stewart et al. 1975); even when a normal diet was restored, the rats' growth and development did not normalize until 3 generations had passed. Moreover, as we will discuss in Chapter 7, rats exposed to low-protein diets in utero have the propensity to develop elevated blood pressure, blood vessel dysfunction, and insulin resistance. Even when a normal diet was fed to subsequent generations, Burdge and colleagues (2007b) found that these adverse effects persisted in the F_1 and F_2 generations (Figure 2.21). Burdge's study demonstrated that the methylation of gene promoters was probably the mechanism for the transgenerational stability of this polyphenism.

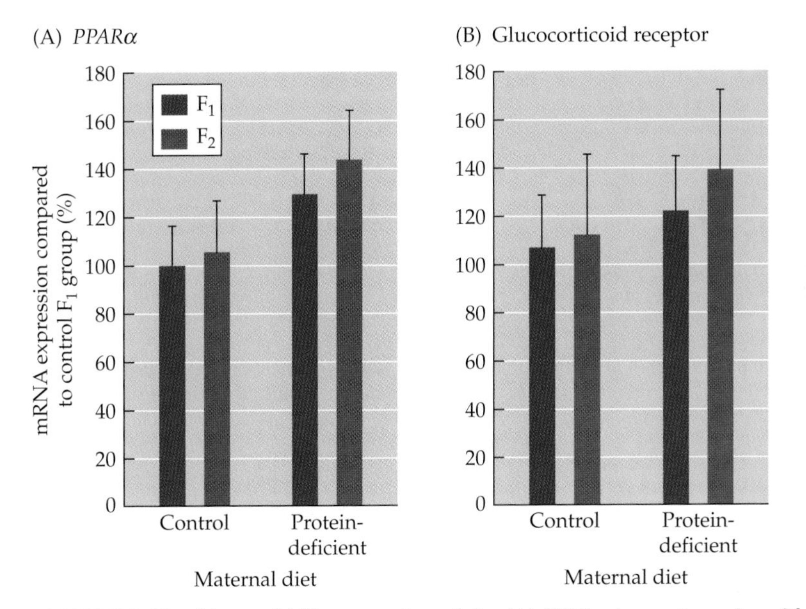

FIGURE 2.21 Liver mRNA expression of the (A) *PPARα* (peroxisomal proliferators-activated receptor alpha) and (B) glucocorticoid receptor (*GR*) genes. Though diet was normalized, altered promoter methylation effects persisted in the F_1 and F_2 generations. (After Burdge et al. 2007b.)

Summary

We now know of several mechanisms whereby the environment can alter development in perfectly understandable ways. Diet, temperature, maternal conditions, and even social interactions can change development. In many instances, this occurs through the altering of gene expression. Gene expression can be altered by direct induction (as when bacteria induce changes in the expression of intestinal genes) or by the neuroendocrine system (as in butterfly wing colors or muscle hypertrophy). There is also evidence that such gene expression differences can be mediated by DNA methylation and by transcription factor activation.

In addition, we now know mechanisms by which polyphenisms can be stably transmitted across generations. DNA methylation has been especially important in explaining how an organism can use the environment to establish a phenotype and then transmit that phenotype stably across several generations, and there are doubtless other epigenetic mechanisms yet to be discovered. We will discuss the evolutionary implications of transgenerational epigenetic transmission later in the book.

References

Agrawal, A. A., C. Laforsch and R. Tollrian. 1999. Transgenerational induction of defences in animals and plants. *Nature* 401: 60–63.

Barbash, J. E., G. P. Thelin, D. W. Kolpin and R. J. Gilliom. 2001. Major herbicides in ground water: Results from the National Water-Quality Assessment. *J. Environ. Qual.* 30: 831–845.

Bard, J. B. L. 1990. Traction and the formation of mesenchymal condensations *in vivo*. *BioEssays* 12: 389–395.

Bates, J. M., E. Mittge, J. Kuhlman, K. N. Baden, S. E. Cheesman and K. Guillemin. 2006. Distinct signals from the microbiota promote different aspects of zebrafish gut differentiation. *Dev. Biol.* 297: 374–386.

Battaglin, W. A., E. T. Furlong, M. R. Burkhardt and C. J. Peter. 2000. Occurrence of sulfonylurea, sulfonamide, imidazolinone, and other herbicides in rivers, reservoirs, and ground water in the midwestern United States 1998. *Sci. Total Environ.* 248: 123–133.

Beldade, P., P. M. Brakefield and A. D. Long. 2002. Contribution of Distal-less to quantitative variation in butterfly eyespots. *Nature* 415: 315–318.

Bilinska, B., M. Kotula-Balak, M. Gancarczyk, J. Sadowska, Z. Tabarowskii and A. Wojtusiak. 2003. Androgen aromatization in cryptorchid mouse testis. *Acta. Histochem.* 105: 57–65.

Bird, A. 2002. DNA methylation patterns and epigenetic memory. *Genes Dev.* 16: 6–21.

Black, M. P., M. Baillien, J. Balthazart and M. S. Grober. 2005. Socially induced and rapid increases in aggression are inversely related to brain aromatase activity in a sex-changing fish, *Lythrypnus dalli*. *Proc. Biol. Sci.* 272: 2435–2440.

Blumenstyk, G. 2003. The story of Syngenta and Tyrone Hayes at UC Berkeley: The price of research. *Chron. Higher Educ.* 50: 110. Available at http://www.mindfully.org/Pesticide/2003/Syngenta-Tyrone-Hayes31oct03.htm.

Boorse, G. C. and R. J. Denver. 2003. Endocrine mechanisms underlying plasticity in metamorphic timing in spadefoot toads. *Integr. Comp. Biol.* 43: 646–657.

Brakefield, P. M. and N. Reitsma. 1991. Phenotypic plasticity, seasonal climate, and the population biology of *Bicyclus* butterflies (Satyridae) in Malawi. *Ecol. Entomol.* 16: 291–303.

Brakefield, P. M. and 7 others. 1996. Development, plasticity, and evolution of butterfly eyespot patterns. *Nature* 384: 236–242.

Bross, R., M. Javanbakht and S. Bhasin. 1999. Anabolic interventions for aging-associated sarcopenia. *J. Clin. Endocrinol. Metab.* 84: 3420–3430.

Bull, J. J. 1980. Sex determination in reptiles. *Q. Rev. Biol.* 55: 3–21.

Bull, J. J., W. H. N. Gutzke and D. Crews. 1988. Sex reversal by estradiol in three reptilian orders. *Gen. Comp. Endocrinol.* 70: 425–428.

Burdge, G. C., M. A. Hanson, J. L. Stater-Jefferies and K. A. Lillycrop. 2007a. Epigenetic regulation of transcription: A mechanism for inducing variations in phenotype (fetal programming) by differences in nutrition during early life? *Brit. J. Nutrit.* 97: 1036–1046.

Burdge, G. C., J. L. Slater-Jefferies, C. Torrens, E. S. Phillips, M. A. Hanson and K. A. Lillycrop. 2007b. Dietary protein restriction of pregnant rats in the F_0 generation induces altered methylation of hepatic gene promoters in the adult male offspring in the F_1 and F_2 generations. *Brit. J. Nutrit.* 97: 435–439.

Burnett, F. M. 1959. *The Clonal Selection Theory of Immunity*. Vanderbilt University Press, Nashville.

Cebra, J. J. 1999. Influences of microbiota on intestinal immune system development. *Am. J. Clin. Nutr.* 69 (suppl.): S1046–S1051.

Coen, E. 1999. *The Art of Genes*. Oxford University Press, Oxford.

Cosgrove, M. S., J. D. Boeke and C. Wolberger. 2004. Regulated nucleosome mobility and the histone code. *Nature Struct. Mol. Biol.* 11: 1037–1043.

Crain, D. A. and L. J. Guillette, Jr. 1998. Reptiles as models of contaminant-induced endocrine disruption. *Anim. Reprod. Sci.* 53: 77–86.

Cubas, P., C. Vincent and E. Coen. 1999. An epigenetic mutation responsible for natural variation in floral symmetry. *Nature* 401: 157–161.

Dalton, R. 2002. Frogs put in the gender blender by America's favorite herbicide. *Nature* 416: 665–666.

Denver, R. J. 1997. Proximate mechanism of phenotypic plasticity in amphibian metamorphosis. *Am. Zool.* 37: 172–184.

Denver, R. J. 1998. Hormonal correlates of environmentally induced metamorphosis in the Western spadefoot toad, *Scaphiopus hammondii*. *Gen. Comp. Endocrin.* 110: 326–336.

Denver, R. J. 1999. Evolution of the corticotropin-releasing hormone signaling system and its role in stress-induced developmental plasticity. *Ann. NY Acad. Sci.* 897: 46–53.

Denver, R. J., N. Mirhadi and M. Phillips. 1998. Adaptive plasticity in amphibian metamorphosis: Response of *Scaphiopus hammondii* tadpoles to habitat desiccation. *Ecology* 79: 1859–1872.

Desvages, G., M. Girondot and C. Pieau. 1993. Sensitive stages for the effects of temperature on gonadal aromatase activity in embryos of the marine turtle *Dermochelys coriacea*. *Gen. Comp. Endocrinol.* 92: 54–61.

Dorizzi, M., G. Richard-Mercier, G. Desvages and C. Pieau. 1994. Masculinization of gonads by aromatase inhibitors in a turtle with temperature-dependent sex determination. *Differentiation* 58: 1–8.

Doumit, M. E., D. R. Cook and R. A. Merkel. 1996. Testosterone up-regulates androgen receptors and decreases differentiation of porcine myogenic satellite cells *in vitro*. *Endocrinology* 137: 1385–1394.

Evans, J. D. and D. E. Wheeler. 2000. Expression profiles during honeybee caste determination. genomebiology.com/2000/2/1/research/000.

Fan, W. and 10 others. 2007. Atrazine-induced aromatase expression is SF1-dependent: Implications for endocrine disruption in wildlife and reproductive cancers in humans. *Environ. Health Persp.* 115: 720–727.

Franke, W. W. and B. Berendonk. 1997. Hormonal doping and androgenization of athletes: A secret program of the German Democratic Republic government. *Clin. Chem.* 43: 1262–1279.

Frost, S. D. W. 1999. The immune system as an inducible defense. In *The Ecology and Evolution of Inducible Defenses*, R. Tollrianand and C. D. Harvell (eds.). Princeton University Press, Princeton, NJ, pp. 104–126.

Fuks, F. 2005. DNA methylation and histone modifications: Teaming up to silence genes. *Curr. Opin. Genet. Devel.* 15: 490–495.

Garma, T., C. Kobayashi, F. Haddad, G. R. Adams, P. W. Bodell and K. M. Baldwin. 2007. Similar acute molecular responses to equivalent volumes of isometric lengthening, or shortening mode resistance exercise. *J. Appl. Physiol.* 102: 135–143.

Gilbert, S. F. 2005. Mechanisms for the environmental regulation of gene expression: Ecological aspects of animal development. *J. Biosci.* 30: 101–110.

Glennemeier, K. A. and R. J. Denver. 2002. Small changes in whole-body corticosterone content affect larval *Rana pipiens* fitness components. *Gen. Comp. Endocrinol.* 127: 16–25.

Godwin, J., J. A. Luckenbach and R. J. Borski. 2003. Ecology meets endocrinology: Environmental sex determination in fishes. *Evol Dev.* 5: 40–49.

Goldspink, G. 2004. Mechanical signals, *IGF-1* gene splicing, and muscle adaptation. *J. Physiol.* 20: 232–238.

Grindstaff, J. L., D. Hasselquist, J.-A. Nilsson, M. Sandell, H. G. Smith and M. Stjernman. 2006. Transgenerational priming of immunity: Maternal exposure to a bacterial antigen enhances offspring humoral immunity. *Proc. Roy. Soc. Lond. B* 273: 2551–2557.

Gruber, A. J. and H. G. Pope, Jr. 2000. Psychiatric and medical effects of anabolic-androgenic steroid use in women. *Psychother. Psychosom.* 69: 19–26.

Gutzke, W. H. N. and D. B. Chymiy. 1988. Sensitive periods during embryology for hormonally induced sex determination in turtles. *Gen. Comp. Endocrinol.* 71: 265–267.

Hägele, B. F., V. Oag, A. Bouaïchi, A. R. McCaffery and S. J. Simpson. 2000. The role of female accessory glands in maternal inheritance of phase in the desert locust *Schistocerca gregaria*. *J. Insect Physiol.* 46: 275–280.

Hayes, T. B. 2005. Welcome to the revolution. Integrative biology and assessing the impact of endocrine disruptors on environmental and public health. *Integr. Comp. Biol.* 45: 321–329.

Hayes, T. B., A. Collins, M. Lee, M. Mendoza, N. Noriega, A. Stuart and A. Vonk. 2002a. Hermaphroditic, demasculinized frogs after exposure to the herbicide atrazine at low ecologically relevant doses. *Proc. Natl. Acad. Sci. USA* 99: 5476–5480.

Tlaskalova-Hogenovava, H. and R. Stepankova. 1980. Development of antibody formation in germ-free and conventionally reared rabbits: The role of intestinal lymphoid tissue in antibody formation to *E. coli* antigens. *Folia Biol.* 26: 81.

Waterland, R. A. and R. L. Jirtle. 2003. Transposable elements: Targets for early nutritional effects of epigenetic gene regulation. *Mol. Cell. Biol.* 23: 5293–5300.

Weaver, I. C. and 8 others. 2004. Epigenetic programming by maternal behavior. *Nature Neurosci.* 7: 847–854.

Weaver, I. C. and 7 others. 2007. The transcription factor nerve growth factor-inducible protein A mediates epigenetic programming: Altering epigenetic marks by immediate-early genes. *J. Neurosci.* 27: 1756–1768.

Weintraub, H. 1985. Assembly and propagation of repressed and derepressed chromosomal states. *Cell* 42: 705–711.

Weismann, A. 1875. Über den Saison-Dimorphismus der Schmetterlinge. In *Studien zur Descendenz-Theorie.* Engelmann, Leipzig.

Wheeler, D. E., N. Buck and J. Evans. 2006. Expression of the insulin pathway genes during the period of caste determination in the honeybee, *Apis mellifera. Insect Mol. Biol.* 15: 597–602.

Williams, S. C., C. R. Altman, R. L. Chow, A. Hemmati-Brivanlou and R. A. Lang. 1998. A highly conserved lens transcriptional element from the *Pax-6* gene. *Mech. Dev.* 73: 225–229.

Wilson J. D. Androgens. In *Goodman and Gilman's Pharmacological Basis of Therapeutics,* 9th ed. J. G. Hardman et al. (eds.). McGraw-Hill, New York, pp. 1441–1457.

Wu, K. C., J. Streicher, M. L. Lee, B. I. Hall and G. B. Müller. 2001. Role of motility in embryonic development: I. Embryo movements and amnion contractions in the chick and the influence of illumination. *J. Exp. Zool.* 291: 186–194.

Yang, S. Y., M. Alnaqeeb, H. Simpson and G. Goldspink. 1996. Cloning and characterization of an IGF-1 isoform expressed in skeletal muscle subjected to stretch. *J. Musc. Res. Cell Motil.* 17: 487–495.

Young, R. L. and A. V. Badyaev. 2007. Evolution of ontogeny: Linking epigenetic remodeling and genetic adaptation in skeletal structures. *Int. Comp. Biol.* 47: 234–244.

Chapter 3

Developmental Symbiosis

Co-Development as a Strategy for Life

There is a tendency for living things to join up, establish linkages, live inside each other, return to earlier arrangements, get along, whenever possible. This is the way of the world.

Lewis Thomas, 1974

Honor thy symbionts.

J. Xu and J. I. Gordon, 2003

As we saw in Chapters 1 and 2, an organism may respond to the presence of other organisms in the environment by altering its development in specific ways to produce a phenotype best suited to survive under the particular circumstances. But what if the "other organisms" are present not just sometimes (as in the case of a predator), but always? Such is the situation when a host responds to cues from its **symbionts**: organisms of different species that live in close association with the host. Indeed, we often find that the host has "outsourced" developmental cues to these expected symbiotic partners and that normal development comprises the interactions of at least two organisms. The presence of the symbiont may be essential for the host's development. This situation—where the complete development of an organism is dependent on symbionts with foreign genomes—is probably not the exception; in fact, it appears to be the rule (McFall-Ngai 2002; Saffo 2006).

Some of the most important examples of symbiosis in general and developmental symbiosis in particular come from the relationships between animals and the microbes that live inside them. One of the environmental factors that is constant is the presence of microbes. We may think in terms of the "The Age of Reptiles" or "The Age of Mammals," but it has always been the microbes' world. We large animals supply food and niches for the enormous microbial population of the planet.

Symbiosis: An Overview

The word **symbiosis** (Greek, *sym*, "together" + *bios*, "life") refers to any close association between organisms of different species (see Sapp 1994). In many symbiotic relationships, one of the organisms is much larger than the other, and the smaller organism may live on the surface or inside the body of the larger. In such relationships, the larger organism is referred to as the **host** and the smaller as the **symbiont**. There are two important categories of symbiosis:

- **Parasitism** occurs when one partner benefits at the expense of the other, as in a tapeworms that lives within and steal nutrients from a human digestive tract. Many important developmental symbioses involve **parasitoid** species that attach to or live within the host, eventually killing and consuming it. This strategy is typified by parasitoid wasp species that lay their eggs inside the larvae of other insects. The wasp larvae eat the insect larvae from within and undergo metamorphosis inside the hollow cuticle of the dead host.
- **Mutualism** is a relationship that benefits both partners. A striking example of this type of symbiosis is that of the Egyptian plover (*Pluvianus aegyptius*) and the Nile crocodile (*Crocodylus niloticus*). Although it regards most birds as lunch, the crocodile allows the plover to roam its body, feeding on the harmful parasites there. Thus the bird obtains food while the crocodile gets rid of its parasites.

A third type of symbiosis, known as **commensalism**, is defined as a relationship that is beneficial to one partner and neither beneficial nor harmful to the other partner. Although many symbiotic relationships may appear on the surface to be commensal, few if any symbiotic relationships in nature are truly neutral with respect to either party. Finally, the term **endosymbiosis** ("living inside") describes the situation where one cell lives inside another cell, a circumstance thought to account for the evolution of the organelles of the eukaryotic cell (see Margulis 1971), and one that describes the *Wolbachia* developmental symbioses discussed at length later in this chapter.

Although this chapter is concerned primarily with the developmental symbioses of animals, it is important to note that animal life as we know it rests on a variety of symbiotic relationships among plants, bacteria, and fungi.

The "Grand" Symbioses

Certain symbioses are essential for maintaining life as we know it on Earth. Some of these "grand" symbioses are responsible for transforming elemental chemicals from the soil or air into biologically useful molecules. Foremost among these reactions is **nitrogen fixation**, in which nitrogen gas (N_2) in the atmosphere is transformed into the biologically useful form NH_4^+.

Nitrogen-fixing nodules

Most of the nitrogen fixation on Earth (about 170 million metric tons of nitrogen per year) is accomplished by living organisms, and much of this amount results from the symbioses between bacterial species of the genus *Rhizobium* and the roots of certain green plants known as the legumes (including peas, soybeans, clover, alfalfa, and numerous tropical shrubs). While smaller amounts of nitrogen (about 20 million metric tons) are converted into ammonia by lightning, volcanic eruptions, and forest fires, and about 80 million metric tons are fixed industrially by the Haber process* (wherein nitrogen gas and hydrogen gas are heated to 500°C at a pressure of 250 atmospheres), *Rhizobium*-legume symbioses accomplish this fixation every day and at normal atmospheric temperatures and pressure. In so doing, they make all other life on Earth possible.

Neither free-living rhizobia nor uninfected legume plants can fix nitrogen; they must be in symbiotic association in order to do so. Moreover, various species of legumes are able to fix nitrogen only when partnered with a particular species of *Rhizobium*. Garden pea plants and soybeans are both legumes, but their nitrogen-fixing symbionts are different species of *Rhizobium*.

The association between the legume root and the rhizobia begins when the root secretes flavonoid compounds that are recognized by the rhizobia (Figure 3.1A). The rhizobia respond by secreting "nod (nodulation) factors" that bind to the root hairs and induce the hairs to grow to the bacteria and curl around them. The rhizobia then create a tunnel that enters the root. Once inside the root, the bacteria release chemicals that activate root cell division, producing nodules (Figure 3.1B). The rhizobia now change shape into a form called a **bacteroid** and express different genes. One of these bac-

*The Haber process is responsible for the production of fertilizers used in gardens and farms throughout the industrialized world. It is one of the most important industrial processes known and has been given major credit for the increase in world population (Smil 2001). The process uses large amounts of energy, and this energy is currently supplied by oil and petroleum. This has important economic, political, and social consequences, in that it now takes oil to grow crops. Thus, Michael Pollan (2002) writes, "Growing the vast quantities of corn used to feed livestock in this country takes vast quantities of chemical fertilizer, which in turn takes vast quantities of oil—1.2 gallons for every bushel. So the modern feedlot is really a city floating on a sea of oil."

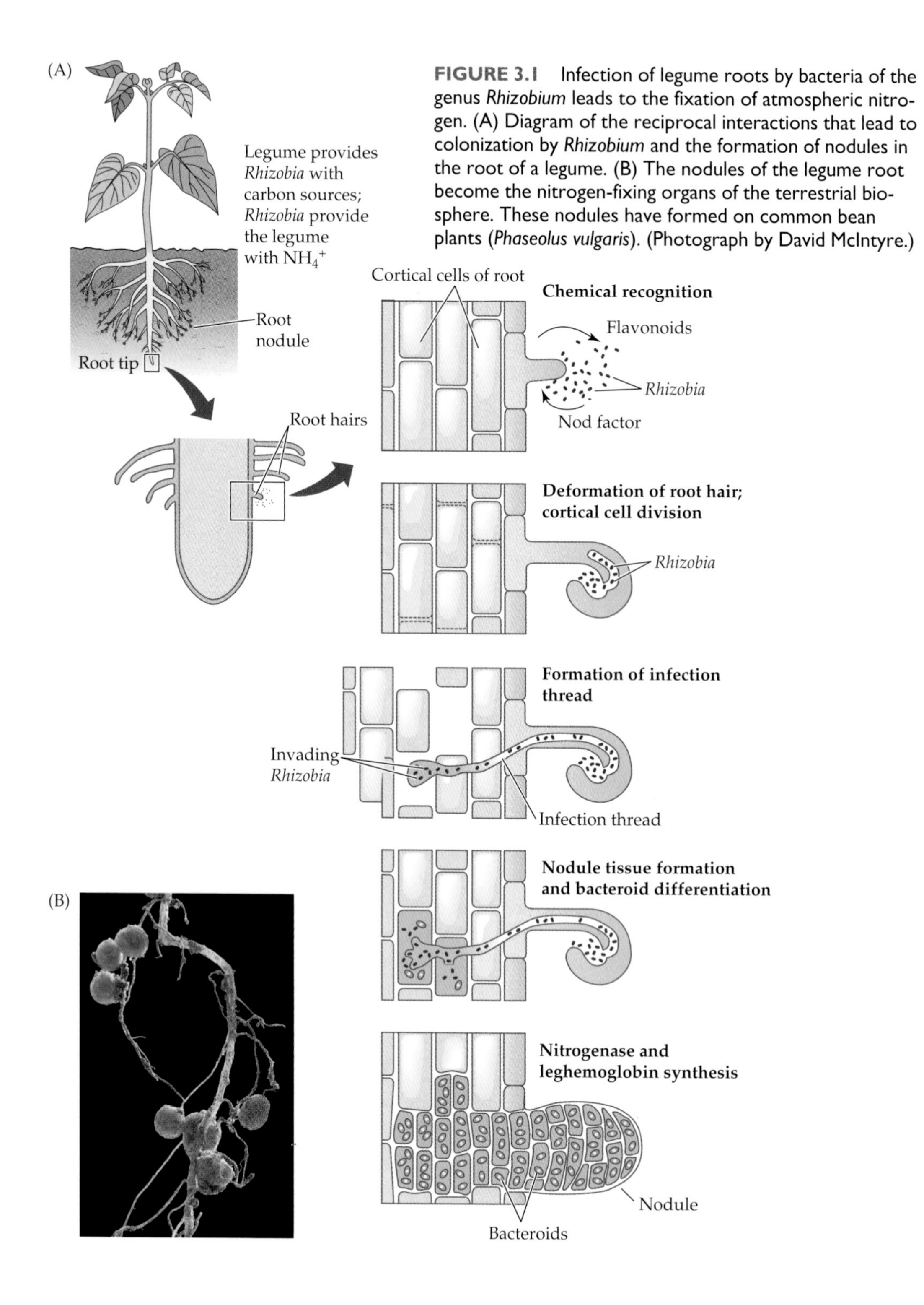

FIGURE 3.1 Infection of legume roots by bacteria of the genus *Rhizobium* leads to the fixation of atmospheric nitrogen. (A) Diagram of the reciprocal interactions that lead to colonization by *Rhizobium* and the formation of nodules in the root of a legume. (B) The nodules of the legume root become the nitrogen-fixing organs of the terrestrial biosphere. These nodules have formed on common bean plants (*Phaseolus vulgaris*). (Photograph by David McIntyre.)

Endophytes

We have long known that lichens are the products of plant-fungal symbioses and have relegated them to a separate place outside the traditional eukaryotic kingdoms of plant, animal, and fungus. They are considered exceptions. However, we now grasp the possibility that most, if not all, plants are products of symbiosis. (Indeed, perhaps all animals are, too.) The idea that each of what we call a "plant" is in fact a lichen-like symbiosis was driven home by the discovery of **endophytes**: fungi living inside the tissues of host plants.

While botanists had been aware of the presence of endophytes in a few species, their ubiquity was discovered accidentally. Evolutionary biologists using the polymerase chain reaction to amplify pine DNA inadvertently sequenced the DNA of fungi living within the pine needles. While at first this may have caused some consternation (since the original data made it seem that pines were really tree-shaped fungi), it soon became apparent that fungi had infiltrated the plant's tissues and become a normal part of its plant's anatomy. In some plant species, fungi have been found to infiltrate seeds, which means they have an immediate host in the next generation. In other cases, fungi generate spores that waft in the wind until they land on a host plant. When they enter a plant, fungi usually establish themselves near the plant's vascular system and live off fluids obtained by the host.

Endophytic fungi have been found in every plant species tested, and some species have numerous fungal symbionts (Vandenkoorhuyse et al. 2002; Milius 2006). Moreover, these fungi may play an important role in plant protection (Saikkonen et al. 2006). A southwestern U.S. rye grass, *Achnatherum robustum*, is commonly called "sleepy grass" because it causes wooziness in horses and other animals that eat it. It turns out that the wooziness is caused by an LSD-like compound produced by a symbiotic fungus—a compound that also prevents insects and nematodes from eating the grass (Faeth et al. 2006). In other species, fungi provide protection against the invasion of neighboring plants.

The plants and their associated fungi generally appear to cooperate for mutual advantage. However, a fungus is a potentially dangerous thing to have around. Tanaka and colleagues (2006) found that a single gene mutation can convert a friendly, mutualistic fungus into a predatory parasite: when the fungus *Epichloë festucae* is mutated such that it cannot induce oxygen radicals in its host rye grass, its proliferation goes unchecked, and it kills the host.

teroidal genes encodes nitrogenase, an enzyme that binds atoms of gaseous nitrogen (N_2) and adds hydrogens to the individual atoms to make NH_4^+.

Another set of rhizobial genes encodes a heme group (similar to that found in mammalian hemoglobin) that attaches to a globin protein made by the legume. This heme-containing protein, **leghemoglobin**, carries oxygen to the bacteroids to sustain their metabolism without interfering with the work of nitrogenase (which is inactivated by oxygen). In this way, atmospheric nitrogen is converted into a form that can be used as the essential material for synthesizing the bases of nucleic acids and the amino acids of proteins.

Mycorrhizae

In addition to nitrogen, plants need mineral nutrients, such as phosphorus, that they must take up from the soil. About 95% of all seed plants have a symbiotic relationship with fungi, which can grow either within the roots

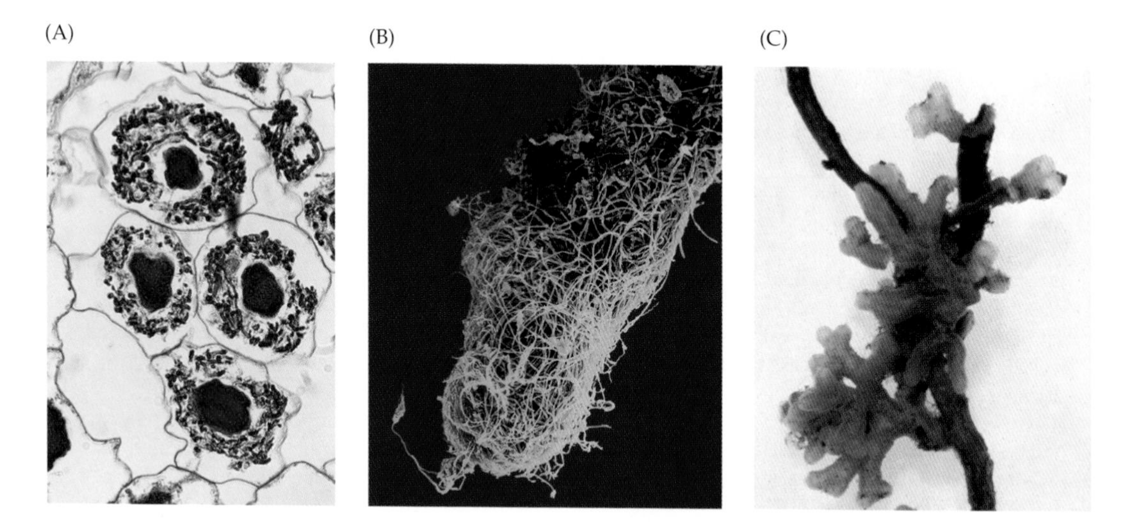

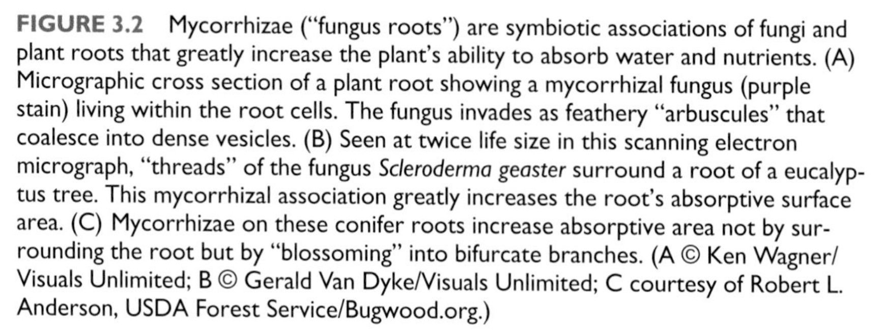

FIGURE 3.2 Mycorrhizae ("fungus roots") are symbiotic associations of fungi and plant roots that greatly increase the plant's ability to absorb water and nutrients. (A) Micrographic cross section of a plant root showing a mycorrhizal fungus (purple stain) living within the root cells. The fungus invades as feathery "arbuscules" that coalesce into dense vesicles. (B) Seen at twice life size in this scanning electron micrograph, "threads" of the fungus *Scleroderma geaster* surround a root of a eucalyptus tree. This mycorrhizal association greatly increases the root's absorptive surface area. (C) Mycorrhizae on these conifer roots increase absorptive area not by surrounding the root but by "blossoming" into bifurcate branches. (A © Ken Wagner/Visuals Unlimited; B © Gerald Van Dyke/Visuals Unlimited; C courtesy of Robert L. Anderson, USDA Forest Service/Bugwood.org.)

(Figure 3.2A) or around the roots, extending the root surface area (Figure 3.2B,C). Such associations are called **mycorrhizae**, which means "fungus roots." The fungal partner receives organic compounds from the plant while it functionally extends the plant's root system, allowing the plant to acquire more water and nutrients. Rain forest soils in particular usually have very low mineral content, and the lush growth of plants in such soil is dependent on mycorrhizae (see Martin et al. 2007; Cameron et al. 2008).

Both mycorrhizal fungi and the nitrogen-fixing bacteria in root nodules induce changes in gene expression in the plant, just as the plant changes gene expression in the fungus or bacterium. Interestingly, in *Medicago trunculata*, a legume species that can associate with both *Rhizobium* and with mycorrhizal fungi, the same gene is used for the establishment of both types of symbiosis. Mycorrhizae can also be important in plant development. Orchid seeds, for example, contain no energy reserves, so a developing orchid plant must acquire carbon from mycorrhizal fungi. All orchids need a mycorrhizal partner for their early development; an orchid may produce thousands of seeds, but only those finding a fungal partner have a chance of germinating (Waterman and Bidartando 2008). This is why orchids grow best in moist tropical environments, where fungi are plentiful.

Mycorrhizae may even constitute a "wood-wide web." Work by Suzanne Simard and colleagues (1997a,b; Selosse et al. 2006) has shown that the mycelia of symbiotic fungi can colonize the roots of different plant species and link them together beneath the soil. Such associations may allow plants to "share the wealth," with nutrient exchange taking place among the individual plants connected by the fungi. Moreover, such mycorrhizal networks may help establish seedlings in understory conditions that would otherwise be hostile (Booth 2004). Thus, below the surface, plants may "fuse" into a cooperative network that modulates competition and provides a structure for an integrated plant community.

Life Cycle Symbioses

The life cycles of most organisms are exquisitely coordinated with the seasons and are intimately connected to their environments. Developmental considerations are critical, since in most organisms, the adult stage of the life cycle is usually very brief and is often completely given over to reproduction. We tend to forget that the beautiful large moths that eclose in the summer live for only one or two days. The males often don't even have mouthparts with which to feed; their bodies have been so formed as to be nothing more (or less) than mobile gonads. The cannibalistic practice of the female praying mantis is readily explainable from the perspective that adult male insects do not survive long, and if providing food for your developing progeny will help them survive, everyone benefits. Indeed, some male mantises die during copulation without female assistance (Johns and Maxwell, 1997; Foellmer and Fairbairn 2003).

Some of the most important life cycle symbioses in nature involve angiosperm (flowering plant) pollination. Here, the male gamete is not a sperm, but rather a cell contained within a pollen grain that must be transported to the female part of the flower by the wind or by an animal pollinator. Although these interactions are usually between the adult plant and an adult pollinator (insect, bird, or mammal), the developmental aspects of these interactions are also crucial. For instance, the maturation of the pollinator has to match the maturation and floral opening of the plant. Global warming has already changed the phenology (timing of cyclic events in a plant's annual cycle) of numerous plants. For example, the cherry trees on Mt. Takao in Tokyo are now flowering 5–6 days earlier than they did 25 years ago. Most cherry tree species flower 3–5 days earlier for each 1°C increase in temperature (and the February-March temperature at Mt. Takao has increased 1.8°C during that time), but some early-blooming varieties flowered as much as 9 days earlier for each 1°C increase in temperature (Miller-Rushing et al. 2007). These altered rates in timing reflect the plasticity of the organism to environmental cues. However, one can imagine that if

Life Cycle Symbioses: The Large Blue Butterfly

The unsuccessful attempt to save the Large Blue butterfly *Maculinea arion* from extinction in England is an example of the need to understand a species' life cycle in all its phases. There were hundreds of thousands of these beautiful butterflies in England during the 1950s, but censuses carried out in the early 1970s found only a few hundred individuals. In an attempt to save the species, conservation biologists tried to protect the thyme plants that were the butterflies' preferred egg-laying sites by eliminating the grazing of cows in these areas. But the butterfly population continued to decline.

It turned out that, although the female *M. arion* does indeed lay her eggs on thyme plants. the larvae do not eat thyme. Instead the caterpillars drop to the ground, where they produce a mixture of volatile chemicals that mimics the smell of larvae of the ant species *Myrmica sabuleti*. The patrolling worker ants mistake the butterfly larvae for their own and carry the caterpillars into the ant nests. Once there, the caterpillars eats young ants until they pupate. They undergo metamorphosis in the ant colony, surfacing as butterflies.

The conservation efforts failed because, while the grazing restrictions allowed pastures to grow thicker and taller, they also changed the soil conditions to favor different species of ants. As a result, the *M. arion* larvae dropped into a predatory field rather than a nutritive one. The species went extinct in Britain in 1979, although a population introduced from Sweden, with the proper conservation management, appears to be growing (Thomas 1995; Nash et al. 2008).

But in the complex web of life histories, the parasitic *Maculinea* larvae can become the parasitized. It seems these caterpillars are the sole food source for the larvae of several species of wasps of the genus *Ichneumon* (Thomas and Elmes 1993; Thomas 2002). A female wasp can detect not only the ant colonies but also the presence of butterfly larva within them. She enters only colonies where caterpillars are present; once there, the wasp emits pheromones that cause the ants to fight among themselves while she goes about laying a single egg in each butterfly larva. Each wasp egg hatches into a larvae that eats the caterpillar as it begins pupation. Eleven months later, the pupal case is shed and there emerges not a butterfly but an adult wasp.

These fascinating interactions are shown in a series of videos on the BBC documentary *Life in the Undergrowth* (Programme 4).

(A)

(B)

(A) A female Large Blue butterfly, *Maculinea arion* from the re-introducton program underway in Somerset, England. (B) Newly hatched caterpillar of a related species (*Maculinea alcon*) being carried by a worker *Myrmica* ant to its nest. (From Nash et al. 2008; photographs courtesy of D. R. Nash.)

insect pollinators of these trees had a different susceptibility to temperature than the plant (or if they were to cycle on a more stable cue, such as photoperiod), they would not eclose from their pupae at a time when the blossoms are open to them (see Stiling 1993).

Photoperiodism (the response to durations of sunlight) is controlled genetically, and there is evidence that some insects and birds in the Northern Hemisphere are evolving photoperiods that resemble those of more southerly members of their species (Bradshaw and Holzapfel 2001; Nussey et al. 2005). Global warming also has a major effect on the prevalence of mites and other predators of pollinating insects. There have been declines in numerous pollinating species, especially bees, due in large part to parasitic predators (NRC, 2007). This is a matter for grave concern because, when there are obligate webs of symbiosis, the elimination of any one of the participants spells disaster for the other members.

The life cycles of many marine invertebrates are regulated by mats of bacteria, called bacterial **biofilms**, that determine where and when the larvae can settle and undergo metamorphosis. Different species can take their cues from biofilms produced by different species of bacteria (Hadfield and Paul 2001; Zardus et al. 2008). These biofilms occur naturally, and they help determine the distribution of the species. However, we humans are changing this distribution by our desire to place large objects into the oceans. Such objects readily acquire biofilms and the resultant marine fauna that attach to them. As early as 1854, Charles Darwin speculated that barnacles were transported to new locales when their larvae settled on the hulls of ships. The ability of biofilms to aid invertebrate larval settlement and colony formation explains the ability of barnacles and tubeworms ("biofouling invertebrates") to accumulate on ships' keels, clog sewer pipes, and deteriorate underwater structures. Such invertebrate colony formation costs billions of dollars a year in prevention, fuel consumption, and structure maintenance (Zardus et al. 2008).

Symbioses in which one organism is essential to another at a particular stage of its life cycle are thus not unusual. However, such life cycle symbioses can be exacting, because they require all the necessary players to be present and participating at precisely the right stages of development.

Getting Symbionts Together with Their Hosts

All symbiotic associations must meet the challenge of maintaining their partnerships over successive generations. In the partnerships that are the main subject here, in which microbes are crucial to the development of their animal hosts, the task of transmission is usually accomplished in one of two ways, by vertical or horizontal transmission.

Vertical transmission refers to the transfer of symbionts from one generation to the next through the germ cells, usually the eggs (Krueger et al.

1996). There are several ways by which embryos can become infected by their mothers. Certain clams (*Solemya reidi, S. velum*, and *Calyptogena soyae*) produce oocytes that contain microbial symbionts (Endow and Ohta 1990; Cary 1994). Bacterial symbionts enter the larval gut of the mollusk *Dentalium* by having ovarian cells infect the egg. These bacteria specifically adhere to those cells destined to become the larval gut. During gastrulation, these cells (originally on the outside of the embryo) become internalized to form the gut lining, and the bacteria ride in with them (Geilenkirchen et al. 1971). In the common composting earthworm (*Eisenia foetida*), symbiotic bacteria from the kidneys (nephridia) are deposited directly onto the capsules inside which the embryos are developing. The developing embryo makes a nephridial duct that recruits only that particular species of bacterium into the kidneys, excluding others. This species is thought to protect the worm's nephridia against colonization by other, more harmful, bacteria (Davidson and Stahl 2008).

And, as we will soon learn, there are many cases in which bacteria of the genus *Wolbachia* reside in the egg cytoplasm of invertebrates and provide important signals for the development of the individuals produced by those eggs. *Wolbachia*, one of the world's most successful bacteria, can prevent development in some species and become an important agent of normal development in others. *Wolbachia* will provide us with fascinating examples of the evolutionary paths that can lead to developmental dependence of a host on its symbiont species.

In **horizontal transmission**, the metazoan host is born free of symbionts but subsequently becomes infected, either by its environment or by other members of the species. One example of horizontal transmission involves aquatic eggs that attract photosynthetic algae. Clutches of amphibian and snail eggs, for example, are packed together in tight masses. The supply of oxygen limits the rate of their development, and there is a steep gradient of oxygen from the outside of the cluster to deep within it; thus embryos on the inside of the cluster develop more slowly than those near the surface (Strathmann and Strathmann 1995). The embryos seem to get around this problem by coating themselves with a thin film of photosynthetic algae, which they obtain from the pond water. In clutches of amphibian and snail eggs, photosynthesis from this algal "fouling" enables net oxygen production in the light, while respiration exceeds photosynthesis in the dark (Bachmann et al. 1986; Pinder and Friet 1994; Cohen and Strathmann 1996). Thus, the algae "rescue" the eggs by their photosynthesis.

Horizontal transmission also appears to be the case in the well-studied association between the bioluminescent bacterium *Vibrio fischeri* and its host, the squid *Euprymna scolopes*. In this relationship, *Vibrio* bacteria provide developmental signals for the formation of the squid's light organ. Nyholm and colleagues (2000) have shown that the newborn squids acquire *Vibrio* by horizontal transmission from seawater; the squid's developing light organ secretes molecules that selectively attract and catch the microbes.

The Squid and the Microbe: A Paradigm of Symbiont Influence

The Hawaiian bobtail squid *Euprymna scolopes* is a tiny (2 inches long) native of the shallow waters off Hawaii (Figure 3.3A). A nocturnal animal, the squid preys on shrimp. Predatory fish may be alerted to the presence of the squid if the moon casts its shadow as it skims across the water overhead. *E. scolopes*, however, has developed a most creative mechanism of dealing with this potential threat: it emits light from its underside, mimicking the moonlight and hiding its shadow from potential predators. The squid cannot accomplish this feat of disguise alone; the presence of the symbiotic bacterium *Vibrio fischeri* in the squid's light organ is required to generate this characteristic glow. Both the squid and the bacterium benefit from this mutualistic relationship—the squid because it gains protection from predators, and the bacterium because it is able to live safely within the

(A)

(B)

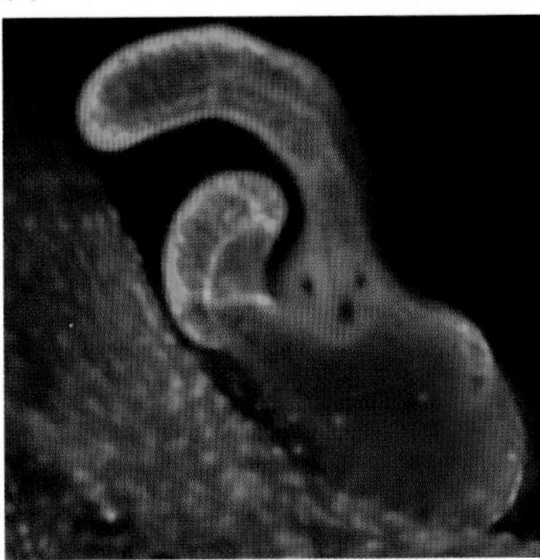

(C)

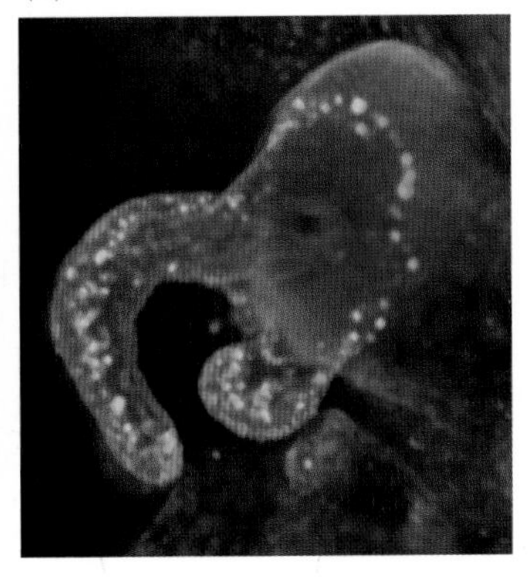

FIGURE 3.3 The *Euprymna scolopes–Vibrio fischeri* symbiosis. (A) An adult Hawaiian bobtail squid (*E. scolopes*) is about 2 inches long. The symbionts are housed in a two-lobed light organ on the squid's belly. (B) The light organ of a juvenile squid is poised to receive *V. fischeri*. Ciliary currents and mucus secretions create an environment (diffuse yellow stain) that attracts seaborne Gram-negative bacteria, including *V. fischeri*, to the organ. Over time all bacteria except *V. fischeri* will be eliminated by mechanisms yet to be exactly elucidated. (C) Once *V. fischeri* are established in the crypts of the light organ, they induce apoptosis of the epithelial cells (yellow dots) and shut down production of the mucosal secretions that attracted other bacteria. (Photographs courtesy of Margaret McFall-Ngai.)

host's light organ, an environment free of predators and adverse environmental conditions.

Vibrio fischeri is transmitted horizontally between host generations. The species is present in low concentrations (between 100 and 1500 *V. fischeri* cells per milliliter of seawater—which is less than 0.1% of all the bacteria present in such a sample). As a newly hatched *E. scolopes* swims through its environment, seawater passes through its immature light organ. The squid captures any *V. fischeri* present in the seawater via the action of a ciliated, mucus-secreting epithelium on the surface of its immature light organ (Figure 3.3B; Nyholm and McFall-Ngai 2004; Goodson et al. 2005). The secretion of this mucus is a response to the common bacterial cell envelope component peptidoglycan. Although both Gram-positive and Gram-negative bacteria contain peptidoglycan in their cell envelopes (and thus can induce mucus secretion), only Gram-negative cells such as *V. fischeri*—which have an extra layer of cell envelope—can successfully aggregate in the squid's mucus. Once aggregated, there is further selection, and as the symbiotic bacteria move into the light organ, all species *except* for *V. fischeri* are eliminated; the mechanism of the species-specific retention of *V. fischeri* is still unknown. One possibility relies on the fact that *V. fischeri* (and not other bacteria) are attracted to *N*-acetylneuraminic acid, a component of the mucus secretions of *Euprymna*. In addition, it seems likely that *V. fischeri* is adapted to adhere to the mucus secreted by its host in the same way mammalian gut bacteria adhere to their hosts' mucus, as detailed later in this chapter.

After being enmeshed by the mucus, the *V. fischeri* cells must find their way into the crypts of the squid's light organ and establish colonies. Once they do this, the bacteria's bioluminescence is stimulated through a quorum-sensing feedback loop. As *V. fischeri* accumulate in the crypts of the squid's light organ in concentrations far exceeding those in the external environment, the bacteria produce two molecules that induce transcription of their own *lux* genes, which are the genes responsible for luminescence (Millikan and Ruby 2001; Visick et al. 2000). Thus, *V. fischeri* is able to bioluminesce *only* after it is established within the light organ of *E. scolopes;* this fundamental attribute of the symbiont is induced only in the presence of its host. It is therefore clear that the host has an important effect on the development of its symbiont; but the symbiont is important to the normal development of its host as well.

When it is first colonized by *V. fischeri, E. scolopes'* light organ is still in a very immature form (Figure 3.3C). The presence of the symbiotic bacteria is in fact necessary to complete the light organ's development. The majority of the changes that occur in the light organ after bacterial colonization have to do with the elimination of the structures that attracted the bacteria in the first place. In particular, the presence of *V. fischeri* induces apoptosis in the light organ's ciliated epithelium, which gradually disintegrates. The presence of *V. fischeri* also shuts down the mucosal secretions that were so important in allowing the bacteria to gain access to the crypts of the light organ. Additionally, the symbiont reduces the squid's production of nitric oxide (NO) by 80%; prior to this reduction, *E. scolope* produces NO in concentrations that, if maintained, would be toxic to the bacteria (Nyholm and McFall-Ngai 2004).

How is *V. fischeri* able to induce all of these changes in its host? Recent studies have shown that the squid is reacting to two specific components expressed on the bacterial cell surface, peptidoglycan and lipopolysaccharide. Both of these molecules are necessary in order to induce loss of the ciliated epithelium. But both peptidoglycan and lipopolysaccharide are expressed by a wide range of bacterial cells, so why does *E. scolopes* undergo morphological changes only in the presence of *V. fischeri*? The answer appears to be that, in order to affect the host's development, the bacterial molecules must be expressed inside the crypts of the light organ—and *V. fischeri* are the only bacteria that are "allowed" to reach these crypts (Koropatnick et al. 2004; Nyholm and McFall-Ngai 2004).

The symbiotic relationship between *E. scolopes* and *V. fischeri* is a complex one. We still do not know how *V. fischeri* is able to attenuate its host's NO production, and many other details about the relationship between the two organisms—such as how the squid is able to expel those bacteria that do not luminesce—have yet to be explained. Even though our knowledge about this excellent model system is as yet incomplete, the relationship between the Hawaiian bobtail squid and its luminescent symbiont can teach us important lessons about environmental influences on the regulation of development. The squid and the bacterium have coevolved such that each plays a fundamental role in the other's development: the squid actively accumulates a high enough population density of *V. fischeri* to allow the bacterium to express its latent bioluminescence, while the bacteria trigger important morphological changes in the light organ of the host. The concrete and well-characterized impact that host and symbiont have on each other's development provides us with an excellent example of how environmental factors can play an integral role in normal development.

Evolution of the Symbiotic Regulation of Development: *Wolbachia*

The relationship between bacteria of the genus *Wolbachia* and their many different arthropod hosts provides clear examples of how symbionts can alter the development of their hosts and how organisms can evolve to depend on their symbionts for normal development. In the many different organisms that *Wolbachia* infects, this symbiont's effects on the sexual development of its host range from transforming genetically male embryos into females to their being necessary for its host's completion of normal oogenesis.

Sex determination by infection

First discovered inhabiting the reproductive organs of the mosquito *Culex pipiens*, the bacterial genus *Wolbachia* includes at least 38 different strains, though the number of species into which these strains should be classified is still uncertain. The bacteria live in the cytoplasm of cells in the ovaries and testes of many different arthropods, including not only mosquitoes but also parasitoid wasps, pillbugs, mites, lepidopterans (butterflies and moths), and *Drosophila*. (Stouthamer et al. 1999). *Wolbachia* can be transmitted both horizontally (from one adult to another) and vertically (from one generation to the next), but vertical transmission is by far the more important mechanism by which *Wolbachia* infects new hosts (Werren 1997).

Wolbachia is transmitted from mother to progeny because it lives within the egg cytoplasm; thus, it is a maternally inherited parasite (Charlat et al. 2003). Because *Wolbachia* can be transmitted only by females, the bacteria benefit if they increase the ratio of females to males among the offspring of the organism that they infect. Indeed, several different methods have evolved by which *Wolbachia* can and does affect this ratio.

The most straightforward way in which *Wolbachia* influences the sex ratio of the host is by killing the male progeny of infected females (Charlat et al. 2003). *Wolbachia* has been found to engage in male killing in both the Asian corn borer moth *Ostrinia furnacalis* and the African butterfly *Acraea encedana* (Kageyama and Traut 2003; Jiggins et al. 2000). In their study of a Ugandan population of *A. encedana*, Jiggins and colleagues (2000) determined that over 95% of the females were infected with *Wolbachia* and that many of these infected females produced no male progeny. They also noted that the broods of infected females (who produced only female progeny) had hatch rates roughly half those of broods from uninfected females, suggesting that *Wolbachia* was actively killing the male progeny during embryogenesis rather than nonlethally altering the sex of the progeny (Figure 3.4). The impact of this male-killing in the population was huge: only 6% of the adult butterflies in the population were male.

Although it is clear that male-killing results in a skewed sex ratio in the host population, exactly how this skew benefits the bacteria is a bit more obscure. After all, *Wolbachia* benefits not by killing off organisms that it can't infect, but by increasing the number of host organisms available to infect; and as seen in Figure 3.4, male-killing doesn't directly influence the absolute number of female progeny. Instead, the female progeny benefit because the elimination of their male siblings means reduced competition. In some species, females benefit further by eating the male embryos that do not survive. Because the death of their male siblings gives the female progeny of infected females a higher survival rate than the female progeny of uninfected females, male-killing does give the infecting *Wolbachia* an advantage: their potential hosts are more likely to survive and reproduce than they would be if their male siblings survived (Charlat et al. 2003). However,

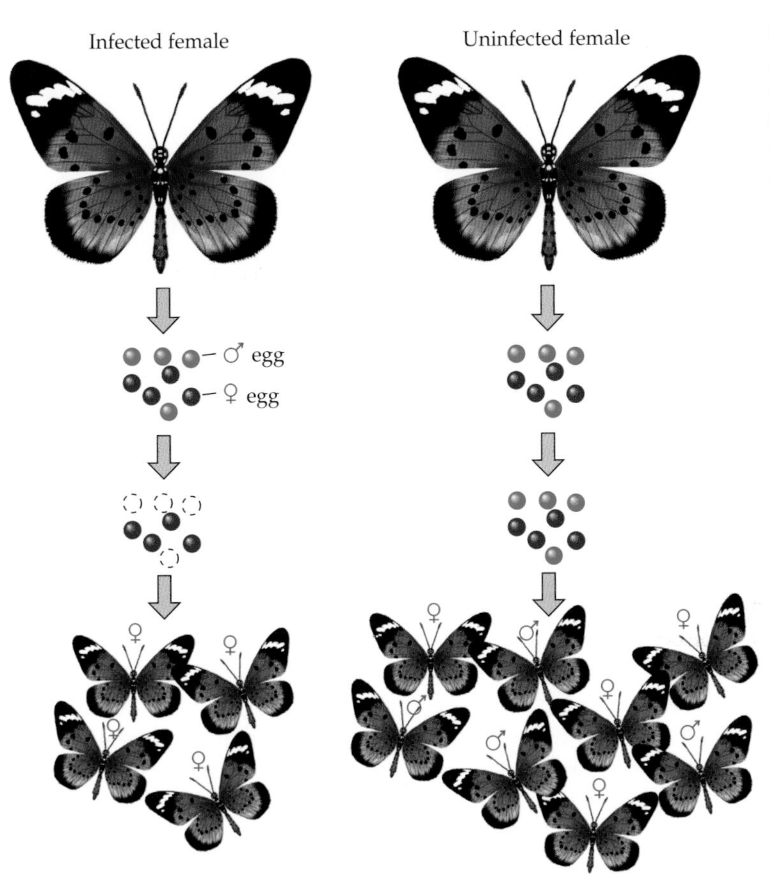

FIGURE 3.4 In *Acraea encedana* infected with *Wolbachia* bacteria, male progeny die during embryogenesis (left). Because all the eggs of uninfected females hatch (right), infected females produce only half as many progeny as uninfected females. However, the females produced have better access to resources.

the absence of male progeny is so deleterious for the host population as a whole that, in insect populations with high incidences of male-killing *Wolbachia* infections, we would expect genes for resistance to either *Wolbachia* infection or to its male-killing effects to be strongly selected for, and indeed there is some evidence for resistance genes in *A. encedana* (Charlat et al. 2003; Jiggins et al. 2000).

A second mechanism by which *Wolbachia* may attempt to skew the sex ratio of the progeny of its host is by feminizing genetically male embryos, resulting in all-female or mostly female progenies without killing embryos. Termed "feminization in diploids," this process has been found to occur in *Wolbachia* infections in many species of isopods, including the common pill bug, *Armadillidium vulgare* (Figure 3.5; Cordaux et al. 2004). In this species, females are usually the heterogametic sex (WZ) while males are homogametic (ZZ); however, when there is *Wolbachia* infection, genetically male ZZ embryos develop as females. This is a striking example of the power of environmental factors to influence development, for the ability of *Wolbachia*

FIGURE 3.5 Male and female *Armadillidium vulgare*. Genetically male pill bugs (right) can be transformed into phenotypic egg-producing females (left) by infection with *Wolbachia* bacteria. (Photograph by David McIntyre.)

to feminize a genetically male host means that, in this case, the presence or absence of the symbiont is a more important factor in sex determination than chromosomal sex.

Like male-killing, feminization of males has a clear advantage for the infecting *Wolbachia* because it leads to an increased number of females, the only sex capable of passing *Wolbachia* on to their progeny. The elimination of male progeny through feminization, however, once again results in a dearth of males for the *Wolbachia*-carrying host females to mate with, and thus it is hypothesized that among arthropods carrying feminizing *Wolbachia*, selection for resistance genes will again occur (Charlat et al. 2003; Cordaux et al. 2004).

For some arthropod species (such as the hymenopteran ants, bees, and wasps), sex is determined not by chromosome composition (whether XY or WZ), but by the number of complete chromosome sets that an organism has. In this scheme, known as *haplodiploidy*, females develop from fertilized eggs and thus are diploid, whereas males develop from unfertilized eggs and are haploid. In several species, including mites and parasitoid wasps, *Wolbachia* can interfere with this sex-determination scheme, turning haploid embryos that normally would become males into diploid females. This phenomenon is known as *parthenogenesis induction*. *Wolbachia* usually induces parthenogenesis (i.e., the ability of a female to produce female progeny without fertilization by a male) by doubling the chromosome number in unfertilized eggs to create a diploid embryo with two identical sets of chromosomes (Zchori-Fein et al. 2001; Charlat et al. 2003). Such an embryo will develop into a female capable not only of passing *Wolbachia* on to its progeny, but of doing so without interaction with a male, making the reduction in male progeny caused by parthenogenesis induction a less severe problem for the host population than the shortage of males caused by male-killing or feminization.

Studies (reviewed in Charlat et al. 2003) have found that if normally haplodiploid species infected with parthenogenesis-inducing *Wolbachia* are treated with antibiotics, haploid male progeny can be produced. These males, however, are often unable to successfully mate with females. A hypothesized reason for this curious fact is that *Wolbachia* infection may also be affecting the evolution of host sex determination. The theory is that, because *Wolbachia*-infected populations are female-biased, any phenotype that promotes the production of more males will be favored evolutionarily. In the case of haplodiploid species, however, a phenotype that favors the production of males is one that somehow prevents or discourages the female from mating with a male, since only her unfertilized eggs will develop into males. Thus, the same genes that might favor the production of males also prevent the females carrying them from mating with males! Such "virginity genes," which in fact have been found in the parasitoid wasp *Telenomus nawai*, represent an important example of how a symbiont can profoundly influence the evolution of development in its host.

Perhaps the most common effect of *Wolbachia* infection, however, is **cytoplasmic incompatibility**, in which embryos resulting from crosses between uninfected males and infected females are viable, but those resulting from crosses between infected males and uninfected females are not (Charlat et al. 2003). Cytoplasmic incompatibility benefits *Wolbachia* because it inhibits the reproduction of uninfected females, thereby increasing the relative reproductive success of infected females and promoting the spread of *Wolbachia* throughout the host population. Because in this instance females become dependent on *Wolbachia* infection for reproductive success, cytoplasmic incompatibility can result in selection for any phenotype that decreases resistance to *Wolbachia* infection, thus making *Wolbachia* a yet more integral component of the host's sexual development.

Evolution of dependence on *Wolbachia* for sexual development

As described above, continued *Wolbachia* infection in an arthropod population may lead to an evolutionary change in sexual development, whether through the induction of parthenogenesis or through an acquired dependence on *Wolbachia* for successful reproduction due to cytoplasmic incompatibility. The ability of *Wolbachia* to rescue sterile *Drosophila melanogaster* females (Starr and Cline 2002) provides some additional insights into how a host might become dependent on its symbiont for reproductive success. In this study, *Wolbachia* was found to eliminate sterility in *D. melanogaster* females who had a particular mutation in the *Sex-lethal* (*Sxl*) gene that resulted in an inability to produce eggs. *Sex-lethal* is a crucial gene in *Drosophila* sex determination, and it must be active for female development.

Although *Wolbachia* was able to alleviate sterility in these particular *Sxl* mutant females, it was unable to do so in flies that were sterile because of a mutation in a different gene, or even in flies with a different *Sxl* mutation. The ability of *Wolbachia* to rescue only females suffering from a certain kind of *Sxl* mutation suggests that there is a very specific interaction between the bacteria and the Sxl protein in the female fly. Such specificity of interaction between host and parasite indicates that the insect and the bacterium have coevolved to give *Wolbachia* an important role in oogenesis in the female fly.

While in this *Sxl* example the female *D. melanogaster* was not dependent on *Wolbachia* infection for fertility (except in the presence of a certain sterilizing mutation), a study on the parasitic wasp *Asobara tabida* found that normal females of this species are dependent on *Wolbachia* infection for successful egg production. In this experiment, *A. tabida* females carrying *Wolbachia* were treated with antibiotics that killed the symbiotic bacteria. The experimenters found this treatment had what they described as a "totally unexpected effect": these female wasps were unable to produce mature oocytes and thus could not reproduce (Figure 3.6; Dedeine et al.

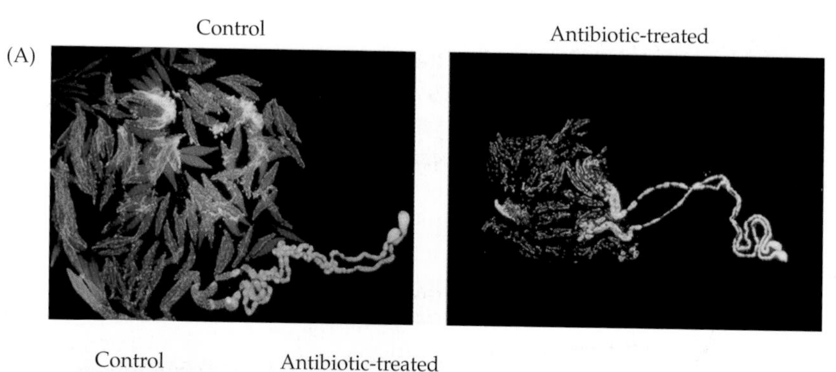

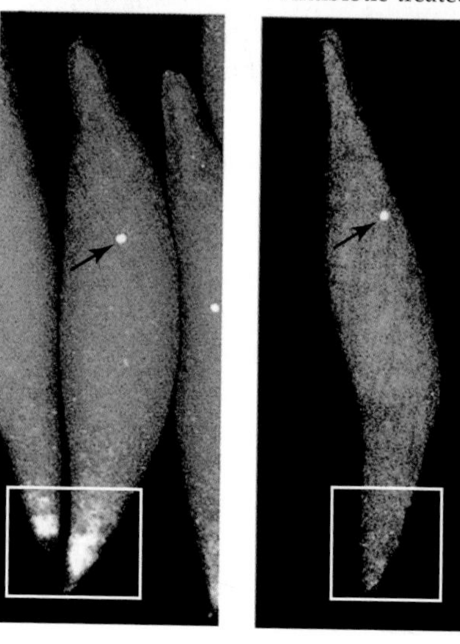

FIGURE 3.6 Comparison of the ovaries and oocytes (immature eggs) of the wasp *Asobara tabida* from control females and females treated with rifampicin antibiotic to remove *Wolbachia*. (A) The ovaries of control wasps had an average of 228 oocytes, while rifampicin-treated females had an average of 36 oocytes as their ovaries had undergone apoptosis in the absence of *Wolbachia* (see Pannebakker et al. 2007). (B) When DNA in the *A. tabida* oocytes was stained, the controls had a nucleus (arrow) as well as a mass of *Wolbachia* at one end (boxed area). Treated oocytes had nuclei but no *Wolbachia*; these eggs were sterile. (From Dedeine et al. 2001.)

2001). The results of this experiment showed that *A. tabida* has become dependent on the presence of *Wolbachia* for successful reproduction, suggesting that the host-*Wolbachia* interactions described earlier also have the

Mutualistic Consortia

While *Euprymna* appears to harbor only one symbiont in its light organ, some animals are home to entire ecosystems of symbionts that can act as organs in an animal. Off the coast of Naples, burrowed in the mud near the island of Elba, resides the oligochaete worm *Olavius algarvensis*. More like a plant than an animal, this species is able to synthesize its own food. Indeed, the worm has neither mouth nor anus nor gut. Even its nephridial excretory pores have mostly degenerated.

What this animal has is an internal ecosystem of sulfur-metabolizing bacteria that can fix carbon dioxide into sugars and manufacture all the amino acids and several vitamins needed by the worm. These bacteria also appear to recycle the animal's waste products. This microbial consortium of four different bacteria species lives just beneath the worm's cuticle. As a group, they are able to fix carbon dioxide through the Calvin-Benson cycle, using the same types of enzymes (including Rubisco) as chloroplasts, but coupled to the oxidation of reduced sulfur compounds. Moreover, by cycling the intermediate sulfur compounds between them, they can supply organic carbon to the worm through three different metabolic pathways. Depending on where the worm sits in the sediment (the deeper layers being more anoxic than the upper layers of mud), different pathways will be utilized (Woyke et al. 2006).

Lynn Margulis (1971, 1994) has written elegantly of symbiosis, and one of her most incredible examples involves the protozoan *Myxotricha paradoxa*, a symbiont of the Australian termite *Mastotermes darwinensis*, without which, the termite would not be able to digest wood, as the termite gut by itself has no enzymes capable of digesting cellulose. But *M. paradoxa* itself is a symbiotic colony, includuing a protist and at least three species of bacteria. One of these bacterial species is a spirochete that attaches to the outer membrane of the protist, forming its locomotor apparatus. A second symbiotic bacterium appears to stabilize this arrangement, and a third lives inside the cytoplasm of the protist. It is the genome of this third bacterium that provides the cellulose-digesting enzyme essential to the termite. All of these bacteria appear to be transmitted through the egg yolk to the next generation (Sacchi et al. 1998).

Such **microbial consortia**—structured arrangements of microbes that function together as a unit—are found in numerous species. Some of the most important microbial corsortia are those established in coral reefs. The photosynthesis of symbiotic algae (dinoflagellates) of the genus *Symbiodinium* provide oxygen and sugars to the coral, and they provide oxygen radicals that, working between the cells, block infection. (The effect is similar to placing hydrogen peroxide on a cold sore, releasing oxygen that destroys bacteria and viruses.) In addition, corals contain a resident microbiological population that include nitrogen-fixing bacteria as well as bacteria that produce antibiotics against other, pathogenic, bacteria. Coral bleaching is caused when this symbiosis is disrupted, and the major cause of this disruption is the pathogenic bacterium *Vibrio shiloi*. Elevated temperatures cause *Vibrio* to express adhesion molecules, which allow the bacteria to bind to the coral's surface, and the substance "toxin P," which blocks photosynthesis in *Symbiodinium*.

Rosenberg and colleagues (2007) have proposed that environmental conditions select for the most beneficial symbiotic relationship between the different coral species and the different bacterial and *Symbiodinium* species. Thus, in this case, the processes of natural selection are operating not on an organism but on a network of relationships among many organisms. This is the so-called *hologenome theory of evolution*, wherein the host and its symbiont population constitute a "hologenome" that can change more rapidly than the host genome alone, thereby providing greater adaptability.

potential to evolve to the point where the host is totally dependent on its symbiont for successful development.

The *Wolbachia* symbiosis systems provide remarkable insights into how symbiont and host might coevolve to allow the symbiont to play an important role in the host's development.

The Mutualistic Bacteria of the Mammalian Gut

Having discussed several examples of how symbionts and symbiotic consortia play important roles in the development of invertebrate host organisms, we will turn to an example much closer to home: the interaction between mammals and the bacteria that reside in the mammalian gut. This symbiotic relationship is a perfect example of mutualism: the bacteria benefit from the safe and nutrient-filled home the host's gut provides, while the host benefits from a range of functional and developmental effects induced by the bacteria. As we shall see, these bacteria not only aid the host's digestion, they also play an important role in the development of the host's intestine, capillary networks, and immune system, as well as having important consequences for human health.

Introduction to the gut microbiota

It is estimated that 90% of the cells in our bodies are prokaryotic. While it is difficult to determine the exact number, it is believed that the human intestinal tract may contain up to 100 trillion (10^{14}) microorganisms—about 10 times as many cells as the eukaryotic "human" cells that make up our bodies (Bäckhed et al. 2005; Ley et al. 2006a). The human gut is home to a consortium of roughly 500–1000 different species of bacteria. It is difficult to obtain a more precise estimate, in part because many human gut bacteria have thus far resisted cultivation in vitro; indeed, only about 7% have been successfully cultured in the laboratory* (Gordon et al. 2005; Rawls et al. 2006).

Compared with the microbial fauna found in many other ecosystems, the bacteria of the mammalian gut are not very diverse. Members of only eight of the known phyla of bacteria are present in the human gut at all, and of these eight, species from only two divisions dominate the gut community (Bäckhed et al. 2005). These two dominant phyla are the Firmicutes

*The major exception, of course, is the Gram-negative proteobacterium *Escherichia coli*, normally found in the distal region of the large intestine. *E. coli* is probably the most cultured organism in all science. A usually harmless bacterium that is easily grown on a variety of media, it is one of the laboratory bench's "model organisms." In particular, *E. coli* strain K-12, isolated at Stanford University in 1922, has been the subject of many groundbreaking molecular genetic studies (see Lederberg 2004).

(e.g., *Bacillus* and *Listeria* spp.) and the Cytophaga-Flavobacterium-Bacteroides complex; the genus *Bacteroides* alone (including *Bacteroides fragilis* and *B. thetaiotaomicron*) accounts for roughly 25% of the bacterial population of the human gut (Xu and Gordon 2003; Bäckhed et al. 2005; Ley et al. 2006a). Although the human gut bacteria are not terribly diverse on the phylum level, their diversity increases dramatically if analyzed on the species or subspecies (strain) level. The high species-level diversity found in the mammalian gut is indicative of a degree of selective pressure for bacteria both to occupy unique niches and to maintain overlapping functions with other members of the gut community; this point will be discussed in more detail later in the chapter.

In addition to aiding their hosts' digestion of polysaccharides, the mutualistic gut bacteria may also play two other major roles in adult mammals: They synthesize vitamins that the adult mammal cannot make (notably vitamins B2 and B6), and they actually protect the intestine against injury. One study showed that the interaction of specific ligands produced by the gut bacteria (particularly lipopolysaccharide and lipoteichoic acid) with the receptors on the gut epithelial cells was necessary for the production of factors that protect against damage to the digestive tract (Rakoff-Nahoum et al. 2004).

Maintaining the gut microbial community: The biofilm model

Just as bacteria affect development and induce gene expression in the mammalian gut, so the gut reciprocates by changing the development of its symbiotic bacteria, which also display phenotypic plasticity. To maintain their presence in the intestinal microbiome, the gut bacteria form **biofilms**, interconnected microbial communities that can resist the shearing forces that threaten to dislodge them and sweep them away (Sonnenburg et al. 2004). Individual bacteria utilize different genes and acquire new properties when they become part of a biofilm (Walter et al. 2003; Macfarlane et al. 2005). These genes are induced by the gut epithelia and mucus, allowing the bacteria to adhere together and to change their nutrient uptake.*

The gut microbes' use of intestinal mucus as an attachment matrix helps to explain the variable composition of the gut bacteria in different regions of the intestine. The intestinal mucus varies in thickness and composition in different regions of the gut, and thus different bacteria are able to form biofilms in different areas. This bacterial segregation allows for specializa-

*Interestingly, the presence of biofilms in the human appendix, and the location and narrow aperture of this structure, have led Bollinger and colleagues (2007) to suggest that the appendix is not a vestigial organ, as has long been supposed, but a very highly adapted structure that serves as a reservoir of symbiotic bacteria. In cases of diarrhea or other severe evacuation of the gut, the appendix would be able to maintain its bacterial population to serve as a culture that can repopulate the bowel during its recovery.

tion due to the spatial separation of bacteria that perform different functions. In one region of the gut, for example, various *Bacteroides* species produce products that are further metabolized by *Escherichia coli*, another gut inhabitant, in a more distal region of the gut (Sonnenburg et al. 2004). Interestingly, the *Bacteroides* themselves are able to metabolize the product they produce, but because the product is washed through the gut lumen along with the rest of the partially digested food while the *Bacteroides* remain attached to the slower-moving mucosal matrix, they are unable to access and metabolize the product. The *E. coli* bacteria farther along in the gut make use of the metabolite instead, thus demonstrating that the differential characteristics of mucus in different regions of the gut help to maintain gut bacterial diversity.

Inheritance of the gut bacteria

Many mammalian gut bacteria cannot survive outside of their host, yet mammalian fetuses maintain a bacteria-free gut until birth. Thus the mutualistic gut bacteria of mammals, like the *Vibrio fischeri* symbionts of the Hawaiian bobtail squid, must be passed to the host through some form of horizontal transmission. A paradox exists, however, since observations and experiments in both mice and humans have indicated that the gut community is inherited. Related mice were found to have more similar gut communities than unrelated mice, and in humans, monozygotic twins, dizygotic twins, and normal siblings were all found to have gut communities that were more similar to each other's than they were to their marriage partner's (reviewed in Bäckhed et al. 2005 and in Ley et al. 2006a).

While inheritance has thus been shown to have a greater impact on the gut community than does the host's environment as an adult, the irrelevance of the type of sibship (monozygotic vs. dizygotic twins vs. non-twin siblings) to the similarity of the siblings' gut communities suggests that it is the early environment of the host, rather than the host's genotype, that has the greatest impact on the makeup of the gut bacterial community. It is currently believed that human babies acquire the beginnings of such communities from the vagina and feces of their mothers. In support of this idea, babies born through Caesarian section have been found to have an altered bacterial colonization pattern early in life compared with vaginally delivered babies (Ley et al. 2006a). It appears, then, that the gut microbiota are an epigenetic inheritance from the mother.

Although the human twin study mentioned above shows that early environment is more important in determining the makeup of the gut community than is host genotype, vast differences in genotype do also influence this composition. A recent transplantation experiment (Rawls et al. 2006) using mice and zebrafish in which a germ-free mouse (i.e., a mouse raised so that it has no microbes of any kind) was inoculated with gut bacteria from a conventionally raised zebrafish, and vice versa, showed that although the lineages that take up residence in the new host are limited to those provided through inoculation, the new host will remodel the proportions of the different lineages present to approach the population normally found in the new host organism. Additionally, bacterial sublineages present in the donor host population but not normally present in the new host are often eliminated. It is thus clear that large differences in genotype do indeed affect the makeup of the gut microorganism community but that the type of bacteria that the host is exposed to also has a vital, and in many cases overwhelming, influence on the community's final composition.

Gut Bacteria and Normal Mammalian Development

In addition to protection against injury and aiding the host in digestion, the enteric bacteria have been shown to induce or regulate the expression of many genes in the gut. In other words, products of the bacterial cells can induce gene expression in the mammalian intestinal cells.* Moreover, this gene expression is essential for the normal development of the gut. Bacteria-induced expression of mammalian genes was first demonstrated by Umesaki (1984), who noticed that a particular fucosyl transferase enzyme characteristic of mouse intestinal villi was induced by bacteria. More recent studies have shown that the intestines of germ-free mice (i.e., mice bred in sterile facilities and having no contact with bacteria or fungi) can initiate, but not complete, their differentiation. For complete development, the microbial symbionts of the gut are needed (Hooper et al. 1998). Microarray analyses of mouse intestinal cells have shown that normally occurring gut bacteria can upregulate the transcription of numerous mouse genes, including those encoding colipase, which is important in nutrient absorption; angiogenin-4, which helps form blood vessels; and Sprr2a, a small, proline-rich protein that is thought to fortify matrices that line the intestine (Figure 3.7; Hooper et al. 2001).

An important role for symbiotic bacteria in the normal development of the host's gut: Angiogenesis induction

Stappenbeck and colleagues (2002) have demonstrated that in the absence of particular intestinal microbes, the capillaries of the small intestinal villi

*The bacterial regulation of host gut gene expression is not limited to mammals. An example from zebrafish was discussed in Chapter 2 (see Figure 2.16), and the proper consortium of gut bacteria is need for the survival of *Drosophila* (see Ryu et al. 2008).

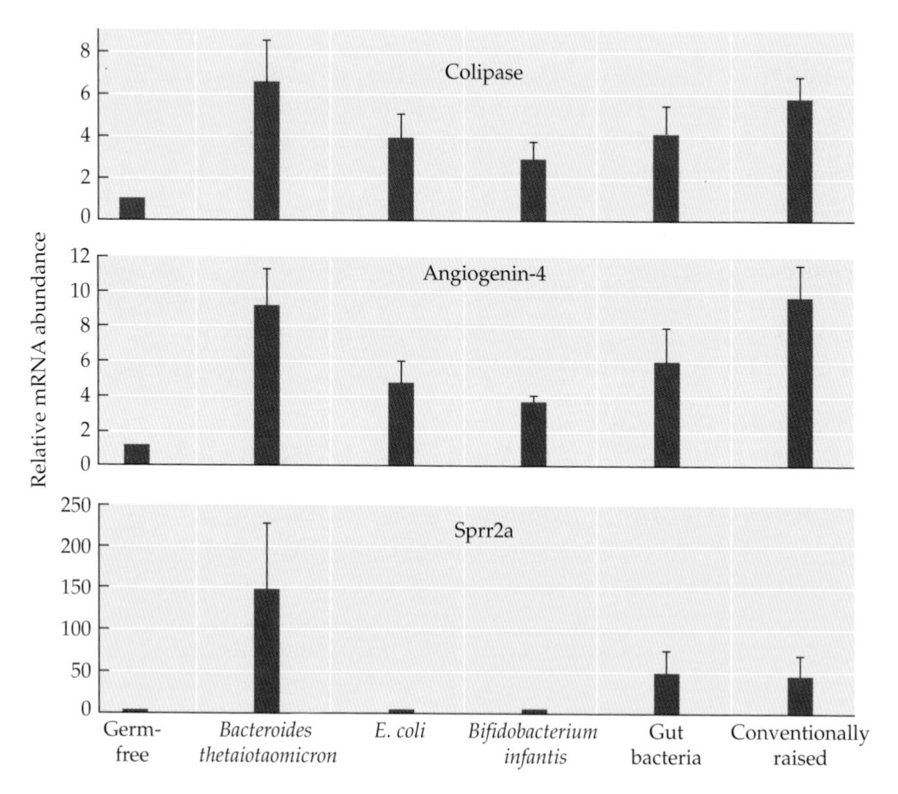

FIGURE 3.7 Induction of mammalian genes by symbiotic microbes. Mice raised in "germ-free" environments were either left alone or inoculated with one or more types of bacteria. After 10 days, their intestinal mRNAs were isolated and tested on microarrays. Mice grown in germ-free conditions had very little expression of the genes encoding colipase, angiogenin-4, or Sprr2a. Several different bacteria—*Bacteroides thetaiotaomicron*, *Escherichia coli*, *Bifidobacterium infantis*, and an assortment of gut bacteria harvested from conventionally raised mice—induced the genes for colipase and angiogenin-4. *B. thetaiotaomicron* appeared to be totally responsible for the 205-fold increase in *sprr2a* expression over that of germ-free animals. This ecological relationship between the gut microbes and the host cells could not have been discovered without the molecular biological techniques of polymerase chain reaction and microarray analysis. (After Hooper et al. 2001.)

fail to develop their complete vascular networks. *Bacteroides thetaiotaomicron* is vital for the induction of host angiogenesis (blood vessel formation). They demonstrated that mice raised so that they have no bacteria in their gut (germ-free mice) have reduced gut capillary network formation compared with normal mice. Colonization of the germ-free mouse intestine either with a sample of bacteria from the gut of a conventionally raised mouse or with

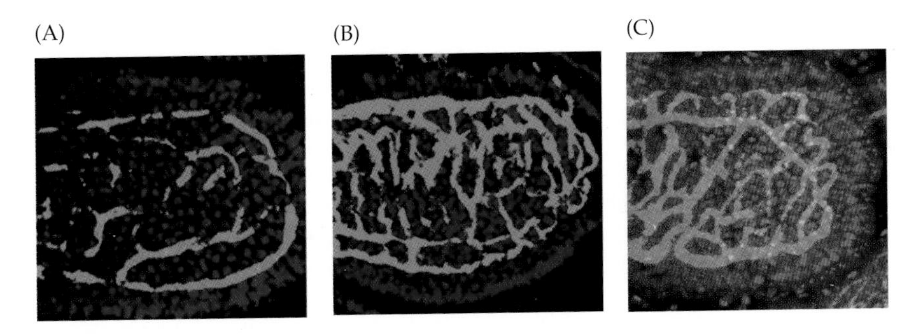

FIGURE 3.8 Gut microbes are necessary for mammalian capillary development. (A) The capillary network (green) of germ-free mice is severely reduced compared with (B) the capillary network in those same mice 10 days after inoculation with normal gut bacteria. (C) The addition of *Bacteroides thetaiotaomicron* alone is sufficient to complete capillary formation. (From Stappenbeck et al. 2002.)

B. thetaiotaomicron alone, however, completed the capillary network formation of the mice within 10 days (Figure 3.8). This result was not dependent on the age of the mouse; even adult mice were able to quickly complete capillary growth in the presence of the required symbiotic bacteria.

The experiment by Stappenbeck and colleagues also investigated the mechanism through which *B. thetaiotaomicron* was able to induce angiogenesis in its host. They determined that the Paneth cells—gut immune system cells found in the intestinal crypts (at the base of the intestinal glands)—were required for the induction of the capillary network. In mice without Paneth cells, the capillary network failed to form properly even after inoculation with *B. thetaiotaomicron* or conventional gut bacteria. Other experiments showed that the Paneth cells were responding to *B. thetaiotaomicron* by transcribing the gene encoding angiogenin-4, a protein known to induce blood vessel formation (Hooper et al. 2001, 2003; Crabtree et al. 2007). These experiments made it clear that the presence of a specific environmental factor (in this case, *B. thetaiotaomicron*) is necessary to induce the expression of important factors necessary for the completion of capillary network formation in the host's gut.

The impact of symbiotic gut bacteria on the development of the host immune system: Antimicrobial secretions

The bacteria of the digestive tract are now known to be fundamental to the proper development of the mammalian immune system. Since a large part of the immune system is localized to the digestive tract, this fact is not terri-

Immunity through Developmental Symbiosis

Symbiosis may be one evolutionary path of protection for eukaryotic organisms. As we will see in Chapter 4, embryos often encounter pathogenic microorganisms before they have developed a functioning immune system. One way for embryos to gain immediate protection from these pathogens is by forming symbiotic relationships with other organisms. For example, symbioses between eggs and bacteria can protect the eggs of aquatic species from fungal pathogens. (As anyone who owns an aquarium knows, uneaten fish food is soon surrounded by a halo of filamentous fungi.)

The outer envelopes (chorions) of several crustacean eggs actually attract bacteria that produce fungicidal compounds. Embryos of the shrimp *Palaeon macrodactylus*, for instance, are extremely resistant to infection by the marine fungus *Lagnidium callinectes*, a known pathogen of many crustaceans. The chorions of these shrimp embryos are consistently found to be infected with the bacterium *Alteromonas*, which produces the antifungal compound 2,3-indolinediol. In laboratory studies, *Palaeon* embryos were bred both with and without the bacterial symbionts and then exposed to the fungal pathogen (Gil-Turnes et al. 1989). When exposed to the fungus, the bacteria-free embryos died rapidly, whereas embryos reinoculated with *Alteromonas* (or treated with 2,3-indoleinedione) survived well. Therefore, it appears that the symbiotic bacteria protect the shrimp embryos from fungal infection.

In some cases, there are entire layers of symbiosis protecting the host. Aphids, for instance, have mutualistic symbioses with numerous bacteria. The pea aphid (*Acyrthosiphon pisum*) is host to the bacterium *Hamiltonella*, a symbiont that gives the aphid protection against the larvae of parasitoid wasps. But the bacterium itself is not what provides this immunity. Rather, *Halmiltonella* cells also harbor a symbiont—a type of bacteriophage (virus) that is critical for their own life cycle. It is a biochemical product of the phage that actually kills the wasp larvae (Moran et al. 2005).

bly surprising. However, we must address an important consideration: How could it possibly benefit the gut bacteria to aid in the development of the host immune system—the very system that is responsible for ridding the host's body of foreign organisms such as bacteria? It would seem that any bacterium would have an interest in inhibiting the host's immune response rather than in promoting it. So why do the gut bacteria play this seemingly counterproductive role?

In fact, playing an active role in immune system development and regulation may allow the bacteria to assert some control over what sorts of things the immune system will attack, thus allowing the bacteria both to prevent the immune system from attacking them, and to upregulate any feature of the immune system that would reduce competition from foreign organisms. Thus, by helping in the development of the host immune system, the symbiotic bacteria are in fact aiding in the construction of their own niche, protecting themselves from both foreign competitors and possible detrimental attacks from their host.

Several different mechanisms have been characterized whereby the symbiotic bacteria reduce competition by inducing the host's cells to produce microbiocidal (antibiotic) compounds. First, in an experiment closely related to those described above as investigating the bacterial influence on angiogenesis, Hooper and colleagues (2003) found that in mice, *B. thetaiotaomicron* has species-specific bactericidal and fungicidal activity. This bactericidal factor appears to be secreted into the gut lumen in response to lipopolysaccharides, a common component of the cell membranes of Gram-negative bacteria such as *B. thetaiotaomicron*. Angiogenin-4 (see pp. 101–103) was found to have bactericidal activity against the pathogenic Gram-positive bacteria *Enterococcus faecalis* and *Listeria monocytogenes*, reducing the populations of each of these bacteria by more than 99% after just 2 hours of Ang4 exposure. By contrast, normal gut inhabitants such as *B. thetaiotaomicron* and *E. coli* were found to be resistant to this microbicidal protein; the *B. thetaiotaomicron* population decreased in size by only 30% after 2 hours of exposure to Ang4. Remarkably, they also found the *non*pathogenic *Listeria innocua* to be Ang4-resistant, in spite of this species' being a close relative of the pathogenic *L. monocytogenes* bacterium so effectively wiped out by Ang4. In this way, microbes may act to structure their populations within the gut.

B lymphocytes and the GALT

In many mammals, the microbial community of the gut appears to play a role not only in the regulation of antimicrobial secretions, but also in the original development and expansion of the host immune system. B-cell

Immune System Cell Types

Many types of cells are involved in the mammalian immune system, and their complex nomenclature can become confusing. A few of the most important and relevant immune-cell types for our discussion here include:

- **B cells (B lymphocytes)**: These cells each produce a unique antibody to recognize foreign substances and mark them for destruction. Once a B cell has successfully bound an antigen, it will replicate itself to create a host of **plasma cells**—short-lived, antibody-producing cells that can bind to and eliminate antigenic substance.
- **Helper T cells**: Also known as CD4$^+$ T cells or T_H cells, these cells tell B cells when to produce antibodies and activate other immune system cells. There are two subtypes of T_H cells, known as **T_H1** and **T_H2**, and each produces a different chemical messenger (a **cytokine**) that stimulates immune system cells. T_H1 cells produce interferon-gamma (IFN-γ) to induce the formation of effector T lymphocytes (including cytotoxic T cells; see below), whereas T_H2 cells produce interleukin-4 (IL4) to induce B cells to divide and form plasma cells.
- **Cytotoxic T cells**: Also known as CD8$^+$ T cells, "killer" T cells, or T_C cells, these cells attack and destroy infected body cells.
- **Antigen-presenting cells**: These include the **macrophages** and **dendritic cells**. They are phagocytic cells that ingest antigens and present them on their surfaces so that helper T-cell receptors can bind to them and induce the appropriate immune response. Dendritic cells can activate both T_H and T_C cells.

ation (bacterial flagellin proteins) identified. These proteins are now being tested to see whether they can protect intestinal stem cells against radiation-induced apoptosis such as that caused by cancer treatments (Abrev 2008; Burdelya 2008).

From the many studies on the role of the mammalian gut bacteria in the development and regulation of the host immune system, it is clear that without these bacteria, many defects and imbalances occur in mammalian immune system cells and tissues. The importance of the gut bacteria for host immune system development is thus an incredibly convincing example of how symbionts can play an integral regulatory role in the development of their host, once again demonstrating the importance of environmental factors for normal development.

The role of the gut bacteria in fat storage: Implications for human obesity

In the United States, 64% of the adult population is overweight or obese (Makdad et al. 2004). Clearly, there are many factors, both innate and external, that contribute to this obesity epidemic; some of the obvious ones include diet, reduced physical activity, and genes. Recently, however, yet another important environmental contributor to obesity has been discovered: the microbes living in our gut. A dramatic study by Bäckhed et al. (2004) showed that, in one mouse strain, adult male germ-free mice had a stunning 42% less body fat than their conventionally raised counterparts, even though the conventionally raised mice ate 29% less food per day. Furthermore, when adult germ-free mice were colonized with conventional mouse gut bacteria, the animals underwent a whopping 57% increase in body fat in just 14 days, even though they were eating 27% less food than they had been eating prior to colonization.

Intriguingly, this huge increase in body fat content was not associated with differential weight gain in the mice, but instead with a decrease in the "lean" (nonfat) body mass; a greater percentage of the new body weight was fat. In addition, conventionalized mice (germ-free mice inoculated with normal gut bacteria) had increased metabolic rates—27% higher than that of their germ-free counterparts (Bäckhed et al. 2004; Wolf 2006).

The mechanisms by which the gut bacteria induce fat synthesis are just beginning to be elucidated, but at least three pathways have been found. The first of these involves the common gut bacterium *B. thetaiotaomicron*. Colonization of germ-free mice with *B. thetaiotaomicron* alone results in about 40% of the total body fat gain associated with colonization by a sample of normal mouse intestinal bacteria (Bäckhed et al. 2004). This bacterium is somehow inducing the expression of the transcription factors ChREBP (carbohydrate response element binding protein) and SREBP1 (sterol response element binding protein-1) in the mammalian liver. These transcription factors in turn activate the genes encoding two enzymes that are important in the biosynthesis of fatty acids: acetyl-CoA carboxylase and fatty acid synthase (Bäckhed et al. 2004). Transcription of several other genes encoding lipid-storage proteins is also induced by the bacteria, including the genes encoding colipase, pancreatic lipase-related protein-2, a fatty acid–binding protein (L-FABP), and apolipoprotein A-IV (Hooper et al. 2001). Upregulation of these proteins leads to an increased capacity for fat storage.

A second component of increased fat synthesis concerns interactions between the symbiotic microbes. For instance, certain Archaea found in the human gut are able to increase the efficiency of *Bacteroides* metabolism, enabling that bacterium to digest complex carbohydrates better and to produce acetate, which can be made into fatty acids that stimulate fat production in the liver (Samuel and Gordon 2006.)

It was also found that genetically obese mice have more Archaea living in their guts; these organisms can increase the efficiency of bacterial fermentation, leading to greater nutrient absorption and fat storage (Samuel and Gordon 2006). Strikingly, when the altered microbial contents of obese mice guts are transplanted into germ-free mice, these mice gain 20% more weight than do germ-free mice colonized with normal gut microflora, demonstrating that the gut microbiota do play a causative role in obesity (Turnbaugh et al. 2006).

A third component involves the interaction of the microbes with diet. In fat mice, the relative abundance of the two most common bacterial groups in the mammalian gut, *Bacteroides* and Firmicutes, was found to be altered such that the *Bacteroides* population decreased by 50% while that of Firmicutes increased by the same amount in obese mice as compared to normal mice (Figure 3.12; Ley et al. 2005; Turnbaugh et al. 2006). A similar correspondence was noted in dieting humans: as individuals lost weight, the relative abundance of members of the *Bacteroides* lineage in their guts increased (Ley et al. 2006b). When mice were made obese by feeding the a

FIGURE 3.12 Mice with a mutation in their leptin genes are genetically obese. Their gut microbes are also different from those of wild-type mice, having 50% more firmicutes and 50% less *Bacteroides*. The mix of microbes in the intestines of the obese mice is more effective at releasing calories from food. (Photograph © Science VU/Visuals Unlimited.)

"prototypical Western diet," the high-calorie food caused a "bloom" in a single, formerly unknown, group of Firmicutes; low-calorie diets decreased the numbers of this species.

Moreover, when bacteria from the intestines of mice fed high-calorie diets were transferred into lean germ-free recipients, these conventionalized mice stored more fat than did mice receiving microbe transplants from mice fed low-calorie diets (Turnbaugh et al. 2008).

Further implications of the enteric gut bacteria for human health

GASTRIC CANCER One of the most surprising stories of modern cancer biology (and one for which Barry Marshall and J. Robin Warren received a Nobel Prize in 2005) is that human gastric cancer can be initiated by bacterial infection. Gastric cancer ranks second only to lung cancer in oncological deaths worldwide, and it is often caused by *Helicobacter pylori*, a bacteria found in the stomachs of nearly half of all humans. When certain strains of this bacteria adhere to gastric cells (which they do by releasing ammonia to neutralize stomach acids), they can cause gut epithelial cells to become malignant. *H. pylori* appears to regulate certain genes in the stomach epithelial cells, causing the overexpression of paracrine factors that lead to cell growth (e.g., epithelial growth factor and its receptor) and the invasion of blood vessels (vascular endothelial growth factor). In addition, the tumor-suppression genes *Runx2* and *MLH1* become hypermethylated and thereby inactivated (Tsuji et al. 2006; Nardone et al. 2007). As we will see in Chapter 7, the resulting lack of tumor-suppressor proteins can allow carcinogenic mutations to accumulate.

ASTHMA AND ALLERGIES While most infectious diseases have been brought under control in industrialized countries, cases of allergies and asthma have actually increased with the advent of modern medicine (Braun-Fahrländer et al. 2002; MacDonald and Monteleone 2005). One widely disseminated hypothesis is that the increased prevalence of these conditions is in fact connected to the developments in medicine. This idea, known as the "hygiene hypothesis," states that we put ourselves at risk for allergies and asthma by removing bacteria from our environment. In addition, the hygiene hypothesis proposes that the level of antibiotic use prevalent in many industrialized nations may permanently disrupt the proportions of bacteria in the gut and thereby alter the direction of the host's immune responses (Noverr and Huffnagle 2004). The hygiene hypothesis is supported by studies showing that infants who develop allergies later in life have statistically different enteric bacterial profiles than infants who do not develop allergies, and that early exposure to certain microbes provides protection against having allergic responses to related microbes later in life (see Noverr and Huffnagle 2004; Debarry et al. 2007).

Interestingly, Chen and Blaser (2007) have shown that the childhood acquisition of the *Helicobacter pylori* bacterium (see above) appears to *lower* the risk of that person getting asthma or allergies. Blaser and colleagues (2006, 2008) have shown that *Helicobacter* species are relatively host-specific, and they hypothesize that for most of human history, *H. pylori* has been a normal, coevolved human gut symbiont (and that its loss in people in industrial countries is associated with acid reflux disease). Normal *H. pylori* (or a similar bacterium) might be a bacterial component regulating the immune system to block the conditions necessary for allergy and asthma induction.

Breast-feeding might also play a role in reducing both the prevalence and severity of asthma. Kull and colleagues (2004) have shown that exclusive breast-feeding for 4 months or longer significantly reduces the risk of a child's getting asthma during the next 4 years. Similarly, a study of some 1000 children in Qatar showed significant decreases in skin and respiratory allergies in children who had been continually breast-fed (Bener et al. 2007). This effect might be brought about in many ways. Mother's milk may support the growth of symbiotic bacteria, and it may transfer potential allergens in a manner that will induce tolerance rather than an immune response (Kull et al. 2004; Verhasselt et al. 2008).

PROBIOTICS The research reported here concerning obesity, allergies, asthma, and ulcerative colitis indicates that dysbioses may be responsible for a variety of human diseases. One possible way to offset such imbalances might be the use of **probiotics**, defined as "live microorganisms which when administered in adequate amounts confer a health benefit on the host" (FAO/WHO 2001). The notion of probiotics dates back to Eli Metchnikoff, one of the founders of the modern science of immunology, who in 1907 suggested that yogurt cultures from rural Russia might have properties that promoted good health and longevity.

There are many potential uses for probiotic therapy. One is the administration of probiotic supplements in early childhood to prevent autoimmune conditions such as asthma, eczema, and allergies (see Savilahti et al. 2008). Another potential use is to treat diarrhea and urinary tract infections in adults (Hatakka and Saxelin 2008). Still another use of probiotics would be to reconstitute a person's gut microbes after colon surgery, where appropriate fears of infection cause surgeons to deplete as many of the gut microbes as possible (Correia and Nicoli 2006; Mogilner et al. 2007).

Many questions remain, however, about probiotic formulas such as those currently found in commercial yogurts or pills (Farnsworth 2008; Morrow and Kollef 2008). First, these products are not marketed as pharmaceuticals and thus do not have to undergo testing as drugs. In addition, there are no standardization requirements. Results from clinical trials are still controversial; the freshness and mode of entry of the product may be very important. Some "hybrid" procedure involving antibiotics followed by probiotics might have advantages (Anukam et al. 2006). And finally, even the strains of bacteria used may act differently in different people. Still, as we discover the mechanisms by which bacteria are used to develop and maintain bodily functions, probiotics may become part of our normal medical care.

Summary

It should be clear by now that most organisms are not "individuals," nor do they develop as individuals. It might be better to think of each of us as a team, or as a little ecosystem. In *At Home in the Universe*, Stuart Kauffman (1995) stated, "All evolution is co-evolution." We may now have to conclude that "all development is co-development."

It is safe to say that the maintenance of life is predicated on symbiotic relationships. Mycorrhizal, endophytic, and nitrogen-fixation symbioses are probably responsible for the existence of life on land. Moreover, symbiosis is an essential element of the normal development of many animals. Molecular biology has revealed that what used to be thought of as exceptions are actually the rule. Whereas microbially induced development was once seen as the province of strange arthropods and a few squids, it is now seen to be a common scheme of development. Even we mammals have "outsourced" the signals that induce normal intestinal gene expression to our gut microbes.

Environmental polyphenisms show important interactions-at-a-distance between our developing selves and the environment; developmental symbioses demonstrate that some of these environmental factors can be closely associated with our developing cells. We are not adults entering into symbiotic relationships; rather, the processes that made us adults included the interactions between us and our symbionts. Thus, the notion of the "individual" becomes semipermeable. Mark Twain is quoted as saying that the only people who can use the "Royal We" are kings, editors, and people with tapeworms. However, it now looks as if each of us develops as a community and can embrace that royal privilege. In *The Lives of a Cell*, Lewis Thomas (1974, p. 142) eloquently points out the lessons of such discoveries: "This is, when you think about it, really amazing. The whole dear notion of one's own Self—marvelous, old free-willed, free-enterprising, autonomous, independent island of a Self—is a myth."

References

Abrams, G. D., H. Bauer and H. Spring. 1963. Influence of the normal flora on mucosal morphology and cellular renewal in the ileum: A comparison of germfree and conventional mice. *Lab. Invest.* 12: 355–364.

Abrev, M. T. 2008. Harnessing the power of bacteria to protect the gut. *New Eng. J. Med.* 359: 756–759.

Anukam, K. C., E. O. Osazuwa, I. Ahonkhai, M. Ngwu, G. Osemenem A. W. Bruce and G. Reid. 2006. Augmentation of antimicrobial metronidazole therapy of bacterial vaginosis with oral probiotic *Lactobacillus rhamnosus* GR-1 and *Lactobacillus reuteri* RC-14: Randomized, double-blind, placebo-controlled trial. *Microbes Infect.* 8: 1450–1454.

Bachmann, M. D., R. G. Carlton, J. M. Burkholder and R. G. Wetzel. 1986. Symbiosis between salamander eggs and green algae: Microelectrode measurements inside eggs demonstrate effects of photosynthesis on oxygen concentrations. *Can. Zool.* 64: 1586–1588.

Bäckhed, F. and 7 others. 2004. The gut microbiota as an environmental factor that regulates fat storage. *Proc. Natl. Acad. Sci. USA* 101: 15718–15723.

Bäckhed, F., R. E. Ley, J. L. Sonnenburg, D. A. Peterson and J. I. Gordon. 2005. Host-bacterial mutualism in the human intestine. *Science* 307: 1915–1920.

Bener, A., M. S. Ehlayel, S. Alsowaidi and A. Sabbah. 2007. Role of breast feeding in primary prevention of asthma and allergic diseases in a traditional society. *Eur. Ann. Allergy Clin. Immunol.* 39: 337–343.

Blaser, M. J., Y. Chen and J. Reibman. 2008. Does *Helicobacter pylori* protect against asthma and allergy? *Gut* 57: 1178–1179.

Bollinger, R., A. S. Barbas, E. L. Bush, S. S. Lin and W. Parker. 2007. Biofilms in the large bowel suggest an apparent function of the human vermiform appendix. *J. Theoret. Biol.* 249: 826–831.

Booth, M. G. 2004. Mycorrhizal networks mediate overstorey-understorey competition in a temperate forest. *Ecol. Lett.* 7: 538–546.

Bradshaw, W. E. and C. M. Holzapfel. 2001. Genetic shift in photoperiodic response correlated with global warming. *Proc. Natl. Acad. Sci USA* 98: 14509–14514.

Braun-Fahrländer, C. and 14 others. 2002. Environmental exposure to endotoxin and its relation to asthma in school-age children. *N. Engl. J. Med.* 347: 869–877.

Burdelya, L. G. and 11 others. 2008. An agonist of toll-like receptor 5 has radioprotective activity in mouse and primate models. *Science* 320: 226–230.

Cameron, D. D., I. Johnson, D. J. Read and J. R. Lenke. 2008. Giving and receiving: Measuring the carbon cost in the green orchid, *Goodyera repens*. *New Phytol.* doi:10.1111/j. 1464–8137. 2008. 02533.x

Cary, S. C. 1994. Vertical transmission of a chemoautotrophic symbiont in the protobranch bivalve *Solemya reidi*. *Mol. Marine Biol. Biotech.* 3: 121–130.

Charlat, S., G. D. D. Hurst and H. Merçot. 2003. Evolutionary consequences of *Wolbachia* infections. *Trends Genet.* 19: 217–223.

Chen, Y. and M. J. Blaser. 2007. Inverse associations of *Helicobacter pylori* with asthma and allergy. *Arch. Int. Med.* 167: 821–827.

Cohen, C. S. and R. R. Strathmann. 1996. Embryos at the edge of tolerance: Effects of environment and structure of egg masses on supply of oxygen to embryos. *Biol. Bull.* 190: 8–15.

Cordaux, R., A. Michel-Salzat, M. Frelon-Raimond, T. Rigaud and D. Bouchon. 2004. Evidence for a new feminizing *Wolbachia* strain in the isopod *Armadillidium vulgare*: Evolutionary implications. *Heredity* 93: 78–84.

Correia, M. I. and J. R. Nicoli. 2006. The role of probiotics in gastrointestinal surgery. *Curr. Opin. Clin. Nutrit. Metab. Care* 9: 618–621.

Crabtree, B., D. E. Holloway, M. D. Baker, K. R. Archava and B. Subramanian. 2007. Biological and structural features of murine angiogenin-4, an angiogenic protein. *Biochemistry* 46: 2431–2443.

Darwin, C. R. 1854. *Living Cirripedia, The Balanidæ, (or sessile cirripedes); the Verrucidæ. etc. etc. etc.* Vol. 2. The Ray Society, London.

Davidson, S. K. and D. A. Stahl. 2008. Selective recruitment of bacteria during embryogenesis of an earthworm. *ISME Journal* 2: 510–518.

Debarry, J. and 10 others. 2007. *Acinetobacter lwoffii* and *Lactococcus lactis* strains isolated from farm cowsheds possess strong allergy-protective properties. *J. Allergy Clin. Immunol.* 119: 1514–1521.

Dedeine, F., F. Vavre, F. Fleury, B. Loppin, M. E. Hochberg and M. Boulétreau. 2001. Removing symbiotic *Wolbachia* specifically inhibits oogenesis in a parasitic wasp. *Proc. Natl. Acad. Sci. USA* 98: 6247–6252.

Endow, K. and S. Ohta. 1990. Occurrence of bacteria in the primary oocytes of the vesicomyid clam *Calyptogena soyae*. *Mar. Ecol. Prog. Ser.* 64: 309–311.

Faeth, S. H., D. R. Gardner, C. J. Hayes, A. Jani, S. K. Wittlinger and T. A. Jones. 2006. Temporal and spatial variation in alkaloid levels in *Achnatherum robustum*, a native grass infected with the endophyte *Neotyphodium*. *J. Chem. Ecol.* 32: 307–324.

FAO/WHO. 2001. Report of a joint FAO/WHO expert consultation on evaluation of health and nutritional properties of probiotics in food including powder milk with live lactic acid bacteria.

Farnsworth, E. R. 2008. The evidence to support health claims for probiotics. *J. Nutrit.* 138: S1250–S1254.

Foellmer, M. W. and D. J. Fairbairn. 2003. Spontaneous male death during copulation in an orb-weaving spider. *Proc. Biol. Sci.* 270: S183–S185.

Gil-Turnes, M. S., M. E. Hay and W. Fenical. 1989. Symbiotic marine bacteria chemically defend crustacean embryos. *Science* 246: 116–118.

Gordon, J. I., R. E. Ley, R. Wilson, E. Mardis, J. Xu, C. M. Fraser, and D. A. Relman. 2005. Extending our view of self: The human gut microbiome initiative (HGMI). http://www.genome.gov/10002154.

Hadfield, M. G. and V. J. Paul. 2001. Natural chemical cues for settlement and metamorphosis of marine invertebrate larvae. In *Marine Chemical Ecology*, J. B. B. McClintock and B. J. Baker (eds.). CRC Press, Boca Raton, FL, pp. 431–461.

Hatakka, K. and M. Saxelin. 2008. Probiotics in intestinal and non-intestinal infectious diseases: Clinical evidence. *Curr. Pharm. Des.* 14: 1351–1367.

Hooper, L. V., T. S. Stappenbeck, C. V. Hong and J. I. Gordon. 2003. Angiogenins: A new class of microbicidal proteins involved in innate immunity. *Nature Immunol.* 4: 269–273.

Hooper, L. V., M. H. Wong, A. Thelin, L. Hansson, P. G. Falk and J. I. Gordon. 2001. Molecular analysis of commensal host-microbial relationships in the intestine. *Science* 291: 881–884.

Jiggins, F. M., G. D. D. Hurst, C. E. Dolman and M. E. N. Majerus. 2000. High-prevalence male-killing *Wolbachia* in the butterfly *Acraea encedana*. *J. Evol. Biol.* 13: 495–501.

Johns, P. M. and M. R. Maxwell. 1997. Sexual cannibalism: Who benefits? *Trends Ecol. Evol.* 12: 127–128.

Kageyama, D. and W. Traut. 2003. Opposite sex-specific effects of *Wolbachia* and interference with the sex determination of its host *Ostrinia scapulalis*. *Proc. Biol. Sci.* 271: 251–258.

Kauffman, S. A.. 1995. *At Home in the Universe: The Search for the Laws of Self-Organization and Complexity*. Oxford University Press, New York.

Koropatnick, T. A., J. T. Engle, M. A. Apicella, E. V. Stabb, W. E. Goldman, and M. J. McFall-Ngai. 2004. Microbial factor-mediated development in a host-bacterial mutualism. *Science* 306: 1186-1188.

Krueger, D. M., R. G. Gustafson and C. M. Cavanaugh. 1996. Vertical transmission of chemoautotrophic symbionts in the bivalve *Solemya velum* (Bivalvia: Protobranchia). *Biol. Bull.* 190: 195–202.

Kulberg, M. C. 2008. Soothing intestinal sugars. *Nature* 453: 602–604.

Kull, I., C. Almqvist, G. Lilja, G. Pershagen and M. Wickman. 2004. Breast feeding reduces the risk of asthma during the first four years of life. *J. Allergy Clin. Immunol.* 114: 755–760.

Lanning, D. K., K. Rhee, and K. L. Knight. 2005. Intestinal bacteria and development of the B-lymphocyte repertoire. *Trends Immunol.* 26: 419–424.

Lederberg, J. 2004. *E. coli* K-12. *Microbiology Today* 31 (Aug): 116.

Ley, R. E., F. Bäckhed, P. Turnbaugh, C. A. Lozupone, R. D. Knight and J. L. Gordon. 2005. Obesity alters gut microbial ecology. *Proc. Natl. Acad. Sci. USA* 102: 11070–11075.

Ley, R. E., D. A. Peterson and J. I. Gordon. 2006a. Ecological and evolutionary forces shaping microbial diversity in the human intestine. *Cell* 124: 837–848.

Ley, R. E., P. J. Turnbaugh, S. Klein and J. L. Gordon. 2006b. Human gut microbes associated with obesity. *Nature* 444: 1022–1023.

Lorber, B. 2005. Infection and mental illness: Do bugs make us batty? *Anaerobe* 11: 303–307.

MacDonald, T. T. and G. Monteleone. 2005. Immunity, inflammation, and allergy in the gut. *Science* 307: 1920–1925.

Macfarlane, S., E. J. Woodmansey and G. T. Macfarlane. 2005. Colonization of mucin by human intestinal bacteria and establishment of biofilm communities in a two-stage continuous culture system. *Appl. Environ. Microbiol.* 71: 7483–7492.

Makdad, A. H., J. S. Marks, D. F. Stroup and J. L. Gerberding. 2004. Actual causes of death in the United States, 2000. *J. Amer. Med. Assoc.* 291: 1238–1245.

Margulis, L. 1971. *Origin of Eukaryotic Cells*. Yale University Press, New Haven.

Margulis, L. 1994. Symbiogenesis and symbioticism. In *Symbiosis as a Source of Evolutionary Innovation*, L. Margulis and R. Fester (eds.). MIT Press, Cambridge, MA, pp. 1–14.

Martin, F., A. Kohler and S. Duplessis. 2007. Living in harmony in the wood underground: Ectomycorrhizal genomics. *Curr. Opin. Plant Biol.* 10: 204–210.

Mazmanian, S. K., C. H. Liu, A. O. Tzianabos and D. L. Kasper. 2005. An immunomodulatory molecule of symbiotic bacteria directs maturation of the host immune system. *Cell* 122: 107–118.

Mazmanian, S. K., J. L. Round and D. L. Kasper. 2008. A microbial symbiosis factor prevents intestinal inflammatory disease. *Nature* 453: 620–625.

McFall-Ngai, M. J. 2002. Unseen forces: The influence of bacteria on animal development. *Dev. Biol.* 242: 1–14.

Metchnikoff, E. 1907. *Essais optimistes*. Paris. (Trans. and edited by P. C. Mitchell as *The Prolongation of Life: Optimistic Studies*. Heinemann, London.)

Miller-Rushing, A. J., T. Katsuki, R. Primack, H. Higuchi, Y. Ishii, S. D. Lee and H. Higuchi. 2007. Impact of global warming on a group of related species and their hybrids: Cherry tree flowering at Mt. Takao, Japan. *Amer. J. Bot.* 94: 1470–1478.

Millikan, D. S. and E. G. Ruby. 2001. Alterations in *Vibrio fischeri* motility correlate with a delay in symbiosis initiation and are associated with additional symbiotic colonization defects. *Appl. Env. Microbiol.* 68: 2519–2528.

Milius, S. 2006. They're all part fungus. *Science News* 169: 231.

Mogilner, J., I. Srugo, M. Laurie, R. Shaoul, A. G. Coran, E. Shiloni and I. Suhkotnik. 2007. Effects of probiotics on intestinal regrowth and bacterial translocation after small bowel resection in a rat. *J. Pediatr. Surg.* 8: 1365–1371.

Moran, N., P. H. Degnan, S. R. Santos, H. E. Dunbar and H. Ochman. 2005. The players in a mutualistic symbiosis: Insects, bacteria, viruses, and virulence genes. *Proc. Natl. Acad. Sci. USA* 102: 16919–16926.

Morrow, L. E. and M. H. Kollef. 2008. Probiotics in the intensive care unit: Why controversies and confusion abound. *Crit. Care* 12 (in press).

Nardone, G., D. Compare, P. De Colibus, G. de Nuzzi and A. Rocco. 2007 *Helicobacter pylori* and epigenetic mechanisms underlying gastric carcinogenesis. *Digest. Disord.* 25: 225–229.

Nash, D. R., T. D. Als, R. Maile, S. R. Jones and J. J. Boomsma. 2008. A mosaic of chemical coevolution in a large blue butterfly. *Science* 319: 88–90.

National Research Council. 2007. *Status of Pollinators in North America*. National Academies Press, Washington, DC.

Noverr, M. C. and G. B. Huffnagle. 2004. Does the microbiota regulate immune responses outside the gut? *Trends Microbiol.* 12: 562–568.

Nussey, D. H., E. Postma, P. Gienapp and M. E. Visser. 2005. Selection on heritable phenotypic plasticity in a wild bird population. *Science* 310: 304–306.

Nyholm, S. V. and M. J. McFall-Ngai. 2004. The winnowing: Establishing the squid-*Vibrio* symbiosis. *Nature Rev. Microbiol.* 2: 632–642.

Nyholm, S. V., E. V. Stabb, E. G. Ruby and M. J. McFall-Ngai. 2000. Establishment of an animal-bacterial association: Recruiting symbiotic *Vibrios* from the environment. *Proc. Natl. Acad. Sci. USA* 97: 10231–10235.

Pannebakker, B. A. B. Loppin, C. P. H. Elemains, L. Humblot and F. Vavre. 2007. Parasitic inhibition of cell death facilitates symbiosis. *Proc. Natl. Acad. Sci. USA* 104: 213–215.

Pinder, A. W. and S. C. Friet. 1994. Oxygen transport in egg masses of the amphibians *Rana sylvatica* and *Ambystoma maculatum:* Convection, diffusion, and oxygen production by algae. *J. Exp. Biol.* 197: 17–30.

Pollan, M. 2002. Power steer. *New York Times* March 31 2002. Available at http://www.michaelpollan.com/article.php?id=14.

Rakoff-Nahoum, S., S. Paglino, F. Eslami-Varzaneh, S. Edberg and R. Medzhitov. 2004. Recognition of commensal microflora by Toll-like receptors is required for intestinal homeostasis. *Cell* 118: 229–241.

Rawls, J. F., M. A. Mahowald, R. E. Ley and J. I. Gordon. 2006. Reciprocal gut microbiota transplants from zebrafish and mice to germ-free recipients reveal host habitat selection. *Cell* 127: 423–433.

Rhee, K. J., P. Sethupathi, A. Driks, D. K. Lanning and K. L. Knight. 2004. Role of commensal bacteria in development of gut-associated lymphoid tissue and preimmune antibody repertoire. *J. Immunol.* 172: 1118–1124.

Rosenberg, E., O. Koren, L. Reshef, R. Efrony and I. Zilber-Rosenberg. 2007. The role of microorganisms in coral health, disease, and evolution. *Nature Rev. Microbiol.* 5: 355–362.

Ryu, J. H. and 9 others. 2008. Innate immune homeostasis by the homeobox gene *Caudal* and commensal gut mutualism in *Drosophila. Science* 319: 777–782.

Sacchi, L. and 8 others. 1998. Some aspects of intracellular symbiosis during embryo development of *Mastotermes darwiniensis* (Isoptera: Mastotermitidae). *Parasitology* 40: 309–316.

Saffo, M. B. 2006. Symbiosis: The way of all life. In *Life As We Know It*, J. Seckbach (ed.). Springer, NY, pp. 325–339.

Saikkonen, K., P. Lehtonen, M. Helender, J. Koricheva and S. H. Faeth. 2006. Model systems in ecology: Dissecting the endophyte-grass literature. *Trends Plant Sci.* 11: 428–433.

Samuel, B. S. and J. I. Gordon. 2006. Humanized gnotobiotic mouse model of host-archaeal-bacterial mutualism. *Proc. Natl. Acad. Sci. USA* 103: 10011–10016.

Sapp, J. 1994. *Evolution by Association: A History of Symbiosis*. Oxford University Press, New York.

Savilahti, E., M. Kuitunen and . Vaarala. 2008. Pre- and probiotics in the prevention and treatment of food allergy. *Curr. Opin. Allergy Clin. Immunol.* 8: 243–248.

Selosse, M. A., F. Richard, X. He and S. W. Simard. 2006. Mycorrhizal networks: Les liaisons dangereuses? *Trends Ecol. Evol.* 21: 621–628.

Simard, S. W., D. A. Perry, M. D. Jones, D. D. Myrold, D. M. Durall and R. Molina. 1997a. Net transfer of carbon between ectomycorrhizal trees in the field. *Nature* 388: 579–582.

Smil, V. 2001. *Enriching the Earth: Fritz Haber, Carl Boschm and the Transformation of World Food Production*. MIT Press, Cambridge, MA.

Sonnenburg, J. L., L. T. Angenent, and J. I. Gordon. 2004. Getting a grip on things: How do communities of bacterial symbionts become established in our intestine? *Nature Immunol.* 5: 569–573.

Stappenbeck, T. S., L. V. Hooper and J. I. Gordon. 2002. Developmental regulation of intestinal angiogenesis by indigenous microbes via Paneth cells. *Proc. Natl. Acad. Sci. USA* 99: 15451–15455.

Starr, D. J. and T. W. Cline. 2002. A host-parasite interaction rescues *Drosophila* oogenesis defects. *Nature* 418: 76–79.

Stiling, P. 1993. Why do natural enemies fail in biological control campaigns? *Amer. Entomol.* 39: 31–37.

Stouthamer, R., J. A. J. Breeuwer and G. D. D. Hurst. 1999. *Wolbachia pipientis:* Microbial manipulator of arthropod reproduction. *Annu. Rev. Microbiol.* 53: 71–102.

Strathmann, R. R. and M. F. Strathmann. 1995. Oxygen supply and limits on aggregation of embryos. *J. Mar. Biol. Assoc. UK* 75: 413–428.

Tanaka, A., M. J. Christensen, D. Takemoto, P. Park and B. Scott. 2006. Reactive oxygen species play a role in regulating a fungus-perennial ryegrass mutualistic interaction. *Plant Cell* 18: 1052–1066.

Thomas, J. A. 1995. The ecology and conservation of *Maculinea arion* and other European species of large

blue butterfly. In *Ecology and Conservation of Butterflies*, A. S. Pullin (ed.). Chapman and Hall, New York.
Thomas, J. A. and G. W. Elmes. 1993. Specialized searching and the hostile use of allomones by a parasitoid whose host, the butterfly *Maculinea rebeli*, inhabits ant nests. *Anim. Behav.* 45: 593–602.
Thomas, J. A. and 7 others 2002. Parasitoid secretions provoke ant warfare. *Nature* 417: 505.
Thomas, L. 1974. *The Lives of a Cell: Notes of a Biology Watcher*. Viking, New York.
Tsuji, S. and 9 others. 2006. *Helicobacter pylori* eradication to prevent gastric cancer: Underlying molecular and cellular mechanisms. *World J. Gastroent.* 12: 1671–1680.
Turnbaugh, P. J., R. E. Ley, M. A. Mahowald, V. Magrini, E. R. Mardis and J. I. Gordon. 2006. An obesity-associated gut microbiome with increased capacity for energy harvest. *Nature* 444: 1027–1031.
Turnbaugh, P. J., F. Bäckhed, L. Fulton and J. I. Gordon. 2008. Diet-induced obesity is linked to marked but reversible alterations in the mouse distal gut microbiome. *Cell Host Microbiol.* 3: 213–223.
Umesaki,Y. 1984. Immunohistochemical and biochemical demonstration of the change in glycolipid composition of the intestinal epithelial cell surface in mice in relation to epithelial cell differentiation and bacterial association. *J. Histochem. Cytochem.* 32: 299–304.
Vandenkoorhuyse, P., S. L. Baldauf, C. Leyval, J. Straczek, and J. P. Young. 2002. Extensive fungal diversity in plant roots. *Science* 295: 2051.
Verhasselt, V. and 7 others. 2008. Breast milk-mediated transfer of an antigen induces tolerance and protection from allergic asthma. *Nature Medicine* 14: 170–175.
Visick, K. L., J. Foster, J. Donio, M. McFall-Ngai, and E. G. Ruby. 2000. *Vibrio fischeri lux* genes play an important role in colonization and development of the host light organ. *J. Bacteriol.* 182: 4578–4586.
Walter, J., N. C. K. Heng, W. P. Hammes, D. M. Loach, G. W. Tannock and C. Hertel. 2003. Identification of *Lactobacillus reuteri* genes specifically induced in the mouse gastrointestinal tract. *Appl. Environ. Microbiol.* 69: 2044–2051.
Waterman, R. J. and M. I. Bidartando. 2008. Deception above, deception below: Linking pollination and mycorrhizal biology of orchids. *J. Exp. Botany* 59: 1085–1096.
Werren, J. H. 1997. Biology of *Wolbachia*. *Annu. Rev. Entomol.* 42: 587–609.
Wolf, G. 2006. Gut microbiota: A factor in energy regulation. *Nutr. Rev.* 64: 47–50.
Woyke, T. and 17 others. 2006. Symbiosis insights through metagenomic analysis of a microbial consortium. *Nature* 443: 950–955.
Xu, J. and J. I. Gordon. 2003. Honor thy symbionts. *Proc. Natl. Acad. Sci. USA* 100: 10452–10459.
Zardus, J. D., B. T. Nedved, Y. Huang, C. Tran and M. G. Hadfield. 2008. Microbial biofilms facilitate adhesion in biofouling invertebrates. *Biol. Bull.* 214: 91–98.
Zchori-Fein, E. and 6 others. 2001. A newly discovered bacterium associated with parthenogenesis and a change in host selection behavior in parasitoid wasps. *Proc. Natl. Acad. Sci. USA* 98: 12555–12560.

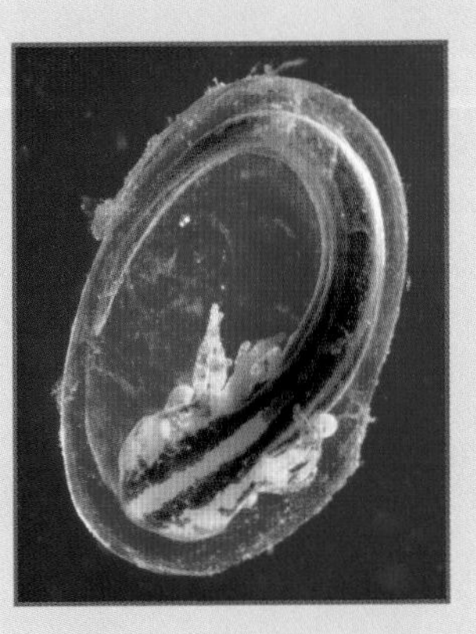

Chapter 4

Embryonic Defenses

Survival in a Hostile World

Self-defense is Nature's eldest law.

John Dryden, 1681

Developmental biologists rightly focus on the mechanisms by which fertilized eggs become new and highly complex individuals. What is not appreciated, however, is that there are two systems operating to ensure the forward movement of development: one system that ensures the fidelity of development, and another that increases an embryo's chances of surviving through its developmental stages. Fidelity is achieved by a buffering, or robustness, which assures small environmental or genetic perturbations do not alter development. Survival is achieved by a defense physiology that allows the fertilized egg to stay alive and continue its development in an environment teeming with predatory animals, toxic compounds, and potentially lethal radiation (Figure 4.1).

In addition, when we look at the diverse reproductive patterns in the animal kingdom, we find that the vast majority of organisms develop as **orphan embryos**, with little or no parental protection during their development (Figure 4.2). In many of these organisms, the adults can be viewed as players in a giant lottery, producing millions of embryos on the chance that one or two or a few will survive. In fact, most embryos share the fate of becoming food for other organisms. And the embryos and larvae that manage to escape predation must still cope with a constantly changing environment, including such challenges as variations in temperatures and humid-

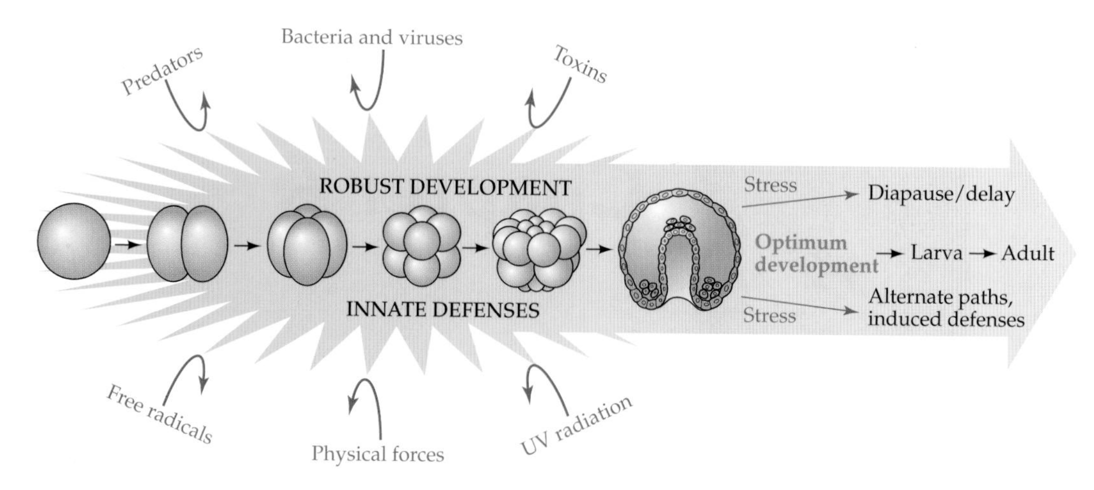

FIGURE 4.1 The embryo has not only a plan for development, but also a battery of defenses in order to survive in its anticipated environment. In early development of embryos with rapid early cleavages (e.g., most "orphan" embryos), the strategy is to have high levels of innate defenses already present when environmental stress is encountered. Later in development, the embryo can respond to the environment by inducing defenses, altering developmental pathways, or temporarily stopping development. These alternate paths can represent a "lottery approach," producing phenotypes which may or may not be adaptive. (After Hamdoun and Epel 2007.)

ity, infection by bacteria and fungi, and irradiation by solar wavelengths that can damage DNA.

But whatever the level of parental protection, the embryo's own protective mechanisms must be at the level of the cell, since the tissues and organs that provide protection for the mature individual haven't yet developed. And even at the cellular level, embryonic cells and adult cells do things very differently. Adult cells are sophisticated, with a strategy similar to that of wealthy nations: they maintain a large arsenal of defense mechanisms that are always available and can be quickly called up when danger is perceived. These adult cells can respond by initiating synthesis of stereotypic defense proteins immediately upon infusion of some toxicant from a recent meal, upon detecting a bacterial infection, or upon sensing denatured proteins from an increase in temperature.

By contrast, embryonic cells seem to be on a fast track, with seemingly no time or no capacity to divert their metabolism to defense. It appears that the protein-generating system of embryonic cells is taken up with instructing the cells how to divide, diversify, and form the basic plan of the adult organism; there is not much left for them to spend on defense until after gastrulation. Given these constraints, we may wonder how any embryo makes it to adulthood. But in fact a wide range of embryo survival mecha-

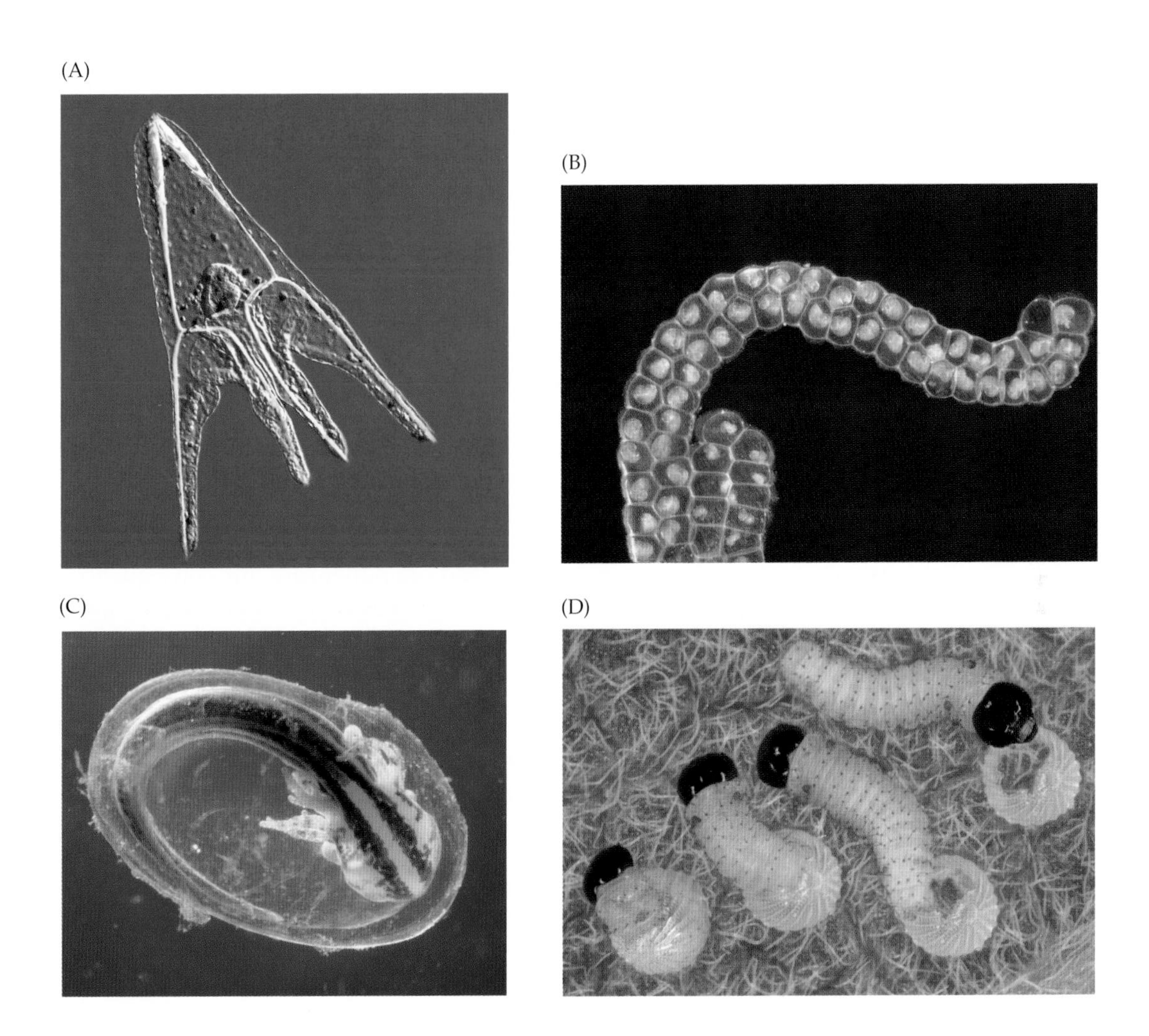

FIGURE 4.2 Most animals begin life as "orphan embryos" with no parental protection. (A) Pluteus-stage larva of a sea urchin. Sea urchins are prototypical of the myriad marine species whose embryos develop unaided in the ebbs and flows of the ocean currents. (B) Gastopod embryos such as these of the pond snail *Limnaea stagnalis* have been the source of much interesting data regarding the defense mechanisms of orphan embryos. (C) Most amphibian species lay fertilized eggs that develop and hatch unaided, as this European crested newt (*Triturus cristatus*) will do. (D) Larvae of the monarch butterfly (*Danaus plexippus*) get early nutrition from eating their egg cases immediately after hatching. (A © Hervé Conge/Photolibrary.com; B,C © Blinkwinkel/Alamy; D © Cathy Keifer/ShutterStock.)

nisms are in place; indeed, in the logical scheme of things, adult survival mechanisms could never have evolved if the embryo had not survived first.

Characteristics of Embryo Defense

Developmental robustness: A necessary but paradoxical defense

The developmental process is precise and, under the expected and anticipated conditions, proceeds in a predictable way. A *Drosophila melanogaster* egg fertilized in Los Angeles will develop identically to the same egg fertilized in New Delhi, even under a variety of environmental conditions such as different temperatures or humidity. Such **robustness**—also called **canalization**—is an essential property of developing systems (Nijhout 2002).

Canalization can also buffer against minor genetic perturbations such that the same phenotype is always produced. Researchers can experimentally "knock out" genes in embryos, and often there are dire consequences to such manipulations. But just as often there will be no phenotypic difference. It is as if the gene isn't even needed. This has led to the idea that robustness or buffering ensues from the presence of many genes controlling the same output, so removing one has no obvious effect on the outcome. It's as if you had ten people carrying a heavy box; removing one of them would not make a perceptible difference (see Cooke et al. 1997).

On one level, canalization is the *opposite* of phenotypic plasticity, because it assures that he same phenotype is produced regardless of environmental or genetic perturbations. At the molecular level, however, such robustness can be considered to be a *product* of plasticity, since the developmental interactions can adjust to compensate for genetic or environmental differences.

TWO EXAMPLES OF DEVELOPMENTAL ROBUSTNESS Major insights into the canalization of developmental pathways have come from studies of the fruit fly *Drosophila melanogaster*. These studies have focused on the stability of the insect's body segmentation pattern, which is determined in large part by the distribution of the Bicoid protein. The mRNA for Bicoid is maternal (i.e., it is carried in the egg cytoplasm) and is localized to the anterior region of the future embryo during oogenesis. Fertilization and egg laying initiate mRNA translation, and the newly translated Bicoid protein diffuses from the anterior end to establish a concentration gradient across the embryo. In the fruit fly embryo, DNA synthesis and mitosis occur without cell division for the first 12 nuclear divisions such that thousands of nuclei are present in what is effectively a large, single-celled embryo. Bicoid protein extends through this cell, forming a gradient that is highest at the anterior and lowest at the posterior of the embryo. The resultant reading of this gradient

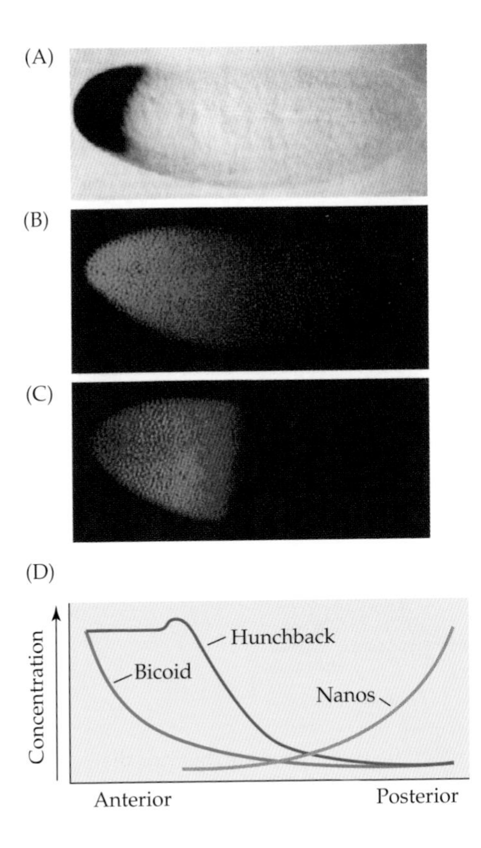

FIGURE 4.3 Bicoid and Hunchback proteins establish the *Drosophila* body axis. (A) The localization of *bicoid* mRNA (black) in the anterior pole of the *Drosophila* egg. (B) Later in development, the *bicoid* mRNA is translated into a Bicoid protein gradient, visualized here by a fluorescent green antibody. (C) Bicoid protein is a transcription factor that induces the expression of the Hunchback protein (red fluorescent antibody). (D) Summary of the interactions producing the Hunchback protein gradient (From http://www.princeton.edu/~wbialek/rome/lecture1.htm.)

establishes the precise distribution of the body segments (see Gilbert 2006).

Among other functions, Bicoid can act as a transcription factor, inducing the expression of the *Hunchback* gene. This causes Hunchback protein to be produced in a gradient from anterior to posterior of the embryo. Conversely, Nanos protein, which is synthesized in the posterior of the embryo, *inhibits* the translation of *Hunchback* mRNA, steepening the Hunchback protein gradient (Figure 4.3). Depending on the concentration of Hunchback present, a given region of the embryo is instructed to form the head (high concentration), the thorax (medium concentration), or the abdomen (low concentration).

Several investigators have shown that when *Drosophila* embryos are raised at different temperatures, the gradient of Bicoid expression is altered but the phenotype of the fly remains the same (see Houchmandzadeh et al. 2002; Lucchetta et al. 2005). In some as yet unknown manner, the regulatory mechanisms instructing *Hunchback* gene expression are buffered such that the Hunchback protein gradient from anterior to posterior remains the same even though the Bicoid gradient differs (Figure 4.4). This may be due

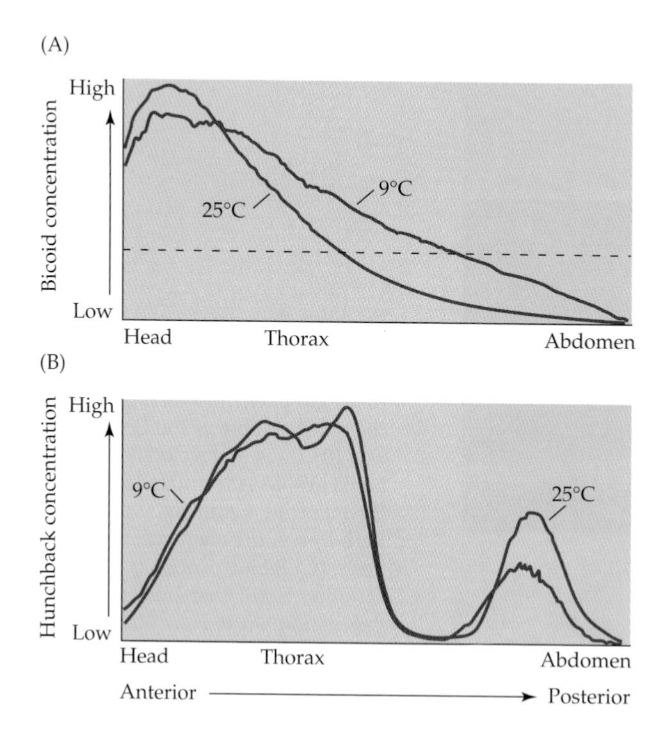

FIGURE 4.4 Changes in temperature affect the Bicoid gradient, but not the Hunchback gradient. (A) The Bicoid gradient is altered in *Drosophila* eggs incubated at different temperatures such that the posterior (abdominal) region of an embryo incubated at 9°C has a Bicoid concentration similar to that of the midsection (thorax) of an embryo incubated at 25°C (dashed line). (B) The Hunchback protein gradient is very similar at either incubation temperature. (After Houchmandzadeh et al. 2002.)

to the fact that Bicoid can be both a transcription factor and a translational regulator, or it might be due to Hunchback's being synthesized both by maternal genes (during egg formation) and by zygotic genes (in the embryo). Several mathematical models have been formulated to explain this precision (see Gregor et al. 2007a,b; Jaeger et al. 2007; Lepzelter and Wang 2008). These models suggest that the stability of the body plan ensues just from interaction of many components of the system. The idea here is that the robustness is an inherent outcome of the complexity of the developmental process.

Similarly, the formation of the vulva in the nematode *Caenorhabditis elegans*—a developmental process that has been extensively studied and is known to proceed usually from a specific and identifiable central cell—shows remarkable robustness in the presence of genetic and environmental variation. Recent research has shown that underlying developmental plasticity allows such robustness (Gleason et al. 2002; Chen and Grunwald 2004; Braendle and Felix 2009). In cases of starvation, for instance, the formation of the vulva will proceed from a different cell, indicating functional redundancy of vulval precursor cells. Furthermore, at least three signaling systems determine which differentiating cells become central to the structure and which are peripheral. If one of these signals fails to function, the other two will still persist.

THE PARADOX OF ROBUSTNESS If development is so robust, so canalized, how does the variation that is a hallmark of natural selection arise? If development was solely exact replication in every generation, then life would be static, there would be no diversity, and life as we know it could not have evolved, since there would have been be no variants for natural selection to act upon.

There are two general solutions to this paradox. The first is that robustness provides a threshold for phenotypic change, so that only large genetic or environmental perturbations produce phenotypic variants. The second answer is that robustness allows the accumulation of mutations that can later be expressed as an ensemble, especially under stressful conditions (Gibson and Dworkin 2004). The importance of this "cryptic variation" to evolution will be discussed in Chapter 10.

Early embryonic cells differ from adult cells

Another paradox of protecting the embryo is that there seems to be a relaxation of cellular defenses in the earliest stages of development. This is most evident in the cleavage period, the interval between fertilization and gastrulation. Cleavage-stage cells, often called **blastomeres**, are not like adult cells. They are not even like late embryonic cells. When teaching biology, we often highlight the remarkable similarities between different cell types. In this case, however, we must highlight the differences.

CELL CYCLE DIFFERENCES The cell cycle of early embryonic cells differs from the adult somatic cell cycle in at least two important ways. First, there is usually no G phase in the blastomere cell cycle. Instead of having a four-part cell cycle, early embryos often cycle from S phase (DNA synthesis) to M phase (mitosis) without G1 or G2 stages (Figure 4.5). In the fruit fly *Drosophila* and the frog *Xenopus*, the G stages are added only after the twelfth division. The blastomere cell cycle appears to have been selected for speed, especially in those marine orphan embryos that must passively ride the currents before acquiring larval organs.

Second, in adult cells, there is a highly orchestrated series of checkpoints by which DNA synthesis, chromosome movement, and cytokinesis (cell division) are coordinated (Murray and Hunt 1993). These checkpoints

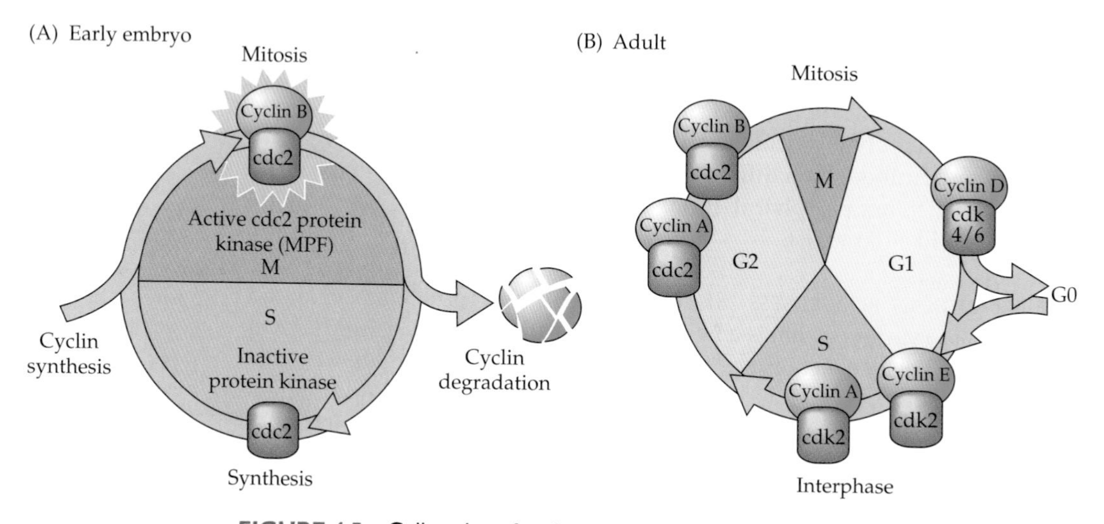

FIGURE 4.5 Cell cycles of embryonic blastomeres differ from those of adult somatic cells. (A) The biphasic cell cycle of animal blastomeres alternates between only two states, S and M. Cyclin B synthesis allows the progression from S to M, while degradation of cyclin B allows the progression of M to S. (B) In adult cells (or embryonic cells later in development), mitosis (M) is followed by a gap (G1), and synthesis is also followed by a gap (G2). Growth usually occurs during these gap phases. Cells that have differentiated are removed from the cell cycle and enter a G0 phase that resembles an extended G1. (After Gilbert 2006.)

ensure "quality control" of DNA replication and the fidelity of chromosome distribution during mitosis. In contrast to adult cells, however, in blastomeres there is often very little linkage between DNA synthesis, chromosome movement, and cytokinesis. For example, one can block cytokinesis and still have DNA synthesis (Alexandre et al. 1982; Whittaker 1973). One can even block DNA synthesis in blastomeres and still have normal cytokinesis (Nagano et al. 1981; Raff and Glover 1988). It seems that the programs for cell division, DNA synthesis, and early differentiation can be separated in the early embryo. As with the lack of G phases, this lack of coordination suggests an embryonic cell cycle that has evolved in response to selection for speed of development.

DIFFERENCES IN STRESS PROTEIN SYNTHESIS The adult cell, when experiencing stresses such as starvation or temperature shock, can synthesize numerous protective **stress proteins** (often called molecular chaperones) and initiate specific signaling cascades that protect the cell. Among the most prevalent are the **heat-shock proteins** that help prevent heat-induced protein denaturation and refold damaged proteins (Lindquist 1986). In some cases, the

Speculations on Cell Defenses in Early Development

One would think that the sorts of protection mechanisms found in adult cells would also be useful in early embryonic cells. But if they are not used, what could be the reasons? One possible explanation is the requirement for precise timing and positioning of cells during the cleavage stage of development. Anything that upsets the temporal sequence or the spatial consequences of cleavage must be avoided. So the embryo foregoes transcription of the genes needed for the stress responses and foregoes many checkpoint delays—all of which would slow down cell division. In addition, the early embryo foregoes apoptosis, as removal of injured cells would upset the spatial map needed to coordinate later development.

This requirement of precise coordination of time and space arises from the nature of the map of the future embryo that is present in the unfertilized or just-fertilized egg. Most eggs have some sort of polarity, such that the cells arising at one end of the zygote become the anterior portion of the embryo and cells at the other end become the posterior portion (see Gilbert 2006). Within these broad distinctions, the interaction of cells with each other allows neighboring cells to exchange information so that the right cells develop in the right place. If even a few cells in the early embryo experience serious damage, their removal via apoptosis could prevent their transmission of a critical message or signal. The solution is to keep the messenger (i.e., the cell) alive until the signal can be sent.

Even if some cells experience minor stress, the mounting of gene transcription as part of the stress response could delay cell division. Such delays would also happen if the checkpoint response to DNA damage delayed completion of the cell cycle. These holdups could mean cells do not arrive at the right place on time, resulting in missing messages or in transmission of the message to the wrong cell or at the wrong moment. Once again, the solution is to keep the messenger alive and on time so that the message can be delivered at the right time and place.

An alternate solution, you might say, is simply to allow more time for development so that stress responses and checkpoints can be used. This solution has indeed evolved among the mammals, where the time between cell divisions is often many hours, allowing ample time for transcription and checkpoint monitoring. Consistent with this, mammalian embryos can mount a stress response as early as the second cell division, and they have efficient mitotic checkpoints (Bensaude et al. 1983; Artus et al. 2006).

But slowing down the tempo of development does not seem to be a practical solution for species that produce orphan embryos. Rather, the selective process seems to have favored rapid development (Staver and Strathmann 2002). As an unprotected orphan embryo, every minute you remain an embryo is an extra minute for a predator to eat you or for an ocean current to carry you farther away from your preferred habitat. So the trade-off in orphan organisms seems to be speed of development—no stopping for repair of DNA, no apoptosis, and brief time windows for proper signaling. Damage can be taken care of after this signaling has been completed.

response to temperature is so extreme that the cell shifts the bulk of its protein synthesis to the manufacture of these heat-shock proteins. In addition to heat, free radicals (which can damage DNA), toxic chemicals, and osmotic changes each induce their own specific set of stress proteins, as well as inducing a general set of protective factors that can repair cellular damage or, in the worst-case scenario, direct the damaged cell to commit suicide (Feder and Hofmann 1999; Kultz 2005; Bukau et al. 2006).

Given these potent stress defenses in adults, it is paradoxical that most embryos do not show any stress response during the period between fertilization and gastrulation when their blastomeres are rapidly dividing (Lang et al. 1999; Heikkila et al. 1997). These cleavage-stage cells are remarkably unresponsive to heat stress, and there is no induction of synthesis of the heat-shock proteins until the embryo enters the **midblastula transition**, a point in development when the tempo of cell division begins to slow down and the G phases of the cell cycle first appear (Newport and Kirschner 1982). Therefore, when discussing the defenses of the zygote and the early embryo, we must remember that early embryonic cells cannot call forth the arsenal of protective agents used by adult (or even later embryonic) cells.

CELL DEATH DIFFERENCES Should the proteins or DNA of an adult cell become badly damaged, the cell sacrifices itself. The two main methods of cellular suicide are referred to as **apoptosis** (programmed cell death) and **autophagy** (the cell "eats" itself in a manner similar to the way in which it clears bacteria out of its cytoplasm). The two processes proceed by different mechanisms, but in both cases, externally induced damage causes the cell to synthesize proteins that digest its own constituents. In the end, the cell is physically removed, with only limited consequences to the surrounding tissue.

Embryonic cells can undergo apoptosis when they are severely damaged, but the timing of such cell death is usually delayed until after gastrulation (Figure 4.6; Hensey and Gautier 1998; Ikegami et al. 1999; Thurber and Epel 2007). So if, during the cleavage stage, embryos are exposed to severe stresses such as radiation or DNA-damaging drugs, the cells continue to divide even though their DNA may be irreparably damaged. Because

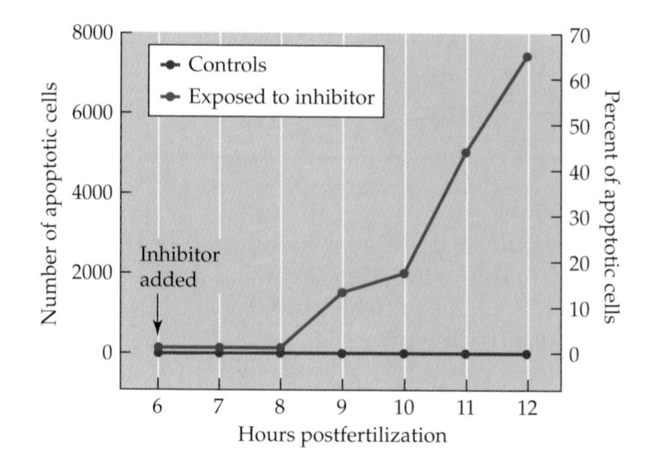

FIGURE 4.6 Exposure of a fish embryo to a DNA-synthesis inhibitor at 6 hours after fertilization does not induce apoptosis until the midblastula stage (at 9 hours). Controls not exposed to the inhibitor show no apoptosis. (After Ikegami et al. 1999.)

every one of the few cells present in an early embryo carries so much information, the loss of even one cell can affect development (see box). Embryonic cells thus appear to "postpone" apoptosis until around the time of the midblastula transition—the same point at which the heat-shock response is first seen.

Strategies for Embryo Defense

The consequences of experiments that subject embryos to diverse insults such as heat shock or immersion in DNA-synthesis inhibitors might suggest that the embryo is poorly prepared for the stresses it will encounter during its development. But these experiments involve all sorts of unnatural exposures; the paradox here is that, in fact, most embryos *are* well prepared for the stresses of the environment in which they normally develop. It seems that they have adapted to the *anticipated, expected, historical* environment and that these adaptations allow normal development to proceed.

The protective strategies found among animal embryos can be grouped into four major categories: induced polyphenism, parental protection, dormancy and diapause, and cellular stress physiologies. These are described briefly below; the remainder of the chapter then details some of the mechanisms by which embryos implement these four strategies.

Strategy 1: Induced polyphenism

The first type of protective program, which was described in detail for a variety of organisms in Chapters 1 and 2 and will be detailed later for early embryonic cells, is induced protection. Here, an embryo can search the environment for cues indicating potential dangers and then alter its developmental trajectory so that the embryo can survive these dangers. The developmental changes might be alterations in structure to foil predators, or changes in physiology and morphology in response to environmental conditions.

Strategy 2: Parental protection

There are many forms of parental protection. In some instances, especially in mammals, the embryo forms the placenta, which provides a partial barrier against pathogens and toxins. The placenta also implants into the mother's uterus, enabling the mother to provide nutrients and oxygen to the embryo. The mother's immune system also protects the placenta-embryo complex from some (although not all; see Chapter 5) environmental bacteria and viruses (Glezen and Alpers 1999).

In most organisms, the oocyte or cells associated with the growing oocyte prepare a protective egg coat—variously named the shell, chorion, egg envelope, egg case, or sheath—that provides a physical barrier around the oocyte. Such barriers can comprise a large portion of the embryo mass and, as we describe later, provide protection against physical trauma as well as bacterial and viral infections.

In many non-mammalian species, the mother prepares protective compounds that she transports into the egg. The hen, for instance, makes antibodies against common barnyard pathogens, and these antibodies get transported into the yolk of the egg. In this way, the embryo and hatchling have a passively acquired immunity against common bacteria, viruses, and fungi before they are able to make their own antibodies. The hen also synthesizes several proteins that prevent bacterial growth and which are transported into the egg albumen. These include lysozyme (which prevents bacteria from making their cell walls) and avidin (which binds and inactivates biotin, a bacterial growth factor).

Vertebrates aren't the only animals whose mothers (and in some cases fathers) provide protective compounds for the egg and early embryo. The eggs of the arctiid moth *Utetheisa ornatrix* get a heavy dose of a foul-tasting alkaloid from both the mother and the father (Dussourd et al. 1988). The parents acquire these bitter alkaloids while still in the larval stage by eating the alkaloid-laden plants of the genus *Crotalaria*. These alkaloids do not harm the larvae, and they continue giving protection to the adult (Figure 4.7). The adult female puts these compounds into her eggs, and the father places the alkaloids into his semen. The female stores the sperm for later fertilization, and the alkaloids in the seminal fluid are transferred to the eggs. Both the male and female contributions to the egg inhibit the predation of the egg by beetles (Dussourd et al. 1988).

Strategy 3: Dormancy and diapause

When the going gets tough, the embryo sleeps through it all. Like the seeds of many angiosperm plants, the embryos and larvae of many animals can curtail development, becoming metabolically inert and entering a state of **dormancy** when conditions are not favorable for growth and survival. They are able to restore normal metabolism and continue growing when favorable conditions return.

For example, in overly dry conditions, brine shrimp embryos and certain fly larvae get rid of all water. This anhydrous condition would normally destroy the cell's proteins, but in these larvae the proteins are kept intact and in a functional (although inert) state by the larval synthesis of high concentrations of the protective sugar trehalose. The dormancy ends when rainfall brings the larvae "back to life." Another well-studied example of dormancy is that of the nematode *Caenorhabditis elegans*, which in the absence of normal food supply can form a **dauerlarva** after its second metamorphic molt (see box). This non-feeding larval stage can be terminated by

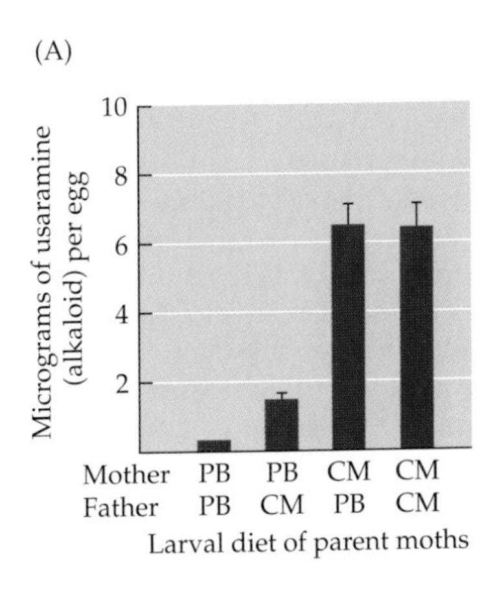

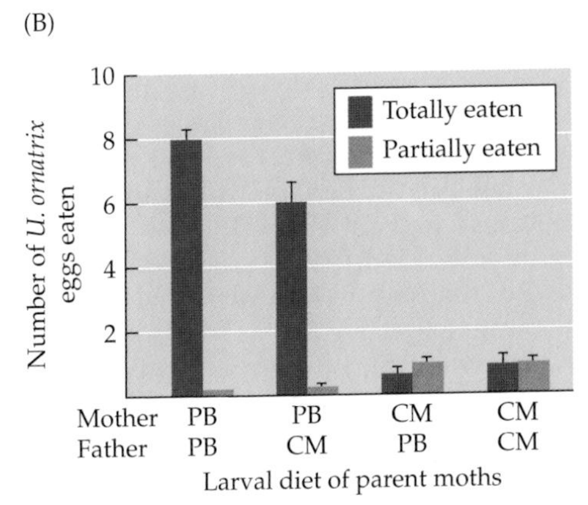

FIGURE 4.7 Eggs and embryos of the moth *Utetheisa ornatrix* acquire plant alkaloids such as usaramine from their parents. These chemicals appear to protect the embryos from being eaten by predators. (A) Alkaloid content of the eggs made by parents raised on diets containing alkaloids (CM) or no alkaloid (PB). (B) The presence of alkaloid is a major determinant of whether or not the eggs are eaten or not. (After Dussourd et al. 1988.)

providing the larva with adequate food (Cassada and Russell 1975). Thus, one way for an embryo to protect itself is to form a state that can sleep through the bad times and wait for better days.

In many animal species, a variant of the dormancy strategy, called **diapause**, has evolved. Diapause is a genetically programmed suspension of development that can occur at the embryonic, larval, pupal, or adult stage, depending on the species. In most cases, diapause is not a physiological response brought about by harsh conditions. Rather, it is brought about by token stimuli that presage a change in the environment; thus diapause usually begins before the actual severe conditions arise.

In some species, a period of diapause has become an obligatory part of the life cycle. This is often seen in temperate-zone insects that overwinter, where diapause is induced by changes in the photoperiod (the relative lengths of day and night). The point at which 50% of the population has

The Dauerlarva of *C. elegans*

Studies on the nematode worm *Caenorhabditis elegans* have provided insights on how the availability of food governs whether, at the second molt stage, this animal continues its metamorphosis and gives rise to a reproductive adult or whether it becomes a dormant larval form known as the dauerlarva. Genetic analysis reveals a complex of interacting sensory systems that determine this decision. The input to these systems comes from chemosensory neurons that detect the "state of the environment" through sensing the concentration of specific chemicals in the environment. One sensory system "measures" the density of the worm population, using a pheromone produced by other *C. elegans* individuals in the immediate environment. Another sensory system reports the availability of food, and a third is sensitive to high temperatures (Bargmann 2006). The worm responds to this sensory input with the activation of signaling cascades that result in the production of numerous neuropeptides, including an insulin-like growth factor and a paracrine factor called transforming growth factor-β (Hu 2007).

These neuropeptides can signal other cells to turn on the *DAF9* gene, which encodes a P450-type cytochrome that converts cholesterol into a sterol-like compound called dafachronic acid (Yabe et al. 2005; Motola et al. 2006). Dafachronic acid is a ligand for the transcription factor DAF12, and binding to DAF12 initiates a cascade of events directing the larva to become a reproductive adult. If food is limiting, however, the neuropeptides are not made, *DAF9* is not active, and in the absence of dafachronic acid the larva enters the dauer stage.

This dissection of the dauer response points to the common mechanisms used in environmental sensing during development. In the other well-studied cases, such as photoperiod or temperature control of phenotype in butterflies or the temperature control of gender in reptiles, the sensing of the environment is ultimately tied to production of a hormone, which then directs a suite of events leading to the new phenotype (see Chapter 2). The sensing mechanism, interestingly, is most likely also through neurons. This is obviously the case with photoperiod. The mechanism of temperature-sensitive hormone production is still not known, but the temperature-sensitive ion channels are an interesting candidate target (Rosenzweig et al. 2005; Voets et al. 2005).

entered diapause—an event that is usually quite sudden—is called the **critical day length** (Figure 4.8; Danilevskii 1965; Tauber et al. 1986).

Embryos of the silkworm moth (*Bombyx mori*) overwinter as embryos, entering diapause just before segmentation. The gypsy moth (*Lymantia dispar*) initiates its diapause as a fully formed larva and is thus ready to hatch as soon as diapause ends. Hormones appear to be a determining factor in the stage at which an insect enters diapause. Diapause in the embryonic stage appears to be regulated in some insects by increased levels of PTTH (prothoracicotropic hormone), whereas in other insects larval diapause appears to be the result of inhibition of this hormone, which prevents the larvae from molting and entering pupation. In the laboratory, diapausing pupae can be reactivated by exposure to 20-hydroxyecdysone. However, under normal conditions, the brains of diapausing pupae (such as those of the moth *Hyalophora*) are activated by the exposure to cold weather for a particular duration. Moth pupae kept in warm conditions will remain in

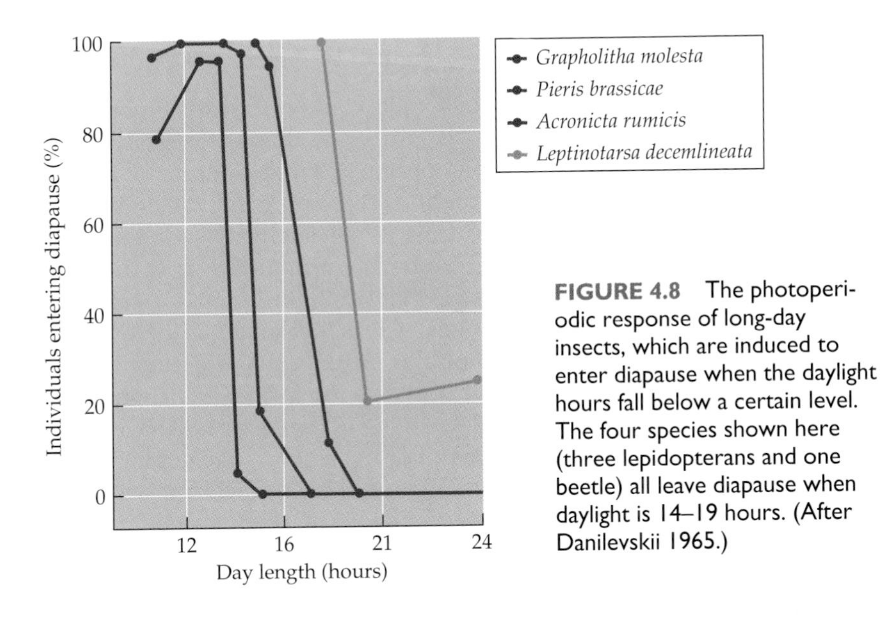

FIGURE 4.8 The photoperiodic response of long-day insects, which are induced to enter diapause when the daylight hours fall below a certain level. The four species shown here (three lepidopterans and one beetle) all leave diapause when daylight is 14–19 hours. (After Danilevskii 1965.)

diapause until they die (see Nijhout 1994). The mechanism by which changes in temperature and day length regulate hormone production remains to be elucidated.

Diapause isn't only for invertebrates. More than a hundred mammalian species also have this ability. For example, some female mammals can delay the implantation of the embryo, keeping it suspended in the uterus. Further development does not proceed until the embryo has attached itself to the uterine lining. In this way, these mammals can mate in the winter and yet "time" their pregnancies such that the young are born in the spring, when food is abundant (Renfree and Shaw 2000).

Strategy 4: Defense physiologies

A range of physiological strategies allows the embryo to cope with everyday stresses. These stresses include heat that can damage proteins, toxic chemicals and radiation that can damage DNA, and physical stresses that can damage cell membranes. These defense physiologies apparently do not interact with the developmental program, but insure that the embryo can survive while carrying out its development. The heat-shock proteins mentioned earlier, as well as molecular mechanisms for removing toxic substances from cells and the provision of "sunscreen" chemicals—both described in detail in the next section—are among the prime examples of embryonic defenses against stress.

A General Strategy: "Be Prepared"

Many of the mechanisms that have evolved to protect the early embryo are versions of the Boy Scout injunction "Be prepared." Just as the scout is prepared to survive emergencies that might come up while camping or hiking, so has there been selection for adaptations that "prepare" the embryo to survive traumas it might encounter during its development (see Figure 4.1). Like the scout's preparedness, the embryo's preparedness is directed toward *potential* problems, not toward *all* problems. If historically the embryo develops in a cold environment such as the Antarctic, there will be adaptations (such as unique microtubular proteins that are functional at –2°C) to allow the embryo to survive and develop in a frigid habitat (Detrich et al. 2000). Cells of an embryo developing in the desert will have adaptations against heat (David et al. 2004). For the embryo developing in the full glare of sunlight on a beach (which would fry our DNA if we forgot sunscreen), embryonic cells have ways of avoiding or repairing UV damage. The protective and defensive evolutionary adaptations found in embryonic cells provide effective and in many cases novel means to survive in remarkably rigorous environments (see Hamdoun and Epel 2007).

Mechanisms of Embryo Defense

How do embryos defend themselves before the differentiation of the adult tissues and organs that normally provide protection? How does the early embryo handle toxic substances, whether natural or man-made? How does it keep its DNA intact in the face of such stresses as UV radiation or the free radicals produced by normal metabolism? How can it defend itself against physical damage? Is the membranous envelope that surrounds the embryo equivalent to skin or skeleton? And, finally, what adaptations are there to ward off pathogens without an immune system, or predators without some sort of escape or behavior response?

The conclusion, as you will see, is that there is a remarkable constellation of defenses that can work well at this few-celled level. Their presence indicates the environmental pressures that have operated over the eons and the remarkable selection that must have taken place to provide these defenses.

We start with how embryos thwart exposure to toxic substances while simultaneously carrying out the developmental program. Remember that these defenses have evolved to defend the embryo in its anticipated, historical milieu. But that milieu, that environment, is being changed rapidly by human activities. Can embryos evolve fast enough to withstand the effects of new chemicals and environmental conditions? Some chemicals will get through the embryos' defenses and block development. These agents are called teratogens and will be discussed in Chapter 5. Some chemicals will affect developmental pathways in ways that won't become apparent until later in life; such agents will be considered in Chapter 6. And some conditions, such as climate change, will result from indirect activities of a rapidly growing and consuming human population. A question to consider, therefore, is whether these ancient embryonic defenses will prove adequate in this rapidly changing world (see Hamdoun and Epel 2007).

Protection against Toxic Substances

An embryo, whether protected by a parent or developing independently, can be exposed to toxic substances from its mother's diet or from the environment. Some of these toxic substances are from other organisms (such as defensive chemicals produced by prey), some are from the abiotic environment (such as heavy metals resulting from volcanic eruptions or industrial spillage), and some are from processes used in manufacturing (such as plasticizers or the lead used in paints and fuel additives).

The general plan: "Bouncers," "chemists," and "policemen"

Whereas the adult animal has several sophisticated enzyme pathways for inactivating, metabolizing, and removing poisons (in vertebrates, these reactions are generally carried out in the liver), embryonic cells generally use a three-tier scheme to rid themselves of poisons:

1. "Bouncers" that prevent the entry of toxic substances into the cell
2. A set of "chemists" that modify (detoxify) the dangerous molecule and then put some sort of chemical tag on the modified molecule.
3. A set of "policemen" that then escort the tagged toxicants out of the cell.

The "bouncer" and "policemen" functions are carried out by **efflux transporters**, which are members of the ATP-binding cassette, or **ABC**, family of proteins, one of the largest protein families known. ABC efflux transporters use the energy of ATP hydrolysis to carry compounds across cell membranes. These proteins can carry compounds into the cell or transport endogenous compounds out of the cell (Figure 4.9).

The "bouncer" function, which is the most efficient way of dealing with toxic compounds, is to not let the compounds even get into the cell. This function is handled by members of the P-glycoprotein subfamily of ABC transporters. These glycoproteins interact with the toxicants within the plasma membrane and actively prevent their entry into the cell (see reviews by Higgins 1992; Cole and Deeley 1998; Epel et al. 2008).

The "chemist" role is carried out by suites of enzymes that modify the putative toxicant, usually by adding groups that make the toxic substance more water-soluble. Typically this can involve oxidation by the enzymes of

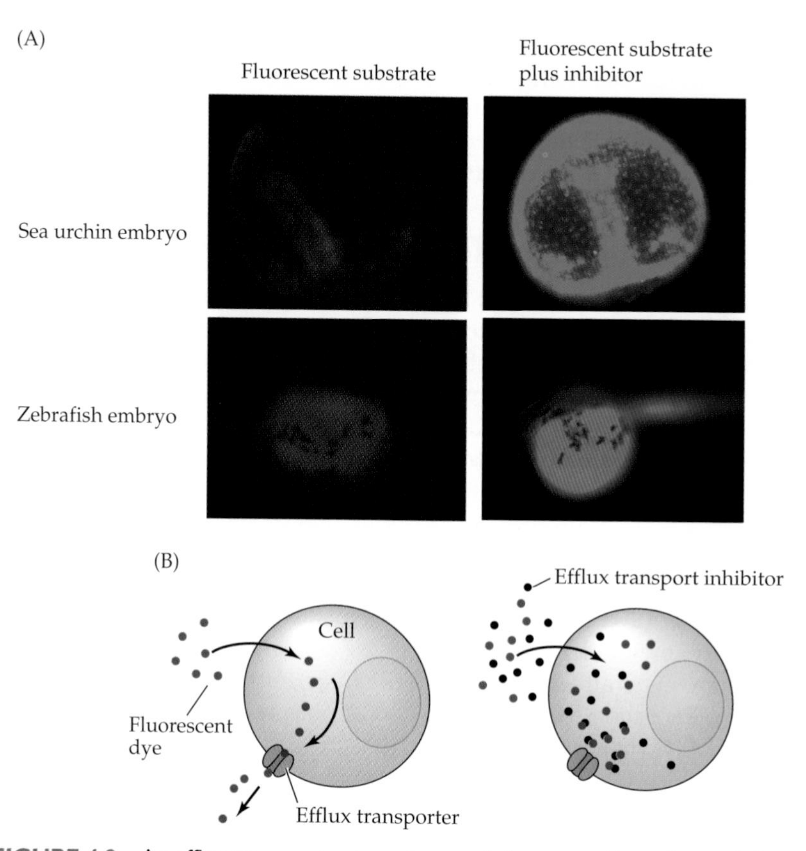

FIGURE 4.9 An efflux transporter in embryos of the sea urchin and zebrafish prevents entry of a fluorescent dye. (A) Micrographs on the left show effective exclusion of the fluorescent dye calcein AM. Micrographs on the right depict the consequences of inhibiting transport activity in the same material. As seen, there is little entry of the dye unless the transporter is inhibited. (B) The diagram shows what happens when transport activity is inhibited. (From Epel et al. 2008.)

the cytochrome P450 system, followed by direct conjugation of a polar (hydrophilic) group to the oxidized toxicant. This polar group can be a glutathione (a tripeptide containing a cysteine), a sugar, or even a sulfate group. If the toxicant can be conjugated directly, the P450 step need not take place.

The next step is carried out by the "policeman." This step involves binding any toxicants that entered the cell and were modified and escorting them out of the cell. This is carried out by the **MRP** (multidrug-resistance proteins) family of ABC transporters.

The properties of the efflux transporters in embryos have been best described in the sea urchin embryo (see Hamdoun et al. 2004; Goldstone et al. 2006). The unfertilized sea urchin egg has low ABC activity, but this activity increases markedly after fertilization. It appears that an ABC transporter belonging to the P-glycoprotein family (which primarily prevents toxicant entry into the cell) and a transporter belonging to the MRP family (which pumps toxicants out of cells after they have entered the cytoplasm and been modified) are already present in the egg in an inactive form. Fertilization initiates steps that move the transporter to the plasma membrane and tips of the microvillus (Hamdoun, unpub. obs.), and this translocation is somehow tied to an almost 200-fold increase in activity that begins about 25 minutes after fertilization.

These transporters defend the embryo against toxic chemicals. As shown in Figure 4.10, cell division in the sea urchin embryo is inhibited by vinblastine, a microtubule-disrupting chemical. The concentration of vinblastine needed to inhibit 50% of the embryos from dividing is 2.5 μ*M*. However, if the transporter activity is inhibited, the effective concentration drops to 0.1 μ*M*. In other words, the transporter activity is so effective that 25 times more drug is required to inhibit mitosis when the transporter is functioning than when it isn't. (Vinblastine is itself a protective alkaloid made by plants of the periwinkle family.)

The limited work on other embryos suggests drug transporters might be general defenses in all embryos. They are present in the embryos of several other marine organisms, such as the echiuroid worm *Urechis caupo* (Toomey and Epel 1993) and the mussel *Mytilus galloprovincialis* (McFadzen et al. 2000), and they are also found in embryos of the soil nematode *Caenorhabditis elegans* (Broeks et al. 1995). All of these organisms develop in environ-

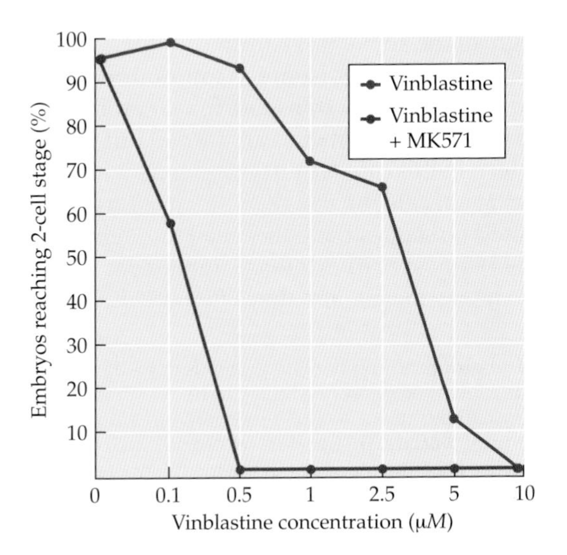

FIGURE 4.10 An efflux transporter protects sea urchin embryos from the effects of a cell division inhibitor. Incubation of *Urechis* embryos in vinblastine destroys microtubules and inhibits mitosis. The 50% concentration necessary for inhibition is about 2.5 μ*M*. If the MRP transporter inhibitor MK571 is added, the effective concentration goes down to 0.1 μ*M*. (After Hamdoun et al. 2004.)

ments that can be very toxic, and the titre of ABC proteins found in *U. caupo* (which lives in mudflats) are higher even than in drug-resistant cancer cells that overexpress ABC proteins (Toomey and Epel 1993). *C. elegans* uses an ABC transporter to export dafachronic acid, the signal regulating dauer formation (see p. 132). Among mammals, protective ABC transporters have been found in early mouse embryos (Ebling et al. 1993), as well as in the murine placenta. Smit and colleagues (1999) showed that these transporters can play important roles in preventing toxins from entering and accumulating in the mouse embryo.

The transporter defense, however, is of no use if the transporter protein does not recognize a specific dangerous chemical, as is presumed to have happened in the case of the drug thalidomide. As will be discussed in Chapter 5, thalidomide is a mild sedative that, in the 1950s and early 1960s, was sometimes prescribed for pregnant woman and was responsible for crippling birth defects. We now know that part of thalidomide's effectiveness as a teratogen (substance that causes developmental defects) was due to the drug's not being recognized by the ABC efflux transporter and thus having unfettered access to the developing fetus (Zimmerman et al. 2006).

Toxic metals

As will be described in Chapter 5, some metal ions—particularly the heavy metals such as mercury, cadmium, and lead and the transition metals such as copper or even iron—can be toxic. Their toxicity can ensue from (1) interfering with enzyme activity, as when mercury or cadmium bind to a protein (often to a critical cysteine residue on the protein), (2) preventing closure of a calcium ion channel, so toxic levels of calcium can enter a cell, or (3) generating reactive oxidative intermediates, as can occur with the transition metals. Some metals can be toxic through all the above mechanisms.

Whereas the protection against many poisonous compounds was through the "bouncers" (i.e., the family of efflux transporters that prohibited entry into the cell), the defenses against heavy metals are carried out by the "chemists" and "policemen," which in this case are enzymes along with specialized peptides and proteins that shackle the metal and then escort it out of the cell. These defenses are based on sequestering the metal through binding to sulfhydryl groups (–SH) on specific defense molecules.* The sulfhydryl groups bind avidly to heavy metals, and they often enclose them,

*We have also learned to use sulfhydryl groups to combat heavy metals in medical practice. British anti-Lewisite (BAL) is a 3-carbon backbone linked to two sulfhydryl groups. BAL was developed in response to the arsenic-containing compound Lewisite (the "dew of death"), an agent of chemical warfare. BAL and its related sulfhydryl-containing compound, DMSA, have since become important in the treatment of people who have been exposed to heavy metals.

thereby preventing the metals from doing damage. Sulfhydryl groups do double duty in cell protection, since they are also antioxidants (i.e., they reduce oxidant molecules, as discussed further later in this chapter).

One of the major defense molecules against heavy metals is the tripeptide glutathione, composed of glutamate-cysteine-glycine, which, with its sulfhydryl-containing cysteine group, binds with high affinity to heavy metals such as mercury and cadmium. It is very abundant in cells (in the m*M* range) and, as noted, provides critical defenses against oxidant damage. The cells get rid of the metal-glutathione conjugate by pumping it out of the cytoplasm using members of the MRP family of efflux transporters, as described above (Cole and Deeley 1998).

Variants of the glutathione defense are the phytochelatins, which are polymers of glutamate-cysteine and similarly bind to heavy metals. Phytochelatins, as the name implies, had been assumed to be plant-specific, but related phytochelatin synthase genes have also been found in animals, suggesting their use may be more widespread (Goldstone et al. 2006).

The second defense against heavy metal toxicity is a special class of metal-binding proteins, the metallothioneins (MTs). These proteins are cysteine-rich (i.e., at least 30% of their component amino acids are cysteines), and as in glutathione, the –SH group of the cysteines binds to the metals (Palmiter 1998). This class of protein also has antioxidant activity.

If the organism is exposed to heavy metals, its cells respond by increasing *MT* gene transcription and translation, which provides further protection. The genes encoding MTs (and other metal transport proteins) have an enhancer region containing a metal-sensing regulatory element, or MRE. The MRE sequence binds the transcription factor MTF-1; MTF-1, however, will only induce *MT* gene expression if it has bound a heavy metal ion. The heavy metal ion appears to alter the shape of MTF-1 so that it can recruit a histone acetyltransferase capable of activating gene expression (Li et al. 2008). Thus, the genes encoding MT and other proteins that protect against heavy metals are activated by the heavy metals themselves. MTF-1 also regulates the transcription of antioxidant enzymes, thus providing multiple defenses against the consequences of metal exposure. Indeed, like the heat-shock protein scenario, a previous exposure of cells to low levels of a toxic metal will induce the synthesis of more protective molecules, in this case more MTs, thus providing protection against subsequent escalation of the metal concentration (Andrews 2000).

The defense against toxic metals in the cleavage stages of orphan embryos utilizes the endogenous glutathione and MTs to sequester metals. If exposed to heavy metals during cleavage, these embryos do not initiate additional transcription of MTs. Like the situation with heat-shock proteins, the induction of new MT occurs only after gastrulation, meaning that the basal levels of protective molecules in the egg are the only protection against any heavy metal exposure (Scudiero et al. 1994). This level is presumably based on the anticipated exposure to heavy metals and will be effective as long as there is no rapid change in the embryos' environment. After gastrulation, when the pace of cell division slows, elective synthesis of MTs can occur (Nemer et al. 1991). Prior to this, the embryo depends on its "be prepared" strategy for the anticipated environment.

In the slower developing (and non-orphan) mammalian embryos, there is a different scenario, most likely related to the long duration of the cell cycle in mammals, which then allows time for new transcription in response to environmental stresses. Here induction of MT by heavy metals can occur during cleavage, by the 4-cell stage (Andrews et al. 1991). This behavior is similar for the induction of heat-shock proteins, which are also inducible in the mouse embryo during cleavage. Work on the mouse embryo, however, indicates that MT is not solely for protection from toxicants; MTs may have other roles, perhaps in protecting from oxidative damage or for the metal homeostasis needed for normal cell functioning (Andrews et al. 1991; Palmiter 1998).

Problems with metal detoxification

The trouble with metal-binding defenses is getting rid of the bound metal. In plants, the metal-loaded phytochelatins are secreted into the plant cell vacuole, which essentially isolates the toxicant from the cell (Cobbett 2000). A similar mechanism may exist in animal cells via transport into lysosomal-type vesicles (Berry 1996). However, it is thought that the bulk of the metals bound to glutathione are ejected via the efflux transporters, which pump the metal conjugate into the extracellular medium, although this method merely displaces the problem to adjacent cells or tissues (since the metal-loaded glutathione is still in the embryo or organism).

There is a similar problem with the metallothioneins. The metal-loaded MTs can stay in the cell a while but eventually are leaked or diffuse from the cell. In vertebrates—both larvae and adults—the metal-loaded MT travels to the kidneys, where hydrolysis of the MT leads to the release of toxic metals; if toxin levels are too high, there can be kidney damage (Klaassen and Liu 1997).

The defense provided by the MTs (and probably by glutathione as well) may actually be a sort of timing game. The immediate toxic effect of high metal levels is ameliorated by sequestration, but eventually the metal is released. The benefit may be that the subsequent release takes place over a longer time period, so the concentration is smaller and hence less toxic. This scheme works for limited exposure to metals but would not be effective if the exposure were chronic.

Protection against Physical Damage

Shells and extracellular coats

Fragile embryos are often protected from dessication and infection by becoming encased in protective layers; however, these layers are themselves often fragile. Indeed, as will be described in Chapter 6, one effect of the spread of DDT throughout the food chain was this pesticide's interference with the normal thickening and calcification of bird eggshells (see Figure 6.1). The situation was so severe for peregrine falcons that, in some cases, their eggshells would break when the parent birds sat on them for normal incubation.

Certainly bird eggs and the embryos within them seem vulnerable, but does this fragility extend to other types of embryos? The intuitive answer is "yes," since unlike the adult protected inside its skin or exoskeleton, most eggs and embryos are out in the open and subject to predation, breakage, and desiccation. But in fact the answer is more often "no"; many animal eggs have protective features that allow the embryo to survive in environments that could easily be expected to convert them into scrambled eggs.

A general feature of the egg is a typically spheroid shape and large size (as compared with other cells). In very large amphibian eggs, the actin cytoskeleton stabilizes the cortex. It is as if the egg is surrounded by an actin "girdle" that maintains its spherical structure and rigidity (Elinson and Houliston 1990; Ryabova and Vassetzky1997). Actin is similarly enriched in the cortex in other eggs, such as the sea urchin (Spudich 1992). If one depolymerizes the actin, as with the microfilament-destroying drug cytochalasin, the egg loses its spherical shape and begins to ooze or flow onto whatever surface it is on.

The other obvious feature in most eggs is some sort of extracellular envelope. In the familiar avian egg, this is the shell. Eggs of almost all species have a similar sort of functional cover that not only provides physical protection but also acts as a "fertilization filter," insuring that only sperm of the same species can bind to the outer envelope or pass through it on the way to fusing with the egg plasma membrane (Vacquier 1998). The avian egg is different from most in that the outer egg coat surrounds the egg *after* fertilization has taken place. The avian egg also has another layer, the albumen, or "egg white." Albumen serves as a shock absorber as well as a source of antibodies and other compounds that prevent infections.

THE MAMMALIAN EXTRACELLULAR COAT The extracellular coat of the mammalian egg is more typical than avian eggshells of the egg coatings seen in the animal kingdom. The components of this thick extracellular matrix, called the **zona pellucida**, are synthesized by the developing oocyte. The zona pellucida surrounds the unfertilized egg and, as noted, is a physical barrier that challenges sperm on their way to the plasma membrane of the oocyte.

The zona pellucida of the mouse egg is the best studied example of this structure. It has three main structural proteins: ZP1, ZP2, and ZP3. If a sperm is of the correct species, and if it is able to bind to ZP3, it can pass through the zona. When ZP3 binds the sperm, an **acrosome reaction** is activated in the sperm head, whereby the sperm releases enzymes that can digest a path through the zona. ZP2 acts as a guide, allowing the sperm to stay in the path that the acrosomal enzymes are digesting, while ZP1 holds these proteins together (Prasad et al. 2000; Jovine et al. 2005).

As the embryo undergoes cleavage inside the zona pellucida, the enclosing zona allows the embryo to reach the uterus without attaching to the Fallopian tubes. If such attachment does occur, it results in an ectopic, or "tubal," pregnancy that can kill the mother as well as the embryo. In addition, the zona apparently protects the embryo from viruses and bacteria that might have evaded the mother's immune system. Mateusen and colleagues (2004) found that if pig embryos were incubated with certain porcine viruses, none of the embryonic cells became infected as long as the zona pellucida was present. When the zona was removed, more than 50% of the cells were rapidly infected.

THE SEA URCHIN EXTRACELLULAR COAT The egg of the sea urchin is unusual in that it does not have a visible external coat. However, it elevates a fertilization envelope within a few seconds after sperm entry. This fertilization envelope arises from the conversion of a thin, ephemeral precursor layer adhering closely to the egg surface—the vitelline layer—into the substantial fertilization envelope (Figure 4.11). The vitelline layer is similar to the zona pellucida in that it also has sperm binding sites, and only sperm of the same species can attach to these sites (Vacquier 1998). The fertilizing sperm digests a hole through the vitelline layer, followed by fusion of the sperm with the egg plasma membrane (Vacquier 1998; Briggs and Wessel 2006).

The elevation of the fertilization envelope is the result of a cascade of events that ultimately results in the exocytosis of thousands of protein-containing **cortical granules** embedded in the egg cortex. In sea urchins, this cortical granule exocytosis releases a large number of proteins and enzymes whose action simultaneously prevents further sperm-egg binding and causes the detachment and subsequent elevation of the vitelline layer from the egg surface.* This fertilization membrane is initially quite soft, but within a few minutes, the membrane undergoes major structural changes to become

*Similarly in the mouse, the bursting of the cortical granules releases enzymes that modify the ZP3 and ZP2 proteins mentioned earlier, preventing any other sperm from entering into the egg (Sun 2003; Wortzman-Show et al. 2007).

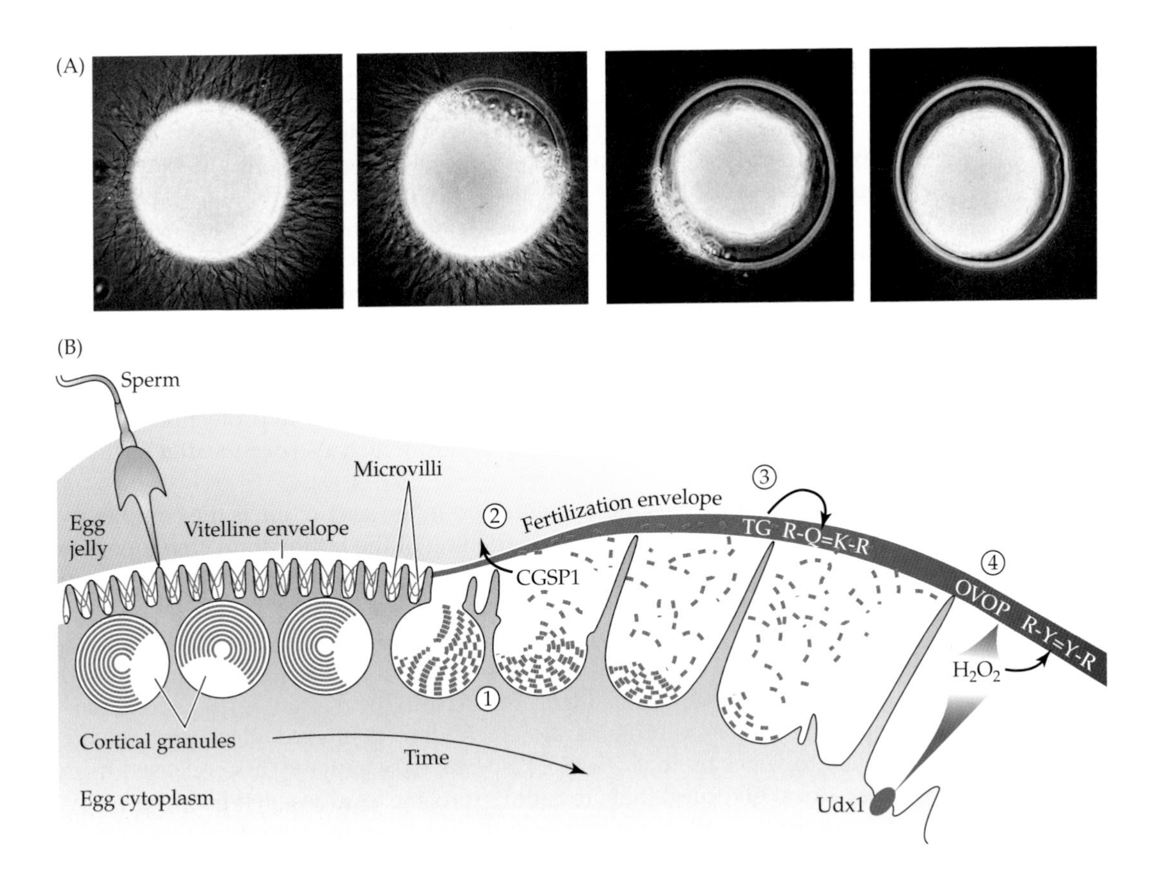

FIGURE 4.11 Formation of the sea urchin fertilization envelope. (A) A series of micrographs taken in the first minute following sperm-egg fusion shows the formation of the rigid fertilization envelope and the concomitant removal of sperm from the vicinity. The fertilization envelope protects the early embryo from shear forces and also blocks other sperm from entering the egg (which would have lethal consequences). (B) Steps leading to the elevation and hardening of the fertilization envelope. (1) Signals arising from the contact or fusion of the sperm and egg cause the membranes of cortical granules to fuse with the egg cell membrane. This exocytosis releases the protein-digesting enzyme CGSP1; the ovoperoxidase (OVOP) enzyme; and several structural proteins targeted to the fertilization envelope. (2) CGSP1 cleaves the protein binding the vitelline envelope to the cell membrane, thereby freeing the vitelline envelope. Mucopolysaccharides from the cortical granules bring in water that separates the cell membrane from the vitelline envelope (which from this point on is called the fertilization envelope). (3) CGSP1 activates a transglutaminase enzyme, present on the tips of the egg's microvilli, which cross-links glutamines and lysines on adjacent proteins. (4) The peroxidase enzymes (UDX1 on the cell membrane and ovoperoxidase) make hydrogen peroxide, which cross-links adjacent tyrosine residues, thereby hardening the fertilization envelope such that it becomes resistant to shear forces. (A from Vaquier and Payne1973, photographs courtesy of V. D. Vaquier; B from Wong and Wessel 2008.)

the **fertilization envelope**—a rigid physical barrier that protects the newly formed embryo (Wong and Wessel 2006, 2008).

The major role of the fertilization envelope is to protect the embryo from physical damage. Its efficacy is best demonstrated by a simple but elegant experiment of Miyaki and McNeil (1998). They compared the resistance of unfertilized eggs (without fertilization envelopes) and zygotes (with fertilization envelopes) to the shear forces generated when an egg or embryo suspension is forced through a syringe and narrow needle. They found that the fertilized eggs were much more resistant to lysis than were unfertilized eggs, supporting the idea that the elevation of the fertilization envelope provides physical protection against shear stresses that might be encountered in the environment. There may also be additional protection from an increase in polymerized actin in the cortex, which also occurs after fertilization (Spudich 1992).

A surprising aspect, best understood in the sea urchin, is the complexity of the assembly, elevation, and hardening of the fertilization envelope (see Figure 4.11). As noted, a "template" of the envelope is present in the form of the vitelline layer, which is closely apposed to the egg surface and lies just above the cortical granules. The constituents of the cortical granules cause the vitelline envelope to elevate, and many of the proteins released from the granules become incorporated into the nascent fertilization envelope. These released proteins, along with the vitelline envelope proteins, are linked together by the actions of a peroxidase enzyme (also released from the cortical granules) that cross-links tyrosine residues on adjacent proteins to convert the soft, membranous vitelline layer into the rigid fertilization envelope (Somers and Shapiro 1991; Wong and Wessel 2008).

Besides protecting the egg from physical stresses such as wave forces, there could be many additional roles for this robust membrane. This area is ripe for ecological analysis. For example, the fertilization envelope might also provide protection against predators, including organisms such as zooplankton that feed on embryos; or, like the zona pellucida, it could protect the embryo from pathogenic bacteria and fungi by forming a physical barrier to their entry.

GASTROPOD EGG CAPSULES Forming a protective physical barrier is also the case with the elaborate egg capsules housing large numbers of gastropod embryos, including snails in the genus *Nucella*. *Nucella* lays large egg capsules that are often attacked by intertidal isopods (Figure 4.12). The larvae develop completely inside these capsules, hatching out as fully formed small snails. The thickness of the *Nucella* capsule is variable, and Rawlings (1994, 1995) found that thin capsules were more vulnerable to predation from these isopods than thick capsules. He also found that the capsule wall protected against pathogens; if the wall was damaged, protozoa and bacteria entered, with larval death rapidly ensuing.

FIGURE 4.12 The dogwinkle snail *Nucella lapillus* is seen surrounded by its oval yellow egg capsules (Photograph © Juniors Bildarchiv/AGE Fotostock.)

The protection provided by extracellular coatings comes at significant expense. Perron (1981) calculated that as much as 50% of the weight of the embryo mass in cone snails is in their large protective capsule. There are between 10,000 and 15,000 cortical granules in sea urchin eggs, comprising 5–10% of the egg proteins (Vacquier 1975). That's a lot of energy for a structure that is not used after development is complete. But the coatings persist until the larva hatches (about 12–24 hours in sea urchins and up to 3 weeks in cone snails), so its cost may be small relative to its protective value. (See also Strathmann 2007 for a discussion of reproductive strategies.)

Cytoplasmic sealing

When a tire is punctured, compressed air leaks out, and usually the tire collapses. However, some tires have an extra layer inside, containing a sealant that wraps around the inserted object and then fills in the gap when the object is removed. The plasma membrane has such self-sealant properties and can quickly repair the membrane and prevent cell lysis.

Historically, such a mechanism could have been inferred from experiments that started in the late 1800s. At that time, embryologists trying to understand the organization of the egg were busily cutting regions out of eggs and embryos and transplanting other regions back in. In some instances, eggs were cut in half with glass needles, and they were found to heal. Later embryologists often injected dyes into eggs, and recently scientists began studying the electrical properties of eggs by inserting microelectrodes (see Gilbert 2006 for a review). In each case, the egg membrane was able to heal, to reseal.

Although it was obvious that embryos could survive these assaults, few people paid attention to the question of how such physical insult could be tolerated. Unique were the 1930 studies of the cell physiologist Lewis Heilbrunn, who noted a requirement for calcium in order for the embryos to survive mechanical damage (Heilbrunn 1943). He suggested that calcium ions (Ca^{2+}) were required for "clotting" the cytoplasm, noting the similarity to the then recently discovered requirement of Ca^{2+} for blood clotting.* Heilbrunn's work, however, was largely forgotten until the question of how cells survive such damage was asked again in the 1990s.

Steinhardt and his colleagues (1994) examined this by studying how sea urchin eggs and embryos sealed themselves after insertion of a microneedle. They injected fluorescent dyes and determined how quickly the dyes diffused from the cell. If the cell was able to seal itself, the dye remained in the cell and there was no change in fluorescence. However, if the cell could not seal itself, the dye flowed out into the extracellular medium and was detected as a loss of fluorescence.

These experiments confirmed Heilbrunn's findings of a calcium requirement for cell healing. But, surprisingly, they found that the calcium concentration required was quite high. Instead of being in the submicromolar range seen in intracellular signaling events, the calcium requirement for sealing was between 1 m*M* and 10 m*M*. This level is more akin to the extracellular calcium concentration in the blood (as in mammals) or seawater (for marine embryos) than to intracellular free calcium concentration.

This work revealed that the repair was initiated by an influx of calcium ions from the outside media, which entered the cell when it was wounded. So this was the signal. This entry induced a calcium-mediated fusion of intracellular vesicles with each other and with the plasma membrane. The fusion generated new membrane material, which then sealed together at the edges of the torn membrane (McNeil and Terasaki 2001). This sealing phenomenon appears to be a general property of cells; for example, a similar calcium-induced fusion of intracellular vesicles seems to operate in muscle cells that are torn during heavy exercise (McNeil and Steinhardt 2003).

That calcium-sensitive vesicles are present throughout the cytoplasm and that their fusion is triggered by calcium is illustrated in Figure 4.13. This work, however, does not account for the ability of severely damaged eggs (such as those cut in half) to heal and to continue developing. This

*Some embryologists took pains to get special water in which to do their dissections. Viktor Hamburger (1988, p. 22) tells of Hans Spemann's ordering water from Würzburg to be shipped to his laboratory because his dissected amphibians did better in that water than in any other. The secret ingredient of the "Würzburger Wasser"? Calcium ions.

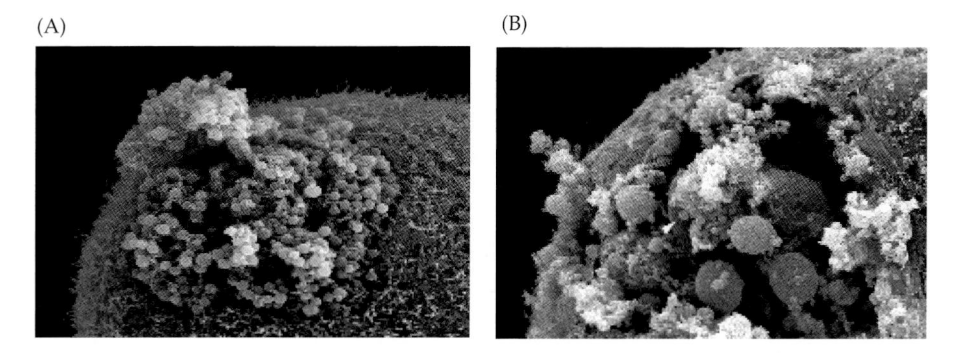

FIGURE 4.13 Calcium is required for cell vesicle fusion. In this experiment, the surfaces of sea urchin eggs were sheared to tear the plasma membrane and then immediately placed into a fixative, which stopped any further fusion. In these micrographs, the membrane has not had time to seal and is still open. (A) In cells sheared in the absence of Ca^{2+}, the intracellular vesicles are of small and uniform size. (B) In cells sheared in the presence of Ca^{2+}, the vesicles are larger and of varying sizes, suggesting that vesicle fusion took place. (From McNeil and Baker 2001; photographs courtesy of Paul McNeil.)

question has been studied by Terasaki et al. (1997), who found that the mechanism for larger-scale wound repair also relied on a calcium-induced fusion of cell vesicles but that the mechanism was more similar to that of a tire patch being added to heal a hole in a pneumatic tire.

An intriguing question is, "How did this calcium-mediated protective sealing mechanism evolve?" One possibility is that it is a by-product of membrane changes seen during mitosis, during normal exocytosis events (such as the acrosome reaction and the bursting of cortical granules), and during fertilization (fusion of the sperm and egg membranes). But what sort of stresses was protective sealing an adaptation to? (Certainly it is not a response to the experiments of embryologists or cell biologists.) A possibility is that cell sealing protects embryos from predation by zooplankton in aquatic environments or by insects in the terrestrial environment. These potential predators, if their mouthparts are small, could kill the early embryo if there were no mechanism to repair this damage. This type of predation has not been demonstrated directly in the field and is another area calling for further work.

Protection against Oxidative Damage

A major stressor for all cells is oxidative damage, caused by reactive oxygen species (ROS) that are produced as a by-product of normal metabolism but which can also arise from exposure to metal ions, ultraviolet radiation, and certain pollutants (whose metabolism generates ROS). Many biologists believe that a large part of aging ensues from a lifetime's accumulation of exposure to these ROS, which can damage all cell constituents, including membranes and DNA (Kujoth et al. 2005). Somatic cells have mechanisms to deal with ROS, including the use of antioxidant molecules such as glutathione, ascorbic acid and the metallothioneins, as well as enzymes that degrade ROS such as catalase and superoxide dismutase. Embryos similarly produce ROS from normal metabolism and use protective strategies that are similar to those used in somatic cells, but they also have some powerful embryo-specific protective mechanisms.

Paradoxically, the ROS implicated in aging and death are also produced at the beginning of development at fertilization. This production is so extensive that additional and novel defenses are needed to counter this potential threat to normal cell function. Sea urchin embryos, for example, have a burst of oxidative metabolism and ROS production at fertilization (Wong et al. 2004; Wong and Wessel 2008) and may use ROS as part of developmental signaling later in development (Finkel 2003; Coffman and Denegre 2007).

The sea urchin zygote generates large amounts of hydrogen peroxide for the peroxidase reactions used for hardening the fertilization envelope (Wong et al. 2004). These produce reactive oxygen species that could damage proteins and DNA of the embryos and wreak havoc with normal cell physiology. Fertilization also results in the production of signaling molecules that are themselves free radicals or that produce free radicals as byproducts of their formation. One of these molecules is nitric oxide, a free radical involved in regulating calcium levels after fertilization (Kuo et al. 2000; Leckie et al. 2003). Arachidonic acid derivatives, arising from the activation of a lipoxygenase and the products of this enzyme, can lead to reactive lipid intermediates that in turn can lead to reactive oxygen and damage to cell components (Perry and Epel 1981).*

So how does the embryo deal with this sudden bolus of ROS at fertilization? One coping mechanism is to use the standard defenses seen in other cells: molecules that scavenge ROS, such as ascorbate and glutathione, and the neutralization of ROS by such enzymes as superoxide dismutase and catalase (Dickinson and Forman 2002). Another way to cope is to use some embryo-specific compounds that seem to have evolved for these precarious times. This other defense, studied extensively in the sea urchin embryo, is a sulfhydryl-containing antioxidant that is a modified histidine (Turner et al. 1987; Shapiro and Hopkins 1991). This antioxidant is called ovothiol, acknowledging its uniqueness to eggs (*ovo*) and its sulfhydryl (*thiol*) group.

*These events do not appear to be unique to sea urchin eggs, as such ROS-creating reactions have been detected in insect eggs (Li and Li 2006) and in starfish eggs (Brash et al. 1991). Most eggs (including those of mammals) have not been studied for these reactions.

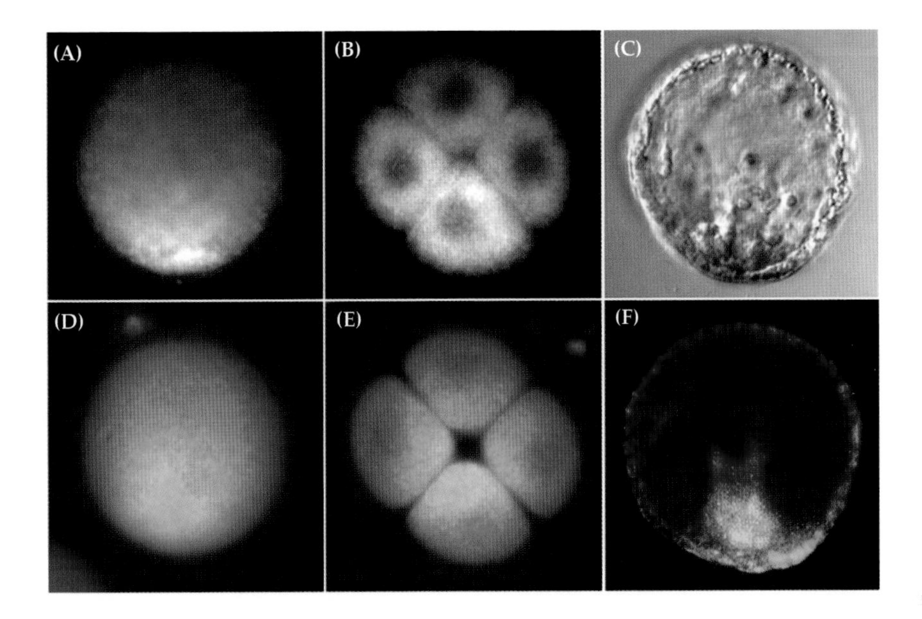

FIGURE 4.14 Oral-aboral axis specification in the sea urchin embryo is determined by the mitochondria and resultant redox state. The positions of the mitochondria were experimentally manipulated by centrifugation (A–C) or by microinjection of extra mitochondria (D–F). (A–C) Mitochondria were centrifuged to the bottom half of the egg (A), and when the egg was fertilized, the future anus/archenteron developed in that area (B, C). (D–F) Mitochondria injected into the unfertilized egg are stained green (D). Following fertilization, the anus/archenteron similarly forms at the site of the extra mitochondria. (From Coffman et al. 2004.)

Ovothiol concentration in sea urchin eggs is high, about 5 m*M* in the cytoplasm (similar to glutathione concentrations). It is also found in the eggs of several other invertebrate organisms (Shapiro and Hopkins 1991).

Why couldn't the sea urchin simply use glutathione? Why go to the trouble of producing another antioxidant compound? Shapiro and colleagues, who first discovered ovothiol in sea urchins, found that this molecule was even more effective than glutathione for the types of oxidants produced at fertilization (Turner et al. 1987). This increased effectiveness may be important in sea urchin embryos, in which ROS are also used as signaling molecules in later development: mitochondrial gradients and a resultant gradient of ROS appear to have important roles in determining the animal's oral-aboral (mouth-posterior) axis (Figure 4.14; Coffman et al. 2004; also see Coffman and Denegre 2007). The regulation of ROS therefore must be highly precise and must not allow high levels of ROS "noise" to interfere with the crucial signaling function of these free radicals.

Protection against Damage to DNA

We are constantly told that exposure to sunlight increases our risk of getting cancer and that we should put on sunscreens to avoid the DNA-damaging effects of ultraviolet (UV) light. But as we lie on the sunlit beach, embryos are floating in tidepools only a few feet away. How are they protected? It may seem as if their mothers put sunscreen on them, too.

Sunscreens prevent DNA damage

The problem for many aquatic embryos is that their physiology and habitat puts them at the air-water interface. Fish and frog embryos, for example, have lipid reserves that can make them so buoyant that they float at the surface. Some tunicate embryos have extraembryonic cells attached to their chorions that are less dense than seawater, which also results in floating eggs and embryos. (This is made possible by the unique adaptation of substituting ammonium ion for sodium ion in the cytoplasm of the extraembryonic cells. Ammonium is lighter than sodium, so these cells act like life jackets that carry the eggs to the surface; see Lambert and Lambert 1978.)

A floating lifestyle can result in exposure to UV radiation and resultant DNA and protein damage. Damage from UV radiation can arise from the direct absorption of the UV by proteins and DNA, which have absorption peaks at 280 nm and at 260 nm, respectively.* Thus, UV light might not seem to be a problem, because the UV radiation that passes through Earth's ozone layer is in the range of 300–400 nm. However, even though the UV radiation that reaches Earth is away from the peak absorption spectra of DNA and proteins, these molecules absorb sufficient light in the 300–320 nm range to damage them. In DNA, this level of UV absorption can generate cross-linking between adjacent thymine nucleotides; this **dimer formation** prevents the proper replication of the DNA (Franklin and Haseltine 1986). Another form of damage, which is indirect, is the generation of ROS by radiation in the 320–400 nm range. This radiation by itself can generate oxidative damage to membranes, proteins, and even DNA (Heck et al. 2003).

One adaptation to avoid UV damage is to spawn at night. This is the situation in Australia's Great Barrier Reef, where more than 100 species of coral spawn on the same night each year (Babcock et al. 1986). This mass spawning event is thought to maximize survival, since the embryos' predators rapidly become satiated by the large amount of food suddenly provided. But the nocturnal spawning also means that the embryos pass

*The UV portion of the light spectrum is divided into UV-B, which is the light at wavelengths between 290 nm and 320 nm, and UV-A, with wavelengths between 320 nm and 400 nm.

through most vulnerable cleavage stages before their first exposure to UV radiation at dawn.

In the waters of the Arctic and Antarctic, embryos face an even more severe problem. The embryos of the Antarctic sea urchin *Sterechinus neumayeri*, for example, experience extreme exposure to UV, since they develop during the polar summer and thus are exposed to 24 hours of sunlight a day. This exposure is compounded by the fact that in the frigid Antarctic water (–1.9°C), development is extremely slow—it takes 3 weeks for an *S. neumayeri* embryo just to reach the pluteus stage. Further problems could arise from increased UV exposure due to the thinning ozone layer in the Antarctic, which would allow even more damaging wavelengths to pass through the atmosphere. Karentz and colleagues (2004) showed that the strategy here is for the embryo to develop deep in the water column, where UV penetration is low. At these depths there is no UV damage to DNA, as estimated by measuring thymine dimers. However, if the embryos are placed in containers near the surface, there is a significant increase in dimer formation.

Most aquatic embryos will be exposed to at least some UV radiation. Even if they don't float at the surface, their eggs contain enough lipid to provide moderate to extensive buoyancy, so they will be found near the water's surface. Even well below the surface, a small proportion of the UV wavelengths penetrate the water column. And, because developmental times to a swimming larvae are more than 12 hours for most species, some portion of development will inevitably take place in daylight, when UV radiation is present.

The solution most often used is a preventive one, sometimes whimsically referred to as the "Coppertone defense." The idea here is for the egg to accumulate chemical sunscreens that absorb UV radiation and prevent harmful wavelengths from penetrating deep into the cytoplasm. A potent sunscreen used in marine embryos (as well as in adult organisms) is provided by a group of chemicals called **mycosporines**, which have high absorption in the UV part of the spectrum. Mycosporine concentrations can reach the m*M* range, and their attenuation of UV can be especially potent at these high concentrations (Shick and Dunlap 2002).

Work by Adams and Shick (2001) provides the best evidence for the role of mycosporines in protecting embryos from UV radiation. Their studies took advantage of the finding that animals cannot synthesize these compounds, but must get them from plants that they eat. The researchers used this dietary requirement to manipulate the mycosporine content in sea urchin embryos by feeding the adult female urchins, during the period that they were making eggs, algae that varied in the amount of mycosporine pigments. Sure enough, there was a protective effect in the embryos of only those animals that were fed algae containing mycosporines, and the degree of UV protection was correlated with the amount of mycosporine present in the embryos (Figure 4.15)

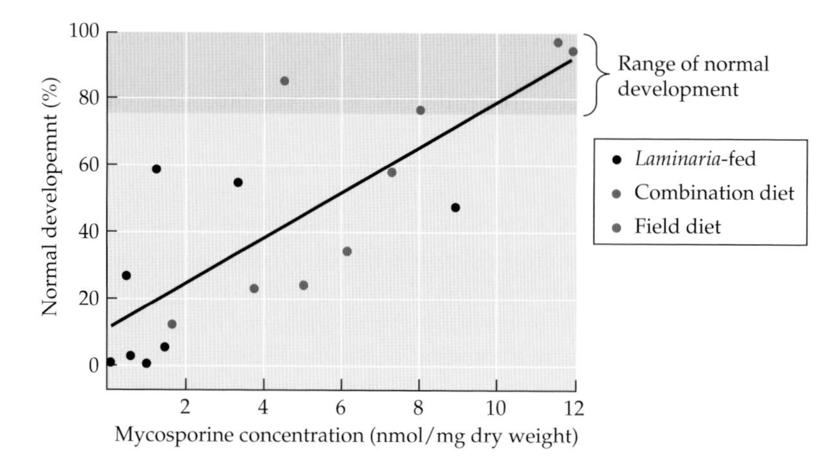

FIGURE 4.15 Mycosporines protect embryos from UV irradiation. Northern sea urchins (*Strongylocentrotus droebachiensis*) were fed algae that produce no sunscreen-containing mycosporines (*Laminaria*-fed); *Laminaria* plus *Chondrus crispus*, a green macroalga high in mycosporines (combination diet); and a control diet of algae collected in the field. Eggs laid by adults in these three experimental groups were then fertilized and exposed to UV. The extent of normal development (shaded bar at top of graph) achieved under conditions of adverse UV irradiation was seen to be related to the mycosporine content in the parental diet. (After Adams and Shick 2001.)

An interesting variant of the mycosporine defense is seen in tunicate embryos, where Epel and colleagues (1999) found that the major attenuation of UV radiation is provided by a layer of extraembryonic "test cells" that surround the egg and embryo. These test cells have such a large amount of the mycosporine pigment that they can absorb 90% of the ambient UV radiation, with a peak at around 360 nm.

A problem with this defense is that the mycosporines are akin to vitamins. As noted, animals can't make these chemicals themselves and depend on eating algae, bacteria, and fungi that make these molecules. Thus, there is a question of reliability: can the animals depend on being able to consume the algae that make these sunscreens? Studies on bacteria and algae show that mycosporine synthesis is induced in response to sunlight (Sinha et al. 2001), which means that when UV radiation is high, the algal diet of animals will also have high levels of mycosporines. The animals do not seem to digest or degrade these chemicals and must have mechanisms to transport them to their ovaries and associated oocytes. These mechanisms remain to be discovered (see Mason et al. 1998).

Repairing damaged DNA

The above protective mechanisms provide some help, but the final, critical defense is having a good DNA damage repair system. The repair system for UV damage is very effective. As noted above, UV causes the dimerization of adjacent thymidines in DNA, which results in misreading during replication and hence in mutation. Cells have an extremely effective repair system carried out by a flavoprotein enzyme called photolyase. Photolyase absorbs light in the yellow range of the visible spectrum and uses the energy from this light absorption to reverse thymidine dimerization and restore the DNA to its original state (Sinha and Häder 2002; Häder and Sinha 2005). This is a very clever mechanism, since the "healing" visible light and the damaging UV part of the light spectrum are both present in sunlight.

This seems like an efficient mechanism, but the few quantitative studies that have been done on UV radiation damage in embryos indicate that not all the dimers are returned to their original state. Also, the level of the photolyase might be linked to an anticipated level of UV radiation, and if that level is exceeded, repair will not be as efficient. Blaustein and colleagues (1994) looked at the levels of photolyase in different amphibian eggs and oocytes. Levels of photolyase varied 80-fold among the tested species, and the differences correlated with the site of egg laying: those species typically laying eggs in sites exposed to full sunlight had the highest levels of photolyase (Table 4.1).

TABLE 4.1 Photolyase activity levels and UV exposure in the eggs of some amphibians

Species	Photolyase activity (10^{11} CBPD/hr/μg)[a]	Egg-laying mode	UV exposure
Caudates (urodeles)			
Plethodon dunni	<0.1	Hidden	Not exposed
Taricha granulosa	0.2	Hidden	Limited exposure
Rhyacotriton variegatus	0.3	Hidden	Not exposed
Plethodon vehiculum	0.5	Hidden	Not exposed
Ambystoma macrodactylum	0.8	Open water	Limited exposure
Ambystoma gracile	1.0	Open, shallow water	Exposed
Anurans			
Xenopus laevis	0.1	Under vegatation	Limited exposure
Bufo boreas	1.3	Open, shallow water	Exposed
Rana cascadae	2.4	Open, shallow water	Highly exposed at high altitudes
Hyla regilla	7.5	Open, shallow water	Highly exposed

Source: After Blaustein et al. 1994.

[a]CBPD, cyclobutane pyrimidine dimer. Refers to the rate at which the photolyase enzyme reverses thymidine dimerization (see text).

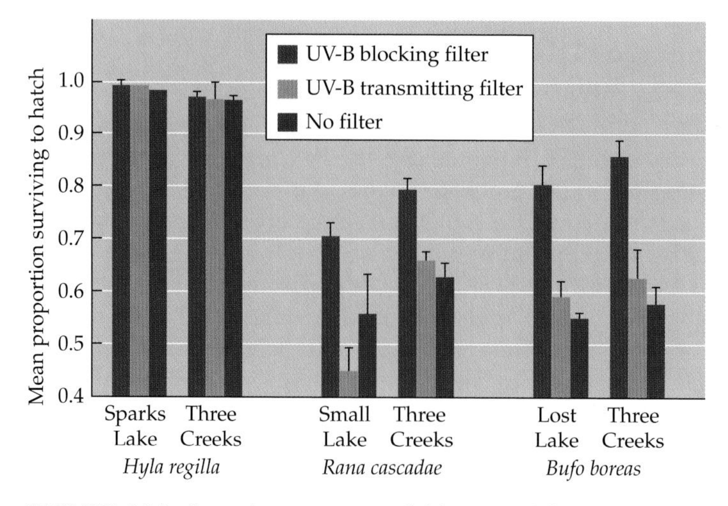

FIGURE 4.16 In studies at separate field sites at different elevations, the effect of UV-B radiation on tadpole hatching was noted for three amphibian species. Blocking UV-B radiation had no effect on *Hyla regilla* but significant effect on the other two species. (After Blaustein et al. 1994.)

In the same study, Blaustein and colleagues tested whether or not UV-B could be a factor in lowering the hatching rate of tadpoles. At two field sites, they divided the eggs of the three anuran species studied into three groups. The first group developed under a filter that blocked UV-B from reaching the eggs. The second group developed under a filter that allowed UV-B to pass through. The third group developed without any sun filter (Figure 4.16). For *Hyla regilla*, the filters had no effect, and hatching was excellent in all three conditions. For *Rana cascadae* and *Bufo boreas*, however, the UV-B–blocking filter raised the percentage of hatched tadpoles from about 60% to close to 80%.

Protection against Pathogens

Pathogens such as bacteria, fungi, and viruses pose a special problem for embryos. The difficulty is that the embryo does not have an adult-type immune system until late in differentiation. As described earlier in the chapter, one embryo-unique protective mechanism against pathogens is provided by the egg coats. Another protective mechanism is transfer of maternal antibodies to the oocyte. A third mechanism, of course, is provided by internal development such as in mammals, where the mother's immune system can protect the developing embryo.

Parental behavior

Defense against pathogens can also arise from parental behavior. An example is the midshipman fish, which comes in from the deep ocean to the shallow waters of Monterey Bay, where the female sheds her eggs and the male fertilizes them, attaches them to rocks, and then remains with the embryos (Crane 1981). Male guarding includes constant grooming of the surfaces of the embryo aggregate. If the males are removed, a fungal growth soon takes over the sessile embryos and destroys them (Epel, unpublished observations). A similar grooming behavior is seen in some crustaceans. In these cases the eggs are carried by the mother, who has a specialized appendage for grooming the eggs; if this appendage is removed, there is increased embryo mortality (Fisher 1983; Förster and Baeza 2001).

Chemical protection

In some species, antibacterial and antifungal chemicals are associated with the egg surface. Kudo (2000) found that carp eggs contain antibacterial substances in the cortical granules, those vesicles described above that are embedded in the cortex of the egg and that undergo exocytosis upon fertilization. In this fish, one of the constituents of the cortical granules is lysozyme, an enzyme that acts on glycoproteins that are specific to the bacterial surface, which results in death of the bacteria. Here the lysozyme is transferred to the fertilization envelope after fertilization, and it is hypothesized to prevent bacterial attachment to this envelope. Lysozyme is also present in a high abundance in the whites of chicken eggs.*

Cortical granules also contain lectins, and these can become part of the egg surface after fertilization (Chang et al. 2004). Lectins are large molecules that bind specifically to glycoproteins that may be important in later development (see Evanson and Milos 1996). However, they also protect against pathogens by binding to the surface of bacteria and fungi (Van Die et al. 2004). This may be a general way that eggs can thwart bacterial and viral pathogens.

Embryonic immune responses

Another strategy is to have components of the innate immune system present in the eggs or embryos. Unlike the adaptive immune system (which uses antibodies), the innate immune system is not very specific. One previously mentioned example seen in eggs and embryos is lysozyme, which

*Lysozyme is also used in adult organisms. For example, it is found in human pores and is an important ingredient in human tears, where it helps keep the eye free of bacterial infections (see Root-Bernstein 1997).

can kill numerous types of bacteria. Another is the use of antibacterial peptides, referred to as defensins. Such molecules have been found in shrimp eggs (Bachère et al. 2000) and embryos of the tobacco hornworm, *Manduca sexta* (Gorman et al. 2004).

A novel variant of the innate immune system repertoire is seen in the purple-and-orange–pigmented eggs and embryos of the sea urchin *Arbacia* and related species. This pigment, a napthoquinone, is stored in cytoplasmic granules that move to the egg surface after fertilization. These granules can be released when the egg is disturbed, emptying napthoquinone into the surrounding seawater. Napthoquinone reacts with the calcium in the seawater to produce hydrogen peroxide, a potent antimicrobial agent (Perry and Epel 1981).

This is very similar to the attack mechanism provided by the white blood cells (especially the eosinophils) in mammals: these immune system cells also produce hydrogen peroxide to kill pathogens, albeit via a completely different mechanism (Moqbel and Lacy 2000). Missing from the sea urchin story, however, is proof of the antimicrobial role of this pigment release. If the hypothesis is proved correct, it would explain at least one function for these odd granules.

Symbiosis and protection from fungi

Fungal infection can be a major pathogen in embryos. A number of embryo-specific fungi exist, but because fungal growth is usually sluggish, these primarily threaten slowly developing embryos. Embryos on the "fast track" will be out of their egg coats before a pathogenic fungus has been able to attach and pass through the chorion. Some susceptible embryos, however, fight fungal pathogens using an approach that will be familiar to some readers from Chapter 3.

The American lobster (*Homarus americanus*) has a months-long developmental period, during which the embryos are brooded in the mother's tail region. In spite of this parental protection, lobster embryos are vulnerable to fungal infection. Such fungal attacks, although quite common in aquaculture facilities, appear to be rare in nature. This is because lobster and shrimp (and perhaps other embryos) have a unique defense against fungal attack: populations of apparently protective symbiotic bacteria are maintained in the outer coats surrounding the crustacean embryos (Fisher 1983; Gil-Turnes et al. 1989). Culturing the bacterial population revealed that these symbionts produce a fungicide. In experiments in which the symbiotic bacteria were removed by incubating the mothers in an antibiotic solution, the embryos were much more susceptible to fungal attack (Figure 4.17).

A similar situation may take place in the embryo of the squid *Loligo opalescens*, whose reproductive lifestyle is quite dramatic. Large numbers of

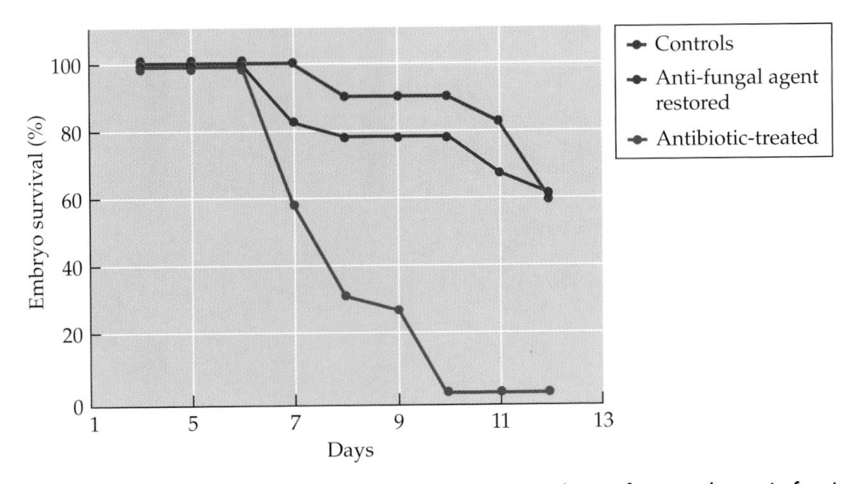

FIGURE 4.17 Symbiotic bacteria protect shrimp embryos from pathogenic fungi. Embryo masses were treated with antibiotics to kill the symbiotic bacteria, and over 12 days none of these embryos survived. However, if the antifungal agent produced by the bacteria was added back, the survival was not significantly different from the controls. These results indicate that the symbionts are providing protection against fungal infection. (After Gil-Turnes et al. 1989.)

these squid migrate into the shallow waters off the California coast, where a mating frenzy ensues. Millions of eggs are deposited—so many that the ocean floor is literarily covered with the egg capsules. Intuitively, one would think that this would provide a major source of food for fungi and bacteria as well as for fish, crustaceans, and other predators. Yet there is very little attrition of these embryos. Examination of the egg cases reveals that, once again, these embryos come surrounded by very dense colonies of symbiotic bacteria (Biggs and Epel 1991). It is assumed (but not proven) that the bacterial consortium provides some sort of protection to the embryos, perhaps similar to the antifungal activity seen in crustacean embryos. Another possibility, however, is that the bacteria simply prevent colonization by other microorganisms by straightforward competition.

There is a surprisingly large number of species associated in consortium with the egg cases of *L. opalescens* (Barbieri et al. 2001). In the female adult squid, these bacteria are stored inside the accessory nidamental gland attached to the oviduct. When the eggs are laid, the bacteria in the nidamental gland are presumably "squeezed out" and become embedded in the thick, translucent layer that covers the egg.

As each embryo in the egg case is surrounded by the bacteria, a first guess would be that the newly hatched larva would become infected when it left the egg case. It turns out, however, that the organ that stores these bac-

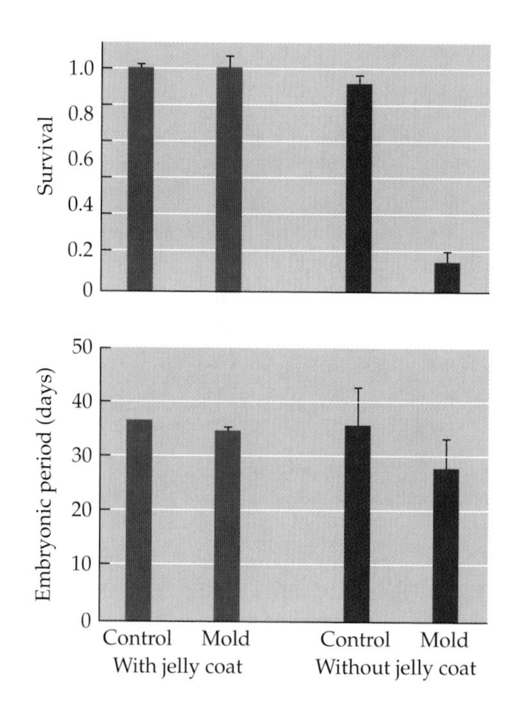

FIGURE 4.18 The jelly coat of the salamander egg *Ambystoma maculatum* provides protection against pathogenic fungi. If the embryos are exposed to a water mold, the control and mold-exposed embryos survive and develop at similar rates. If the jelly coat is removed, survival is drastically reduced; however, those few eggs that do survive the mold are somehow induced to develop more rapidly. (After Gomez-Mestre et al. 2006.)

teria is not present in the newly hatched larvae, and indeed it does not differentiate until 3 months later (Kaufman et al. 1998). Somehow the right bacteria are present in the environment and are sensed, collected, and retained by the gland and stored there for later use. This cycle, remarkably similar to what happens in infection of the light organ of *Euprymna* (see Chapter 3), begins again when the squid returns to the shores and lays her eggs.

Amphibian eggs, which are laid in water, are also susceptible to water molds and have developed several means of protection. The eggs of the spotted salamander (*Ambystoma maculatum*) are protected by a thick layer of jelly that repels water molds (Figure 4.18). The wood frog (*Rana sylvetica*) is protected because the adult females lay eggs very early in the spring, while the pond water is still cold and mold growth is slow. The eggs of the American toad (*Bufo americanus*) are highly susceptible to water molds, having only a thin jelly coating and being laid after the pond water has warmed. However, if a *B. americanus* egg becomes infected, it will hatch more rapidly and develop immune responses that can repel the mold. Interestingly, *R. sylvetica* tadpoles (which hatch earlier in the season) can have a positive effect on *B. americanus* eggs—the feeding *Rana* tadpoles eat the mold off infected *Bufo* egg clutches (Gomez-Mestre et al. 2006).

Protection from Predation

Another problem for embryos is that they often serve as food for other animals. Predatory *Homo sapiens*, for example, consume chicken eggs. Or go to a Japanese restaurant and you may find *uni*, the roe of sea urchins wrapped in rice and seaweed; or eat in an upscale restaurant and order caviar, the coveted roe of sturgeon. In many areas of the world, the nesting sites of endangered turtles have to be protected from the various primate and canine species who eat their eggs.

Leaving human predation aside, are there defenses against other predators? Some defenses are general, such as rapid development into a motile larva or small adult. Indeed, in most orphan embryos, the progression through cleavage and gastrulation to the point of larval motility is extremely rapid. This common feature suggests that the fast track has been selected for in part because it allows the embryo to get off the ocean floor (or the base of a leaf, or the surface of a pond) and at least become a moving target instead of (so to speak) a sitting duck (Strathmann et al. 2002). Indeed, some orphan embryos become ciliated and motile while still in the early cleavage period.

Another approach, mentioned earlier, is to make the embryos taste bad (Lindquist 2000). This can be assessed very dramatically by offering a test egg to a potential predator, such as a fish. The fish will take the egg into its mouth and will immediately swallow it (tastes all right) or almost immediately spit it out (tastes bad). The experimental test of egg palatability, which avoids the problem of the predator visually recognizing the egg, is to make an egg extract and then put the extract into an innocuous food pellet, such as one made of agar. This pellet is then fed to the fish, and if the pellet was made from a bad-tasting egg, it is similarly spit out.

Not all eggs or embryos taste bad; its provision to the egg is an example of a specific adaptation. We earlier mentioned the alkaloids put into moth eggs by both the father and mother. Bryozoans (ectoprocts) and tunicates are particularly good chemists, depositing all sorts of exotic protective compounds into their eggs. In the bryozoan *Bugula neritina*, symbiotic bacteria produce a toxic chemical called bryostatin. This chemical is not present in adult tissues; it is transferred to the eggs and embryos, which it renders unpalatable (Figure 4.19; Lopanik et al. 2004).

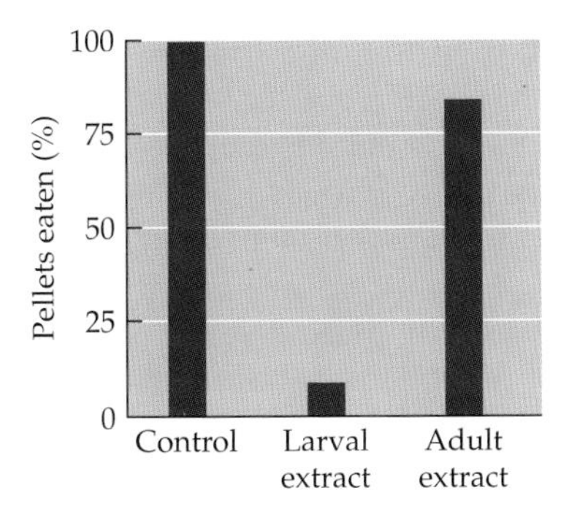

FIGURE 4.19 Embryos (but not adults) of *Bugula neritina* contain substances that make them unpalatable. Here artificial food pellets contain either larval extract or adult extract and are offered to a pinfish. Rejection of the pellet is seen only for the larval extract. (After Lopanik et al. 2001.)

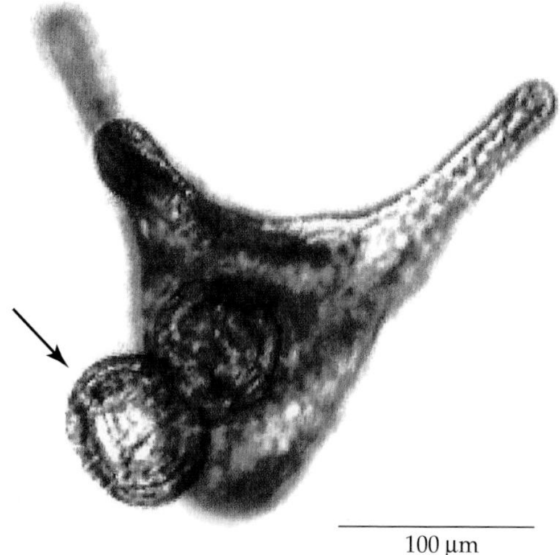

FIGURE 4.20 Exposure of a sand dollar larvae (*Dendraster* sp.) to mucus of a fish predator induces the larvae to form a clone of itself (arrow). The "clone," which looks like a gastrula-stage embryo, develops into a smaller larva that can escape predatory fish. (From Vaughn and Strathmann 2008.)

Recently, a new predator avoidance mechanism was discovered in sand dollars. When exposed to predator cues (the mucus from fish), sand dollar larvae cloned themselves by detaching buds from the region opposite the mouth. Slow-moving larvae have no way to escape predatory fish, and such cloning may provide a means for them to reproduce and become smaller than the fish can readily see (Figure 4.20; Vaughn and Strathmann 2008).

Summary

Development is precise and robust, with buffering and canalization of the embryonic programs to ensure normal development and with embryonic defense mechanisms to ensure survival of the embryo while it develops. Such survival is challenging, especially since the embryos of most animal species develop as orphans, with no or little parental protection. This problem is further exacerbated by the fact that the embryo cannot use adult defense systems until differentiation of these systems has taken place.

This vulnerability is compensated for by a wide range of defenses operating at the cellular level. These defenses are not directed to all eventualities or to every possible danger. Rather, they are a "be prepared" approach tailored to the evolutionary history of the embryo and, as a result, to those challenges the embryo might be expected or anticipated to face. These include ABC transporter proteins that prevent toxins from accumulating in cells; thiol-containing compounds that block heavy-metal toxicity; extracellular coats to shield against physical damage and pathogens; and a series of enzymes, sunscreens, and ROS scavengers to protect proteins and nucleic acids from UV radiation damage.

Some chemicals, however, are not recognized by these defense mechanisms, and sometimes the concentrations of pathogens or harmful chemicals can overwhelm them. Part 2 of this book will investigate various ways in which exposure to chemicals and pathogens can alter development and cause abnormal phenotypes.

References

Adams, N. L. and J. M. Shick. 2001. Mycosporine-like amino acids prevent UVB-induced abnormalities during early development of the green sea urchin *Strongylocentrotus droebachiensis*. *Mar. Biol.* 138: 267–280.

Alexandre, H., B. De Petrocellis and J. Bracket. 1982. Studies on differentiation without cleavage in *Chaetopterus:* Requirement for a definite number of DNA replication cycles shown by aphidicolin pulses. *Differentiation* 22: 1–3, 132–135.

Andrews, G. K., Y. Huet-Hudson, B. C. Paria, M. T. McMaster and S. K. Dey. 1991. Metallothionein gene expression and metal regulation during preimplantation mouse embryo development (*MT* mRNA during early development). *Dev. Biol.* 145: 13–27.

Andrews, G. K. 2000. Regulation of metallothionein gene expression by oxidative stress and metal ions. *Biochem. Pharmacol.* 59: 95–104.

Artus, J., C. Babinet and M. Cohen-Tannoudji. 2006. The cell cycle of early mammalian embryos: Lessons from genetic mouse models. *Cell Cycle* 5: 499–502.

Babcock, R. C., G. D. Bull, P. L. Harrison, A. J. Heyward, J. K. Oliver, C. C. Wallace and B. L. Willis. 1986. Synchronous spawning of 105 scleractinian coral species on the Great Barrier Reef. *Mar. Biol.* 90: 379–394.

Bachère, E., D. Destoumieuxa and P. Bulet. 2000. Penaeidins, antimicrobial peptides of shrimp: A comparison with other effectors of innate immunity. *Aquaculture* 191: 71–88.

Barbieri, E., B. J. Paster, D. Hughes, L. Zurek, D. P. Moser, A. Teske and M. L. Sogin. 2001. Phylogenetic characterization of epibiotic bacteria in the accessory nidamental gland and egg capsules of the squid *Loligo pealei* (Cephalopoda: Loliginidae). *Environ. Microbiol.* 3: 151–167.

Bargmann, C. I. 2006. Chemosensitization in *C. elegans*. The *C. elegans* Research Community Wormbook. www.wormbook.org, doi/10.1895/wormbook.1.123.1.

Bensaude, O., C. Babinet, M. Morange and F. Jacob. 1983. Heat-shock proteins: First major products of zygotic gene activity in mouse embryo. *Nature* 305: 331–333.

Berry, J. P. 1996. The role of lysosomes in the selective concentration of mineral elements: A microanalytical study. *Cell. Mol. Biol. (Noisy-le-grand)* 42: 395–411.

Biggs, J. and D. Epel. 1991. Egg capsule sheath of *Loligo opalescens* Berry: Structure and association with bacteria. *J. Exp. Zool.* 259: 263–267.

Blaustein, A. R., P. D. Hoffman, D. G. Hokit, J. M. Kiesecker, S. C. Walls and J. B. Hays. 1994. UV repair and resistance to solar UV-B in amphibian eggs: A link to population declines? *Proc. Natl. Acad. Sci. USA* 91: 1791–1795.

Braendle, C. and M.-A. Felix. 2009. The other side of phenotypic plasticity: How a developmental system generates an invariant phenotype despite environmental variation. *J. Biosciences*, in press.

Brash, A. R., M. A. Hughes, D. J. Hawkins, W. E. Boeglin, W. C. Song and L. Meijer. 1991. Allene oxide and aldehyde biosynthesis in starfish oocytes. *J. Biol. Chem.* 266: 22926–22931.

Briggs, E. and G. M. Wessel. 2006. In the beginning: Animal fertilization and sea urchin development. *Dev. Biol.* 300: 15–26.

Broeks, A., H. W. Janssen, J. Calafat and R. H. Plasterk. 1995. A P-glycoprotein protects *Caenorhabditis elegans* against natural toxins. *EMBO J.* 14: 1858–1866.

Bukau, B., J. Weissman and A. Horwich. 2006. Molecular chaperones and protein quality control. *Cell* 125: 443–451.

Cassada, R. C. and R. L. Russell. 1975. The dauerlarva, a postembryonic developmental variant of the nematode *Caenorhaditis elegans*. *Dev. Biol.* 46: 326–342.

Chang, B. Y., T. R. Peavy, N. J. Wardrip and J. L. Hedrick. 2004. The *Xenopus laevis* cortical granule lectin: cDNA cloning, developmental expression, and identification of the eglectin family of lectins. *Comp. Biochem. Physiol. A: Mol. Integr. Physiol.* 17: 115–129.

Chen, N. and I. Grunwald. 2004. The lateral signal for LIN12/Notch in *C. elegans* vulval development comprises redundant secreted and transmembrane DSL proteins. *Dev. Cell* 6: 183–192.

Cobbett, C. S. 2000. Phytochelatins and their roles in heavy metal detoxification. *Plant Physiol.* 123: 825–832.

Coffman, J. A. and J. M. Denegre. 2007. Mitochondria, redox signaling, and axis specification in metazoan embryos. *Dev. Biol.* 308: 266–280.

Cole, S. P. and R. G. Deeley. 1998. Multidrug resistance mediated by the ATP-binding cassette transporter protein MRP. *Bioessays* 20: 931–940.

Cooke, J., M. A. Nowak, M. Boerlijst and J. Maynard Smith. 1997. Evolutionary origins and maintenance of redundant gene expression during metazoan development. *Trends Genet.* 13: 360–364.

Crane, J. M. 1981. Feeding and growth by the sessile larvae of the teleost *Poricthys notatus*. *Copeia* 1981: 895–897.

Danilevskii, A. S. 1965. *Photoperiodism and Seasonal Development of Insects*. Oliver and Boyd, Edinburgh.

David, J. R., R. Allemand, P. Capy, M. Chakir, P. Gibert, G. Pétavy and B. Moreteau. 2004. Comparative life histories and ecophysiology of *Drosophila melanogaster* and *D. simulans*. *Genetica* 120: 151–163.

Detrich III, H. W., S. K. Parker, R. C. Williams, Jr., E. Nogales and K. H. Downing. 2000. Cold adaptation of microtubule assembly and dynamics: Structural interpretation of primary sequence changes present in the alpha- and beta-tubulins of Antarctic fishes. *J. Biol. Chem.* 275: 37038–37047.

Dickinson, D. A. and H. J. Forman. 2002. Cellular glutathione and thiols metabolism. *Biochem. Pharmacol.* 64: 1019–1026.

Dussourd, D. E., K. Ubik, C. Harvis, J. Resch, J. Meinwald and T. Eisner. 1988. Biparental defense endowment of eggs with acquired plant alkaloid in the moth *Utetheisa ornatrix*. *Proc. Natl. Acad. Sci. USA* 85: 5992–5996.

Ebling, L., W. Berger, A. Rehberger, T. Waldhor and M. Micksche. 1993. P-glycoprotein regulates chemosensitivity in early developmental stages of the mouse. *FASEB Journal* 7: 1499–1506.

Elinson, R. P. and E. Houliston. 1990. Cytoskeleton in *Xenopus* oocytes and eggs. *Semin. Cell Biol.* 1: 349–357.

Epel, D. 1998. Use of multidrug transporters as first lines of defense against toxins in aquatic organisms. *Comp. Biochem. Physiol. A: Mol. Integr. Physiol.* 120: 23–28.

Epel, D., T. Luckenbach, C. N. Stevenson, L. A. MacManus-Spencer, A. Hamdoun and T. Smital. 2008. Efflux transporters: Newly appreciated roles in protection against pollutants. *Environ. Sci. Tech.* In press.

Epel, D., K. Hemela, M. Shick and C. Patton. 1999. Development in the floating world: Defenses of eggs and embryos against damage from UV radiation. *Amer. Zool.* 39: 271–278.

Evanson, J. E. and N. C. Milos. 1996. A monoclonal antibody against neural crest-stage *Xenopus laevis* lectin perturbs craniofacial development of *Xenopus*. *J. Cranio. Genet. Dev. Biol.* 16: 74–93.

Feder, M. E. and G. E. Hofmann. 1999. Heat-shock proteins, molecular chaperones, and the stress response: Evolutionary and ecological physiology. *Annu. Rev. Physiol.* 61: 243–82.

Finkel, T. 2003. Oxidant signals and oxidative stress. *Curr. Opin. Cell Biol.* 15: 247–254.

Fisher, W. F. 1983. Eggs of *Palaemon macrodactylus*. III. Infection by the fungus *Lagenidium callinectes*. *Biol. Bull.* 164: 214–226.

Förster, C. and J. A. Baeza. 2001. Active brood care in the anomuran crab *Petrolisthes violaceus* (Decapoda: Anomura: Porcellanidae): Grooming of brooded embryos by the fifth periopods. *J. Crust. Biol.* 21: 606–615.

Franklin, W. A. and W. A. Haseltine. 1986. The role of the (6–4) photoproduct in ultraviolet light-induced transition mutations in *E. coli*. *Mutat. Res.* 165: 1–7.

Gibson, G. and I. Dworkin. 2004. Uncovering cryptic genetic variation. *Nature Rev. Genet.* 5: 681–690.

Gilbert, S. F. 2006. *Developmental Biology*, 8th Ed. Sinauer Associates, Sunderland, MA.

Gil-Turnes, M. S., M. E. May and W. Fenical. 1989. Symbiotic marine bacteria chemically defend crustacean embryos from a pathogenic fungus. *Science* 246: 116–118.

Gleason, J. E., H. C. Koswagen and D. M. Eismann. 2002. Activation of Wnt signaling bypasses requirement for RTK/Ras signaling during *C. elegans* vulval induction. *Genes Dev.* 16: 1281–1290.

Glezen, W. P. and M. Alpers. 1999. Maternal immunization. *Clin. Infect. Dis.* 28: 219–224.

Goldstone, J. V. and 9 others. 2006. The chemical defensome: Environmental sensing and response genes in the *Strongylocentrotus purpuratus* genome. *Dev. Biol.* 300: 366–384.

Gomez-Mestre, I., J. C. Touchon and K. M. Warkentin. 2006. Amphibian embryo and parental defenses and a larval predator reduce egg mortality from water mold. *Ecology* 87: 2570–2581.

Gorman, M. J., P. Kankanala and M. R. Kanos. 2004. Bacterial challenge stimulates innate immune responses in extraembryonic tissues of tobacco hornworm eggs. *Insect Mol. Biol.* 13: 19–24.

Gregor, T., D. W. Tank, E. F. Wieschaus and W. Bialek. 2007a. Probing the limits to positional information. *Cell* 130: 153–164.

Gregor, T., E. F. Wieschaus, A. P. McGregor, W. Bialek and D. W. Tank. 2007b. Stability and nuclear dynamics of the bicoid morphogen gradient. *Cell* 130: 141–152.

Häder, D. P. and R. P. Sinha. 2005. Solar ultraviolet radiation-induced DNA damage in aquatic organisms: Potential environmental impact. *Mutat. Res.* 571: 221–233.

Hamburger, V. 1988. *The Heritage of Experimental Embryology: Hans Spemann and the Organizer*. Oxford University Press, Oxford.

Hamdoun, A. and D. Epel. 2007. Embryo stability and vulnerability in an always changing world. *Proc. Natl. Acad. Sci. USA* 104: 745–750.

Hamdoun, A., G. N. Cherr, T. A. Roepke, K. R. Foltz and D. Epel. 2004. Activation of multidrug efflux transporter activity at fertilization in sea urchin embryos (*Strongylocentrotus purpuratus*). *Dev. Biol.* 276: 413–423.

Heck, D. E., A. M. Vetrano, T. M. Mariano and J. D. Laskin. 2003. Unexpected role for catalase. *Biol. Chem.* 278: 22432–22436.

Heikkila, J. J., N. Ohan, Y. Tam and A. Ali. 1997. Heat-shock protein expression during *Xenopus* development. *Cell Mol. Life Sci.* 53: 114–121.

Heilbrunn, L. V. 1943. *An Outline of General Physiology*, 2nd Ed. W. B. Saunders Co., Philadelphia.

Hensey, C. and J. Gautier. 1998. Programmed cell death during *Xenopus* development: A spatio-temporal analysis. *Dev. Biol.* 203: 36–48.

Higgins, C. F. 1992. ABC transporters: From microorganisms to man. *Annu. Rev. Cell Biol.* 8: 67–113.

Houchmandzadeh, B., E. Wieschaus and S. Leibler. 2002. Establishment of developmental precision and proportions in the early *Drosophila* embryo. *Nature* 415: 798–802.

Hu, P. J. 2007. Dauer. The *C. elegans* Research Community Wormbook, doi/10.1895/wormbook.1.144.1. http://www.wormbook.org.

Ikegami, R., P. Hunter and T. D. Yager. 1999. Developmental activation of the capability to undergo checkpoint-induced apoptosis in the early zebrafish embryo. *Dev. Biol.* 209: 409–433.

Ivarie, R. 2006. Competitive bioreactor hens on the horizon. *Trends Biotechnol.* 24: 99–101.

Jaeger, J., D. H. Sharp and J. Reinitz. 2007. Known maternal gradients are not sufficient for the establishment of gap domains in *Drosophila melanogaster*. *Mech. Dev.* 124: 108–128.

Jovine, L., C. C. Darie, E. S. Litscher and P. M. Wassarman. 2005. Zona pellucida domain proteins. *Annu. Rev. Biochem.* 74: 83–114.

Karentz, D., I. Bosch and D. M. Mitchell. 2004. Limited effects of Antarctic ozone depletion on sea urchin development. *Mar. Biol.* 145: 277–292.

Kaufman, M., Y. Ikeda, C. Patton, G. Van Dijkhausen and D. Epel. 1998. Bacterial symbionts colonize the accessory nidamental gland of *Loligo opalescens* via horizontal transmission. *Biol. Bull.* 194: 36–43.

Klaassen, C. D. and J. Liu. 1997. Role of metallothionein in cadmium-induced hepatotoxicity and nephrotoxicity. *Drug Metab. Rev.* 29: 79–102.

Kudo, S. 2000. Enzymes responsible for the bactericidal effect in extracts of vitelline and fertilization envelopes of rainbow trout eggs. *Zygote* 8: 257–265.

Kujoth, G. C. and 16 others. 2005. Mitochondrial DNA mutations, oxidative stress, and apoptosis in mammalian aging. *Science* 309: 481–484.

Kultz, D. 2005. Molecular and evolutionary basis of the cellular stress response. *Annu. Rev. Physiol.* 67: 225–257.

Kuo, R., G. Baxter, S. H. Thompson, S. A. Stricker, C. Patton, J. Bonaventura and D. Epel. 2000. Nitric oxide is both necessary and sufficient for activation of the egg at fertilization. *Nature* 406: 633–636.

Lambert, C. C. and G. Lambert. 1978. Tunicate eggs utilize ammonium ions for flotation. *Science* 200: 64–65.

Lang, L., D. Miskovic, P. Fernando and J. J. Heikkila. 1999. Spatial pattern of constitutive and heat-shock induced expression of the small heat-shock protein gene family, hsp30, in *Xenopus laevis* tailbud embryos. *Dev. Genet.* 25: 365–374.

Leckie, C., R. Empson, A. Becchetti, J. Thomas, A. Galione and M. Whitaker. 2003. The NO pathway acts late during the fertilization response in sea urchin eggs. *J. Biol. Chem.* 278: 12247–12254.

Lepzelter, D. and J. Wang. 2008. Exact probabilistic solution of spatial-dependent stochastics and associated potential landscape for the bicoid protein. *Physical Rev. E* 77: 041917.

Li, Y., R. W. Huyck, J. H. Laity and G. R. Andrews. 2008. Zinc-induced function of a co-activator complex containing the zinc-sensing transcription factor MTF-1, p300/CBP, and SP1. *Mol. Cell Biol.* In press.

Lindquist, N. 2000. Chemical defense of early life stages of benthic marine invertebrates. *J. Chem. Ecol.* 28: 1987–2000.

Lindquist, S. 1986. The heat-shock response. *Annu. Rev. Biochem.* 55: 1151–1191.

Lopanik, N., N. Lindquist and N. Targett. 2004. Potent cytotoxins produced by a microbial symbiont protect host larvae from predation. *Oecologia* 139: 131–139.

Lucchetta E. M., J. H. Lee, L. A. Fu, N. H. Patel and R. F. Ismagilov. 2005. Dynamics of *Drosophila* embryonic patterning network *per* turned in space and time using microfluidics. *Nature* 434: 1134–1138.

Mason, D. S., F. Schafer, J. M. Shick and W. C. Dunlap. 1998. Ultraviolet radiation-absorbing mycosporine-like amino acids (MAAs) are acquired from their diet by medaka fish (*Oryzias latipes*) but not by SKH-1 hairless mice. *Comp. Biochem. Physiol. A* 120: 587–598.

Mateusen, B., R. E. Sanchez, A. Van Soom, P. Meerts, D. G. Maes and H. J. Nauwynck. 2004. Susceptibility of pig embryos to porcine circovirus type 2 infection. *Theriogenology* 61: 91–101.

McFadzen, I., N. Eufemia, C. Heath, D. Epel, M. Moore and D. Lowe. 2000. Multidrug resistance in the embryos and larvae of the mussel *Mytilus edulis*. *Mar. Environ. Res.* 50: 319–323.

McNeil, P. L. and M. M. Baker. 2001. Cell surface events during resealing visualized by scanning electron microscopy. *Cell Tiss. Res.* 304: 141–146.

McNeil, P. L. and R. A. Steinhardt. 2003. Plasma membrane disruption: Repair, prevention, adaptation. *Annu. Rev. Cell Dev. Biol.* 19: 697–731.

McNeil, P. L. and M. Terasaki. 2001. Coping with the inevitable: How cells repair a torn surface membrane. *Nature Cell Biol.* 3: E124–E129.

Miyake, K. and P. L. McNeil. 1998. A little shell to live in: Evidence that the fertilization envelope can prevent mechanically induced damage of the developing sea urchin embryo. *Biol. Bull.* 195: 214–215.

Moqbel, R. and P. Lacy. 2000. Molecular mechanisms in eosinophil activation. *Chem. Immunol.* 78: 189–198.

Motola, D. L. and 10 others. 2006. Identification of ligands for DAF-12 that govern dauer formation and reproduction in *C. elegans*. *Cell* 124: 1209–1223.

Nagano, H., S. Hirai, K. Okano and S. Ikegami. 1981. Achromosomal cleavage of fertilized starfish eggs in the presence of aphidicolin. *Dev. Biol.* 85: 409–415.

Nemer, M., R. D. Thornton, E. W. Stuebing and P. Harlow. 1991. Structure, spatial, and temporal expression of two sea urchin metallothionein genes, SpMTB1 and SpMTA. *J. Biol. Chem.* 266: 6586–6593.

Newport, J. and M. Kirschner. 1982. A major developmental transition in early *Xenopus* embryos. I. Characterization and timing of cellular changes at the midblastula stage. *Cell* 30: 675–686.

Nijhout, H. F. 1994. *Insect Hormones.* Princeton University Press, Princeton, NJ.

Nijhout, H. F. 2002. The nature of robustness in development. *BioEssays* 24: 553–563.

Palmiter, R. D. 1998. The elusive function of metallothioneins. *Proc. Natl. Acad. Sci. USA* 95: 8428–8430.

Perron, F. E. 1981. The partitioning of reproductive energy between ova and protective capsules in marine gastropods of the genus *Conus*. *Am. Nat.* 118: 110–118.

Perry, G. and D. Epel. 1981. Ca^{+2}-stimulated production of H_2O_2 from napthoquinone oxidation in *Arbacia* eggs. *Exp. Cell Res.* 114: 65–72.

Raff, J. W. and D. M. Glover. 1988. Nuclear and cytoplasmic mitotic cycles continue in *Drosophila* embryos in which DNA synthesis is inhibited with aphidicolin. *J. Cell Biol.* 107: 2010–2019.

Rawlings, T. A. 1994. Encapsulation of eggs by marine gastropods: Effect of variation in capsule form on the vulnerability of embryos to predation. *Evolution* 48: 1301–1313.

Rawlings, T. A. 1995. Shields against ultraviolet radiation: An additional protective role for the egg capsules of benthic marine gastropods. *MEPS* 136: 81–95.
Renfree, M. B. and B. Shaw. 2000. Diapause. *Annu. Rev. Physiol.* 62: 353–375.
Root-Bernstein, R. S. 1997. *Discovering*. Harvard University Press, Cambridge, MA.
Rosenzweig, M., K. M. Brennan, T. D. Tayler, P. O. Phelps, A. Patapoutian and P. A. Garrity. 2005. The *Drosophila* ortholog of vertebrate TRPA1 regulates thermotaxis. *Genes Dev.* 19: 419–424.
Ryabova, L. V. and S. G. Vassetzky. 1997. A two-component cytoskeletal system of *Xenopus laevis* egg cortex. *Int. J. Dev. Biol.* 41: 843–851.
Scudiero, R., C. Capasso, P. P. De Prisco, A. Capasso, S. Filosa and E. Parisi. 1994. Metal-binding proteins in eggs of various sea urchin species. *Cell Biol. Internatl.* 18: 47–53.
Shapiro, B. M. and P. B. Hopkins. 1991. Ovothiols: Biological and chemical perspectives. *Adv. Enzymol. Relat. Areas Mol. Biol.* 64: 291–316.
Shick, J. M. and W. C. Dunlap. 2002. Mycosporine-like amino acids and related gadusols: Biosynthesis, accumulation, and UV-protective functions in aquatic organisms. *Annu. Rev. Physiol.* 64: 223–262.
Sinha, R. P. and D. P. Häder. 2002. UV-induced DNA damage and repair: A review. *Photochem. Photobiol. Sci.* 1: 225–236.
Sinha, R. P., M. Klisch, E. W. Helbling and D. P. Hader. 2001. Induction of mycosporine-like amino acids (MAAs) in cyanobacteria by solar ultraviolet-B radiation. *J. Photochem. Photobiol. B* 60: 129–135.
Somers, C. E. and B. M. Shapiro. 1991. Functional domains of proteoliaisin, the adhesive protein that orchestrates fertilization envelope assembly. *J. Biol. Chem.* 266: 16870–16875.
Spudich, A. 1992. Actin organization in the sea urchin egg cortex. *Curr. Top. Dev. Biol.* 26: 9–21.
Steinhardt, R. A., G. Bi and J. M. Alderton. 1994. Cell membrane resealing by a vesicular mechanism similar to neurotransmitter release. *Science* 263: 390–393.
Strathmann, R. S., J. M. Staver and J. R. Hoffman. 2002. Risk and the evolution of cell cycle durations of embryos. *Evolution* 56: 708–720.
Strathmann, R. R. 2007. Three functionally distinct kinds of pelagic development. *Bull. Mar. Sci.* 81: 167–179.
Sun, Q. Y. 2003. Cellular and molecular mechanisms leading to cortical reaction and polyspermy block in mammalian eggs. *Microsc. Res. Tech.* 61: 342–348.
Tauber, M. J., C. A. Tauber and S. Masaki. 1986. *Seasonal Adaptations of Insects*. Oxford University Press, Oxford.
Terasaki, M., K. Miyake and P. L. McNeil. 1997. Large plasma membrane disruptions are rapidly resealed by Ca^{2+}-dependent vesicle-vesicle fusion events. *J. Cell Biol.* 139: 63–74.
Thurber, R. V. and D. Epel. 2007. Apoptosis in early development of the sea urchin, *Strongylocentrotus purpuratus*. *Dev. Biol.* 303: 336–346.
Tilney, L. G. and L. A. Jaffe. 1980. Actin, microvilli, and the fertilization cone of sea urchin eggs. *J. Cell Biol.* 87: 771–782.
Toomey, B. H. and D. Epel. 1993. Multixenobiotic resistance in *Urechis caupo* embryos: Protection from environmental toxins. *Biol. Bull.* 185: 355–364.
Turner, E., R. Klevit, L. J. Hager and B. M. Shapiro. 1987. Ovothiols, a family of redox-active mercaptohistidine compounds from marine invertebrate eggs. *Biochemistry* 26: 4028–4036.
Vacquier, V. D. 1998. Evolution of gamete recognition proteins. *Science* 281: 1995–1998.
Vacquier, V. D. 1975. The isolation of intact cortical granules from sea urchin eggs: Calcium ions trigger granule discharge. *Dev. Biol.* 43: 62–74.
Vacquier, V. D. and J. E. Payne. 1973. Methods for quantitating sea urchin sperm in egg bining. *Exp. Cell Res.* 82: 227–235.
Van Die, I., A. Engering and Y. Van Kooyk. 2004. C-type lectins in innate immunity to pathogens. *Trends Glycotechnol.* 16: 265–279.
Voets, T., K. Talavera, G. Owsianik and B. Nilius. 2005. Sensing with TRP channels. *Nature Chem. Biol.* 1: 85–92.
Whittaker, J. R. 1973. Segregation during ascidian embryogenesis of egg cytoplasmic information for tissue-specific enzyme development. *Proc. Natl. Acad. Sci. USA* 70: 2096–2100.
Wong, J. L. and G. M. Wessel. 2006. Defending the zygote: Search for the ancestral animal block to polyspermy. *Curr. Top. Dev. Biol.* 72: 1–151.
Wong, J. L. and G. M. Wessel. 2008. Free-radical crosslinking of specific proteins alters the function of the egg extracellular matrix at fertilization. *Development* 135: 431–440.
Wong, J. L., R. Créton and G. M. Wessel. 2004. The oxidative burst at fertilization is dependent upon activation of the dual oxidase Udx1. *Dev. Cell.* 7: 801–814.
Wortzman-Show, G. B., M. Kurokawa, R. A. Fissore and J. P. Evans. 2007. Calcium and sperm components in the establishment of the membrane block to polyspermy: Studies of ICSI and activation with sperm factor. *Mol. Hum. Reprod.* 13: 557–565.
Xu, W. H., Y. Sato, M. Ikeda and O. Yamashita. 1995. Stage-dependent and temperature-controlled expression of the gene encoding the precursor protein of diapause hormone and pheromone biosynthesis activating neuropeptide in the silkworm, *Bombyx mori*. *J. Biol. Chem.* 270: 3804–3808.
Yabe, T., N. Suzuki, T. Furukawa, T. Ishihara and I. Katsura. 2005. Multidrug resistance-associated protein MRP-1 regulates dauer diapause by its export activity in *C. elegans*. *Development* 132: 3197–3207.
Zimmermann, C., H. Gutmann and J. Drewe. 2006. Thalidomide does not interact with P-glycoprotein. *Cancer Chemother. Pharmacol.* 57: 599–606.

PART 2

Ecological Developmental Biology and Disease States

Chapter 5

Teratogenesis

Environmental Assaults on Development

It is all of a piece, thalidomide and pesticides. They represent our willingness to rush ahead and use something without knowing what the results will be.

Rachel Carson, 1962

According to the Surgeon General, women should not drink alcohol during pregnancy because of the risk of birth defects.

United States Surgeon General's Office warning on wine bottles

The environment is a source not only of instructive signals for normal development, but also of signals that can disrupt normal development. Between 2% and 5% of human infants are born with an observable anatomical abnormality (Thorogood 1997). These abnormalities may include missing limbs, missing or extra digits, cleft palate, eyes that lack certain parts, hearts that lack valves, and so forth. These anomalies can be caused by genetic influences, by environmental agents, or by the interactions between environmental factors and the inherited abilities of the embryo to deal with these agents.

Experimental embryology originated in attempts to determine the causes of such birth defects in human populations (Oppenheimer 1968). By growing embryos at different temperatures or in different salt solutions, the early experimental embryologists attempted to reproduce in chicks, amphibians, and marine invertebrates anatomical abnormalities similar to those seen in anomalous human development. Physicians need to know the causes of specific birth defects in order to counsel prospective parents. In addition, the study of birth defects can help us understand how the human body is normally formed. In the absence of experimental data on human embryos, we often must rely on nature's "experiments" to learn how the human body becomes organized.* In the eighteenth century, the physician Johann Friedrich Meckel was probably the first to realize that parts of the body that were affected together in abnormal **syndromes** (Greek, "happening together") must have some common developmental origin or mechanism that was being affected (Opitz et al. 2006).

The modern medical term for birth defects is **congenital anomalies** (congenital, "at birth"; anomaly, "not normal"). There are two main classes of congenital anomalies:

- Abnormalities caused by genetic events (gene mutations, chromosomal aneuploidies, and translocations) are usually called **malformations**.
- Abnormalities that are the direct result of exogenous agents (including certain chemicals, viruses, radiation, and excessive heat) are usually called **disruptions**.

In some cases, the same condition can be the result of either a malformation or a disruption. For example, chondrodysplasia punctata is a congenital syndrome of bone and cartilage characterized by abnormal bone mineralization, underdevelopment of nasal cartilage, and short fingers. It is caused by a defective gene on the X chromosome. An identical phenotype is produced by exposure of fetuses to the anticoagulant compound warfarin. It appears that, in the malformation, the defective gene (*CDPX2*) fails to produce a specific enzyme necessary for cartilage growth. In the disruption, the gene produces the correct enzyme, but the warfarin compound inhibits the enzyme's action (Franco et al. 1995).

Some agents in the environment can be responsible for the genetic damage that results in malformations; such agents are referred to as **mutagens**. This chapter, however, is concerned with the exogenous agents responsible for disruptions; these factors are known as **teratogens** (Greek, "monster-formers"), and the study of how environmental agents disrupt normal development is called **teratology**.

Medical Embryology and Teratology

At one time, doctors and biologists believed the fetus to be protected from the environment by the mother's body and the placenta (see Dally 1998). But in 1941, Norman Gregg, an Australian ophthalmologist, documented that women who contracted rubella (German measles) during the first trimester of pregnancy had a 1 in 6 chance of giving birth to an infant with

*Our ability to learn about normal development from abnormal occurrences is seen in the language once used to describe birth defects. The word *monster*, frequently encountered in medical textbooks prior to the mid-twentieth century to describe malformed infants, comes from the Latin *monstrare*, "to show" or "to point out." This is also the root of the English word *demonstrate*.

eye cataracts, heart malformations, and/or deafness. This study provided the first evidence that the mother could not fully protect the fetus from the outside environment. Twenty years later, the U.S. rubella epidemic of 1963–1965 resulted in over 10,000 fetal deaths and the births of some 20,000 infants with birth defects (CDC 2002). Children and most adults infected by the rubella virus show relatively mild symptoms (indeed, some are unaware of even being sick), but infants born to mothers who contract rubella during early pregnancy are likely to be born blind, deaf, or both. Many are also born with heart defects or mental retardation. The epidemic underscored the fact that embryonic and fetal development can be disrupted by environmental factors.

Wilson's principles of teratology

In 1956, James Wilson put forth six principles that (with some modernization) are still applied to nearly all discussions of teratogenesis:

1. Susceptibility to the teratogenic effect of an agent depends on the genotype of the embryo, the genotype of the mother, and the ways in which their genotypes allow mother and fetus to interact with the adverse environmental factors.
2. There are critical periods of development when embryos are susceptible to being disrupted by teratogenic agents. These windows of susceptibility are when different organs are forming.
3. Teratogenic agents act in specific ways on genes, cells, and tissues in the developing organism to disrupt normal sequences of developmental events.
4. Several conditions affect the ability of a teratogen to disrupt normal development. These conditions include the nature of the agent itself, the route and degree of maternal exposure, the ability of the mother to detoxify or block the agent, the rate of transfer through the placenta, the rate of fetal absorption, and composition of the maternal and embryonic/fetal genotypes.
5. There are four manifestations of disrupted development: death, malformation, growth retardation, and functional defects.
6. Manifestations of deviant development increase in frequency and degree as the dosage of teratogens increases.*

Wilson also noted (1961, p. 191) that "an agent which is very damaging to the embryo may be relatively harmless to the mother."

*As we will see in Chapter 6, this tenet must be modified when dealing with certain compounds that elicit responses that are not linear to dose. For instance, there are several compounds that disrupt development by impairing the endocrine system. These *endocrine disruptors* produce disease states at moderate concentrations, but (probably due to feedback regulation in the organism) not at higher concentrations.

Thalidomide and the window of susceptibility

In 1961, the subject of teratogens was brought to public attention in dramatic fashion. Two researchers working independently, Widukind Lenz and William McBride, accumulated evidence that the drug thalidomide, prescribed as a mild sedative to many pregnant women, caused an enormous increase in a previously rare syndrome of congenital anomalies. The most noticeable of these anomalies was *phocomelia*, a condition in which the long bones of the limbs are deficient or absent (Figure 5.1A). Worldwide, more than 7000 affected infants were born to women who took thalidomide, and a woman need only have taken one tablet to produce a child with deformities of all four limbs (Lenz 1962, 1966; Toms 1962). Other abnormalities induced by the ingestion of this drug included heart defects, absence of the external ears, and malformed intestines. Thalidomide was withdrawn from the market in November 1961.

We now know that one major way that thalidomide causes its damage is to induce oxidative stress in certain embryonic cells. Using cultured human skin cells and chick embryos, Knobloch and colleagues (2007) showed that embryonic cells respond to this thalidomide-induced stress by upregulating the BMP signaling pathway and downregulating the Wnt signaling pathway. Both the BMP and Wnt proteins are secreted by cells to regulate the cell division and differentiation of their neighboring cells. The Wnt pathway is especially critical for limb development, and if thalidomide is countered by agents that promote Wnt signaling, limb formation continues (Knobloch et al. 2007). Another mechanism whereby thalidomide might disrupt development is by inhibiting blood vessel formation (D'Amato et al. 1994; Yabu et al. 2005). Both these effects are allowing thalidomide to be prescribed once more—although not to pregnant women—as a potential anti-tumor and anti-autoimmunity drug (Raje and Anderson 1999).

Nowack (1965) documented a **window of susceptibility** during which thalidomide caused abnormalities. The drug was found to be teratogenic only during days 34–50 after the last menstruation (20–36 days after conception). This specificity of thalidomide action is shown in Figure 5.1B. From day 34 to day 38, no limb abnormalities are seen. During this period, thalidomide can cause the absence or deficiency of ear components. Malformations of upper limbs are seen before those of the lower limbs, since the arms form slightly before the legs during development.

The concept of a window of susceptibility for teratogens is an important one; most teratogens produce their harmful effects only during certain critical periods of development. Human development is usually divided into two periods, the **embryonic period** (from conception to the end of week 8, or the first 2 months of gestation) and the **fetal period** (the remaining time in utero). It is during the embryonic period that most of the organ systems

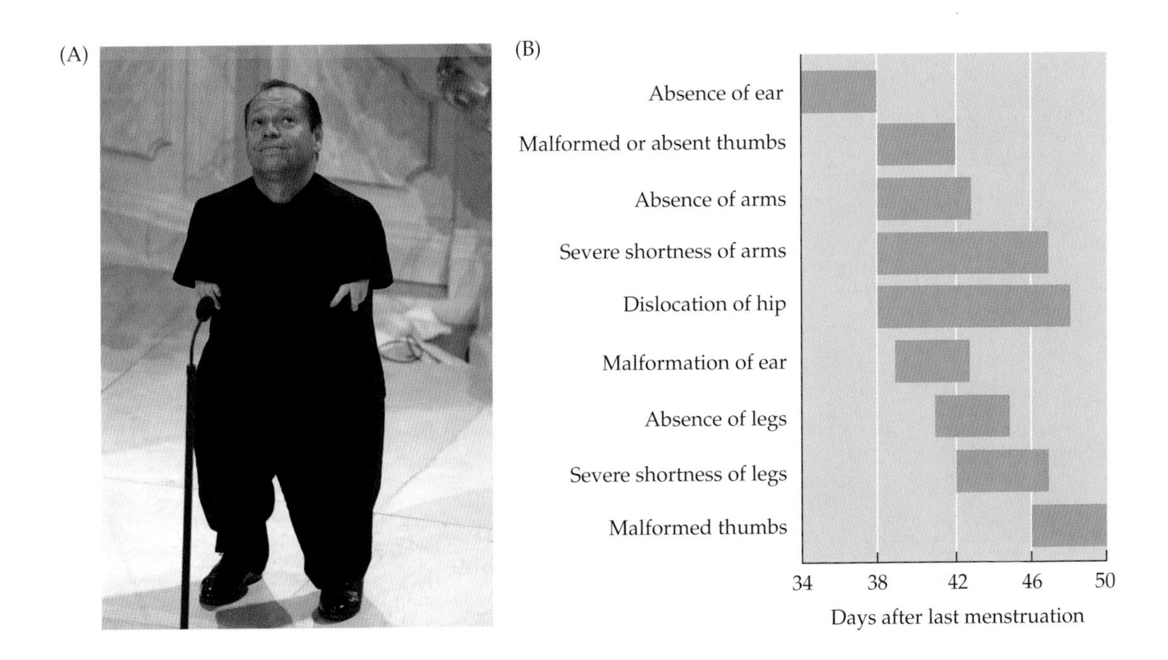

FIGURE 5.1 During the early 1960s, doctors (primarily in Europe) began prescribing the drug thalidomide to pregnant women as an effective mild sedative and remedy for morning sickness. The resulting epidemic of a specific syndrome of birth defects soon identified the drug as a teratogen. (A) German singer Thomas Quasthoff, Grammy-nominated performer of classical and jazz music, was born with phocomelia (improper limb development). Phocomelia, in this case most pronounced in the arms, is the most visible of the birth defects resulting from the ingestion of thalidomide during pregnancy. (B) Thalidomide disrupts different structures at different times of human development. (Photograph © AP Photo; B after Nowack 1965.)

form; the fetal period is generally one of growth and modeling. Figure 5.2 indicates the time frames during which various organs are most susceptible to teratogens.

Prior to week 3, exposure to teratogens does not usually produce congenital anomalies because a teratogen encountered at this time either (1) damages most or all the cells of an embryo, resulting in its death, or (2) damages only a few cells, allowing the embryo to recover (since at this stage many cells remain substantially totipotent). The period of maximum susceptibility to teratogens is between weeks 3 and 8 because that is when cell types are differentiating and most organs are formed. The nervous system, however, is forming continually throughout gestation (and, indeed, even after birth) and thus remains susceptible to teratogenic agents throughout infancy.

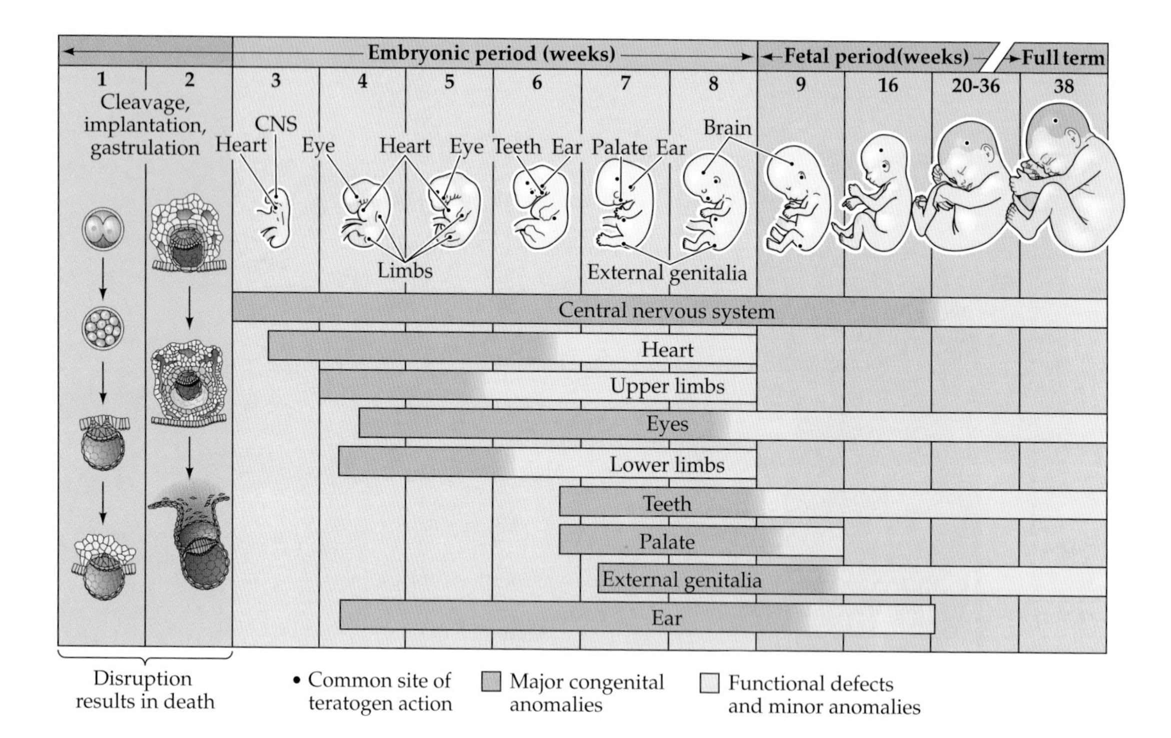

FIGURE 5.2 Periods (weeks of gestation) and degrees of sensitivity of embryonic organs to teratogens. (After Moore and Persaud 1993.)

Teratogenic Agents

The largest class of human teratogens includes drugs and chemicals (including heavy metals such as lead and mercury). Although teratogenic effects are usually associated with anthropogenic chemicals, some chemicals found naturally in the environment can also cause birth defects. Quinine and alcohol, two common substances derived from plants, also cause developmental disruptions. Quinine ingested by a pregnant mother can cause deafness, and (as we will describe shortly) alcohol can cause physical and mental retardation in the infant.

Viruses, radiation, hyperthermia, and metabolic conditions in the mother can also act as teratogens. A partial list of agents that are teratogenic in humans is given in Table 5.1.

Chemical teratogens: Industrial mercury and Minamata disease

As modern agriculture has become dependent on pesticides, herbicides, and fertilizers, and as chemical and mining industries have expanded their interests across the globe, the teratogenic effects of anthropogenic chemicals have traveled with them. The unregulated "industrial production at all

TABLE 5.1 Some agents thought to disrupt human fetal development[a]

DRUGS AND CHEMICALS	IONIZING RADIATION (X-RAYS)
Alcohol	HYPERTHERMIA (FEVER)
Aminoglycosides (Gentamycin)	INFECTIOUS MICROORGANISMS
Aminopterin	Coxsackie virus
Antithyroid agents (PTU)	Cytomegalovirus
Bromine	Herpes simplex
Cortisone	Parvovirus
Diethylstilbesterol (DES)	Rubella (German measles)
Lead	*Toxoplasma gondii* (toxoplasmosis)
Methylmercury	*Treponema pallidum* (syphilis)
Penicillamine	METABOLIC CONDITIONS IN THE MOTHER
Retinoic acid (Isotretinoin, Accutane)	Autoimmune disease (including Rh incompatibility)
Streptomycin	Diabetes
Tetracycline	Dietary deficiencies, malnutrition
Thalidomide	Phenylketonuria
Trimethadione	
Valproic acid	
Warfarin	

Source: Adapted from Opitz 1991.

[a]This list includes known and possible teratogenic agents and is not exhaustive.

costs" approach of the former Soviet Union left a legacy of soaring birth defect rates. In some regions of Kazakhstan, there are high concentrations of heavy metals in the drinking water, vegetables, and air. In such locations, nearly half the people tested have extensive chromosome breakage, and in some areas the incidence of birth defects doubled between 1980 and 1994 (Edwards 1994). Before the 1960s, however, there were very few warnings that embryos might be endangered by the progress of human technologies. Other than Rachel Carson, whose book *Silent Spring* warned that our songbird population was being wiped out by DDT, few people thought that plastics, pesticides, or the chemicals that comprise them were dangerous. Perhaps the first disaster to raise public awareness of the danger from industrial chemicals occurred in Japan, where methylmercury was found to be the cause of neurological deformities in nearly 10% of the children born near the area of Minamata Bay.

In the years after World War II, the Minamata plant of Japan's Chisso Corporation manufactured numerous organic chemicals, including acetaldehyde, an intermediate in the manufacture of consumer products from plastics to paints to perfumes. By 1951, the plant was producing more than 6000 tons of acetaldehyde a year. A by-product of the process (in which mercuric sulfate is used to oxidize acetylene into acetaldehyde) is mercury, which the company simply dumped into Minamata Bay. Microbes in the water of the bay converted the mercury into an organic form, methylmercury. Fish and shellfish—dietary staples for the townspeople in the area—consumed and concentrated this methylmercury.

As the 1950s progressed, cats in the bay area began dying amidst fits of convulsions, crows were seen falling from the sky, and fish floated dead on the bay's surface. Some villagers began having problems speaking and walking. Eventually, these same people had difficulty seeing, hearing, and swallowing. By October 1956, 40 such patients had been documented, of which 14 died of convulsions. A research team from Kumamoto University concluded that the symptoms were due not to contagious microorganisms, but to a heavy metal (Figure 5.3). When mercury levels in the area were measured, the amounts seen were shocking.* The Chisso wastewater canal contained over 2 kg of mercury in every ton of sludge—so much that it was actually profitable to mine it!

One of the most important findings of Masazumi Harada and his team of physicians (see Harada 1972; Ui 1992) was that children born to unaffected mothers could be severely damaged by methylmercury. Simply by eating their normal diet, pregnant women were inadvertently exposing their fetuses to high doses of this compound. Brain and eye deficiencies could be caused both by transmission of mercury across the placenta and by its transmission through mother's milk. It was once thought that the placenta protected the fetus from such substances; in fact, the placenta probably *concentrates* methylmercury and presents it to the fetus. Mercury is selectively absorbed by regions of the developing cerebral cortex (Eto 2000; Kondo 2000; Eto et al. 2001), and when pregnant mice are given mercury on day 9 of gestation, nearly half of the pups are born with small brains or small eyes (O'Hara et al. 2002).

This finding forced a shift in thinking about the protection of the fetus against environmental chemicals. It also set in motion a change in attitude, to one that held corporations legally responsible for their actions, and governments to be legally responsible for policing such corporations. Despite this, in the United States, industrial dumping of mercury, cadmium, and

*The Chisso management's response to the situation was to change the site where it dumped mercury (thereby killing the fish in the tributary Minamata River and spreading Minamata disease to new areas). The company prevented its own investigator from releasing his findings that water from the factory caused neurological problems in previously healthy cats, and instead it published its own research indicating that factors other than its wastes might be the cause of the disease. Moreover, government authorities—whose investigations revealed extremely large amounts of mercury in the bodies of the Minamata victims—failed to publish their findings (see Harada 1972; Ui 1992). Other companies have placed heavy metal effluents into rivers where the townspeople have little power. Minamata syndrome has been documented in indigenous peoples in Ontario, Canada, and in Brazil's Amazon Basin where new chemical plants have been established (Harada et al. 1976, Harada 1996; Gilbertson 2004).

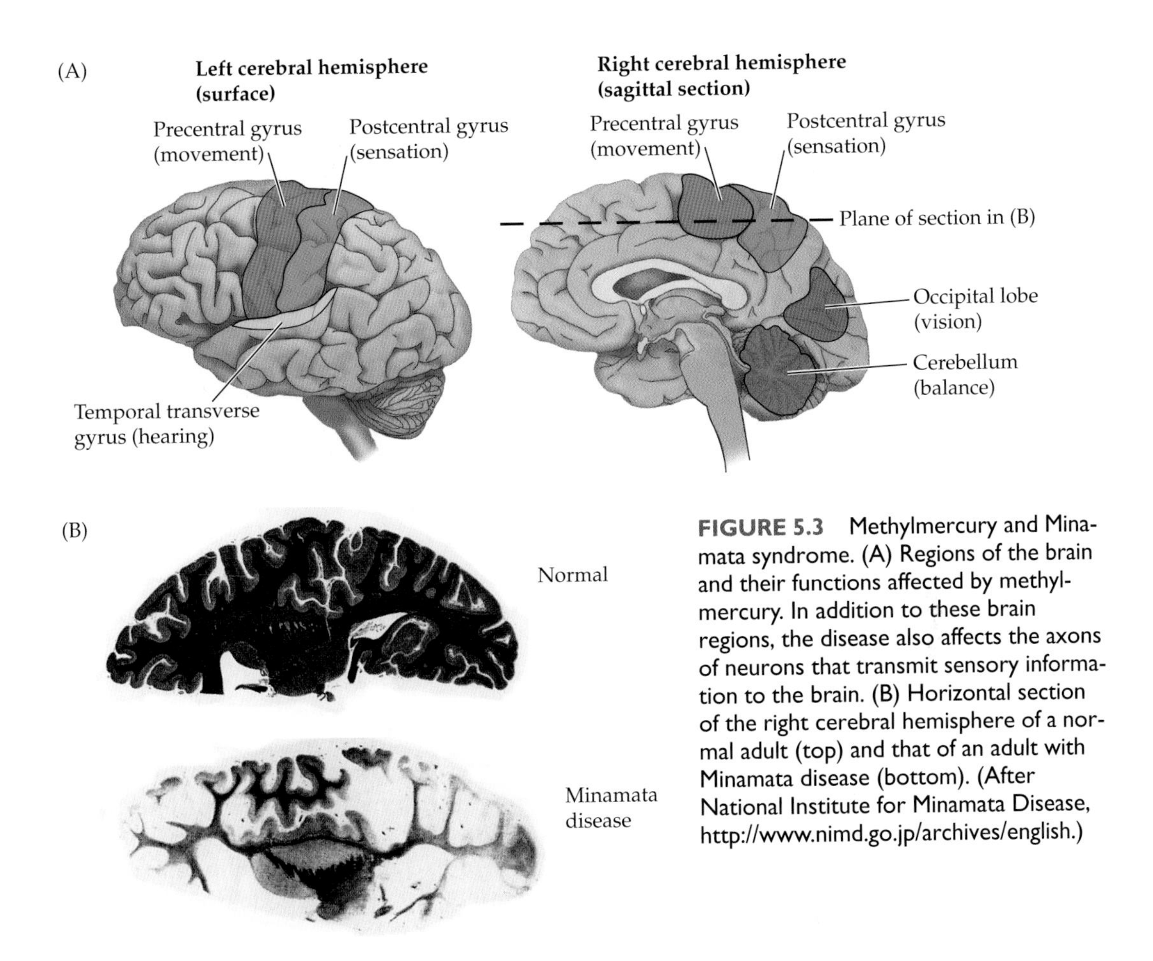

FIGURE 5.3 Methylmercury and Minamata syndrome. (A) Regions of the brain and their functions affected by methylmercury. In addition to these brain regions, the disease also affects the axons of neurons that transmit sensory information to the brain. (B) Horizontal section of the right cerebral hemisphere of a normal adult (top) and that of an adult with Minamata disease (bottom). (After National Institute for Minamata Disease, http://www.nimd.go.jp/archives/english.)

lead and the lax enforcement of antipollution laws have created a situation where lakes throughout the country have warnings against eating fish caught therein. The International Joint Commission of the U.S. and Canada (2000) warns that "eating Great Lakes sport fish may lead to birth anomalies and serious health problems for children and women of childbearing age."

The risk from anthropogenic chemical teratogens increases as more and more untested compounds enter the environment. Over 50,000 artificial chemicals are currently used in the United States, and between 200 and 500 new compounds are manufactured each year (Johnson 1980). Most industrial chemicals have not been screened for teratogenic effects. Standard screening protocols are expensive, take a long time, and are subject to the differences in metabolism between humans and the test animals used in screenings. There is still no consensus on how to accurately test a substance's teratogenicity for human embryos.

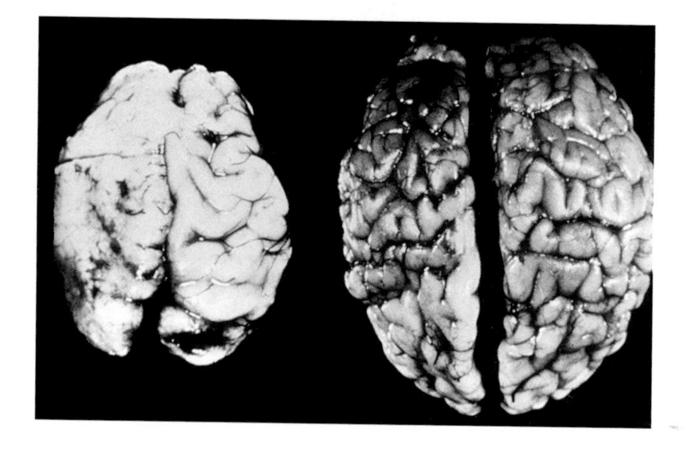

FIGURE 5.4 Comparison of a brain from an infant with fetal alcohol syndrome (FAS) with a brain from a normal infant of the same age. The brain from the infant with FAS is smaller, and the pattern of convolutions is obscured by glial cells that have migrated over the top of the brain. (Photographs courtesy of S. Clarren.)

Alcohol as a teratogen

In terms of the frequency of its effects and its cost to society, the most devastating human teratogen is undoubtedly ethanol. In 1968, Lemoine and colleagues noticed a syndrome of birth defects in the children of alcoholic mothers. This **fetal alcohol syndrome**, or **FAS**, was confirmed by Jones and Smith (1973). Babies with FAS are characterized by small head size, an indistinct philtrum (the pair of ridges that runs between the nose and mouth above the center of the upper lip), a narrow upper lip, and a low nose bridge. The brain of such a child may be dramatically smaller than normal and often shows defects in neuronal and glial migration (Figure 5.4; Clarren 1986). There is also prominent abnormal cell death in the frontonasal process and the cranial nerve ganglia (Sulik et al. 1988). Fetal alcohol syndrome is the third most prevalent type of mental retardation (following fragile-X syndrome and Down syndrome) and affects about 1% of liveborn infants (May and Gossage 2001). It is also thought that lower amounts of alcohol ingestion (as little as a single drink during a particularly susceptible time in pregnancy) can lead to *fetal alcohol effect*, a condition that does not cause the distinct facial appearance of FAS but which nevertheless lowers the functional and intellectual abilities of the affected person.

Some alleles of alcohol-metabolizing enzymes (e.g., alcohol dehydrogenase) appear to be better than others at detoxifying ethanol, and thus there is great variability in the ability of both mothers and fetuses to metabolize ethanol (Warren and Li 2005). It is believed that 30–40% of the children born to alcoholic mothers who drink during pregnancy will be affected by FAS. Children with fetal alcohol syndrome are developmentally and mentally retarded, with a mean IQ of about 68 (Streissguth and LaDue 1987). FAS patients with a mean chronological age of 16.5 years were found to have the functional vocabulary of 6.5-year-olds and to have the mathematical abili-

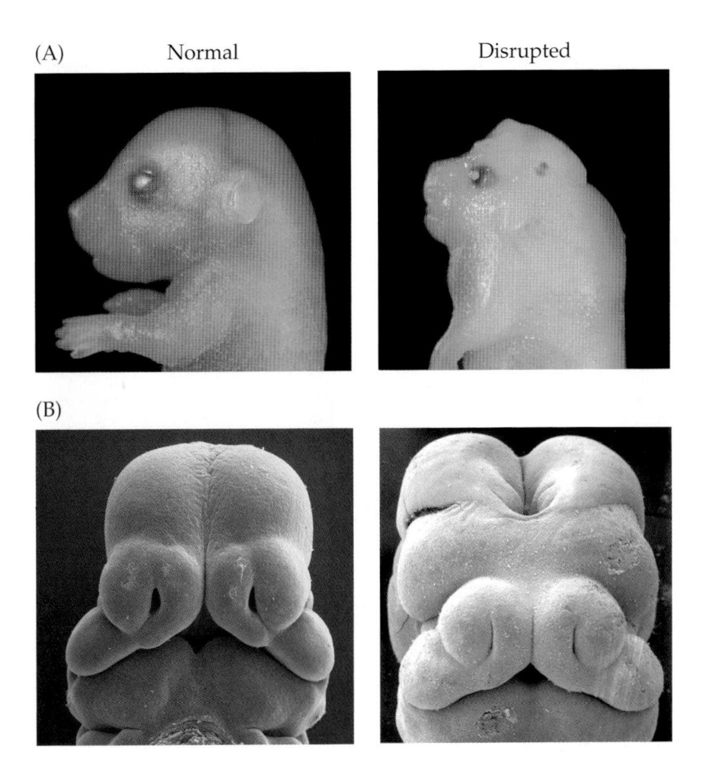

FIGURE 5.5 Alcohol-induced craniofacial and brain anomalies in mice. (A) Control mouse at embryonic day 17 compared with an anencephalic (brain lacking) mouse from mother who ingested alcohol during pregnancy. (B) Normal brain and facial development in a control mouse at embryonic day 11 (left) compared with aberrant development resulting from maternal alcohol ingestion. (From Green et al. 2007; A courtesy of T. Knudsen; B courtesy of K. Sulik.)

ties of fourth-graders. Most adults and adolescents with FAS cannot handle money, and they have difficulty learning from past experiences.*

A mouse model system has been used to explain the effects of ethanol on the face and nervous system. When mice are exposed to ethanol at the time of gastrulation, it induces the same range of developmental defects as in humans. As early as 12 hours after the mother ingests alcohol, developmental abnormalities are observed in her embryos. Their midline facial structures fail to form, causing the abnormally close proximity of the medial processes of the face. Forebrain anomalies are also seen, and the most severely affected fetuses lack a forebrain entirely (Figure 5.5; Sulik et al. 1988). These mouse studies suggest that ethanol may induce its teratogenic effects by several mechanisms. First, anatomical evidence suggests that neural crest cell migration is severely impaired. Neural crest cells usually migrate from the dorsal region of the neural tube (which will form the brain) to generate the bones of the face. Instead of migrating and dividing,

*For remarkable accounts of raising children with fetal alcohol syndrome, read Michael Dorris's *The Broken Cord* (1989) and Liz and Jodee Kulp's *The Best I Can Be* (2000). For an excellent account of the debates within the media and the medical profession about FAS, see Janet Golden's *Message in a Bottle* (2005).

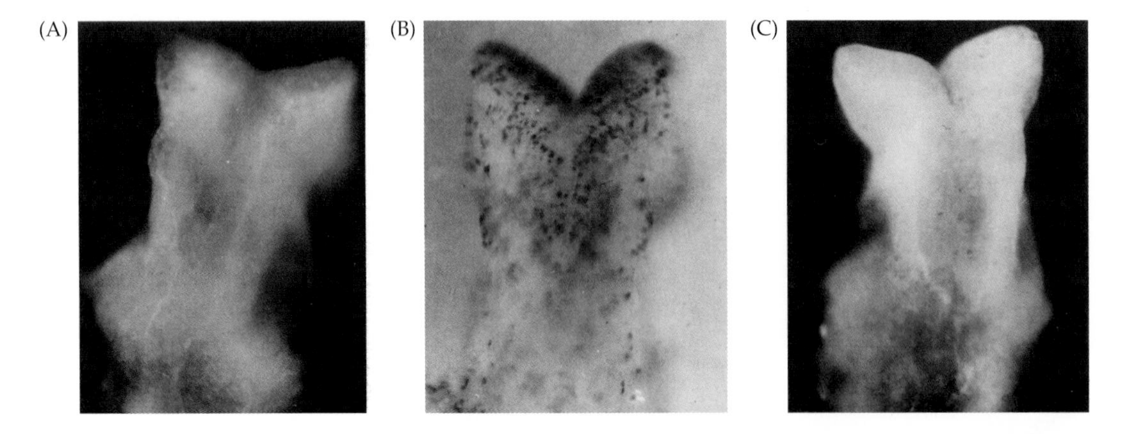

FIGURE 5.6 Cell death caused by ethanol-induced superoxide radicals is a possible mechanism producing fetal alcohol syndrome. Staining with Nile blue sulfate shows areas of cell death. (A) Control 9-day mouse embryo (head region). (B) Head region of ethanol-treated 9-day embryo. (C) Head region of embryo treated with both ethanol and superoxide dismutase, an inhibitor of superoxide radicals. (From Kotch et al. 1995; photographs courtesy of K. Sulik.)

ethanol-treated neural crest cells prematurely initiate their differentiation into facial bones (Hoffman and Kulyk 1999). Similarly, ethanol injected into the amnion of a chick embryo resulted in the apoptosis of cranial neural crest cells and failure to form the frontonasal process that generates the facial skeleton (Ahlgren et al. 2002). This apoptosis was correlated with a loss of *sonic hedgehog* gene expression in the pharyngeal arches. The sonic hedgehog protein is critical for normal formation of the facial skeleton. (It is the protein whose signaling is inhibited by cyclopamine, resulting in the formation of a single eye in the center of the head; see Figure 5.9). Sonic hedgehog-secreting cells placed in the head mesenchyme at this time prevented the ethanol-induced apoptosis of the cranial neural crest cells. Ethanol has recently been found to downregulate *sonic hedgehog* expression in the mouse embryo as well (Chrisman et al. 2004; Aoto et al. 2008). By interfering with the synthesis of sonic hedgehog protein, ethanol can produce facial skeletal defects.

Second, ethanol-induced apoptosis can delete millions of neurons from the developing forebrain and cranial nerve ganglia (Figure 5.6). One reason for this nerve cell death is that alcohol generates superoxide radicals that can oxidize cell membranes and lead to cytolysis (Davis et al. 1990; Kotch et al. 1995; Sulik 2005). Those neurons that do remain have impaired development of their mitochondria (Xu et al. 2005).

Third, alcohol may directly interfere with the ability of the cell adhesion molecule L1 to hold cells together. In this case, ethanol is acting on the cell's

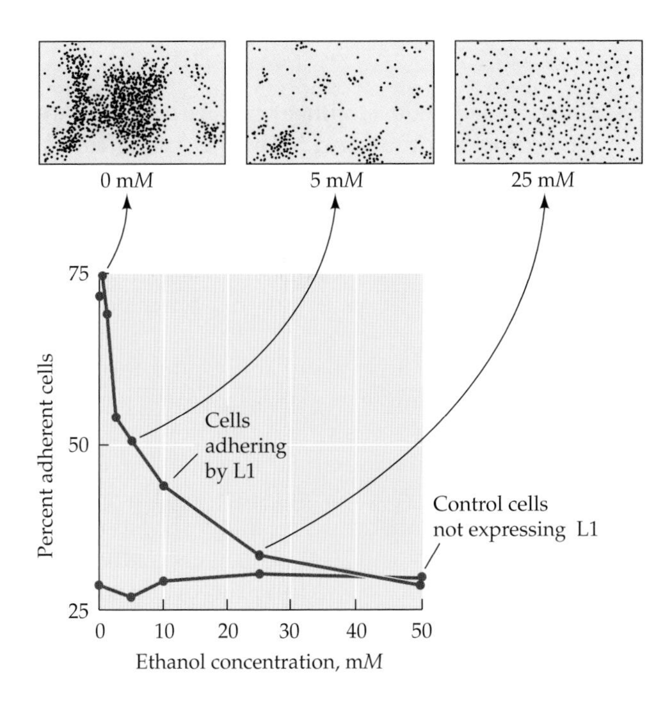

FIGURE 5.7 The inhibition of L1-mediated cell adhesion by ethanol is another possible factor in fetal alcohol syndrome. (From Ramanathan et al. 1996.)

proteins and not on genes that encode them. Ramanathan and colleagues (1996) have shown that, in vitro, ethanol can block the adhesive function of the L1 protein at levels as low as 7 m*M*—a concentration that can be produced in the blood or brain by a single drink (Figure 5.7). Moreover, mutations in the human L1 gene cause a syndrome of mental retardation and malformations similar to that seen in severe cases of fetal alcohol syndrome.

Ethanol ingestion by the mother also causes disruption in those genes whose functions include organizing the cytoskeleton and maintaining mitochondrial function (Green et al. 2007). Taken together, the above studies show that ethanol is exceptionally dangerous to the developing vertebrate because its effects are so numerous, because it can travel across the placenta, and because it can operate in several ways. For this reason, the United States Surgeon General's office recommends that pregnant women (and those who are contemplating pregnancy) refrain from drinking alcoholic beverages.*

*Men should also be warned. Alcohol (like marijuana and tobacco) is believed to cause reduced sperm counts and increases the incidence of defective sperm. It also results in impotence and testosterone deficiency (Emanuele and Emanuele 1998; Battista et al. 2008). Interestingly, the notion that alcohol was bad for developing embryos appears to have been known for a long time, and then the knowledge was lost. Philip Pauly (1996) has documented that in some instances, the harmful effects of ethanol on an individual were seen as doing a public service, getting rid of weak embryos and allowing only the strong ones to survive.

Retinoic acid

Even a compound normally involved in development can disrupt development if it is present in the wrong amounts and/or at the wrong times. Retinoic acid (RA) is important for the formation of the anterior-posterior axis of the mammalian embryo, as well as for proper jaw formation. In normal development, RA is secreted from discrete cells and works in circumscribed areas of the embryo. However, if RA is present in large amounts, cells that normally would not receive high concentrations of this molecule are exposed to it and will respond to it.

13-*cis*-retinoic acid (also called isotretinoin and sold under the trade name Accutane) has been useful in treating severe cystic acne and has been available for this purpose since 1982. Because the deleterious effects of administering large amounts of retinoids to pregnant animals have been known since the 1950s (see Cohlan 1953; Giroud and Martinet 1959; Kochhar et al. 1984), the drug carries a label warning that it should not be used by pregnant women. However, about 160,000 women of childbearing age (15–45 years) have taken this drug since it was introduced, and some of them have used it during pregnancy. Lammer and his coworkers (1985) studied a group of women who inadvertently exposed themselves to retinoic acid and who elected to remain pregnant. Of their 59 fetuses, 26 were born without any noticeable anomalies, 12 aborted spontaneously (miscarried), and 21 were born with obvious anomalies. The affected infants had a characteristic syndrome, including absent or defective ears, absent or small jaws, cleft palate, aortic arch abnormalities, thymic deficiencies, and abnormalities of the central nervous system.

This pattern of multiple congenital anomalies is similar to that seen in rat and mouse embryos whose pregnant mothers were given retinoic acid. Goulding and Pratt (1986) placed 8-day mouse embryos in a solution containing 13-*cis*-retinoic acid at a very low concentration ($2 \times 10^{-6}\ M$). Even at this concentration, approximately one-third of the embryos developed a very specific pattern of anomalies, including dramatic reduction in the size of the first and second pharyngeal arches (Figure 5.8). In normal mice, the first arch eventually forms the maxilla and mandible of the jaw and two ossicles of the middle ear, while the second arch forms the third ossicle of the middle ear, as well as other facial bones.

The basis for this developmental disruption appears to be RA's ability to alter the expression of the Hox genes. Hox genes are critical in specifying which part of the embryo is anterior, which is posterior, and which region lies between them. Hox genes are also critical in specifying the fates of certain cells, and they inhibit neural crest cells from forming cartilage and bone. By altering the expression of Hox genes, RA can re-specify portions of the anterior-posterior axis in a more posterior direction and can inhibit neural crest cells from migrating to form facial cartilage (Moroni et al. 1994;

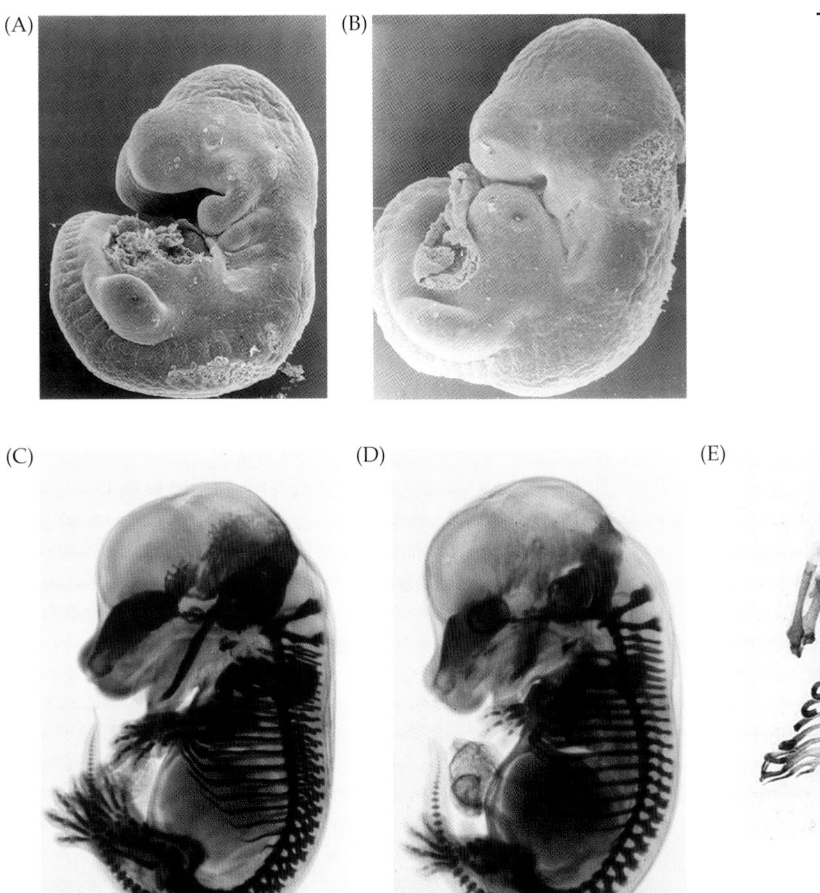

FIGURE 5.8 Effects of retinoic acid on mouse embryos. (A,B) Embryos cultured in control medium (A) or in medium containing retinoic acid (B), seen on day 10. The first pharyngeal (jaw) arch of the treated embryo is malformed and has fused with the second arch. The ossification of some of the skull has failed, and there are limb abnormalities. (C,D) Skeletal formation in control embryo (C) and an embryo exposed to retinoic acid in utero (D), seen at day 17. Craniofacial malformations are seen as the neural crest-derived cartilage of the jaw and of the middle ear failed to form properly. (E) In some cases, exposure to retinoic acid results in severe loss of posterior vertebrae (arrow). (A–D courtesy of G. Morriss-Kay; C,D from Morriss-Kay 1993; E from Kessel 1992, courtesy of M. Kessel.)

Studer et al. 1994). Radioactively labeled RA can be seen to bind to the cranial neural crest cells and arrests both their proliferation and their migration (Johnston et al. 1985; Goulding and Pratt 1986). This binding seems to be specific to the cranial neural crest cells, and the teratogenic effect of the drug is confined to a specific developmental period (days 8–10 in mice; days 20–35 in humans). Animal models of retinoic acid teratogenesis have been extremely successful in elucidating the mechanisms of teratogenesis at the cellular level.

Retinoic acid is a critical public health concern because there is significant overlap between the population using acne medicine and the population of women of childbearing age, and because it is estimated that half of the pregnancies occurring in the United States are unplanned (Nulman et al. 1997). Vitamin A, the precursor of retinoic acid that is transformed into RA in the body, is itself teratogenic in megadose amounts. Rothman and colleagues (1995) found that pregnant women who took more than 10,000 international units of vitamin A per day (in the form of vitamin supplements) had a 2% chance of having a baby born with disruptions similar to those produced by retinoic acid. According to the rules of the U.S. Food and Drug Administration, every patient using isotretinoin, every physician prescribing it, and every pharmacy selling it must sign a registry. Moreover, women who use this drug are expected to take a pregnancy test within 7 days before filling their prescription and to agree to use two methods of birth control and adhere to pregnancy testing on a monthly basis.

Teratogens and Cognitive Function

Although birth defects are usually defined anatomically, recent studies indicate that some teratogens cause defects in neural patterning that are first detected by changes in normal cognitive or behavioral abilities. One possible cognitive alteration is lowered intelligence and, indeed, we saw that the intelligence of people affected by fetal alcohol syndrome can be dramatically below normal expectations. And, while there are many causes for mental dysfunction, environmental teratogenesis is gaining prominence as a possible cause of at least some types of **autism**, a spectrum of behavioral and cognitive changes that includes repetitive behaviors and impaired communication and social interactions.

There are probably genetic causes for autism, environmental causes for autism, and interactions between some genotypes and some environmental factors that can lead to autism (see Autism Genome Consortium 2007; Morris et al. 2008). The first clue that autism might be caused by specific environmental agents came from studies of children with thalidomide syndrome. While autism is found in 0.1–0.2% of the general population, it was present in 5% of the Swedish children having the thalidomide syndrome described earlier in the chapter. Moreover, it was found specifically in those children whose mother had taken thalidomide between days 21 and 24 of embryonic development (Miller et al. 2005).

Valproic acid (VA) is also known to disrupt normal embryonic development. Mice and humans whose mothers took VA during pregnancy often have defects of neural tube closure, ribs, and facial formation, and may have cognitive disorders as well. The IQs of children whose mothers took VA during pregnancy is significantly (10–20 points) below that of their siblings (see Meador et al. 20006), and epidemiological studies suggest that as many as 11% of children born with valproic acid syndrome have autism (Moore et al. 2000; Arndt et al. 2005). It has been suggested that VA might cause autism by affecting the brain development.

Although it is impossible to diagnose autism in rats, recent studies suggest that VA causes the same spectrum of neurological alterations in rodents and humans. These included the reduction in the numbers of neurons in several cranial nerve ganglia, as well as markedly reduced numbers of Purkinje cells in the cerebellum (see Ingram 2000; Arndt et al. 2005). Neurotransmitter abnormalities in these rats were also similar to those seen in human patients

Other teratogenic agents

CIGARETTE SMOKE Cigarette smoke retards the growth of human fetuses and gives a heightened risk of fetal and newborn death. Indeed, compared with nonsmoking women having their first birth, women who smoked one or more packs of cigarettes a day had a 56% greater risk of having the fetus or newborn infant die (Kleinman et al. 1988; Werler 1997). There is also evidence that nicotine, an important component of cigarette smoke, may damage the fetal brain and lungs during development. Dwyer and colleagues (2008) showed that nicotine induces abnormalities of synapse formation and cell survival in the developing brain. Maritz (2008) showed that nicotine can induce lung cells to have an altered metabolism that ages them more rapidly, and prenatal exposure to cigarette smoke is associated with increased risks of impaired lung functions later in life (Wang and Pinkerton 2008). Smoking also significantly lowers the number, quality, and motility of sperm in the semen of males who smoke at least four cigarettes a day (Kulikauskas et al. 1985; Mak et al. 2000; Shi et al. 2001).

(Narita et al. 2002). One recent study (Rinaldi et al. 2008) found that a single exposure to valproic acid in utero caused changes in the rat neocortex, including a weaker excitatory synapse response, diminished contact between neuronal layers, and more autonomic neural activity. This fits in with models of human autism that predict greater activity of local neural modules at the expense of globally coordinated activity.

Some of the above-mentioned anomalies may be caused by the misregulation of *Hoxa1* in the embryonic rats. This gene is critically important in forming the cerebellum. In human embryos, exposure to thalidomide in the time period 20–24 days after fertilization causes hindbrain anomalies that mimic loss-of-function *HOXA1* mutations (Miller and Strömland 1999; Tishfield et al. 2005). In mice, valproic acid exposure in utero induces the overexpression of *Hoxa1*, as well as its expression at the wrong time during development (Stodgell et al. 2006). While *HOXA1* mutations are extremely rare in humans and are probably not a major cause of autism, it appears that teratogen-induced misregulation of this gene can lead to altered neural patterning and perhaps to the cognitive anomalies seen in autistic children (Miller and Strömland 1999; Tischfield et al. 2005).

Other drugs, especially widely used recreational drugs and antidepressants, are now being scrutinized to see if they cause subtle cognitive or behavioral changes in the offspring of women who use them. Although these studies are not complete, it appears that antidepressants may cause transient, but not permanent, changes in newborn behavior (Koren et al. 2001; Kalra et al. 2005). Maternal marijuana and cocaine use may delay the onset of neurological milestones (that is, the age at when certain normal behaviors are seen) (Handu et al. 2008; Kunert et al. 2008).

The effects of antidepressants and recreational drugs are still being studied, but they are not nearly as obvious as those of ethanol. It is difficult to extrapolate possible effects on humans from animal studies for these drugs, and there are often confounding factors (e.g., the effects of diet and the use of other drugs) that make it difficult to associate a particular drug with a particular effect. The March of Dimes, one of the leading agencies attempting to prevent birth defects through treatment and education, keeps a website on this subject at www.marchofdimes.com/professionals/14332_1169.asp.

DRUGS Each year, hundreds of new artificial chemical compounds come into general use in our industrial society, and many of these are found in pharmaceutical products. Some drugs used to control diseases in an adult may have deleterious effects on a fetus (see Table 5.1). The teratogenic effects of the sedative thalidomide were described at the beginning of this chapter. Other such drugs include cortisone (used for rheumatoid arthritis), warfarin (used to prevent blood clots), tetracycline (an antibiotic), and valproic acid (an anticonvulsant used to treat epilepsy).

Valproic acid (VA), as an example, is known to disrupt development in several ways that can result in major and minor spinal defects in the developing fetus. Finnell and colleagues (1997) showed that VA blocked folate* from being absorbed by the embryo and thereby led to neural tube defects. Barnes and colleagues (1996) have shown that valproic acid decreases the level of *Pax1* transcription in chick somites. This decrease causes malformation of the somites and corresponding malformations of the vertebrae and ribs. Recent findings even implicate VA in certain cases of autism (see box on previous page). Recent research (Menegola et al. 2005; Eikel et al. 2006) indicates that valproic acid inhibits the activity of a class of enzymes called histone deacetylases, which are critical regulators of gene expression (see Chapter 2).

PATHOGENS Another class of teratogens includes viruses and other pathogens. At the beginning of this chapter, we described the effects of the rubella (German measles) epidemic of 1963–1965. The rubella virus is able to enter many cell types, where it produces a protein that stops mitosis by blocking kinases that are necessary for the cell cycle to progress (Atreya et al. 2004). Thus, numerous organs are affected, and the earlier in pregnancy the rubella infection occurs, the greater the risk that the embryo will be malformed. The first 5 weeks of development appear to be most critical because that is when the heart, eyes, and ears are formed (see Figure 5.2). Two other viruses, cytomegalovirus and the herpes simplex virus, are also known to be teratogenic. Cytomegalovirus infection of early embryos is nearly always fatal, and infection of later embryos can lead to blindness, deafness, cerebral palsy, and mental retardation.

Bacteria and protists are rarely teratogenic, but at least two of them are known to damage human embryos. *Toxoplasma gondii*, a protist carried by rabbits and cats (and their feces), can cross the placenta and cause brain and eye defects in the fetus. *Treponema pallidum*, the bacterium that causes syphilis, can kill early fetuses and produce congenital deafness and facial damage in older ones.

*Folate, or vitamin B9, is critical for neural tube closure. The U.S. Center for Disease Control has estimated that some 50% of neural tube defects can be prevented by pregnant women taking folate supplements (CDC 1992).

HEAT Hyperthermia, the elevation of core body temperature, is thought to be dangerous to the fetus. An extended maternal temperature of 102°F (38.9°C) or higher during the first 6 weeks of pregnancy is considered detrimental to the closing of the embryonic neural tube (OTIS 2007). Such elevated body temperature can occur as a result of illness (fever), long exposure to unrelieved heat wave conditions, or extensive stays in a hot tub. Proper use of saunas does not appear to be as dangerous because, by causing one to sweat, saunas minimize changes in internal body temperature. A study by Saxén and colleagues (1982) showed that although 98.5% of pregnant Finnish women visited a sauna regularly, Finland had an extremely low incidence of neural tube and other birth defects.

Natural Killers: Teratogens from Plants

Given the voracity of insects and their larvae, it's amazing that any plant survives. Plants can't run away from predators, so they must protect themselves in other ways. Poisoning one's enemies is a time-honored way to rid oneself of them, so perhaps it is not surprising that most of the poisons known to humans are substances that were originally found occurring naturally in plants. A somewhat more subtle way to get rid of a predator is to destroy its offspring, and several teratogenic compounds found among the plants do just this.

Veratrum alkaloids

The plant *Veratrum californicum* (corn lily, or California false hellebore) produces several alkaloid substances, including veratramine, jervine, and cyclopamine, that can block the functions of the Hedgehog family of paracrine factors (Keeler and Binns 1968; Beachy et al. 1997; Chen et al. 2002). One of the major functions of hedgehog protein in vertebrates is to allow the proliferation of cells in the ventral midline of the face. Blocking the *sonic hedgehog* gene signal leads to the failure of the optical field to separate and also prevents the pituitary gland from forming. If a pregnant ewe eats *Veratrum*, her lambs are likely to be born with a single eye ("cyclops") in the center of the head—hence the name "cyclopamine" for one of the compounds causing this effect (Figure 5.9).

Cyclopia is another example of a birth defect that can occur either by environmental or genetic means. In humans, loss-of-function mutations in either one of two separate genes can result in cyclopia. The first of these genes encodes the enzyme sterol-Δ7 reductase, which is critical in cholesterol biosynthesis. Cholesterol is a critical co-factor in allowing the formation and function of Sonic hedgehog protein. The second gene encodes the Sonic hedgehog protein (Roessler et al. 1996; Traiffort et al. 2004). However, a person carrying the mutant *Sonic hedgehog* allele (an autosomal dominant

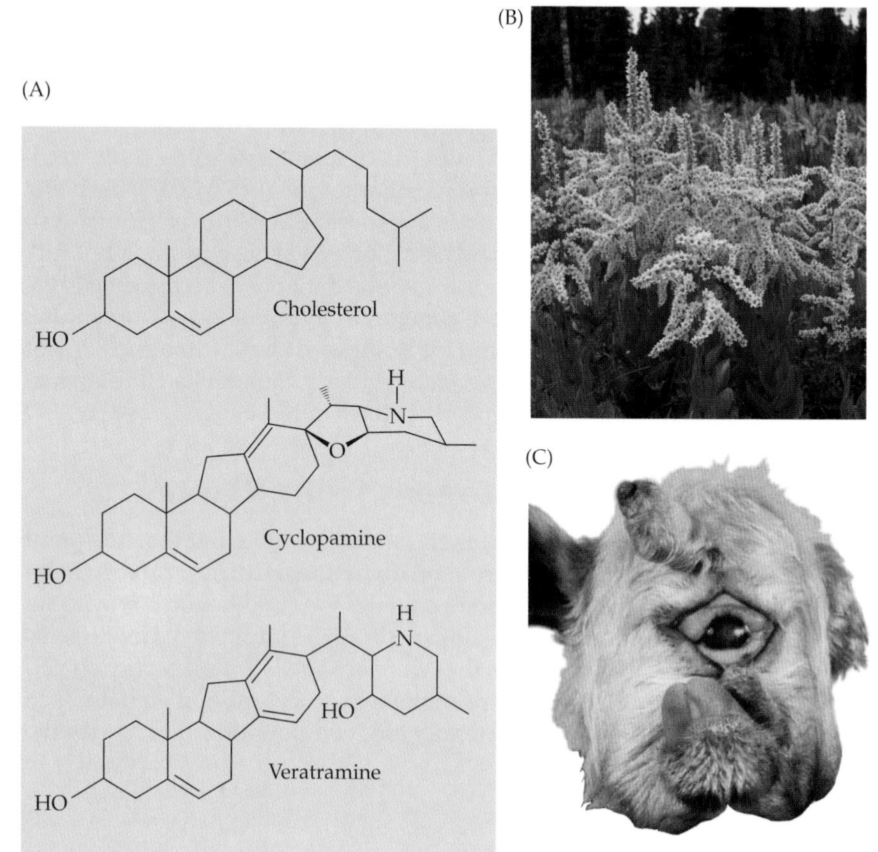

FIGURE 5.9 Cyclopia caused by *Veratrum* alkaloids. (A) Structures of cholesterol and two *Veratrum* alkaloids, cyclopamine and veratramine. (B) *Veratrum californicum*, California false hellbore. (C) Head of a cyclopic lamb stillborn to a ewe who grazed on *Veratrum californicum* early in pregnancy. The cerebral hemispheres fused, forming only one central eye, and the lamb had no pituitary gland. (B Courtesy of D. Powell, USDA Forest Service, Bugwood.org; C courtesy of L. James, USDA ARS Poisonous Plant Research Laboratory.)

mutation) is not necessarily born with lethal cyclopia. Related individuals who carry the *same mutation* of this gene can have *different phenotypes*, ranging from normal to cyclopic (medically, this range is referred to as the *holoprosencephaly spectrum*; Muenke 1995; Cordero 2004). Such **phenotypic heterogeneity** (the ability of one mutant gene to produce a range of phenotypes) is thought to result from the interaction of the genotype with the environment. It is probable that the mutant *Sonic hedgehog* allele causes a deficiency of Sonic hedgehog protein, and that this deficiency is exacerbated by an environmental condition, perhaps a limited amount of cholesterol (Dehart et al. 1997; Edison and Muenke 2003; Cordero et al. 2004).

Plant juvenile hormones

One of the most fascinating examples of teratogenic control of animal pests occurs at the level of juvenile hormone. Juvenile hormone is made in the corpora allata of insect larvae, and its secretion prevents the larvae from undergoing metamorphosis at the next molt; instead, the next molt simply results in a larger larva. Inappropriate amounts of juvenile hormone at the wrong time can delay metamorphosis too long or precipitate it too soon, either one of which can have lethal outcomes for the developing insect.

DELAYED METAMORPHOSIS When Karel Sláma came from Czechoslovakia to work in Carroll Williams's laboratory at Harvard, he brought his chief experimental animal with him—the European plant bug *Pyrrhocoris apterus*. To the consternation of the entire laboratory, his bugs failed to undergo normal metamorphosis, which for this species should take place at the end of the fifth instar.* Rather, they became something never before observed in nature or the laboratory: large, sixth-instar larvae that died before becoming adults. After many variables were tested without finding anything, the papers lining the culture dishes were tested for their effect on the larvae. The results were as conclusive as they were surprising: larvae reared on European paper (including pages of the journal *Nature*) underwent metamorphosis as usual, whereas larvae reared on American paper (such as shredded copies of the journal *Science*) did not undergo metamorphosis. It was eventually determined that the source of the American paper was the balsam fir (*Abies balsamea*), a tree indigenous to the northern United States and Canada. *A. balsamea* synthesizes a biochemical compound that closely resembles juvenile hormone (Bowers et al. 1966; Sláma and Williams 1966; Williams 1970). This compound presumably protects the tree by interfering with the metamorphosis of its insect predators.

The teratogenic effect on insects of some natural plant products has been studied and turned to human advantage in a number of pest-control products. The juvenile hormone mimic methoprene, for instance, is used to control ant and mosquito infestations and is a major ingredient in flea collars (EPA 2007).

PRECOCIOUS METAMORPHOSIS Some plants have compounds that produce the same effect—the death of insect herbivores—but do so by eliciting metamorphosis too early. Two compounds that have been isolated from herbaceous composites (the sunflower family, Compositae), including goatweed and floss flower (*Ageratum conyzoides* and *A. houstonianum*), have been found to cause the premature metamorphosis of certain insect larvae

*__Instar__ is the term used to describe the larva between molts. Thus, the first instar larva is that hatched from the egg. The first instar molts into the second instar, and so forth until the insect forms a pupa and then has a final molt into an adult.

into sterile adults (Bowers et al. 1976). These compounds are called **precocenes**; their chemical structures are shown in Figure 5.10A. When the larvae or nymphs of these insects are dusted with either of these compounds, they undergo one more molt and then metamorphose into the adult form (Figure 5.10B).

Precocenes cause the selective death of the corpus allatum cells responsible for synthesizing juvenile hormone in the immature insect (Schooneveld 1979; Pratt et al. 1980). Without these cells to produce juvenile hormone, the larva commences its metamorphic and imaginal molts. Moreover, juvenile hormone is also responsible for the maturation of the insect egg. Without

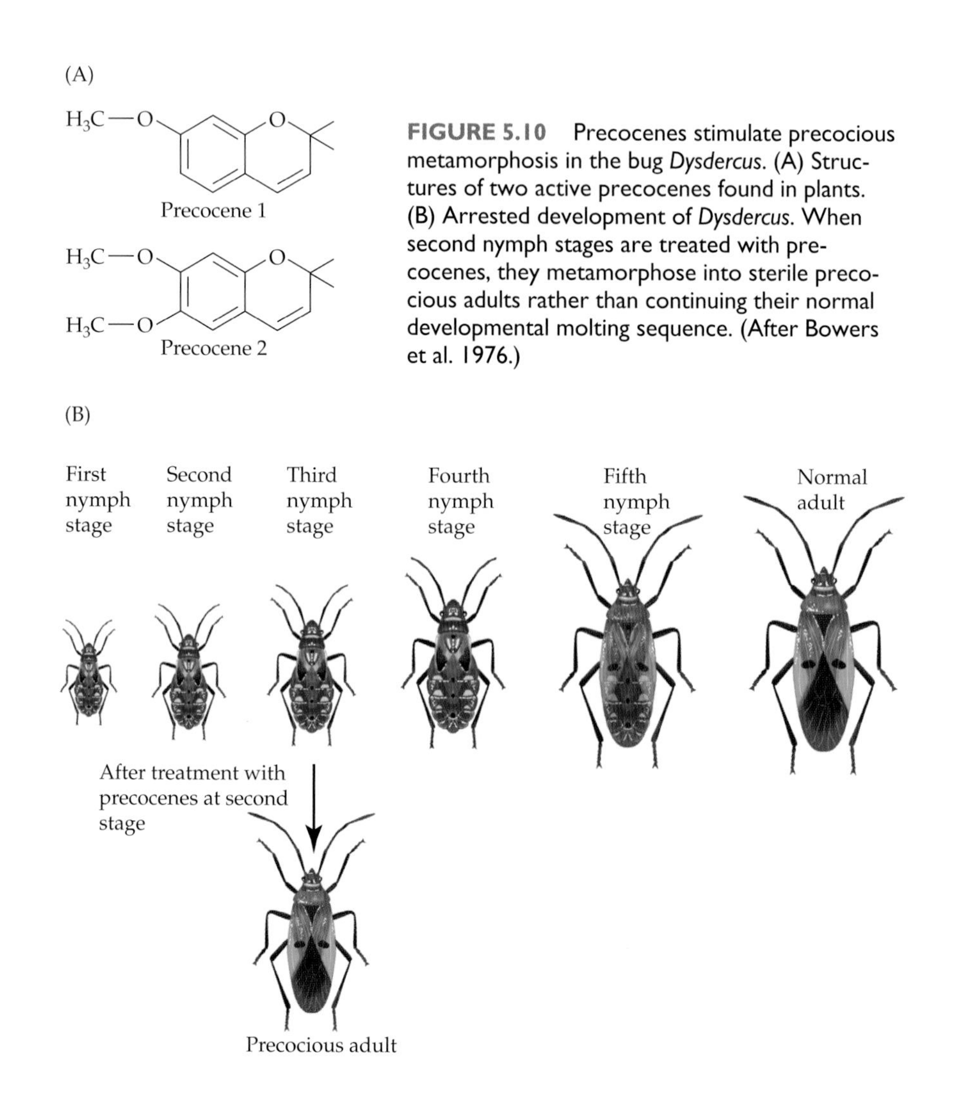

FIGURE 5.10 Precocenes stimulate precocious metamorphosis in the bug *Dysdercus*. (A) Structures of two active precocenes found in plants. (B) Arrested development of *Dysdercus*. When second nymph stages are treated with precocenes, they metamorphose into sterile precocious adults rather than continuing their normal developmental molting sequence. (After Bowers et al. 1976.)

prolonged exposure to this hormone, females are sterile. Thus, precocenes are able to protect the plant by causing the premature metamorphosis of certain insect larvae into sterile adults.

Deformed Frogs: A Teratological Enigma

Over the past two decades, one phenomenon connected with teratogenicity has received a great deal of attention in both the scientific and the national media. Throughout the United States and southern Canada, there has been a dramatic increase in the number of deformed frogs and salamanders observed in what seem to be pristine woodland habitats. In some localized areas, it is estimated that as many as 60% of certain amphibian species have been documented to display visible malformations (Ouellet et al. 1997). These deformities include extra or missing limbs (Figure 5.11), missing or misplaced eyes, deformed jaws, and malformed hearts and guts. We do not know what is causing these abnormalities; David Stocum (2000) has called the situation "an 'eco-devo' riddle wrapped in multiple hypotheses by insufficient data." Numerous hypotheses have been put forward to explain the sudden prevalence of these anomalies (see Lannoo 2008), and in recent years three of these explanations appear to be gaining acceptance.

A combination of factors

The first hypothesis proposes a combination of causes. The proximate cause of many limb deformities may be the infection of the larval amphibian limb buds by the larvae of trematode (flatworm) parasites. Under most conditions, the tadpole's immune system can destroy the trematode larvae. In some ponds, however, it appears that the tadpoles suffer from an acquired

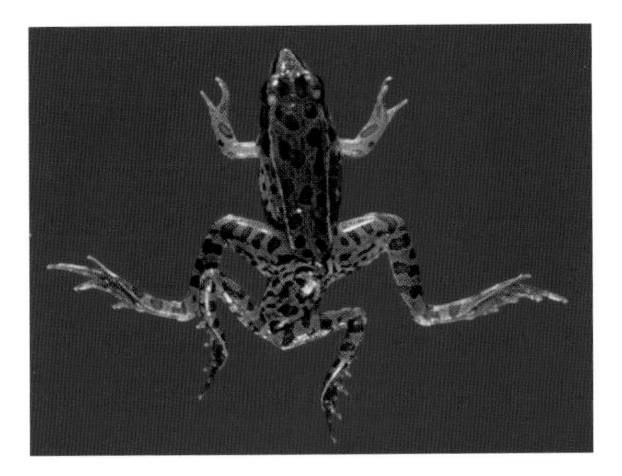

FIGURE 5.11 A northern leopard frog (*Rana pipiens*) from a wild population, suffering from polymelia (an extra set of hindlimbs). (Photograph courtesy U.S. Geological Survey.)

immune deficiency syndrome that leaves them vulnerable to the parasites (Christin et al. 2004). Kiesecker (2002) has shown that frogs living in habitats contaminated by certain pesticides are less able to resist trematode infection. He found that neither pesticide pollution alone nor trematodes alone were sufficient to cause deformities in wild populations of frogs, but that the combination of the two factors resulted in a significant proportion of the frog population showing limb deformities.

At least in some species, then, trematode infestation has to be coupled with pollutants such as pesticides in order for significant cases of limb deformities to be observed. Similar multiagent causation, where one agent is a pesticide, has been shown by Davidson and Knapp (2007) to be responsible for the mortality of some frog populations.

The radiation hypothesis

There are habitats with high proportions of malformed frogs, however, that do not appear to be infested with trematodes. A second hypothesis proposes that ultraviolet (UV) radiation can cause these abnormalities. Ankley and colleagues (2002) have argued that, whereas UV irradiation can kill amphibian embryos (see Chapter 4), those individuals that survive are likely to have limb deformities. The spectrum of limb abnormalities seen in frogs that develop in the laboratory from UV-exposed embryos and tadpoles is somewhat different than that seen in natural populations (Meteyer et al. 2000), but this explanation may account for some of the observed teratogenesis.

Pesticides and herbicides

A third possibility is that pesticidal and herbicidal chemical are the direct cause of physiological disruptions. As mentioned in Chapter 2, the herbicide atrazine is believed to be responsible for gonadal malformations in many frog species. Moreover, atrazine exposure early in development has been seen to cause developmental abnormalities of the limbs, gut, and head, as well as the apoptosis of brain and kidney cells (Lenkowski et al. 2008).

It is also of interest that the spectrum of abnormalities seen in natural populations of deformed frogs resembles some of the malformations that result when tadpoles are exposed to retinoic acid (Crawford and Vincenti 1998), and it is possible that some herbicides and pesticides disrupt development after being converted naturally into chemicals that resemble RA. Experiments in the Gardiner laboratory (Gardiner and Hoppe 1999; Gardiner et al. 2003) have shown that RA can induce all the limb malformations seen in wild populations of some frogs. They have also purified a retinoid compound from the water of one lake that had a high incidence of malformed amphibians. Since no one is suspected of dumping retinoic acid into

Conservation Biology: Saving the Frogs

For some time now, the decline of amphibian populations has been a major cause for alarm among conservation biologists. Interestingly, Lannoo (2008) proposes that we probably know enough to save many of these species even as the specific mechanisms of developmental disruption remain unknown. He points out that nearly all the proposed causes for these anomalies can be prevented simply by controlling runoff from agricultural lands.

Lannoo notes that nutrients entering ponds from runoff contaminated by fertilizers encourage plant growth, which encourages snail population growth. Increased numbers of snails (the vectors of trematode larvae) means more parasitic trematodes to infest the frog larvae and disrupt their development. Moreover, pesticides and herbicides in the runoff may be destroying amphibian immune capabilities and/or directly inducing developmental anomalies.

In all these cases the remedy—stopping erosion from farmlands—is the same. The idea gains support from recent studies by McCoy and colleagues (2008), who quantified the relationship between gonadal anomalies in male toads and the amount of agricultural activity at or near specific ponds. They found statistically significant increases in both gonadal abnormalities and reproductive failures as agricultural activity levels increased. However, as Davidson and Knapp (2007) have pointed out, erosive runoff is only one way chemicals get into the water. Pesticides and herbicides can be blown on the wind and still have severe consequences for the frogs in the ponds in which they finally settle. Still, controlling runoff would probably be a very good start.

the lake, how did such a compound get there? One possibility involves the insectidal chemical methoprene.

Methoprene is an insect juvenile hormone mimic that inhibits mosquito pupae from metamorphosing into adults. Because vertebrates do not have juvenile hormone, it was assumed that this pesticide would be not harm fish or amphibians (or humans). Indeed, this is to some extent the case; methoprene itself has no teratogenic properties. However, on exposure to sunlight, methoprene breaks down into chemical products that have significant teratogenic activity in frogs. These compounds have a structure very similar to that of retinoic acid and will bind to the RA receptor (Harmon et al. 1995). When *Xenopus* eggs are incubated in water containing these compounds, the resulting tadpoles are often malformed and show a spectrum of deformities similar to those found in the wild (La Claire et al. 1998).

Summary

The environment is a source of chemicals that are crucial for normal development, but it is also a source of chemicals that can impair development. Some of these teratogens are naturally occurring, while others are the products of industrial manufacture. Some of these compounds (such as retinoic acid) are used during normal development, and their ingestion in excess can cause the normally acting compound to work in new places or at inappropriate times. Other compounds (such as ethanol) impair development by preventing the synthesis or function of several molecules, interfering with genes and cell membranes. Some teratogens (such as thalidomide) are specific for certain organ systems, while others (such as rubella virus) affect numerous organ systems.

Teratogens are a major source of human congenital malformations, and it has only been recently that many of these compounds had been identified and their mechanisms of tertogenicity delineated. Ethanol, which was not considered to be a teratogen until the 1970s, is particularly damaging, as fetal alcohol syndrome has become the leading cause of congenital mental retardation. Public policy has made the labeling of alcoholic products as teratogenic mandatory, and it has also made it very difficult for women to obtain retinoic acid-containing acne medicines. There is enormous debate about how to test for the teratogenic potential of new substances, because animals metabolize these substances in different ways.

In addition to the established substances discussed in this chapter, a new and widespread class of teratogen has recently been uncovered. These teratogens work by disrupting the endocrine system, and they usually affect the physiological function of the organism more than they disrupt its anatomy. Because of this, their effects are often unnoticed until adulthood. These **endocrine disruptors** are the topic of the next chapter.

References

Ahlgren, S. C., V. Thakur and M. Bronner-Fraser. 2002. Sonic hedgehog rescues cranial neural crest from cell death induced by ethanol. *Proc. Natl. Acad. Sci. USA* 99: 10476–10481.

Aoto, K., Y. Shikata, D. Higashiyama, K. Shiota and J. Motoyama. 2008. Fetal ethanol exposure activates protein kinase A and impairs *Shh* expression in prechordal mesendoderm cells in the pathogenesis of holoprosencephaly. *Birth Def. Res. A: Clin. Mol. Teratol.* 82: 224–231.

Ankley, G. T., S. A. Diamond, J. E. Tietge, G. W. Holcombe, K. M. Jensen, D. L. Defoe and R. Peterson. 2002. Assessment of risk of solar ultraviolet radiation to amphibians. I. Dose-dependent induction of hindlimb malformations in the northern leopard frog (*Rana pipiens*). *Environ. Sci. Tech.* 36: 2853–2858.

Arndt, T. L., C. J. Stodgell and P. M. Rodier. 2005. The teratology of autism. *Int. J. Devl. Neurosci.* 23: 189–199.

Atreya, C. D., K. V. Mohan and S. Kulkarni. 2004. Rubella virus and birth defects: Molecular insights into the viral teratogenesis at the cellular level. *Birth Def. Res. A: Clin. Mol. Teratol.* 70: 431–437.

Autism Genome Consortium. 2007. Mapping autism risk loci using genetic linkage and chromosomal rearrangements. *Nature Genet.* 39: 319–328.

Barnes, G. L., Jr., B. D. Mariani and R. S. Tuan. 1996. Valproic acid-induced somite teratogenesis in the chick embryo: Relationship with *Pax-1* gene expression. *Teratology* 54: 93–102.

Battista, N., C. Rapino, M. DiTomasso, M. Bari, N. Pasquaielo and M. Maccarone. 2008. Regulation of male fertility by the endocannabinoid system. *Mol. Cell Endocrinol.* 286 (suppl.): S17 – S23.

Beachy, P. A. and 7 others. 1997. Multiple roles of cholesterol in Hedgehog protein biogenesis and signaling. *Cold Spring Harb. Symp. Quant. Biol.* 62: 191–204.

Bowers, W. S., H. M. Fales, M. J. Thompson and E. C. Uebel. 1966. Identification of an active compound from balsam fir. *Science* 154: 1020–1021.

Bowers, W. S., T. Ohta, J. S. Cleere and P. A. Marsella. 1976. Discovery of insect anti-juvenile hormones in plants. *Science* 193: 542–547.

CDC (Centers for Disease Control). 1992. Recommendation for the use of folic acid to reduce the number of cases of spina bifida and other neural tube defects. *Morb. Mort. Wkly. Rpt.* 41: 1–7.

CDC (Centers for Disease Control). 2002. Rubella near elimination in the United States. Press release. http://www.cdc.gov/od/oc/media/pressrel/r020429.htm.

Chen, J. K., J. Taipale, M. K. Cooper and P. A. Beachy. 2002. Inhibition of hedgehog signaling by direct binding of cyclopamine to Smoothened. *Genes Dev.* 16: 2743–2748.

Chrisman, K., R. Kenney, J. Comin, T. Thal, L. Suchoki, Y. G. Yueh and D. F. Gardner. 2004. Gestational ethanol exposure disrupts the expression of *Fgf8* and *Sonic hedgehog* during limb patterning. *Birth Def. Res. A: Clin. Mol. Teratol.* 70: 163–171.

Christin, M. S. and 7 others. 2004. Effects of agricultural pesticides on the immune system of *Xenopus laevis* and *Rana pipiens. Aquat. Toxicol.* 67: 33–43.

Clarren, S. K. 1986. Neuropathology in the fetal alcohol syndrome. In J. R. West (ed.), *Alcohol and Brain Development*. Oxford University Press, New York.

Cohlan, S. Q. 1953. Excessive intake of vitamin A as a cause of congenital anomalies in the rat. *Science* 117: 535–537.

Cordero, D., R. Marcucio, D. Hu, W. Gaffield, M. Tapadia and J. A. Helms. 2004. Temporal perturbations in sonic hedgehog signaling elicit the spectrum of holoprosencephaly phenotypes. *J. Clin. Invest.* 114: 485–494.

Crawford, K. and D. M. Vincenti. 1998. Retinoic acid and thyroid hormone may function through similar and competitive pathways in regenerating axolotls. *J. Exp. Zool.* 282: 724–738.

Dally, A. 1998. Thalidomide: Was the tragedy preventable? *Lancet* 351: 1197–1199.

D'Amato, R. J., M. S. Loughnan, E. Flynn and J. Folkman. 1994. Thalidomide is an inhibitor of angiogenesis. *Proc. Natl. Acad. Sci. USA* 91: 4082–4085.

Davidson, C. and R. A. Knapp. 2007. Multiple stressors and amphibian declines: Dual impacts of pesticides and fish on yellow-legged frogs. *Ecol. Appl.* 17: 587–597.

Davis, W. L., L. A. Crawford, O. Cooper, G. R. Farmer, D. Thomas and B. L. Freeman. 1990. Ethanol induces the generation of reactive free radicals by neural crest cells in culture. *J. Craniofac. Genet. Dev. Biol.* 10: 277–293.

Dehart, D. B., L. Lanoue, G. S. Tint and K. K. Sulik. 1997. Pathogenesis of malformations in a rodent model for Smith-Lemli-Opitz syndrome. *Am. J. Med. Genet.* 68: 328–337.

Dwyer, J. B., R. S. Broides and F. M. Leslie. 2008. Nicotine and brain development. *Birth Def. Res. C: Embryo Today* 84: 30–44.

Edison, R. and M. Muenke. 2003. The interplay of environmental and genetic factors in craniofacial morphogenesis: Holoprosencephaly and the role of cholesterol. *Congen. Anom. Kyoto* 43: 1–21.

Edwards, M. 1994. Pollution in the former Soviet Union: Lethal legacy. *Natl. Geog.* 186(2): 70–115.

Eikel, D., A. Lampen and H. Nau. 2006. Teratogenic effects mediated by inhibition of histone deacetylases: Evidence from quantitative structure activity relationships of 20 valproic acid derivatives. *Chem. Res. Toxicol.* 19: 272–278.

Emanuele, M. A. and N. V. Emanuele. 1998. Alcohol's effects on male reproduction. *Alcoh. Health Res. World* 22: 195–201.

EPA (Environmental Protection Agency). 2007. Insect growth regulator fact sheet. http://epa.gov/oppbppd1/biopesticides/ingredients/factsheets/factsheet-igr.htm.

Eto, K. 2000. Minamata disease. *Neuropathology* 20 (suppl.): S14–S19.

Eto, K. and 7 others. 2001. Methylmercury poisoning in common marmosets: A study of selective vulnerability within the cerebral cortex. *Toxicol. Pathol.* 29: 565–573.

Finnell, R. H., B. C. Wlodarczyk, J. C. Craig, J. A. Piedrahita and G. D. Bennett. 1997. Strain-dependent alterations in the expression of folate pathway genes following teratogenic exposure to valproic acid in a mouse model. *Am. J. Med. Genet.* 70: 303–311.

Franco, B. and 12 others. 1995. A cluster of sulfatase genes on Xp22.3: Mutations in chondrodysplasia punctata (*CDPX*) and implications for warfarin embryopathy. *Cell* 81: 15–21.

Gardiner, D. and D. M. Hoppe. 1999. Environmentally induced limb malformations in mink frogs (*Rana septentrionalis*). *J. Exp. Zool.* 284: 207–216.

Gardiner, D., A. Ndayibagira, F. Grün and B. Blumberg. 2003. Deformed frogs and environmental retinoids. *Pure Appl. Chem.* 75: 2263–2273.

Gilbertson, M. 2004. Male cerebral palsy hospitalization as a potential indicator of neurological effects of methylmercury in Great Lakes communities. *Environ. Res.* 95: 375–384.

Giroud, A. and M. Martinet. 1959. Teratogenese pur hypervitaminose A chez le rat, la souris, le cobaye, et le lapin. *Arch. Fr. Pediatr.* 16: 971–980.

Golden, J. 2005. *Message in a Bottle: The Making of Fetal Alcohol Syndrome*. Harvard University Press, Cambridge, MA.

Goulding, E. H. and R. M. Pratt. 1986. Isotretinoin teratogenicity in mouse whole-embryo culture. *J. Craniofac. Genet. Dev. Biol.* 6: 99–112.

Green, M. L., A. V. Singh, Y. Zhang, K. A. Nemeth, K. K. Sulik and T. B. Knudson. 2007. Reprogramming of genetic networks during initiation of the fetal alcohol syndrome. *Dev. Dyn.* 236: 613–631.

Gregg, N. M. 1941. Congenital cataracts following German measles in the mother. *Trans. Opthalmol. Soc. Austral.* 3: 35.

Handu, S. S., H. Datta, A. Sankaranarayanan, H. James, K. Al-Kahja and R. P. Sequeira. 2008. Prenatal exposure to cocaine alters neurobehavioral developmental milestones in rats (Abstract). *Birth Def. Res. A: Clin. Mol. Teratol.* 82: 303.

Harada, M. 1972. *Minamata Disease.* Kumamoto Nichinichi Shinbun Center, Iwanami Shoten, Tokyo. (Trans. 2004 by T. S. George and T. Sachie.)

Harada, M. 1996. Characteristics of industrial poisoning and environmental contamination in developing countries. *Environ. Sci.* 4 (suppl.): S157–S169.

Harada, M., T. Fujino, T. Akagi and S. Nishigaki. 1976. Epidemiological and clinical study and historical background of mercury pollution on Indian reservations in northwestern Ontario, Canada. *Bull. Instit. Constitut. Med.* 26: 169–184.

Harmon, M. A., M. F. Boehm, R. A. Heyman and D. J. Mangelsdorf. 1995. Activation of mammalian retinoid-X receptors by the insect growth regulator methoprene. *Proc. Natl. Acad. Sci. USA* 92: 6157–6160.

Hoffman, L. M. and W. M. Kulyk. 1999. Alcohol promotes in vitro chondrogenesis in embryonic facial mesenchyme. *Internatl. J. Dev. Biol.* 43: 167–174.

Ingram, J. L., S. M. Peckham, B. Tisdale and P. M. Rodier. 2000. Prenatal exposure of rats to valproic acid reproduces the cerebellar anomalies associated with autism. *Neurotox. Teratol.* 22: 319–324.

International Joint Commission of the United States and Canada. 2000. *Tenth Biennial Report of Great Lakes Water Quality*. IJC, Ottawa.

Johnson, E. M. 1980. Screening for teratogenic potential: Are we asking the proper questions? *Teratology* 21: 259.

Johnston, M. C., K. K. Sulik, W. S. Webster and B. L. Jarvis. 1985. Isotretinoin embryopathy in a mouse model: Cranial neural crest involvement. *Teratology* 31: 26A.

Jones, K. L. and D. W. Smith. 1973. Recognition of the fetal alcohol syndrome. *Lancet* 2(7836): 999–1001.

Kalra, S., A. Einarson and G. Koren. 2005. Taking antidepressants during late pregnancy: How should we advise women? *Can. Fam. Physician.* 51: 1077–1078.

Keeler, R. F. and W. Binns. 1968. Teratogenic compounds of *Veratrum californicum* (Durand). V. Comparison of cyclopian effects of steroidal alkaloids from the plant and structurally related compounds from other sources. *Teratology* 1: 5–10.

Kessel, M. 1992. Respecification of vertebral identities by retinoic acid. *Development* 115: 487–501.

Kiesecker, J. M. 2002. Synergism between trematode infection and pesticide exposure: A link to amphibian limb deformities in nature? *Proc. Natl. Acad. Sci. USA* 99: 9900–9904.

Kleinman, J. C., M. B. Pierre, Jr., J. H. Madans, G. H. Land and W. E. Schramm. 1988. The effects of maternal smoking on fetal and infant mortality. *J. Epidemiol.* 127: 274–282.

Knobloch, J., J. D. Shaughnessy, Jr. and U. Ruther. 2007. Thalidomide induces limb deformities by perturbing the Bmp/Dkk1/Wnt signaling pathway. *FASEB J.* 21: 1410–1421.

Kochhar, D. M., J. D. Penner and C. Tellone. 1984. Comparative teratogenic activities of two retinoids: Effects on palate and limb development. *Teratogen. Carcinogen. Mutagen.* 4: 377–387.

Kondo, K. 2000. Congenital Minamata disease: Warnings from Japan's experience. *J. Child Neurol.* 15: 458–464.

Koren, G., A. Pastuszak, S. Jacobson and I. Nulman. 2001. The safety of commonly used antidepressants in pregnancy. In G. Koren (ed.), *Maternal-Fetal Toxicology: A Physician's Guide*. Marcel Dekker, New York, pp. 85–104.

Kotch, L. E., S.-Y. Chen and K. K. Sulik. 1995. Ethanol-induced teratogenesis: Free radical damage as a possible mechanism. *Teratology* 52: 128–136.

Kulikauskas, V., A. B. Blaustein and R. J. Ablin. 1985. Cigarette smoking and its possible effects on sperm. *Fertil. Steril.* 44: 526–528.

Kunert, H. J. and F. Loehrer. 2008. Neuroscientific aspects of chronic cannabis abuse: The impact of age of onset (Abstract). *Birth Def. Res. A: Clin. Mol. Teratol.* 82: 350.

La Claire, J., J. A. Bantle and J. Dumont. 1998. Photoproducts and metabolites of a common insect growth regulator produce developmental deformities in *Xenopus*. *Environ. Sci. Technol.* 32: 1453–1461.

Lammer, E. J. and 11 others. 1985. Retinoic acid embryopathy. *New Engl. J. Med.* 313: 837–841.

Lannoo, M. 2008. *Malformed Frogs: The Collapse of Aquatic Ecosystems.* University of California Press, Berkeley.

Lemoine, E. M., J. P. Harousseau, J. P. Borteyru and J. C. Menuet. 1968. Les enfants de parents alcoholiques: Anomalies observées. *Oest. Med.* 21: 476–482.

Lenkowski, J. R., J. M. Reed, L. Deninger and K. A. McLaughlin. 2008. Perturbation of organogenesis by the herbicide atrazine in the amphibian *Xenopus laevis*. *Environ. Health Persp.* 116: 223–230.

Lenz, W. 1962. Thalidomide and congenital abnormalities. *Lancet* 1: 45. (Reported in a symposium in 1961.)

Lenz, W. 1966. Malformations caused by drugs in pregnancy. *Amer. J. Dis. Child.* 112: 99–106.

Mak, V., K. Jarvi, M. Buckspan, M. Freeman, S. Hechter and A. Zini. 2000. Smoking is associated with the retention of cytoplasm by human spermatozoa. *Urology* 56: 463–466.

Maritz, G. S. 2008. Nicotine and lung development. *Birth Def. Res. C: Embryo Today* 84: 45–53.

May, P. A. and J. P. Gossage. 2001. Estimating the prevalence of fetal alcohol syndrome: A summary. *Alch. Res. Health* 25: 159–167.

McCoy, K. A., L. J. Bortnick, C. M. Campbell, H. J. Hamlin, L. J. Guillette Jr. and C. M. St. Mary. 2008. Agriculture alters gonadal form and function in the toad *Bufo marinus*. *Environ. Health Persp.*, in press.

Meador, K. J. and 10 others. 2006. In utero antiepileptic drug exposure: Fetal death and malformations. *Neurology* 67: 407–412.

Menegola, E., F. Di Renzo, M. L. Broccia, M. Prudenziati, S. Minucci, V. Massa and E. Giavini. 2005. Inhibition of histone deacetylase activity on specific embryonic tissues as a new mechanism for teratogenicity. *Birth Def. Res. B: Dev. Reprod. Toxicol.* 74: 392–398.

Meteyer, C. U. and 9 others. 2000. Hindlimb malformations in free-living northern leopard frogs (*Rana pipiens*) from Maine, Minnesota, and Vermont suggest multiple etiologies. *Teratology* 62: 151–171.

Miller, M. T. and K. Strömland. 1999. Thalidomide: A review, with a focus on ocular findings and new potential uses. *Teratology* 60: 306–321.

Miller, M. T., K. Strömland, L. Ventura, M. Johansson, J. M. Bandim and C. Gillberg. 2005. Autism associated with conditions characterized by developmental errors in early embryogenesis: A minireview. *Internatl. J. Dev. Neurosci.* 23: 201–219.

Moore, K. L. and T. N. N. Persaud. 1993. *Before We Are Born: Essentials of Embryology and Birth Defects.* W. B. Saunders, Philadelphia.

Moore, S. J., P. Turnpenny, A. Quinn, S. Glover, D. J. Lloyd, T. Montgomery and J. C. S. Dean. 2000. A clinical study of 57 children with fetal anticonvulsant syndrome. *J. Med. Genet.* 37: 489–497.

Moroni, M. C., M. A. Vigano and F. Mavilio. 1994. Regulation of human *Hoxd-4* gene by retinoids. *Mech. Dev.* 44: 139–154.

Morriss-Kay, G. M. 1993. Retinoic acid and craniofacial development: Molecules and morphogenesis. *Bioessays* 15: 9–15.

Morrow, E. M. and 24 others. 2008. Identifying autism loci and genes by tracing recent shared ancestry. *Science* 321: 218–223.

Muenke, M. 1995. Holoprosencephaly as a genetic model for normal craniofacial development. *Semin. Dev. Biol.* 5: 293–301.

NARCAM (North American Reporting Center for Amphibian Malformations). 2002. Northern Prairie Wildlife Research Center, Jamestown, ND. http://www.nprwc.usgs.gov/narcam.

Narita, N., M. Kato, M. Tazoe, M. Miyazake, M. Narita and N. Okado. 2002. Increased monoamine concentration in brain and blood of fetal thalidomide- and valproic acid-exposed rat: Putative animal models for autism. *Pediat. Res.* 52: 576–579.

Nowack, E. 1965. Die sensible Phase bei der Thalidomide-Embryopathie. *Humangenetik* 1: 516–536.

Nulman, I. and 8 others. 1997. Neurodevelopment of children exposed in utero to antidepressant drugs. *New Engl. J. Med.* 336: 258–262.

O'Hara, M. F., J. H. Charelap, R. C. Craig and T. B. Knudsen. 2002. Mitochondrial transduction of ocular teratogenesis during methylmercury exposure. *Teratology* 65: 131–144.

Oppenheimer, J. M. 1968. Some historical relationships between teratology and experimental embryology. *Bull. Hist. Med.* 42: 145–159.

Opitz, J., R. Schultka and L. Göbbel. 2006. Meckel on developmental pathology. *Am. J. Med. Genet. A* 140: 115–128.

OTIS (Organization of Teratology Information Specialists). 2007. http://otispregnancy.org/pdf/hyperthermia.pdf.

Ouellet, M., J. Bonin, J. Rodriguez, J. L. DesGanges and S. Lair. 2007. Hindlimb deformities (ectromelia, ectrodactyly) in free-living anurans from agricultural habitats. *J. Wildlife Disord.* 33: 95–104.

Pauly, P. J. 1996. How did the effects of alcohol on reproduction become scientifically uninteresting? *J. Hist. Biol.* 29: 1–28.

Pratt, G. E., R. C. Jennings, A. F. Hammett and G. T. Brooks. 1980. Lethal metabolism of precocene-I to a reactive epoxide by locust corpora allata. *Nature* 284: 320–323.

Raje, N. and K. Anderson. 1999. Thalidomide: A revival story. *New Engl. J. Med.* 341: 1606–1609.

Ramanathan, R., M. F. Wilkemeyer, B. Mittel, G. Perides and M. E. Charness. 1996. Alcohol inhibits cell-cell adhesion mediated by human L1. *J. Cell Biol.* 133: 381–390.

Rinaldi, T., G. Silverberg and H. Markram. 2008. Hyperconnectivity of local neocortical microcircuitry induced by prenatal exposure to valproic acid. *Cereb. Cort.* 18: 763–770.

Roessler, E., E. Belloni, K. Gaudenz, P. Jay, P. Berta, S. W. Scherer, L.-C. Tsui and M. Muenke. 1996. Mutations in the human *Sonic hedgehog* gene cause holoprosencephaly. *Nature Genet.* 14: 357–360.

Rothman, K. J., L. L. Moore, M. R. Singer, U. S. Nguyen, S. Mannino and A. Milunsky. 1995. Teratogenicity of high vitamin A intake. *New Engl. J. Med.* 333: 1369–1373.

Saxén, L., P. C. Holmberg, M. Nurminen and E. Kuosma. 1982. Sauna and congenital defects. *Teratology* 25: 309–313.

Schooneveld, H. 1979. Precocene-induced collapse and resorption of corpora allata in nymphs of *Locusta migratoria*. *Experientia* 35: 363–364.

Shi, Q., E. Ko, L. Barclay, T. Hoang, A. Rademaker and R. Martin. 2001. Cigarette smoking and aneuploidy in human sperm. *Mol. Reprod. Dev.* 59: 417–421.

Sláma, K. and C. M. Williams. 1966. The juvenile hormone. V. The sensitivity of the bug *Pyrrhocoris apterus* to a hormonally active factor in American paper-pulp. *Biol. Bull.* 130: 235–246.

Stocum, D. 2000. Frog limb deformities: An "eco-devo" riddle wrapped in multiple hypotheses surrounded by insufficient data. *Teratology* 62: 147–150.

Stodgell, C. J., J. L. Ingram, M. O'bara, B. K. Tisdale, H. Nau and P. M. Rodier. 2006. Induction of the homeotic gene *Hoxa1* through valproic acid's teratogenic mechanism of action. *Neurotoxicol. Teratol.* 28: 617–624.

Stopper, G. F., L. Hecker, R. A. Franssen and S. K. Sessions. 2002. How trematodes cause limb deformities in amphibians. *J. Exp. Zool.* 294: 252–263.

Streissguth, A. P. and R. A. LaDue. 1987. Fetal alcohol: Teratogenic causes of developmental disabilities. In S. R. Schroeder (ed.), *Toxic Substances and Mental Retardation.* American Association of Mental Deficiency, Washington, DC, pp. 1–32.

Studer, M., H. Pöpperl, H. Marshall, A. Kuroiwa and R. Krumlauf. 1994. Role of a conserved retinoic acid response element in rhombomere restriction of *Hoxb1*. *Science* 265: 1728–1732.

Sulik, K. K. 2005. Genesis of alcohol-induced craniofacial dysmorphism. *Exp. Biol. Med.* 230: 366–375.

Sulik, K. K., C. S. Cook and W. S. Webster. 1988. Teratogens and craniofacial malformations: Relationships to cell death. *Development* 103 (suppl.): S213–S231.

Thorogood, P. 1997. The relationship between genotype and phenotype: Some basic concepts. In P. Thorogood (ed.), *Embryos, Genes, and Birth Defects*. Wiley, New York, pp. 1–16.

Tischfield, M. A. and 10 others. 2005. Homozygous *HOXA1* mutations disrupt human brainstem, inner ear, cardiovascular, and cognitive development. *Nature Genet*. 37: 1035–1037.

Toms, D. A. 1962. Thalidomide and congenital abnormalities. *Lancet* 2: 400.

Traiffort, E. and 7 others. 2004. Functional characterization of *SHH* mutations associated with holoprosencephaly. *J. Biol. Chem.* 279: 42889–42897

Ui, J. 1992. *Industrial Pollution in Japan*. United Nations University Press, Tokyo. http://www.unu.edu/unupress/unupbooks/uu35ie/uu35ie00.htm#Contents.

Wang, L. and K. E. Pinkerton. 2008. Detrimental effects of tobacco smoke exposure during development on postnatal lung function and asthma. *Birth Def. Res. C: Embryo Today* 84: 54 – 60.

Warren, K. R. and T.-K. Li. 2005. Genetic polymorphisms: Impact on the risk of fetal alcohol spectrum disorders. *Birth Def. Res. A: Clin. Mol. Teratol.* 73: 195–203.

Werler, M. M. 1997. Teratogen update: Smoking and reproductive outcomes. *Teratology* 55: 382–388.

Williams, C. M. 1970. Hormonal interactions between plants and insects. In E. Sondheimer and J. B. Simeone (eds.), *Chemical Ecology*. Academic Press, New York, pp. 103–132.

Wilson, J. G. 1961. General principles in experimental teratology. In M. Fishbein (ed.), *Proceedings of the First International Conference on Congenital Malformations*. Lippincott, Philadelphia.

Wilson, J. G. (ed.) 1973. *Environment and Birth Defects*. Academic Press, London.

Xu, Y., P. Liu and Y. Li. 2005. Impaired development of mitochondria plays a role in the central nervous system defects of fetal alcohol syndrome. *Birth Def. Res. A: Clin. Mol. Teratol.* 73: 83–91.

Yabu, T., H. Tomimoto, Y. Taguchi, S. Yamaoka, Y. Igarishi and T. Okazaki. 2005. Thalidomide-induced anti-angiogenic action is mediated by ceramide through depletion of VEGF receptors, and is antagonized by sphingosine-1-phosphate. *Blood* 106: 125–134.

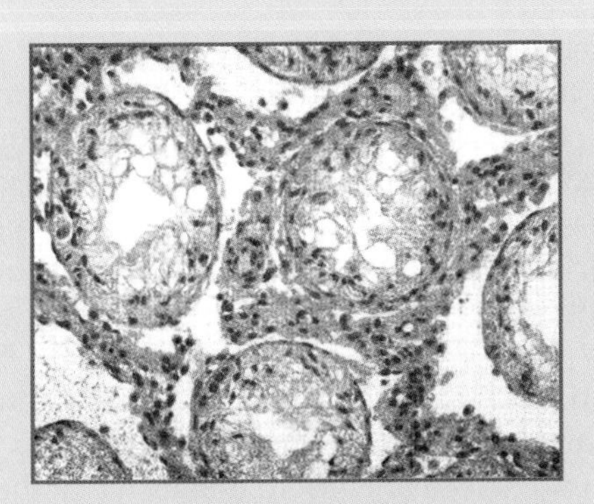

Chapter 6

Endocrine Disruptors

And I brought you into a plentiful land, to eat the fruit thereof and the goodness thereof; but when ye entered, ye defiled my land, and made my heritage an abomination.

Jeremiah 2:7, quoted in Lannoo 2008

Suddenly, I'm not half the man I used to be.

Paul McCartney and John Lennon, 1965

One of the most active and controversial areas of teratology concerns the misregulation of the endocrine system during development. Such studies have led to the **endocrine disruptor hypothesis**, which states that hormonally active molecular compounds in the environment—endocrine disruptors—alter gene expression during early development in ways that have a significant impact on the health of human and wildlife populations. The changes produced by endocrine disruptors during development are not the obvious anatomical malformations classically produced by the teratogens described in Chapter 5; indeed, the anatomical alterations induced by endocrine disruptors are often visible only with a microscope. Rather, the major changes are physiological, and in many cases the aberrant phenotypes are not seen until adulthood. These functional changes are more subtle than the visibly aberrant phenotypes produced by teratogens, but they nevertheless can be extremely important phenotypic alterations. Moreover, in some instances, the alterations can persist in offspring for generations after an organism's exposure to the disruptor.

Within the paradigms of teratology, it was once thought that there were only a few "bad" agents and that people were exposed to dangerous doses of these agents only inadvertently, by occupational exposure, or by accident. However, we now recognize that hormone-disrupting chemicals are everywhere in our technological society. Many ubiquitous products—including plastics, cosmetics, and pesticides—contain endocrine disruptors, and animal studies have shown that even low-dose in utero exposure to these chemicals can produce major disabilities later in life. While their manufacturers did not realize these chemicals would affect organisms in such unanticipated ways, the chemicals' rapid introduction and wide use has involuntarily exposed recent generations of embryos to a wide range of endocrine disruptors. Because these compounds are not as obviously toxic as some of the classic teratogens, it required detective work to establish that they could indeed disrupt the developing endocrine system. We now know that the list of endocrine disruptors includes chemicals that line baby bottles and that leach from the brightly colored plastic of the water containers we drink from while exercising; chemicals found in cosmetics such as sunblocks and hair rinses; chemical coatings that prevent clothing and bed linens from being highly flammable; compounds responsible for the flexible softness of vinyl toys; and even the substances responsible for air freshener fragrances and that "new car smell" (see Aitken et al. 2004; Schlumpf et al. 2004).

The Nature of Endocrine Disruptors

The term "endocrine disruptor" was proposed by Theo Colborn, Frederic vom Saal, and Ana Soto in 1993. These compounds are also known by other names: hormone mimics, environmental signal modulators, environmental estrogens, or hormonally active agents. Endocrine disruptors are exogenous (i.e., arising outside the body) chemicals that interfere with the normal functions of hormones and consequently disrupt development (Colborn et al. 1993, 1996). Endocrine disruptors can interfere with hormonal functions in many ways:

- Endocrine disruptors can be **agonists**, mimicking the effect of a natural hormone and binding to its receptors. An example of an agonist is the paradigmatic endocrine disruptor diethylstilbestrol (DES), which mimics the sex hormone estradiol (a common form of estrogen) and binds to the body's estrogen receptors.
- Endocrine disruptors can act as **antagonists**, either preventing the binding of a hormone to its natural receptor or blocking the hormone's synthesis. DDE, a metabolic product of the insecticide DDT, can act as an anti-androgen (i.e., an antagonist of masculinizing hormones) by binding to the androgen receptor and preventing normal testosterone from functioning properly.
- Endocrine disruptors can increase hormone synthesis. Compounds such as the herbicide atrazine (see Chapter 2 and later in this chapter),

for instance, act by elevating the synthesis of certain hormones (in this case, estrogen via induction of the aromatase enzyme).

- Endocrine disruptors can affect the elimination or transportation of a hormone within the body. One of the ways that the polychlorinated biphenyls (PCBs) disrupt the endocrine system is to interfere with the elimination and degradation of thyroid hormones. Several pesticides may act indirectly to affect hormone levels by inducing the synthesis of liver enzymes involved in the synthesis and degradation of certain hormones (see Guillette 2006).
- Endocrine disruptors can "prime" the organism to be more sensitive to hormones later in life. For instance, when a rat fetus is exposed to bisphenol A (BPA), the embryonic mammary gland is induced to make more estrogen receptors. These extra receptors alter mammary gland growth responses to natural estrogen later in life, predisposing breast tissue to cancer formation (see Wadia et al. 2007; Soto et al. 2008).

The endocrine disruptor hypothesis

The roots of the endocrine disruptor hypothesis can be found in Rachel Carson's 1962 book *Silent Spring*, which documented the effects of DDT and other insecticides on reproductive failures in birds and other wildlife. Within the next decade, endocrine disruption in human populations was documented when diethylstilbestrol (DES), a drug commonly prescribed during pregnancy, was found to cause reproductive abnormalities and also a rare form of carcinoma of the vagina. The drug's effects, however, were seen not in the mothers, but rather in their daughters, who were exposed to DES in utero when the mothers took the drug (see later in this chapter). The widespread occurrence of endocrine disruption was more fully appreciated when endocrinologist John A. McLachlan and wildlife biologist Theo Colborn synthesized the vast literature on the subject and published scientific and popular articles concerning reproductive failures and developmental anomalies (Colborn et al. 1996; Krimsky 2000).

According to Jerrold Heindel of the National Institute of Environmental Health Sciences, the endocrine disruptor hypothesis has had a difficult time being accepted in mainstream medical circles because its paradigms are so different from those of infectious disease, trauma relief, or toxicology—the usual Western approaches to medical care (Heindel 2006). The following insights point to the importance of thinking in terms of ecology and developmental biology in filling the gaps in our medical knowledge:

- Endocrine disruption is a functional change in a tissue that superficially appears normal. The pathology may be evident only upon microscopic examination, or it may manifest solely as a change in gene expression.

- The mechanism of pathology is not infection, injury, or toxicity. Rather, the disease is likely to result from altered gene expression caused by the environmental agent, resulting in altered morphogenesis leading to dysfunctional physiology.
- Sensitivity to disruptive agents will depend on the context of the exposure—it will depend on the stage of development, the dose of the agent, and even the sex of the exposed individual.
- The environmental insult may be additive or synergistic with nutritional influences and is influenced by the exposed organism's genetic background.
- The effect can be transgenerational. That is to say, the agents may cause the changes in gene expression throughout the embryo, including the newly formed germ cells. If the germ cells are altered, the effect can be transmitted to the next generation.

This chapter focuses on a set of endocrine disruptors that interact with the reproductive hormone pathways: these include diethylstilbestrol (DES), bisphenol A (BPA), nonylphenol, atrazine, and the polychlorinated biphenyls (PCBs). This is the area of endocrine disruption that has been best studied and may be the most medically relevant.

DDT: The start of it all

Silent Spring was one of the most influential books of the twentieth century. Its author, fisheries biologist Rachel Carson, warned that pesticides were destroying wildlife, that DDT in particular appeared to be exterminating shorebird populations, and that pesticides were becoming a staple of the American diet. For this she was reviled by the agricultural chemicals industry and called a fanatic, a Communist, and worse (see Lear 1998; Orlando 2002). But subsequent research bore out Carson's claims and revealed the first evidence of endocrine disruption due to exposure to environmental chemicals.

The chemical components of DDT (dichloro-diphenyl-trichloroethane; Figure 6.1A) cannot be broken down and eliminated by vertebrate organisms; it remains in their bodies and builds up, becoming especially concentrated in organisms that feed on the DDT-containing tissues of other animals. This persistence, along with its solubility in lipids, results in tremendous **bioaccumulation** such that even though DDT has not been legally used in the United States since 1972, most of us still have this chemical in our bodies. Indeed, it is so resistant to metabolic degradation, and so much was manufactured, that it remains not only in humans and other terrestrial animals but also in fishes, marine mammals, and seabirds. It has an environmental half-life of about 15 years, which means it can take 100 years or more for the concentration of DDT in the soil to fall below active levels.

Bioaccumulation of DDT was especially pronounced in some birds of prey living at the top of the food chain. Peregrine falcons and bald eagles became endangered because of DDT-induced fragility of their eggshells (Cooke 1973). Even seabirds were affected when DDT in runoff accumulated in the fish they fed on. Fragile eggshells resulted in high mortality as the developing bird embryos became desiccated, were easily preyed upon, and often were not able to withstand even minor physical forces (Figure 6.1B).

Discovering why DDT contamination results in thin eggshells was a formidable research challenge. It turns out that DDT acts as an estrogenic compound (see the next section), while its chief metabolic product, DDE (which lacks one of DDT's chlorine atoms), inhibits androgens such as testosterone from binding to the androgen receptor (Davis et al. 1993; Kelce et al. 1995; Xu et al. 2006). Eggshell thinning is caused by several actions of DDT and DDE. Hens with high DDT levels had poorly developed shell glands, with capillaries deficient in carbonic anhydrase—an enzyme critical for the deposition of shell-strengthening calcium carbonate in the egg (Holm et al.

(A)

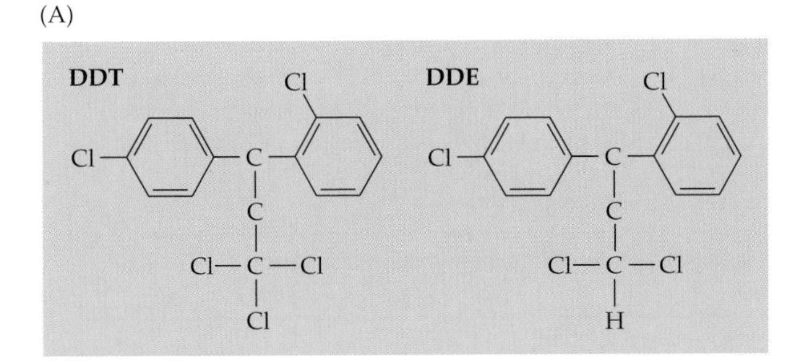

(B)

FIGURE 6.1 Effects of pesticide bioaccumulation. (A) Chemical structures of DDT (dichloro-diphenyl-trichloroethane) and its metabolic by-product, DDE. DDT is an estrogenic compound, while DDE is an androgen inhibitor. (B) One notable effect of environmental bioaccumulation of these chemicals was the prevalence of thin, nonviable eggshells found among many bird species (particularly birds of prey), with subsequent severe population declines. This brown pelican egg cracked open long before the embryo inside was ready to hatch. (Photograph © L. Kiff/Visuals Unlimited.)

Establishing a Chain of Causation

DDT and DDE have been linked not only to fragile eggshells in birds but also to other incidents of reproductive failure in wildlife. In the 1990s, researchers linked a pollutant spill in Florida's Lake Apopka (a discharge including DDT, DDE, and numerous polychlorinated biphenyls) to a 90% decline in the birth rate of alligators and reduced penis size in young males (Guillette et al. 1994; Matter et al. 1998). Indeed, this work revealed how the interaction of many chemicals at once can result in unanticipated consequences on reproduction.

The Lake Apopka study is an example of how environmental toxicology research establishes **chains of causation**. Whether in law or in science, establishing such chains is a demanding and necessary task. In environmental toxicology, numerous end points must be checked and many different levels of causation must be established (Crain and Guillette 1998; McNabb et al. 1999). For example, to establish whether the pollutant spill in Lake Apopka was responsible for the decline in the juvenile alligator population there, researchers first had to establish exactly *why* the population was declining, then follow through with various studies to show how the specific chemicals present in the spill could contribute to the observed reasons for the decline. The table shows the postulated chain of causation, from overall observations of the entire population through the molecular biochemistry.

As seen in the accompanying table, *population level* studies revealed that the Lake Apopka population decline was the result of a decrease in the number of alligators being born (the birth rate). At the *organism level*, researchers found unusually high levels of estrogens in juvenile female alligators and unusually low levels of testosterone in juvenile males. On the *tissue and organ level*, they observed elevated production of estrogens in juvenile male testes, along with penis and testis malformations, and changes in enzyme activity in the female gonads. On the *cellular level*, ovarian abnormalities in females correlated with elevated estrogen levels. These cellular changes could be explained at the *molecular level* by noting that many of the chemicals

Chain of causation linking contaminant spill in Lake Apopka to endocrine disruption in juvenile alligators

Level	Evidence
Population	The juvenile alligator population in Lake Apopka has decreased.
Organism	Juvenile Apopka (JA) females have elevated circulating levels of estradiol-17β.
	JA males have depressed circulating concentrations of testosterone.
Tissue/organ	JA females have altered gonad aromatase activity.
	JA males have poorly organized seminiferous tubules.
	JA males have reduced penis size.
	Testes from JA males have elevated estradiol (estrogen) production.
Cellular	JA females have polyovular follicles that are characteristic of estrogen excess.
Molecular	Many contaminants bind the alligator estrogen receptors and progesterone receptor.
	Many of these contaminants do not bind to the alligator cytosol proteins that blockade excess hormones.

Source: After Crain and Guillette 1998.

in the spill bind to alligator estrogen and progesterone receptors; that these chemicals are able to circumvent the cell's usual defenses against the overproduction of steroid hormones; and that the alligators exposed to these chemicals had altered expression patterns of those genes encoding proteins used in sex determination and steroid hormone production (Crain et al. 1998; Guillette et al. 2007; Kohno et al. 2008). Whereas newly hatched alligators from an uncontaminated lake (the controls) had sexually dimorphic expression of several of these genes, in newly hatched alligators from the contaminated lake, the expression of these genes was the same in both males and females (Milnes et al. 2008). Thus, the pollutant chemicals can be linked by a logical chain of events to reproductive anomalies that could explain the decreased birth rate among alligators in the lake.

In many instances, the "chain" forms a circle, in which endocrine disruptor chemicals from human activities enter the environment, the pollutant chemicals affect the ecosystem, and the environmental pollution returns to affect the human population. Such work shows that even though many of the candidate disruptors are compounds referred to as "pesticides," what constitutes a "pest" is more a value judgment than a scientific designation. We share enzymes and receptor molecules with every animal on the planet, and our mammalian hormones are especially similar to those of other vertebrates.

2006). High DDE levels in the shell gland also prevent calcium carbonate deposition by downregulating the synthesis of prostaglandins (a group of fatty acid derivatives that regulates many vertebrate physiological processes). In birds, one of these prostaglandins is critical for the transport of calcium ions through the shell gland for use in shell formation (Lundholm 1997; Guillette 2006).

DDT use has been banned in the United States since 1972, and is also banned in most of Europe. The Stockholm Convention of 2001, which attempts a worldwide ban on the use of several dangerously persistent organic chemicals, limits the use of DDT to the control of disease-carrying insects (notably mosquitoes), and only in those countries where there is no affordable and/or comparably effective alternative for managing a serious public health threat (primarily malaria; see Chapter 8).

Estrogen and Endocrine Disruptors

Some of the best studied endocrine disruptors are those that mimic or block the actions of **estrogens**, a family of steroid hormones that regulate growth, differentiation, and function of reproductive tissues and of other organs such as the bones, brain, and cardiovascular organs (Brosens and Parker 2003). Although estrogen is popularly known as the female sex hormone, both sexes need estrogens for proper bone and connective tissue development.

We are continuously exposed to estrogen-like compounds in the environment. These compounds are often called **xenoestrogens** since they are foreign to the body (Greek, *xenos*, "foreigner"). The water we drink and the food we eat contain xenoestrogens. Although some of these compounds, such as the phytoestrogens (plant estrogens) in soy products, are from nat-

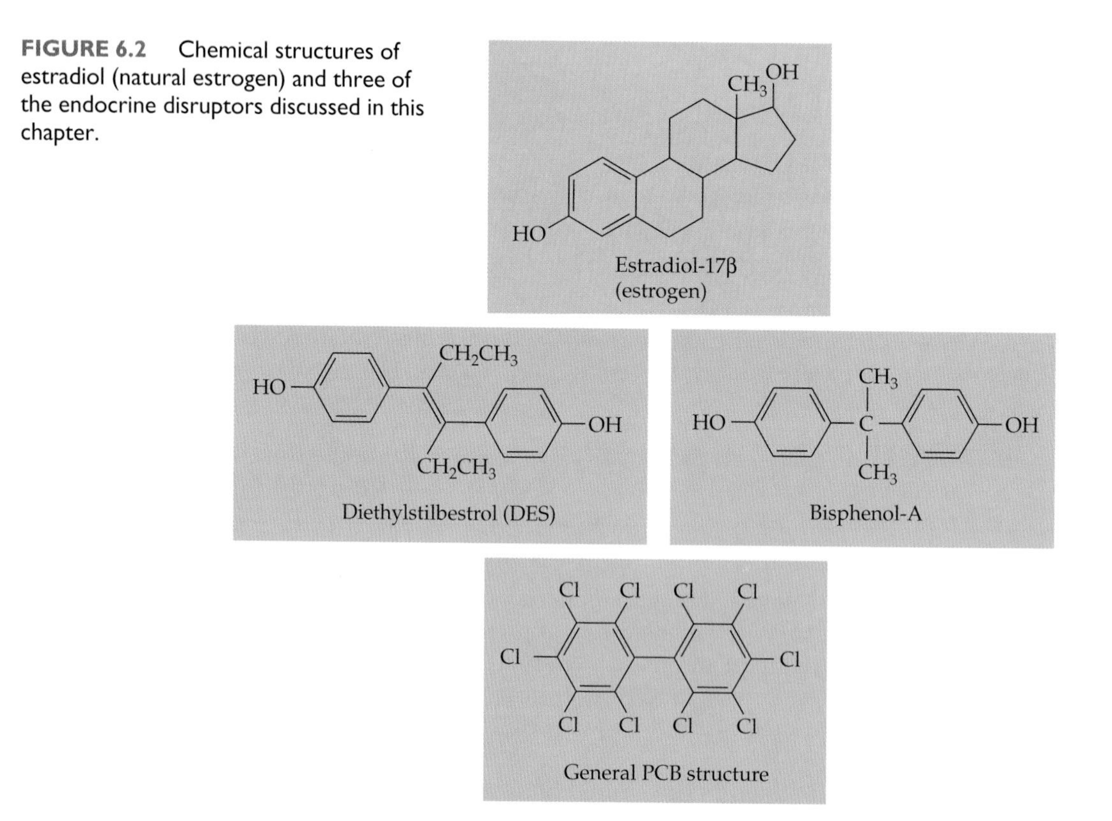

FIGURE 6.2 Chemical structures of estradiol (natural estrogen) and three of the endocrine disruptors discussed in this chapter.

ural sources, modern technological industries have manufactured an estrogenic environment. The plastics that store our water and sodas are slowly leaching these compounds into our drinks. Because the estrogen receptors in our bodies bind to so many of the compounds used in plastics production, we now develop in and live in a world full of unanticipated estrogenic structures (Figure 6.2; Colborn et al. 1996).

When present early in development, xenoestrogens have consequences for health later in life. Indeed, Heindel (2007) has pointed to a "developmental estrogen syndrome" wherein exposure to estrogenic compounds early in development can cause both men and women to experience fertility problems, cancers, and obesity later in life.

The structure and mechanisms of estrogen receptors

The two major estrogen receptor molecules, ERα and ERβ, are nuclear transcription factors that are activated by binding ligands such as 17β-estradiol (the most common estrogen). Once they have bound the estrogen, the estrogen receptors dimerize (pair up and fuse) and bind to the regulatory regions of estrogen-sensitive genes. Here the estrogen-bound receptors interact with other transcription factors to regulate the expression of these genes. A third estrogen receptor is related to ERα and is located in the plasma membrane (Powell et al. 2001), where it binds xenoestrogens at very low (picomolar) concentrations. When this receptor binds estrogens or xenoestrogens, it activates different types of signaling cascades, such as those opening calcium channels (Wozniak et al. 2005; Watson et al. 2007). Recent studies also suggest that this membrane-bound estrogen receptor is capable of binding DDE, suggesting an estrogen-like agonist activity of this chemical in addition to DDE's effect as an anti-androgen (Thomas et al. 2006).

The nuclear estrogen receptor proteins have three major domains. The first is the estrogen-binding domain, which binds the hormone. The second is a protein-interaction domain that allows the estrogen receptor to interact with other proteins (including other estrogen receptors). The third domain of the estrogen receptor binds to specific DNA sequences (the estrogen-responsive elements, or EREs) and promotes the transcription of the genes adjacent to these regions. When an estrogen molecule binds to its receptor protein, the receptor protein breaks its connections to several other proteins and combines with another bound estrogen receptor (dimerizes). The paired and hormone-bound estrogen receptors bind to the estrogen response elements in the enhancers of specific genes, where they control the transcription of these genes and generate a cascade of cellular actions (Figure 6.3; Brosens and Parker 2003).

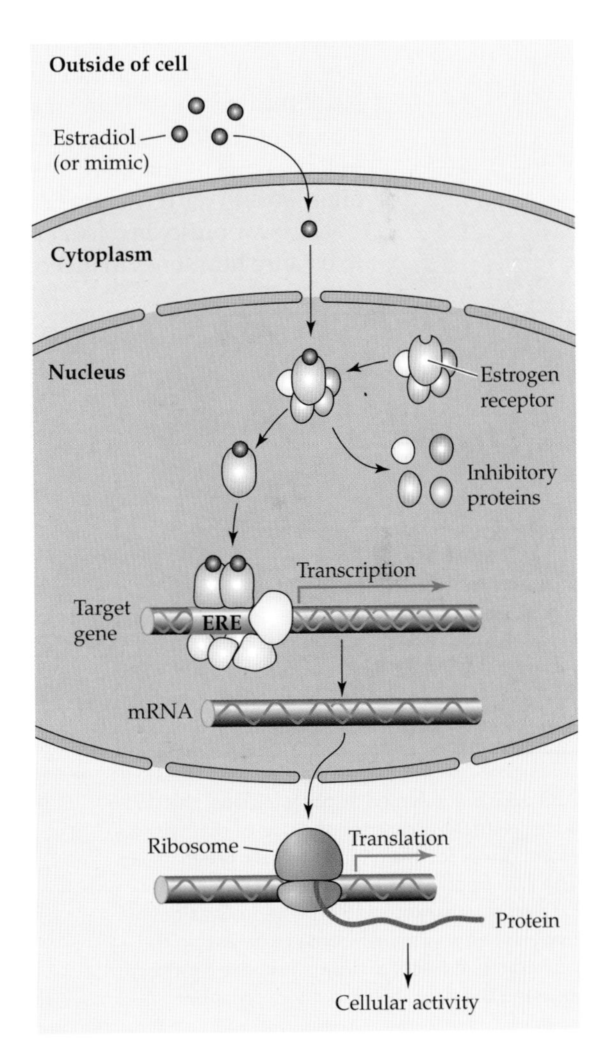

FIGURE 6.3 Mechanism of the estrogen receptor within a general cell nucleus. The estrogen receptor is a protein that can bind estrogen (or compounds resembling estrogen). It is usually bound to numerous other proteins that prevent its binding to DNA. When the receptor is activated by hormonal ligands such as estradiol or an estrogen mimic, the inhibitory proteins leave, and the activated receptors then bind to DNA sequences (estrogen-responsive elements, or EREs). These EREs are in the enhancers of specific genes, where they control the transcription of target genes and generate a cascade of cellular actions.

Some endocrine disruptors can similarly bind to estrogen receptors. This can occur because, unlike most other receptors, the hormone-binding domain of the estrogen receptor can bind numerous differently shaped molecules (see Figure 6.2). Some endocrine disruptors are estrogen agonists and bind in a manner that mimics the effects of estrogen (which is not a good thing if it occurs during times of development when estrogens are not normally produced). Other disruptors are estrogen antagonists and can enter the hormone-binding site of the receptor in a way that prevents it from functioning.

Diethylstilbestrol

One of the best studied of the xenoestrogens is the drug **diethylstilbestrol**, or **DES**. DES provides a paradigmatic example of an endocrine disruptor (see Bell 1986 and Palmund 1996). Indeed, Sir Charles Dodds first synthesized the drug in 1938 in an attempt to obtain inexpensive synthetic estrogens that could be used to study how this hormone acts. It was later marketed as one of the many "miracle drugs" that became available shortly after World War II (Figure 6.4).

DES was prescribed for several reasons; it was used to suppress lactation, to balance hormones in menopausal women, and to rectify hormone imbalances that were thought to contribute to the premature termination of pregnancies (NRC 1999; Krimsky 2000). At the time DES was first marketed, the idea that a drug could damage the fetus, or that it could produce effects that would not show up until decades later, was not seriously considered.

FIGURE 6.4 Aimed at obstetricians, this 1956 advertisement promoted the use of diethylstilbestrol (DES) for maintaining at-risk pregnancies. The U.S. Food and Drug Administration took DES off the market in 1971, when it was definitively linked to a specific type of cancer in the reproductive tracts of women whose mothers ingested the drug during early pregnancy.

It is estimated that in the United States, over a million fetuses were exposed to DES between 1947 and 1971, and this is a small fraction of exposures worldwide. In addition to DES administered to pregnant women, biologically relevant levels of DES were present in meat, since the drug was also used to accelerate livestock growth (Knight 1980). Even though research done in the 1950s showed that in fact DES had *no* effect on the maintenance of a pregnancy, the drug continued to be prescribed right up until the U.S. Food and Drug Administration banned its use in 1971. The ban was instituted when a specific type of tumor—clear-cell adenocarcinoma—was discovered in the reproductive tracts of women whose moth-

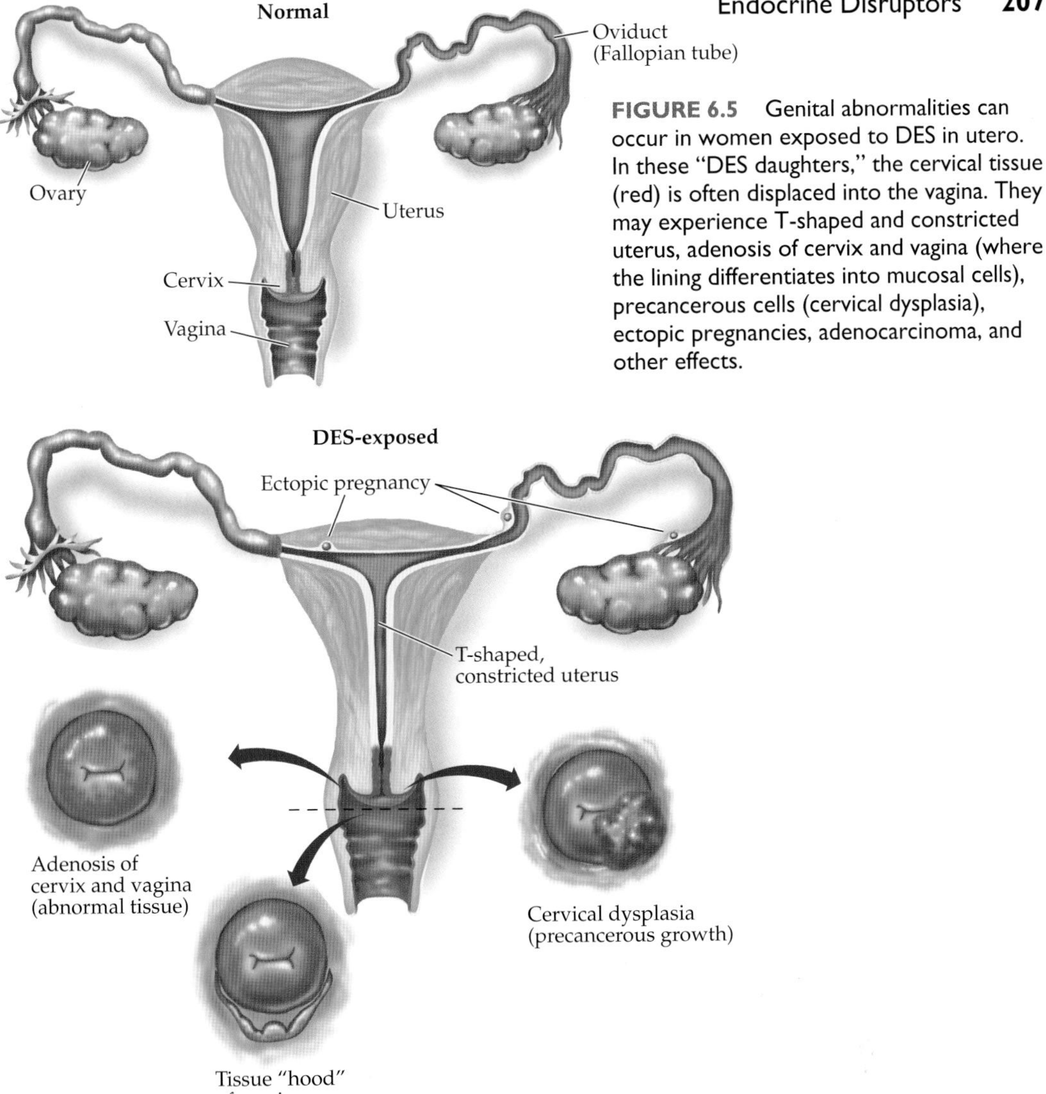

FIGURE 6.5 Genital abnormalities can occur in women exposed to DES in utero. In these "DES daughters," the cervical tissue (red) is often displaced into the vagina. They may experience T-shaped and constricted uterus, adenosis of cervix and vagina (where the lining differentiates into mucosal cells), precancerous cells (cervical dysplasia), ectopic pregnancies, adenocarcinoma, and other effects.

ers had ingested DES prior to week 18 of pregnancy. This is a rare type of tumor, but its epidemiology revealed a clear association with this early exposure to DES in utero. Although the frequency of this cancer in the women studied was in fact low (0.1%), the studies revealed that more than 95% of females exposed to DES in utero had some abnormalities in the structure of their reproductive organs, and the risk of these women getting breast cancer after age 50 increased three-fold (Figure 6.5; Newbold 2004; Palmer et al. 2006).

DES interferes with sexual development by causing cell type changes in the female reproductive tract (the derivatives of the Müllerian duct, which forms the upper portion of the vagina, the cervix, the uterus, and oviducts). In many cases, DES causes the boundary between the oviduct and the uterus (the uterotubal junction) to be lost, resulting in infertility or subfertility (Robboy et al. 1982; Newbold et al. 1983). Moreover, the distal Müllerian ducts often fail to come together to form a single cervical canal (see Figure 6.5). Symptoms similar to the human DES syndrome occur in mice exposed to DES in utero or shortly after birth.

Interestingly, DES is also an **obesigen**—a substance that stimulates the body both to produce fat cells and to accumulate fat within these cells. After studying the endocrine problems of the female mice born of mothers treated with DES, Newbold and her colleagues (2007) noticed that these mice seemed to be fatter than mice born of mothers who were not so treated.* By weighing the different groups of mice, the Newbold laboratory (2004, 2007) confirmed that mice treated shortly after birth with as little as 1 part per billion DES had a marked obesity syndrome later in life. The DES-exposed and control mice were the same until about 8 weeks after birth (roughly the time of "mouse puberty"). After that, the DES-exposed mice grew much fatter than the controls (Figure 6.6). DES had sensitized the mice early in life, and when the large concentrations of estrogen associated with sexual maturity began to be synthesized, the mice rapidly became obese.

Mechanisms of DES action

The particular regions of the female reproductive tract are normally specified by the expression of the *HOXA* genes.† Hox genes are physically linked together on human chromosome 7, and they are expressed in a nested fashion throughout the Müllerian duct. *HOXA9* mRNA is detected throughout the uterus and continues to be found about halfway through the presumptive oviduct. *HOXA10* expression exhibits a sharp anterior boundary at the junction between the future uterus and the future oviduct. *HOXA11* has strong expression in the anterior regions where *HOXA10* is expressed, but its expression weakens in the posterior regions. *HOXA13* expression is seen only in the cervix (Figure 6.7).

Ma and colleagues (1998) showed that the effects of DES on the female mouse reproductive tract could be explained as the result of altered *Hoxa10* expression in the Müllerian duct. They showed that estrogen and proges-

*Obesity may be a common effect of estrogenic compounds. For example, Rubin and colleagues (2001) made observations similar to Newbold's in rats treated with bisphenol A.

†The term "Hox gene" has become common usage in biology, referring to a group of genes widely conserved across many families and phyla and involved in axis specification and pattern formation during animal development. The general term is not italicized, but the specific genes are. Following convention, human genes are named using capital letters, thus differentiating them from mouse genes, which are referred to with initial capital letters only.

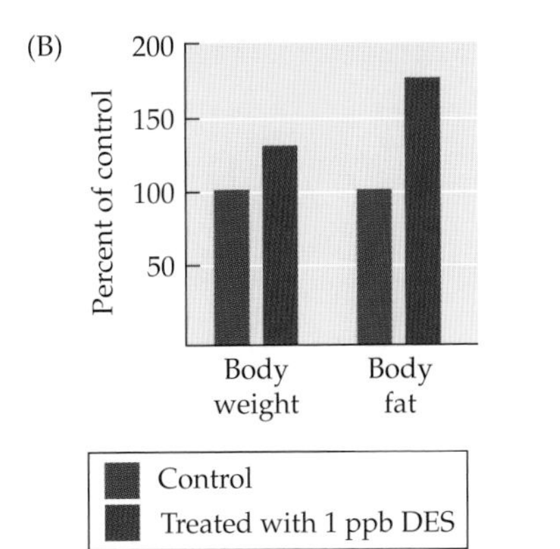

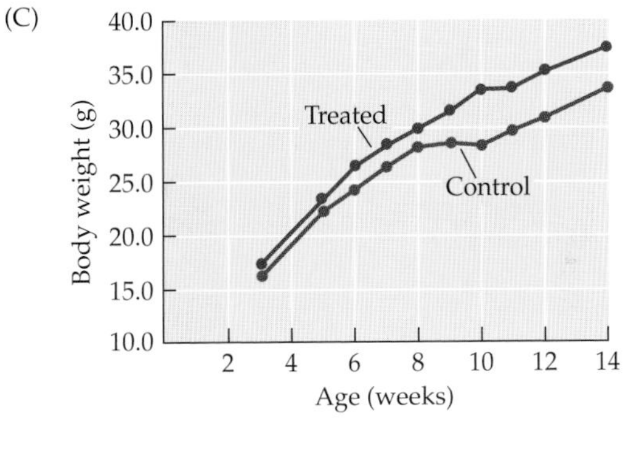

FIGURE 6.6 DES as an obesigen. (A) Exposure to DES in utero causes mice to become fat. (B) As little as 1 part per billion DES can cause the accumulation of fat to nearly double that of control mice who were not exposed to DES. (C) Weight and fat gain occurs after "puberty" in mice. (A © Science VU/Visuals Unlimited; B,C after Newbold 2004.)

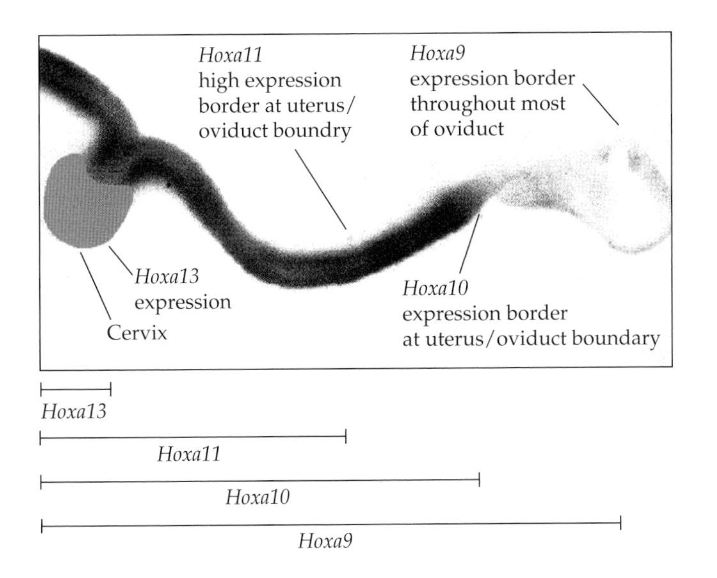

FIGURE 6.7 *Hoxa* gene expression in the reproductive system of a normal 16.5-day embryonic female mouse. *Hoxa9* expression extends from the cervix through the uterus to about halfway up the oviduct. *Hoxa10* expression has a sharp anterior border at the transition between the oviduct and the presumptive uterus. *Hoxa11* has the same anterior border as *Hoxa10*, but its expression diminishes closer to the cervix. *Hoxa13* is expressed only in the cervix and upper vagina. (After Ma et al. 1998.)

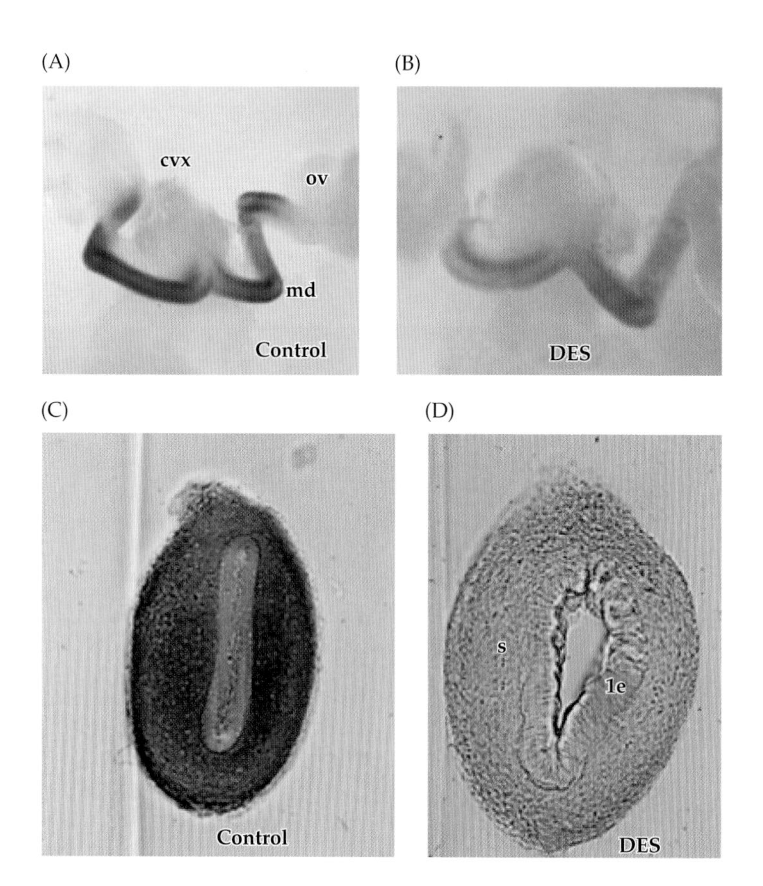

FIGURE 6.8 DES exposure represses *Hoxa10*. (A) Normal 16.5-day embryonic female mice show *Hoxa10* expression from the boundary of the cervix through the uterus primordium and most of the oviduct (compare with Figure 6.7). (B) In mice exposed prenatally to DES, this expression was severely repressed. (C) In control female mice at 5 days after birth (when the reproductive tissues are still forming), a section through the uterus shows abundant expression of the *Hoxa10* gene in the uterine stroma. (D) In female mice given high doses of DES 5 days after birth, *Hoxa10* gene expression in the mesenchyme is almost completely suppressed. Cvx, cervix; md, Müllerian duct; ov, ovary. (From Ma et al. 1998.)

terone are able to regulate certain genes of the HoxA cluster (*Hoxa9*, *Hoxa10*, *Hoxa11*, and *Hoxa13*). To determine whether DES changed Hox gene expression patterns, DES was injected under the skin of pregnant mice, and the fetuses were allowed to develop almost to birth. When the fetuses from the DES-injected mothers were compared with fetuses from mothers that had not received DES, it was seen that DES almost completely repressed the expression of *Hoxa10* in the Müllerian duct (Figure 6.8). This repression was most pronounced in the stroma (mesenchyme) of the duct, the place where experimental embryologists had localized the effect of DES (Boutin and Cunha 1997). The case for DES's acting through repression of *Hoxa10* is strengthened by the phenotype of the *Hoxa10* mutant mouse, in which this gene fails to function (Benson et al. 1996; Ma et al. 1998). In such *Hoxa10*-deficient mutants, there is a transformation of the proximal quarter of the uterus into oviduct tissue, and there are abnormalities at the border of the uterus and oviduct. Similar changes are also seen in Hox gene expression in the human cervical cells exposed to DES (Block et al. 2000.)

The link between Hox gene expression in the stroma and epithelium of the female reproductive tract (i.e., the determination of whether a particular portion of the tract functions as a uterus, an oviduct, or a cervix) is the Wnt proteins. Wnt proteins are associated with cell proliferation and protection against apoptosis (cell death), and the reproductive tracts of DES-exposed female mice also resemble those of mutant mice that lack *Wnt7a* genes. Miller and colleagues (1998) have shown that the Hox genes and the Wnt genes communicate with each other to keep each other activated during the specification and morphogenesis of the reproductive tissues (Figure 6.9). However, DES, acting through the estrogen receptor, represses the *Wnt7a* gene. This repression prevents the maintenance of the Hox gene expression pattern, and it also prevents the activation of another Wnt gene, *Wnt5a*, which encodes a protein necessary for cell proliferation.

The role of Wnt signaling in uterine morphology was confirmed by Carta and Sassoon (2004) when they observed that the *Wnt7a* gene product plays a key role as a cell death suppressor. The uteri of *Wnt7a* mutant mice,

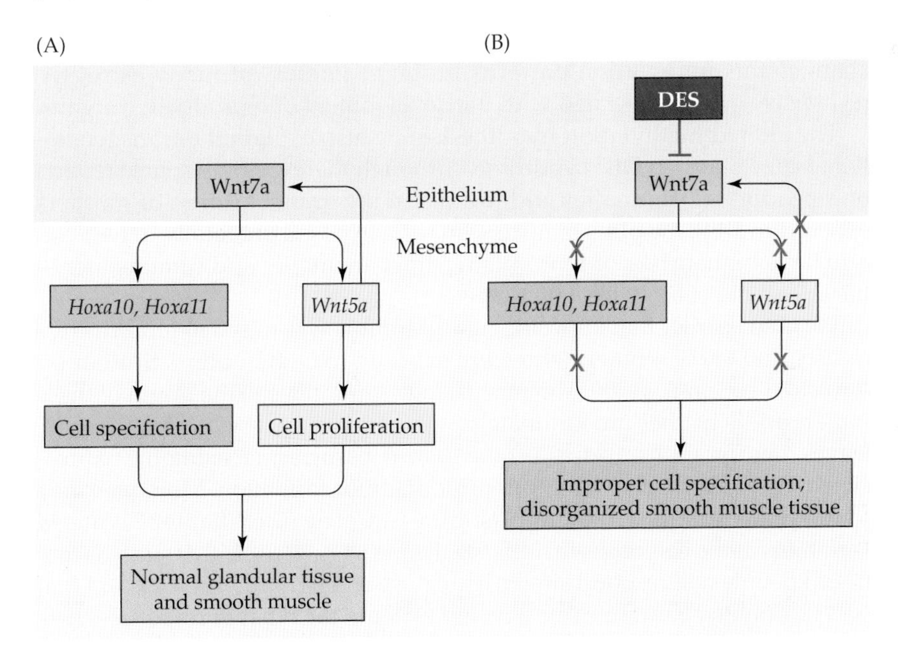

FIGURE 6.9 Misregulation of Müllerian duct morphogenesis due to DES exposure. (A) In normal female morphogenesis, the *Hoxa10* and *Hoxa11* genes in the mesenchyme are activated and maintained by the *Wnt7a* gene product. Wnt7a also induces *Wnt5a* gene expression; the *Wnt5a* gene product both maintains *Wnt7a* gene expression and stimulates cell proliferation and the normal morphogenesis of the uterus. (B) Acting through the estrogen receptor, DES blocks expression of *Wnt7a*. Proper activation of the Hox genes and *Wnt5a* thus does not occur (red X's), resulting in a drastically altered female genital morphology. (After Gilbert 2006.)

like those of DES-exposed mice, displayed high levels of cell death, while the wild-type uteri displayed almost no cell death. The expression pattern of the alpha estrogen receptor, Hox and Wnt genes, and other regulatory genes that guide uterine development reveals either abnormal regulation or atypical spatial distribution of transcripts in response to DES exposure (Carta and Sassoon 2004). This study demonstrates that *Wnt7a* coordinates multiple cell and developmental pathways involved in uterine growth and hormonal responses, and disruption of these pathways by DES results in uterine cell death.

Wnt signaling may also be important in regulating DNA methylation, a mechanism described in Chapter 2. DNA methylation regulates gene expression, and abnormalities of DNA methylation have been observed in many types of human tumors, including those of the uterus and cervix. Li and colleagues (2003) have shown that exposure to a wide variety of endocrine disruptors during critical periods of mammalian development can interfere with normal DNA methylation patterns, leading to atypical gene expression. They demonstrated that exposure of mice to DES around the time of birth (when the mouse reproductive tract is being developed) changes the methylation patterns in the promoters of many estrogen-responsive genes associated with the development of reproductive organs.

The altered methylation hypothesis may also explain the transgenerational effects seen in DES syndrome. Although there are only sparse reports of reproductive anomalies and tumors in the grandchildren of mothers exposed to DES, the pups of mice who were exposed prenatally to DES can also have higher risks of reproductive tract anomalies and tumors later in life (Newbold et al. 1998, 2000, 2006).

We noted earlier that a specific class of tumors was found in reproductive tissues in 0.1% of women who were exposed to DES in utero. Studies by Cook and colleagues (2005) on rats provide some clues to the origin of these tumors and why they are fortunately of low frequency. They found that DES leaves a hormonal "imprint" on the uterine tissues, increasing the expression of estrogen-responsive genes prior to tumor formation. Moreover, they found that variations in a tumor-suppressor gene (see Chapter 7) caused variations in the responses to DES. The lesson suggested by this study is that not all people will respond similarly to these endocrine disruptors, and that genetic polymorphism may be a critical factor in the types of responses seen. In the case of DES, almost all women exposed in utero experienced some sort of developmental modification of the uterus; the polymorphism leading to cancer upon DES exposure, however, was more limited.

Soy estrogens

Genistein, the estrogenic compound found in soy (and in soy products such as tofu) is also being scrutinized for endocrine-disrupting effects (Newbold et al. 2001; Wiszniewski et al. 2005; Jefferson et al. 2007). Newborn mice treated with genistein have reproductive deficiencies and aberrant mammary gland development (Padilla-Banks et al. 2006; Jefferson et al. 2007), and frog tadpoles treated with genistein fail to undergo normal thyroid-dependent metamorphosis (Ji et al. 2007). However, Cabanes and colleagues (2004) find that genistein might reduce mammary tumors, and Cederroth and colleagues (2007) have found that a soy diet (both prenatally and after birth) decreases fat deposition and increases energy usage in mice. As of 2006, the National Institute of Environmental Health Sciences (Rozman et al. 2006) concluded that, while genistein can produce reproductive deficiencies in mice and rats, there are no data available on whether genistein produces a similar set of abnormalities in humans. However, Chavarro and colleagues (2008) found that men with extremely high dietary soy intake had significantly lower sperm counts than men who did not eat large amounts of soy.

Declining sperm counts and testicular dysgenesis syndrome

Females are not the only ones whose reproductive success may be affected by the estrogenic compounds being released into the environment. During the past two decades, there has been an increase in testicular cancers and a decrease in sperm count in men throughout the industrialized world (Carlsen et al. 1992; Aitken et al. 2004). Sperm count (number of sperm per mL of semen) has dropped precipitously throughout much of Europe and the Americas. Skakkebaek and his team at the University of Copenhagen reviewed 61 international studies involving 14,947 men between 1938 and 1992. They found (Carlsen et al. 1992) that the average sperm count had fallen from 113 million per mL in 1940 to 66 million in 1990. In addition, the number of "normal" sperm fell from 60 million per mL to 20 million in the same period.

Subsequent studies have confirmed and extended Skakkebaek's findings. A survey of 1,350 sperm donors in Paris found a decline in sperm counts by around 2% each year over the past 23 years, with younger men having the poorest quality semen (Auger et al. 1995). These men had an average sperm count of 90 million sperm per mL in 1973, but only 60 million sperm per mL in 1992. A study from Finland (Pajarinen et al. 1997) showed a similar decline. Using testicular tissue from 528 middle-aged Finnish men who died suddenly, they showed that among the men who died in 1981, 56.4% had normal sperm production, but by 1991 this figure had dropped to 26.9%. Moreover, the average weight of the men's testes also decreased over the decade, while the proportion of useless fibrous testicular tissue increased at the expense of the sperm-producing seminiferous tubules. A Scottish study (Irvine et al. 1996) showed that men born in the

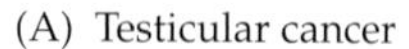

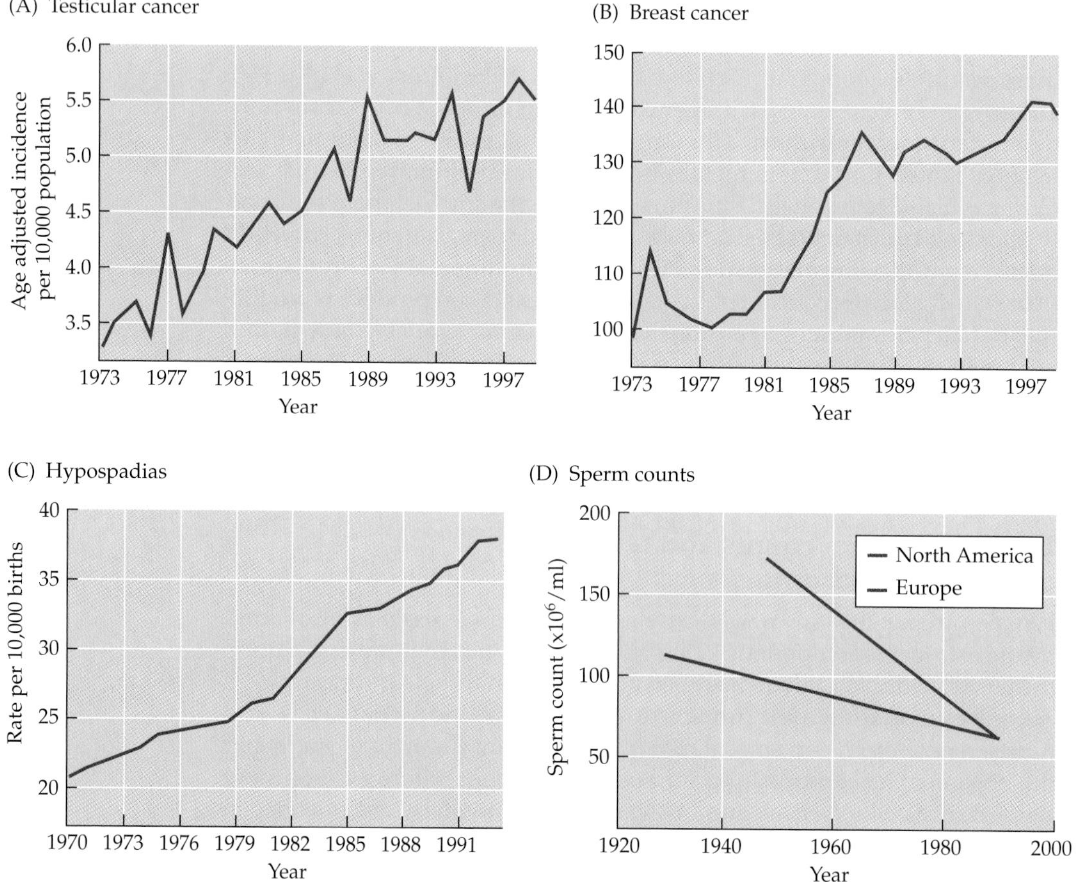

FIGURE 6.10 Developmental estrogen syndrome is manifest in climbing breast cancer rates and testicular dysgenesis. The rise of testicular cancers (A) parallels the rise of breast cancers (B) and abnormalities of penis development (C) such as hypospadias (the failure to completely close the penis). Sperm counts among North American males, moreover, have declined nearly 50% within the past century (D); the decline has been even steeper among European men. (After Sharpe and Irvine 2004.)

1970s were producing some 24% fewer motile sperm in their ejaculate than men who were born in the 1950s. In addition to the drop in sperm count, there has also been an increase in testicular cancers during this time (Figure 6.10)

Sharpe (1994) and Skakkebaek (2004) have analyzed these data and have hypothesized that there exists a **testicular dysgenesis syndrome** (Greek, *dys*, "bad" + *genesis*, "formation") characterized by disorganized testis development, testicular germ cell tumors, and low sperm count. Moreover, Sharpe (1994) has suggested that this syndrome may be due, at least in large part, to endocrine disruptors. While the chain of causation has not been completely established (see Sharpe and Irvine 2004), subsequent studies have provided further evidence that the components of this syndrome can be caused by environmentally relevant concentrations of endocrine disruptors.

Many endocrine disruptors, including dioxins, nonylphenol, bisphenol A, acrylamide, and certain pesticides and herbicides, can adversely affect testes morphology and sperm production (see Aitken et al. 2004). The sunscreen 4-MBC, a camphor derivative, has been found to decrease the size of testes and prostate glands, and it can delay male puberty in rats (Schlumpf et al. 2004.)

Indeed, many aspects of testicular dysgenesis syndrome (including all of the developmental anomalies, but not the testicular tumors) can be induced in laboratory animals by another group of endocrine disruptors, the phthalates. Phthalates are ubiquitous in industrialized societies, being widely used in cosmetics, air fresheners, and vinyl plastics; the "new car smell" consists largely of volatilized phthalates. Many components of testicular dysgenesis syndrome can be induced by administering phthalate derivatives to pregnant rats (Fisher et al. 2003; Mahood et al. 2007). Male rodents exposed in utero to dibutyl phthalate had an extremely high rate (>60%) of cryptorchidism (undescended testes), hypospadias (improperly formed penis), low sperm count, Leydig cells trapped in the seminferous tubules, and other testis abnormalities—conditions very similar to human testicular dysgenesis syndrome. Other evidence, obtained from newborn human males, suggests that those babies exposed in utero to relatively high phthalate levels had morphological changes in their testes (Swan et al. 2005; Huang et al. 2008). In response to these types of findings, there is increasing regulation to limit exposure to the phthalate chemicals. Several types of phthalates are banned in Europe, and a California regulation that will go into effect in 2009 will similarly regulate their use.

Pesticides and infertility in males

The link between pesticides and male infertility has been known for a long time (Carlsen 1992; Colborn et al. 1996). Data from Swan and colleagues (2003, 2006) showed lower sperm counts in men living in agricultural regions, where pesticide and herbicide use is prevalent, as opposed to those living in urban areas. A recent quantitative study of toad populations throughout Florida demonstrated that the frequency of testicular abnormalities and intersex gonads increased linearly with the amount of agricultural activity (McCoy et al. 2008). In highly agricultural sites, testosterone production in male toads had declined so significantly that the sexual dimorphism in skin color had vanished, with males looking the same as the females.

Sensitivity to Disruption: A Genetic Component

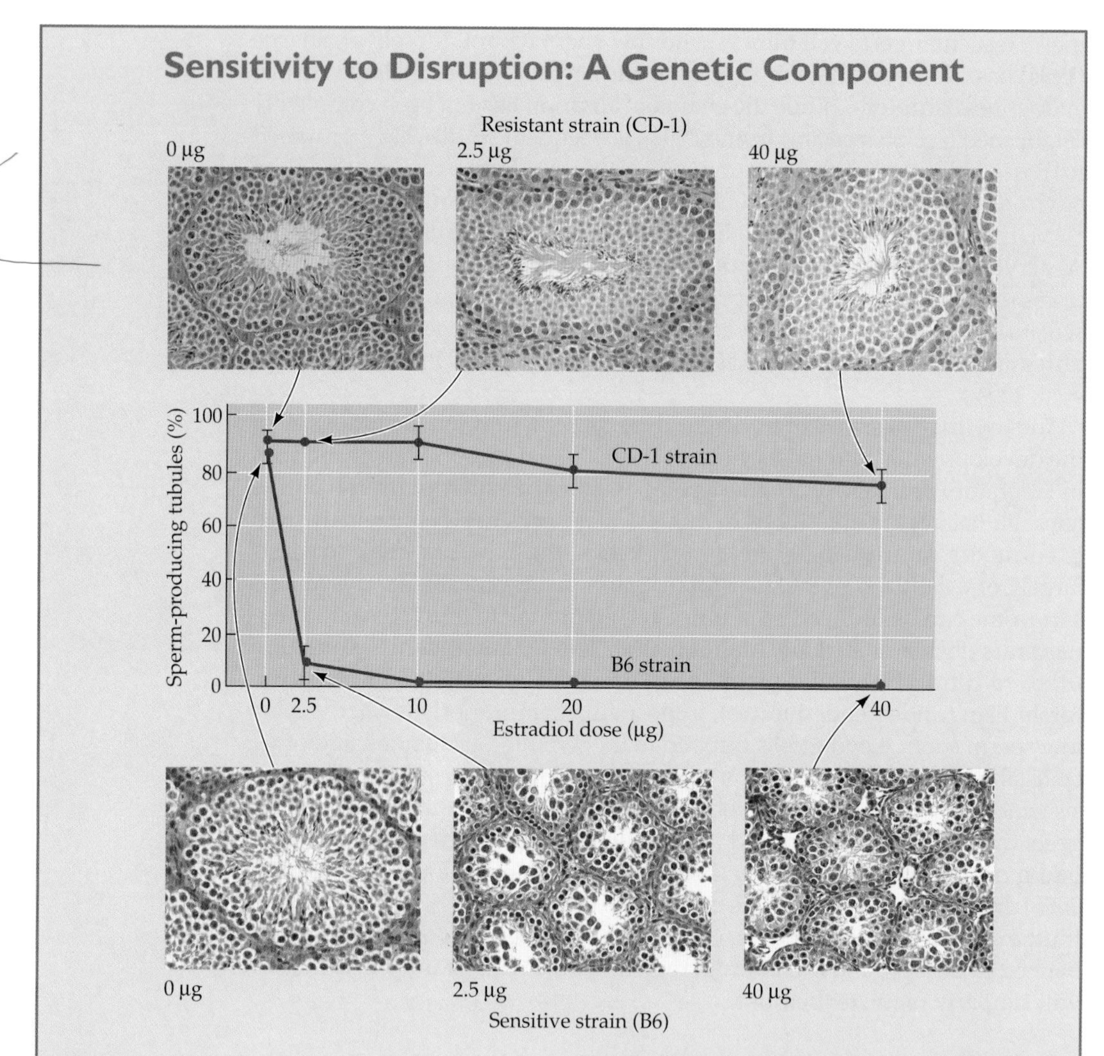

Some scientists have argued that claims about endocrine disruptors affecting reproduction are exaggerated, citing tests on mice that indicate that litter size, sperm count, and gonadal development are not affected by *environmentally relevant* concentrations of environmental estrogens. In fact, it is difficult to determine the concentrations of endocrine disruptors that would cause testicular dysgenesis in a majority of human males. One reason is the genetic heterogeneity of human populations.

Varied effects of estrogen implants on different strains of mice. The graph shows the percentage of seminiferous tubules containing elongated spermatozoa. (The mean ± standard error is for an average of 6 individuals.) The micrographs show cross sections of the testicles (all at the same magnification). Although 40 μg of estradiol did not affect spermatogenesis in the CD-1 mice, as little as 2.5 μg almost completely abolished spermatogenesis in the B6 strain. (After Spearow et al. 1999; photographs courtesy of J. L. Spearow.)

Mice are inbred, and each genetic strain responds differently. Investigations by Spearow and colleagues (1999) have shown a remarkable genetic difference in sensitivity to estrogen among these different strains of mice. The CD-1 strain of laboratory mice used for testing environmental estrogens is at least 16 times more resistant to endocrine disruption than are the more sensitive mouse strains, such as B6. When estrogen-containing pellets were implanted beneath the skin of young male CD-1 mice, very little happened. However, when the same pellets were placed beneath the skin of B6 mice, their testes shrank and the number of sperm seen in the seminiferous tubules dropped dramatically (see figure). This widespread range of sensitivities, which can vary for different tissues in the same strain of mice (see Wadia et al. 2007), has important consequences for determining safety limits for humans. It could also be related to human gene polymorphisms, which would mean that some men will be highly resistant to these chemicals whereas others could be sensitive.

These considerations confirm several important principles of ecological developmental biology. The first is that genetic differences can be critical in how a developing organism responds to environmental cues. The second is that these differences exist (as Darwin said they would) among individuals of the same species. The third is that testing for endocrine disruption is challenging, since different species may have evolved different means for coping with or using environmental cues, and the choice of an experimental model is probably critical in the detective work needed to see if disruption is occurring.

The fungicide vinclozolin, widely used on vineyard grapes, is an androgen antagonist, inserting itself into the androgen (i.e., testosterone) receptor and preventing testosterone from binding there (Grey et al. 1999; Monosson et al. 1999). In experiments, rat embryos exposed to high concentrations of this chemical during the period of spermatogenesis have malformations of the penis, absent sex accessory glands, and very low (around 20% of normal) sperm production (Figure 6.11; Anway et al. 2006). This exposure not only impaired the fertility of males in the generation exposed in utero, but also impaired fertility in males for at least three generations afterward. We will discuss this transgenerational effect in detail later in the chapter. But to end our discussion of agriculturally prominent estrogenics, we return to the discussion of the controversial herbicide atrazine.

Atrazine, again

Discussed in regard to sex determination in Chapter 2, the estrogenic herbicide atrazine is the second-largest-selling weed killer in the world (Miller et al. 2000; Capel and Larsen 2001)—this in spite of the fact that its use is banned in Europe (including in Switzerland, the headquarters of the company that manufactures it). Most sales of atrazine are in the United States, where nearly 80 million pounds are applied to the soil annually (U.S.G.S. Web site). Atrazine is the world's most common herbicidal pollutant of groundwater and surface water, and it is extremely stable. The chemical is still found in French groundwater, even though neither France nor any

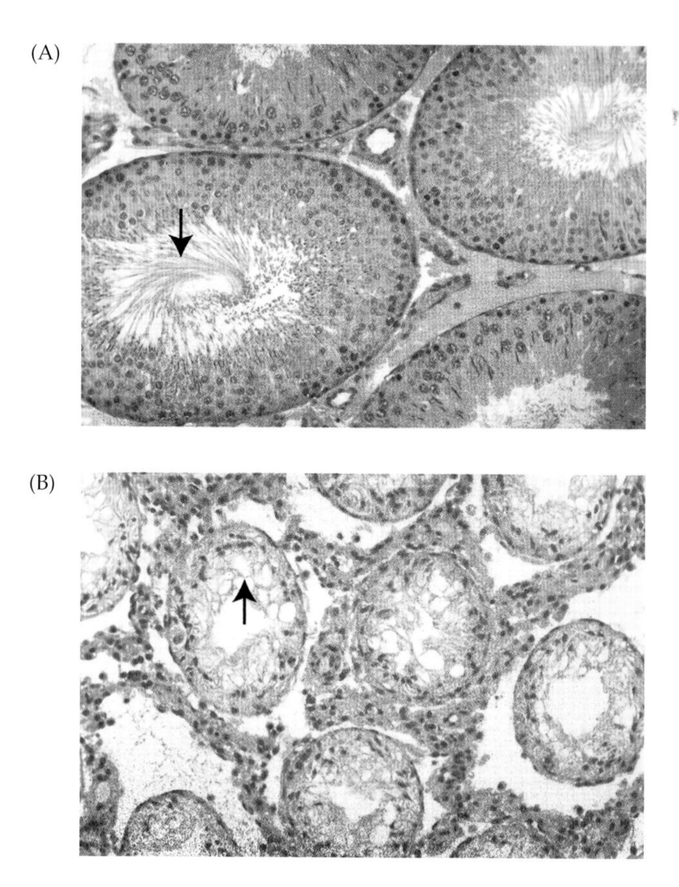

FIGURE 6.11 Cross section of seminiferous tubules from the testes of a control rat and a rat whose grandfather was born from a mother who had been injected with vinclozolin. (A) Normal sperm are present in the control rat, as seen by the large number of sperm tails (arrow). (B) In the vinclozolin-affected rat, the arrow points to the lack of sperm in the much smaller seminiferous tubules. This rat was infertile. (From Anway et al. 2005; photographs courtesy of Michael Skinner.)

neighboring nation has used it for over 15 years (Kolpin et al. 1998; Thurman and Cromwell 2000; Hennion et al. 2004).

REPRODUCTIVE FAILURES As mentioned in Chapter 2, atrazine induces the enzyme aromatase, which converts testosterone into estrogen (estradiol). In several classes of vertebrates—fishes, amphibians, and laboratory mammals—the decrease in testosterone results in "chemical castration" in the form of low sperm count, reduced fertility, and smaller masculine organs. In those animals whose sex determination can be affected by hormones, males are known to have been converted into females or hermaphrodites (organisms having both sexes in the same body). Hayes and colleagues (2003) have documented that atrazine causes the formation of oocytes within what had been the testes (see page 55), and have established a probable chain of causation for the reproductive problems of frog populations exposed to atrazine (Figure 6.12). This chain reflects the difficult issue of how human political and social needs and agendas interact with geological and biological agents.

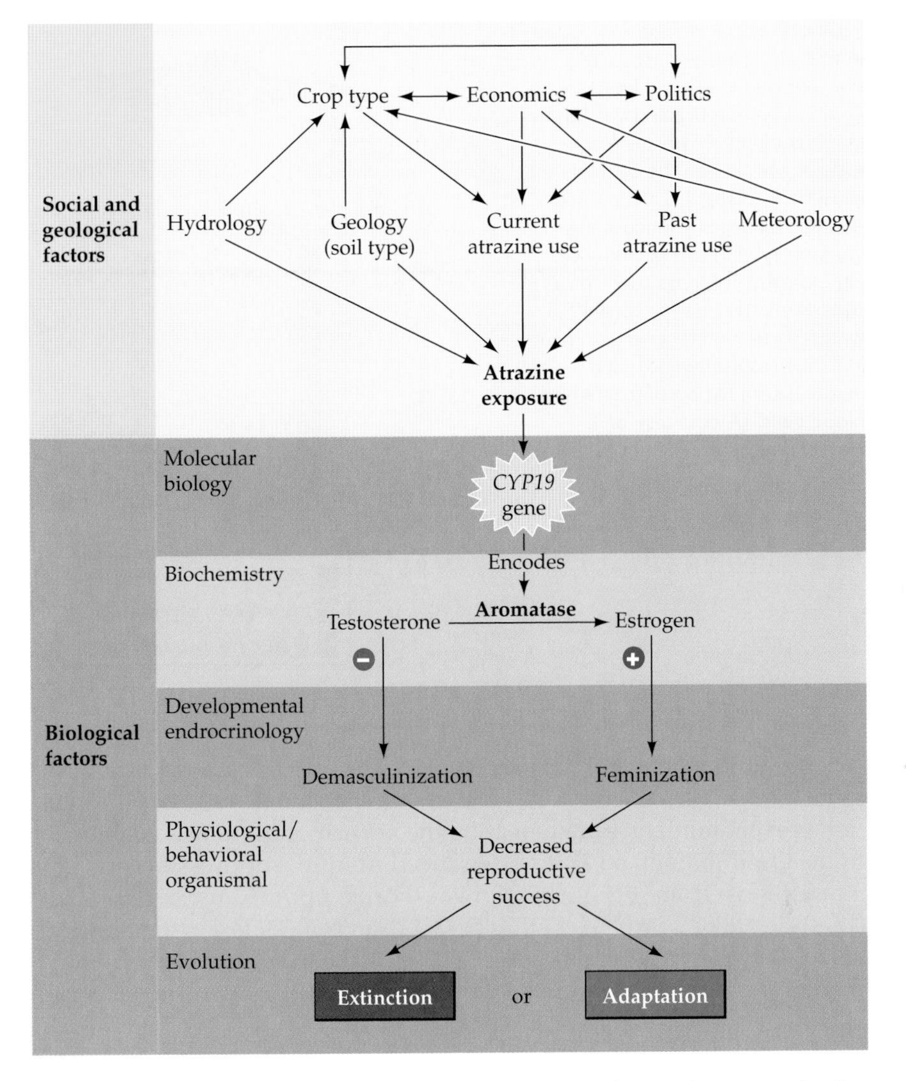

FIGURE 6.12 Possible chain of causation leading to the demasculinization of male frogs and the decline of frog populations in regions where atrazine use is prevalent. Both social and biological agents contribute to the chain. Sanderson and colleagues (2000) have shown that transcription of the human *CYP19* gene that encodes aromatase is induced by herbicides such as atrazine. (After Hayes 2005.)

IMMUNOSUPPRESSION Ecologically, one of the most impressive dangers of atrazine involves its immunosuppressive effects. The chemical has been shown to produce an immunodeficiency syndrome in amphibians. Many amphibians exposed to atrazine (which they absorb through their skin)

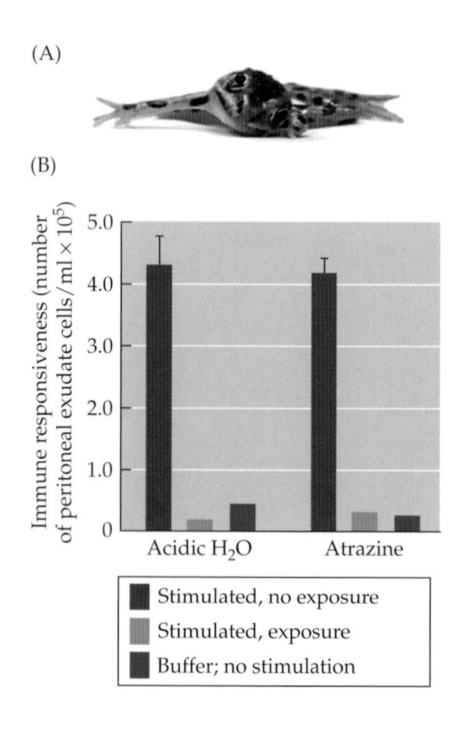

FIGURE 6.13 Atrazine-induced immunodeficiency in frogs. (A) Meningitis in a leopard frog due to immune system breakdown following a "cocktail" of pesticides, including atrazine. The bacteria that causes this disease is always present, but a normally functioning immune system can exclude it. (B) Exposure of frogs to 21 parts per billion atrazine (or to pH 5.5 water, a known immune disruptor in frogs) resulted in reduced immune stimulation of frog white blood cells. (A courtesy of T. Hayes; B after Brodkin et al. 2007.)

lack the ability to fight off common pathogens such as bacteria, fungi, parasites, and viruses (Figure 6.13). Just as cancer chemotherapy or HIV infection prevent humans from defeating the normal opportunistic pathogens in the environment, so atrazine-induced immunosuppression prevents amphibians from fighting off waterborne opportunistic pathogens (Kiesecker 2002; Christin et al. 2003; Gendron et al. 2003; Forson and Storfer 2006a,b; Hayes 2006; Brodkin et al. 2007.) The livers and kidneys of tadpoles in atrazine-contaminated water are often full of parasite larvae (T. Hayes, pers. comm.).

Evidence is also accumulating that exposure to atrazine and other endocrine disruptors causes immune deficiencies in mammals (Filipov et al. 2005; Karrow et al. 2005; Rowe et al. 2006). Here, however, the spectrum of disease includes not only immunosuppression but also hyperpotentiation (allergies). When Rowe and colleagues (2006) exposed mice to atrazine for the last part of their gestation and suckling, their adult immune function included hypersensitivity to certain common environmental bacteria.

Plastics and Plasticity

All human fetuses and adults have substantial levels of estrogenic endocrine disruptors in their circulatory systems. Many of these chemicals probably come from the contamination of food products with material used in the stabilizing or hardening of the plastics in which our foods are stored, cooked, and served.

The discovery of the estrogenic effect of plastic stabilizers was made in a particularly alarming way. Investigators at Tufts University School of Medicine had been studying estrogen-responsive tumor cells. These cells are inhibited from proliferating when exposed to serum proteins; estrogen overcomes this inhibition. The studies were going well until 1987, when the experiments suddenly went awry. The control cells began to show high growth rates suggesting stimulation comparable to that of the estrogen-treated cells. It was as if someone had contaminated the control culture medium by adding estrogen to it. But where could such contamination possibly be coming from? After spending 4 months testing all the components of their experimental system, the investigators found that the source of the estrogenic factor was the plastic containers that held the water and serum. The company that made the containers refused to describe its process for stabilizing the polystyrene plastic, so the researchers had to discover it for themselves.

The culprit turned out to be *p*-nonylphenol, a compound that is used to harden the plastic of the pipes that bring us water and to stabilize the polystyrene plastics that hold water, milk, orange juice, and other common liquid food products (Soto et al. 1991; Colborn et al. 1996). This compound is also a degradation product of detergents, household cleaners, and contraceptive creams. Nonylphenol has been shown to alter reproductive physiology in female mice and to disrupt sperm function. It is also correlated with developmental anomalies in wildlife (Fairchild et al. 1999; Kim et al. 2002; Adeoya-Osiguwa et al. 2003; Smith and Hill 2006; Kurihara et al. 2007).

Bisphenol A

One of the most ubiquitous plasticizing compounds is **bisphenol A (BPA)**. In the early years of hormone research, when the actual steroid hormones were difficult to isolate, chemists looked for synthetic analogues that would accomplish the same tasks. The pioneering British chemist Sir Charles Dodds discovered that BPA was estrogenic in 1936. (Two years later, Dodds synthesized diethylstilbestrol.) Later, polymer chemists realized that BPA could be used in plastic production, and today it is one of the top 50 chemicals in production worldwide. Four corporations in the United States make almost *2 billion pounds* of it each year for use in the resin lining in food cans, the polycarbonate plastic used in bottles and in children's toys, and as dental sealant. Its modified form, tetrabromobisphenol A, is the major flame retardant coating the world's fabrics.

Some of the known effects of in utero exposure to bisphenol A on mice and rats are summarized in Figure 6.14. Data from the vom Saal lab (vom

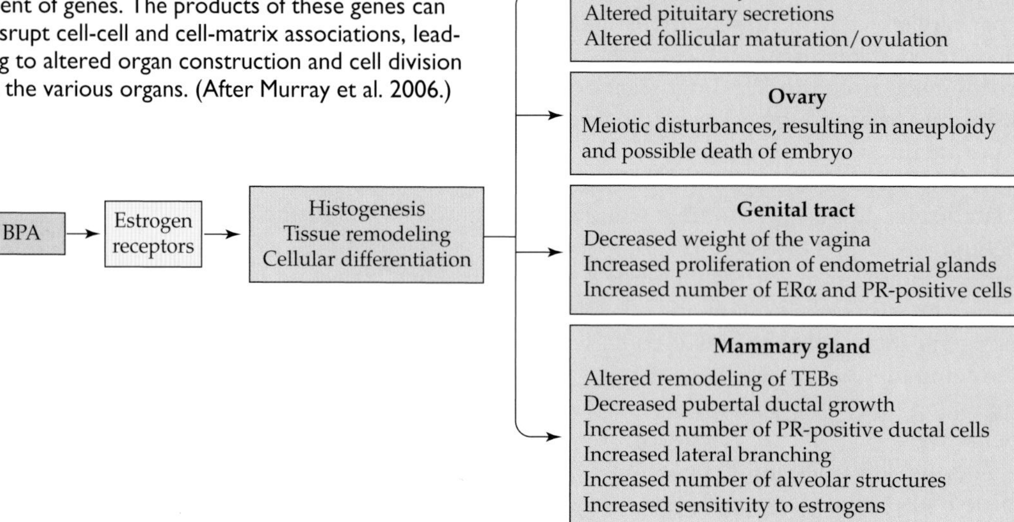

FIGURE 6.14 Some effects of perinatal BPA exposure. Schematic representation of the effects of BPA on adult mice. BPA works through the estrogen receptors, thereby activating an assortment of genes. The products of these genes can disrupt cell-cell and cell-matrix associations, leading to altered organ construction and cell division in the various organs. (After Murray et al. 2006.)

Saal and Hughes 2005) link BPA exposure to an even more far-reaching array of effects that include impaired immune function, reduced sperm count, increased incidence of prostate tumors, and decreased levels of antioxidant enzymes. This data also suggests that BPA may affect behavior, manifest in ways such as hyperactivity, learning impairments, and altered responses to fear- and pain-provoking stimuli.

We have mentioned the ubiquitous presence of BPA in plastics. However, studies by several laboratories have demonstrated BPA is not fixed in plastic forever (Krishnan et al. 1993; vom Saal 2000; Howdeshell et al. 2003). If water sits in an old polycarbonate rat cage at room temperature for a week, the water can acquire levels of BPA that will cause weight changes in the uteruses of young mice. This leaching from plastic can cause chromosome anomalies. When a laboratory technician in Patricia Hunt's laboratory mistakenly rinsed some polycarbonate cages in an alkaline detergent, the female mice housed in these cages had meiotic abnormalities in 40% of their oocytes (the normal amount is about 1.5%). When bisphenol A was administered to pregnant mice under controlled circumstances, Hunt and colleagues (2003) showed that a short, low-dose exposure to BPA was sufficient to cause similar meiotic defects in maturing mouse oocytes (Figure 6.15).

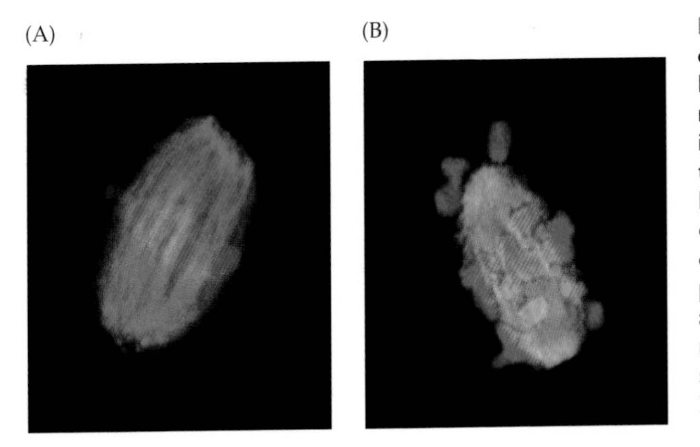

FIGURE 6.15 Bisphenol A causes meiotic defects in maturing oocytes. (A) During the first meiotic metaphase, chromosomes (stained red) normally line up in a paired fashion at the center (equator) of the spindle. (B) Brief exposure to BPA causes chromosomes to align randomly on the spindle. Different numbers of chromosomes then enter the egg and polar body, resulting in chromosomal aneuploidy (incorrect number of chromosomes in the daughter cells of meiosis) and infertility. (From Hunt et al. 2003; photographs courtesy of P. Hunt.)

Studies on laboratory animals indicate that bisphenol A at environmentally relevant concentrations can cause disruptions in the morphology of the fetal sex organs, as well as low sperm counts (vom Saal et al. 1998; Palanza et al. 2002). Moreover, BPA also appears to affect the sexual biology of the rodent brain. In rats and mice, certain areas of the brain are different in males and females, and there are behavioral differences that are more stereotyped than in humans. Kubo and colleagues (2003) demonstrated that low-dose exposure to BPA in utero can destroy the sex differences in the developing rat brain and elicit behavioral changes when these fetuses become adults. These results were duplicated in mice when female mice exposed to BPA in utero showed characteristic masculine behaviors, and prenatal and neonatal exposure of both sexes to BPA made locomotor and reward-response behaviors more sensitive to morphine (Narita et al. 2006; Rubin et al. 2006).

Other work suggests that BPA might be involved in eliciting early puberty. The age at which girls begin to express adult female sexual characteristics has declined over the past hundred years (probably due to nutrition), and new research suggests that female puberty is starting even earlier (Herman-Giddens et al. 1997). Work in vom Saal's laboratory showed that when female mice are exposed in utero to low doses of bisphenol A, they reach sexual maturity faster than unexposed mice (Howdeshell et al. 1999; Howdeshell and vom Saal 2000). Moreover, exposed female mouse embryos showed altered mammary development at puberty, as well as alterations in the organization of their breast tissue and ovaries and altered estrous cyclicity as adults (Markey et al. 2003). Each mammary gland also produced more terminal buds and was more sensitive to estrogen.

Murray and colleagues (2007) showed that fetal exposure to BPA caused the development of carcinoma in situ (early stage cancer) in the mammary glands of 33% of the rats exposed in utero to low levels of BPA. None of the control rats developed such tumors. Moreover, fetal exposure to BPA increased the number of "preneoplastic lesions" (areas of rapid cell growth within the ducts) three- to fourfold. Furthermore, gestational exposure to as little as 25 μg BPA (per kilogram each day), followed at puberty by a "subcarcinogenic dose" of a chemical mutagen, resulted in the formation of tumors only in the animals exposed to the BPA (Durando et al. 2006; Vandenberg et al. 2007). Studies on fetal and adolescent rats showed that perinatal exposure to environmentally relevant doses of BPA causes changes in anatomical patterns of mammary gland development at puberty (Muñoz-de-Toro et al. 2005; Vandenberg et al. 2007). They also found that gestational exposure to BPA induces conditions that can lead to tumors when a second exposure of estrogenic hormones or carcinogens is experienced later in life (Figure 6.16).

Especially noteworthy is the fact that, in all of the above studies, the effective dosage of BPA was extremely low—as much as *2000 times lower* than the dosage set as "safe" by the U.S. government.

Females aren't the only sex affected by BPA. The chemical has also been implicated (along with other estrogenic compounds) in the decline in sperm quality and density. Several of these investigations also suggest that BPA may result in an increased incidence of prostate enlargement in men; in the United States, 65% of men in their 60s have enlarged prostate glands. Studying the mouse prostate gland (whose size is sensitive to estrogens), vom Saal and colleagues (1997) found that when they gave pregnant mice 2 parts per billion BPA—that is, 2 *nanograms* per gram of body weight—for 7 days at the end of pregnancy (equivalent to the period when human reproductive organs are developing), the male progeny showed an increase in prostate size of about 30%. The sperm count of these mice was also lowered. Adult male mice exposed to small amounts of BPA also have enlarged prostates, and BPA increases the rate of mitosis in human prostate cells (Wetherill et al. 2002; Timms et al. 2005).

But is there any evidence that bisphenol A reaches the fetus in concentrations that matter? The BPA ingested by pregnant rats appears to pass readily into the fetus and is not hindered by steroid-binding hormones. Within an hour of ingestion, BPA is found in the fetus at the same doses that it had been in the mother (Miyakoda et al. 1999). The situation may be even more problematic in humans. Recent studies (Ikezuki et al. 2002; Schönfelder et al. 2002) show that BPA in the human placenta is neither eliminated nor metabolized into an inactive compound; rather, it accumulates to concentrations that can alter development in laboratory animals. Moreover, recent survey studies have found that 95% of urine samples of people in the United States and Japan have measurable BPA levels, indicating general exposure of the population to this chemical (Calafat et al. 2005, 2008). Most critically, children have higher concentrations of BPA in their blood than do adolescents or adults.

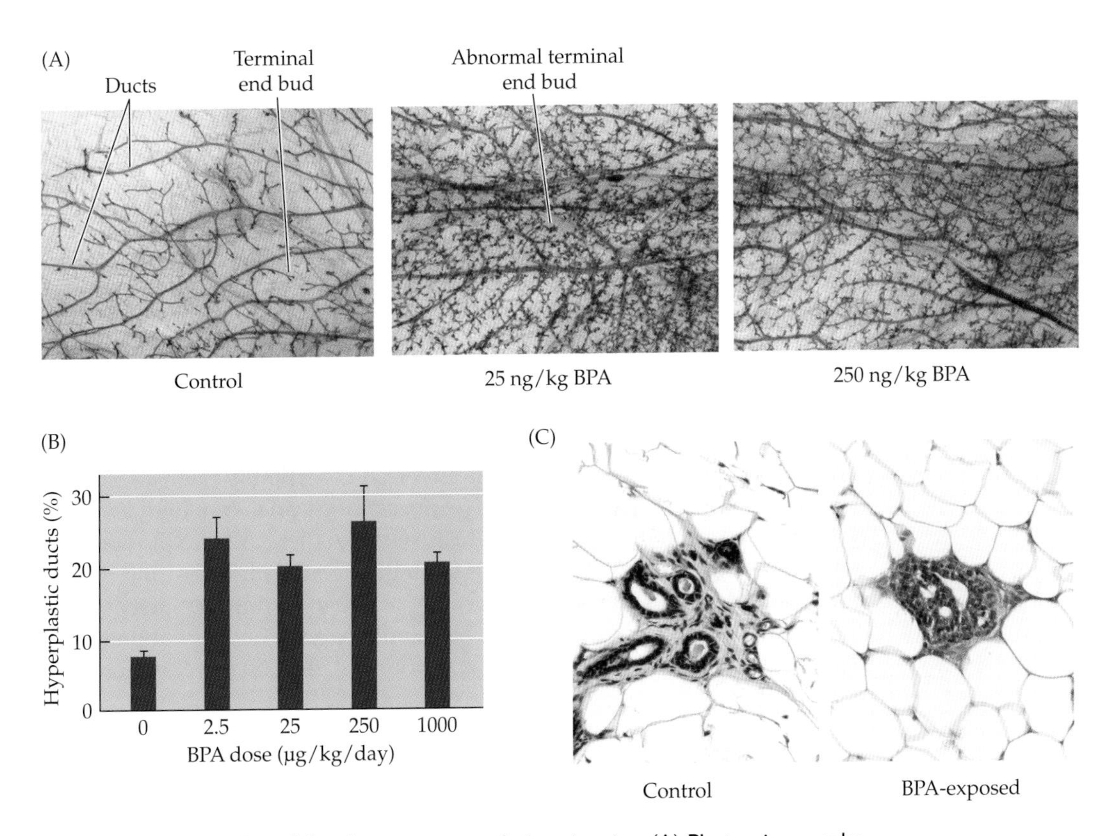

FIGURE 6.16 Bisphenol A induces mammary lesions in mice. (A) Photomicrographs of mammary glands show profound differences between a control animal and animals exposed perinatally to nanogram levels of BPA. The increase in terminal end buds and branching in the BPA-exposed mammary glands predisposes the animal to cancer later in life. (B) The percentage of mammary glands showing intraductal hyperplasia (indicating a cancer-prone state) is significantly increased at postnatal day 50 in BPA-treated animals. (C) Later in life, BPA-exposed animals displayed significant cancer in situ. (From Murray et al. 2007 and Soto et al. 2008; photographs courtesy of Ana Soto.)

The dose-response curve of BPA action

Many of BPA's effects are observed at extremely low concentrations, and higher concentrations often do not exert the same effect. This strange dose-response relationship is probably a function of the nature of endocrine disruptors, which have a different dose-response curve from that of classic teratogens such as alcohol, thalidomide, and heavy metals. Classical teratology, based on the principles of toxicology, operates on the principle that "the dosage makes the poison" and that there is a linear relationship between the dosage of the toxic substance and the severity or the incidence of the abnormality. This linearity is the basis of traditional toxicology, where toxicity is usually reported as an LD50 ("lethal dose," the concentration at which 50% of the animals die) or ED50 ("effective dose," at which 50% of the animals show an effect).

Such linear effects are not seen with most endocrine disruptors. Instead, there is usually an increasing effect from low to medium concentrations (typically the developmental anomalies described above) and then the gradual *disappearance* of this effect at higher concentrations (Figure 6.17A). That is to say, the dosage relationship is not linear, and the abnormalities might not even occur at high concentrations of the chemical. These sorts of responses are shown in the effects of estrogen on cultured human breast cancer cells (Welshons et al. 2003) and in the induction of terminal end buds in the mammary gland (Figure 6.17B; Markey et al. 2001) and prostate tumor proliferation (Figure 6.17C; Wetherill et al. 2002). The curve seen in Figure 6.17A is often referred to as an **inverted U response**. Such a response is expected when there are negative feedback loops, such that high levels of compound can actually inhibit the effect, or when there are two or more different effects, such that the effects at the higher concentrations inhibit each other. The inverted U response is very important in testing the safety of any compound, since testing at high levels (standard in toxicology and teratology) does not necessarily reveal the compound's effect on development (see vom Saal and Welshons 2006).

The molecular biology of the BPA effect

For many years the low concentration effects of chemicals like BPA were dismissed, since its binding to the two forms of the nuclear estrogen receptor, ERα and ERβ, was 10,000 times less potent than estradiol, which is their natural ligand. Recent research, however, indicates three alternate mechanisms of action. The first is presumed interference with normal functioning of ERβ, at least in developing oocytes.

We noted earlier the work of Hunt's lab showing that BPA exposure could lead to a high degree of aneuploidy (an abnormal number of chromosomes in the daughter cells produced by meiosis) in mouse oocytes. Hunt's group has recently shown that a knockout of ERβ gives the same phenotype of aneuploidy as does adding BPA (Susiarjo et al. 2007). Moreover, BPA does not increase aneuploidy in this ERβ knockout, indicating that BPA somehow inactivates ERβ in wild-type mice. This inactivation could be produced by interference with estradiol binding, or perhaps by binding to an as-yet undescribed, meiosis-specific protein that acts through ERβ.

A second target of BPA could be the recently discovered plasma membrane estrogen receptor. Wozniak and colleagues (2005) found that nanomolar levels of estradiol bind to a plasma membrane estrogen receptor; BPA also binds to this receptor, and at similar low concentrations. The consequence of binding to this membrane receptor is the opening of a cal-

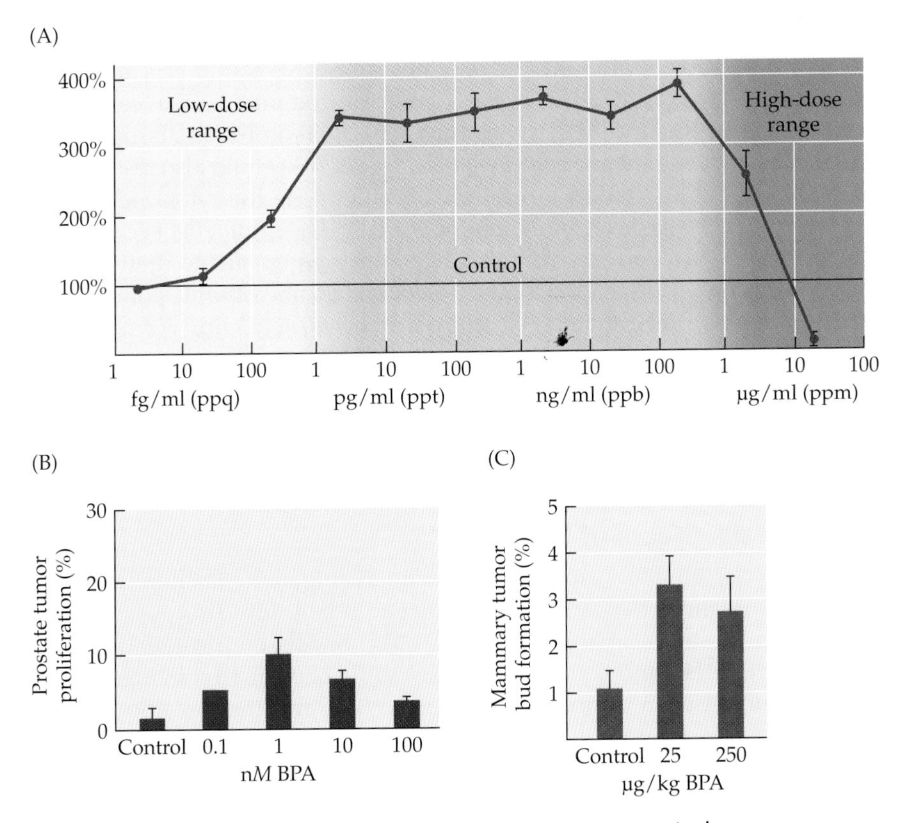

FIGURE 6.17 Endocrine disruptors often have a non-monotonic dose-response curve. (A) The inverted U shape of a typical non-monotonic curve indicates that medium-level dosage has the greatest effect and that neither low nor high doses of the compound produce as striking an effect. (B, C) This type of dose-response curve is revealed in histograms showing the effects of BPA on mammary terminal end bud formation (B) and prostate tumor growth (C). (A after Wellshons et al. 2003; B after Markey et al. 2001; C after Wetherill et al. 2002.)

cium channel, the rise in cytoplasmic calcium levels, and the activation of calcium-dependent responses in the cell. There may even be a fourth estrogen receptor that binds both naturally occurring estrogens and low concentrations of BPA (Matsushima et al. 2007).

A third target could be microtubule function. Microtubules are essential to meiosis and mitosis, being the principal structures forming the meiotic or mitotic apparatus—the structures along which the chromosomes move during cell division. The formation of the apparatus and the functioning of the microtubules depend on the polymerization of the microtubular protein tubulin. George and colleagues (2008) found that concentrations of BPA as low as 0.5 μ*M* disrupt mitosis in sea urchin embryos by altering the normal spatial regulation of microtubule polymerization. Similar effects also were observed in *Xenopus* cell-free extracts and cultured mammalian cells. This could explain the findings of the Hunt lab (2003) on BPA's disruption of meiosis with resultant aneuploidy. This effect of BPA could also result in all sorts of cellular aberrations, since microtubules form the cytoskeleton that determines the shape and function of the cell. The same effect could also explain the neural effects of BPA, since guiding neurons to their proper interconnections during development depends on microtubule polymerization (Bouquet and Nothias 2007). Moreover, because calcium is a regulator of tubulin polymerization (Dent et al. 2003), it is possible that the microtubular effects are derived from the activation of the membrane estrogen receptor. This research, then, provides mechanisms for BPA effects even at the extremely low levels to which many people are exposed daily.

Epigenetic effects of BPA

It is also possible that bisphenol A acts in an epigenetic fashion (through DNA methylation) that could lead to cancer in later life. This possibility is suggested by studies that looked at the effect of BPA on prostate development in mice. Ho and colleagues (2006) published the first evidence of a direct link between perinatal exposure to BPA and the development of prostate cancer later in life. BPA-exposed rats exposed to elevated testosterone and estrogen during adult life were more likely than unexposed animals to develop a type of cancerous lesion in the prostate—prostatic intraepithelial neoplasia (PIN)—that correlates to an early stage of prostate cancer in human males.

Importantly, PIN development may have an epigenetic origin in which BPA (or estrogen) reduces the methylation of specific regions of the *PDE4D4* gene. *PDE4D4* encodes a phosphodiesterase that degrades cyclic AMP (cAMP, a crucial regulatory molecule for normal cell growth and differentiation). It has been proposed that decreased methylation of *PDE4D4* allows for the production of greater than normal amounts of phosphodiesterase, which results in more cAMP degradation and thus less cAMP, with consequent abnormal cell proliferation and development (Ho et al. 2006).

An even more generalized hypomethylating effect of BPA is suggested by Dolinoy and coworkers (2007), who demonstrated that bisphenol A could alter the coat color of mice having the methylation-sensitive *viable Agouti* gene (see Chapter 2, pp. 44–45). Expression of this gene is sensitive to methylation, and Dolinoy and colleagues found that *viable Agouti* mice receiving bisphenol A in utero had decreased methylation at this locus (and at other loci) compared with mice that did not receive BPA.

What is of great concern is that the methylation patterns of the *viable Agouti* gene are stably inherited from one generation to the next. Thus, endocrine disruptors could alter the methylation patterns of germ cells, which could result in the propagation of this methylation pattern from one generation to the next. If so, the next generation would express the effects of the endocrine disruptor without having ever been exposed to the endocrine disruptor.

Polychlorinated biphenyls

Polychlorinated biphenyls (PCBs) are a family of about 200 related chemicals that have been used since 1928 in the manufacture of transformers, capacitors, flame retardants, hydraulic fluids, insulating coolants, and liquid seals (Colborn et al. 1996). Studies with mammals have shown PCBs to be associated with reproductive abnormalities, neurological and cognitive deficits, and thyroid deficits, as well as cancers and immune dysfunctions (Brouwer et al. 1999). PCBs have been associated with abnormalities of neural development in animals, and abnormalities of neural and psychological development were seen in Taiwanese children born to mothers accidentally exposed to high levels of PCBs in cooking oil (Schantz et al. 2003; Guo et al. 2004). Although banned in the 1970s, PCBs still contaminate landfills and the water system (Colborn et al. 1996; Domingo and Bocio 2007).

PCB exposure appears to affect reproductive development and has different impacts on males and females. In male mammals, studies have found that perinatal PCB exposure can cause cryptorchidism (failure of testes to descend), hypospadias (abnormalities of penis formation), testicular cancers, and low sperm count (Facemire et al. 1995; Den Hond and Schoeters 2006). In females, PCB exposure has led to a decrease in the age of menarche (onset of menstruation) and an increased incidence of endometriosis (Den Hond and Schoeters 2006). The highest concentration of PCBs is not found in urban areas but in the Arctic, where the food chain lacks primary producers and the diet is high in animal fat (AMAP 2002). Here, PCB concentrations in polar bears have been linked to malformations of the reproductive and genital organs (Christian et al. 2006).

Studies in mammals also show decreased immune response after consumption of PCB-laden fish oils. In seals, decreased size of the thymus has been documented, which results in a reduced immune response (Yoshimura 2003; Ross et al. 2004). PCB exposure affects thyroid function by influencing the hypothalamus-pituitary-thyroid axis through various mechanisms and leads to a decrease in thymus size in rhesus monkeys and other animals. Wang and colleagues (2005) have shown that in utero exposure to PCBs and dioxins causes feedback alterations in the hypothalamus, causing an overproduction of thyroid hormones. This in turn can cause the PCB-associated neurological disturbances which include deficits in visual recognition, short-term memory, and learning, as well as interference with sexual differentiation of the brain (Ribas-Fito et al. 2001; EPA-PCB).

Possible mechanisms for the effects of PCBs

Polychlorinated biphenyls can affect the estrogen, androgen, thyroid, retinoid, corticosteroid, and other endocrine pathways (Brouwer et al. 1999). Some PCBs interrupt estrogen metabolism through the cytochrome P450 enzyme system, inducing aromatase and thus increasing production of estradiol from testosterone (Brouwer et al. 1999). These same cytochrome P450 enzymes convert adrenal progesterone into cortisol and aldosterone, two steroids that can induce hypertension (Lin et al. 2006). Several PCB metabolites appear to bind to both the estrogen receptor and thyroid hormone receptors (Brouwer et al. 1999). PCBs have a weaker affinity for sex hormone binding globulins (SHBGs), which are proteins that bind excess steroids, than does estrogen. This property leaves PCBs free to bind with estrogen receptors when endogenous estrogen is bound up by SHGB.

Unbound ("free") PCBs often interfere with the production or elimination of naturally occurring compounds, and this can cause damage in several ways. One effect of PCBs is to decrease follicle-stimulating hormone (FSH) levels, thereby inhibiting the production of male Sertoli cells and consequently the production of sperm (Mol et al. 2002). Another detrimental effect of free PCBs is to induce the liver enzymes uridine-5-diphosphate-glucuronyltransferases (UGTs), which promote the elimination and excretion of essential thyroid hormones. Exposure to PCBs results in reduced plasma thyroid hormone (Brouwer et al. 1999). This can cause thyroid dysplasia and tumors (due to the lack of thyroid hormone inhibition on TSH, the hormone that instructs the thyroid follicle cells to divide).

PCBs also may directly affect the brain. Royland and colleagues (2008) showed that in utero exposure of rats to one particular PCB caused dramatic changes in the gene expression pattern in the hippocampus. The genes altered were predominantly those involved in intracellular signaling (such as the genes responsible for normal calcium signaling) and axonal guidance. These changes are consistent with the hypothesis that PCBs change neuronal growth patterns and that existing neurons do not function properly (Royland and Kodavanti 2008).

Transgenerational Effects of Endocrine Disruptors

One of the surprising lessons of the DES affair was that the effects of the endocrine disruptor lasted for at least another generation. Not only were the daughters of the women who took DES at risk for reproductive tract tumors and uterine abnormalities, *their daughters* were as well (Newbold et al. 1998, 2000.) In mice, both the female and the male grandpups of females exposed in utero to DES also had reproductive tract anomalies and tumors (Turusov et al. 1992; Walker and Kerth 1995; Walker and Haven 1997).

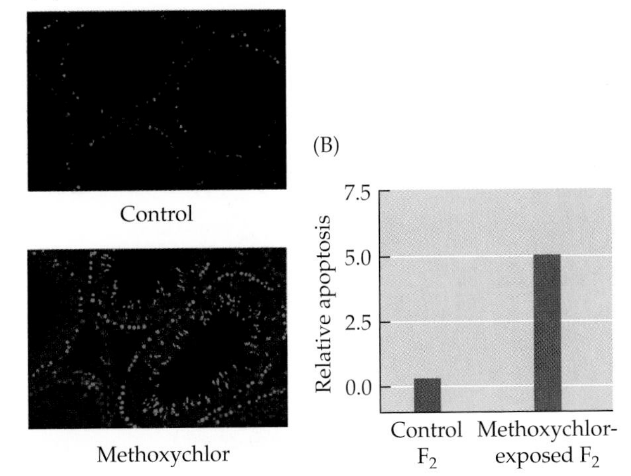

FIGURE 6.18 The effects of methoxychlor include testicular malformations and a drastic reduction of sperm production. (A) When cell death (apoptosis, which shows up here as green fluorescence) was examined, animals exposed in utero to methoxychlor had much greater cell death than controls. (B) A transgenerational effect is seen when the male offspring of a male mouse exposed in utero to methoxychlor are compared with the male offspring of a control (not exposed) male. (From Skinner and Anway 2005; photographs courtesy of Michael Skinner.)

Transgenerational effects have now been documented for certain pesticides and insecticides. Anway and colleagues (2005, 2006) demonstrated that two endocrine disruptors, the fungicide vinclozolin (used in vineyards) and methoxychlor (a widely used insecticide), impaired male fertility not only in the generation initially exposed in utero, but in males born for at least three generations afterward. Both methoxychlor and vinclozolin had been known to cause sterility in male rats born to mothers who had been injected with these pesticides late in pregnancy (Figure 6.18). However, this new study showed that if the pregnant rats were injected at a slightly earlier time, the offspring could still reproduce, but their testis cells underwent greater apoptosis than usual, their sperm count dropped 20%, and the sperm that remained were significantly less motile. Moreover, when affected males were mated with normal females, the resultant male offspring also had the same testicular dysgenesis syndrome. Some of the offspring were sterile and some had reduced fertility. The study ended after the fourth generation, with males continuing to show low sperm count, low sperm motility, prostate disease, and high rates of testicular cell apoptosis (Figure 6.19).

One possible mechanism by which the effect of endocrine disruptors might be transmitted from one generation to the next involves the altered methylation of the male germline cells (Figure 6.20). Chang and colleagues (2006) found several genes whose DNA methylation patterns were changed after exposure to vinclozolin, and these altered patterns were seen in genes expressed in the male germline (i.e., sperm precursor cells); they were also seen in the third generation of mice descended from a male originally exposed in utero to the fungicide.

One of the genes receiving a new methylation pattern is the gene for NCAM1, a cell adhesion molecule necessary for normal testis and brain development. When brains from third-generation mice were compared, the mice with the history of vinclozolin exposure (i.e., mice whose grandfathers were exposed to vinclozolin in utero) had less than 10% of the *NCAM1* gene expression of control mice (Chang et al. 2006). Further studies have shown

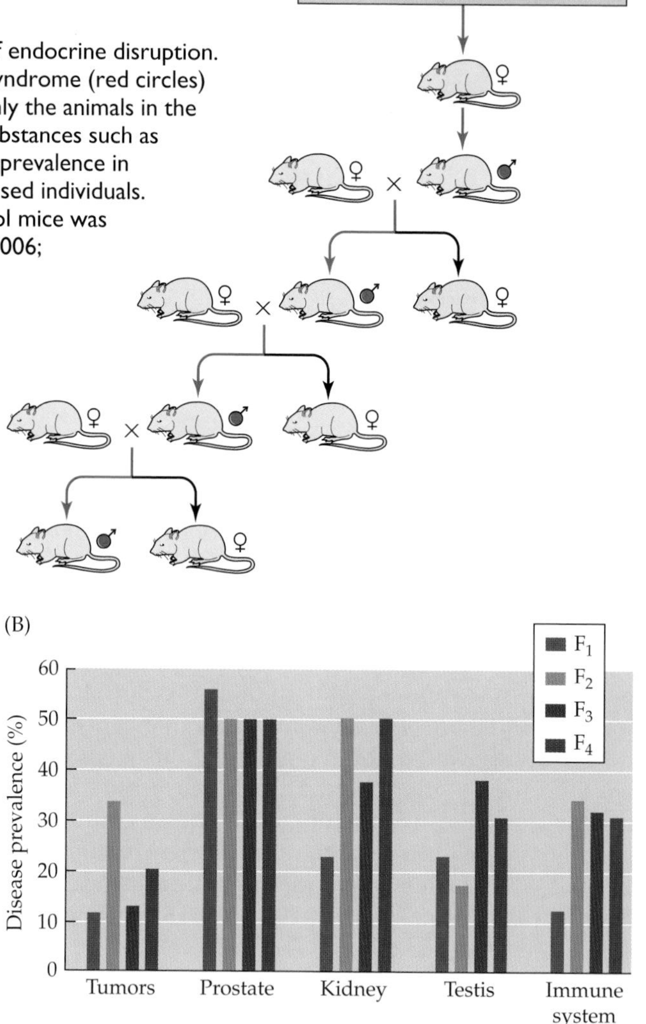

FIGURE 6.19 Epigenetic transmission of endocrine disruption. (A) Transmission of testicular dysgenesis syndrome (red circles) through four generations of male mice. Only the animals in the F_1 generation were exposed in utero to substances such as vinclozolin and methoxychlor. (B) Disease prevalence in multiple generations descended from exposed individuals. The presence of these conditions in control mice was essentially 0%. (After Anway and Skinner 2006; Anway et al. 2006.)

the male offspring of a male mouse exposed to viclozolin in utero to have altered methyltransferase activity in their germlines (Anway et al. 2008). The ability of endocrine disruptors to re-program germ cells has important implications for preventative medicine and public health.

DES may also be causing its transgenerational effects by stably altering the pattern of DNA methylation. As mentioned above, Li and colleagues (2003) have shown that perinatal exposure to DES alters methylation patterns in the promoters of many estrogen-responsive genes associated with the development of reproductive organs. Moreover, Ruden and colleagues (2005) have proposed a model wherein DES causes the demethylation of the c-*fos* gene in the epithelial cells of the Müllerian duct. This gene is a

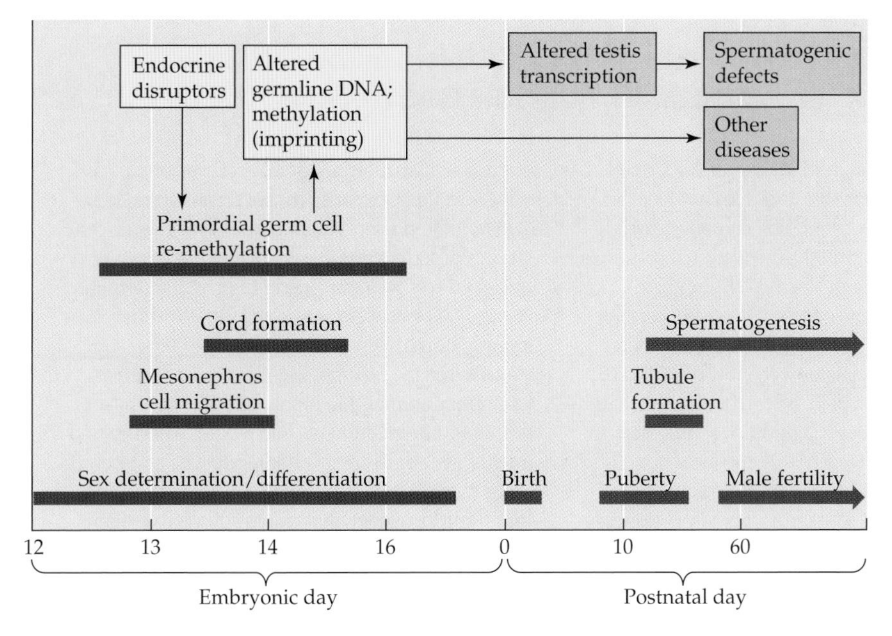

FIGURE 6.20 Possible schema by which endocrine disruptors alter germline methylation, which causes defects in both spermatogenesis (through the germline) and other organs (through the somatic cells derived from the germ cells).

growth promoter that can predispose cells to becoming tumors. Ruden and colleagues propose that alterations in Wnt signaling changes are required for the demethylation of the c-*fos* gene and that this demethylation might be inherited from one generation to the next, thereby predisposing this entire lineage to tumors (Ruden et al. 2005).

Summary

The environment can act in many ways during an organism's development. It can act in its normal role to produce the expected phenotype (as in mammalian/bacterial developmental symbiosis). It can act instructively to enable an organism to develop an alternate phenotype predicted to be especially fit for a given environment (as in polyphenisms). It can act in an overtly deleterious manner, disrupting normal development (as in teratogenesis). And it can act disruptively but covertly, as when endocrine disruptors modify gene expression in a manner that does not become apparent until much later in life.

Endocrine disruptors differ from classical teratogens in numerous ways. First, they can have effects at low doses, rather than having to be present at

Regulatory and Policy Decisions on BPA and Other Endocrine Disruptors

The Food and Drug Administration of the United States holds that BPA is safe and does not recommend that consumers avoid BPA-containing products (FDA 2008a,b). Similarly, the National Institutes of Environmental Health Science (NIEHS 2008), while expressing "some concern" over the possible effects of BPA on the developing brain, has not recommended legislation against it. However, in 2007, a consensus statement of 38 researchers in the field called for an immediate ban on BPA, stating that currently prevalent levels of this chemical in humans are already higher than those shown to cause developmental damage to laboratory animals (vom Saal et al. 2007).

The government agencies base their policy decisions on the ambiguity surrounding the assessments of BPA's safety, and on the fact that no scientific experiments have yet shown that the chemical is harmful to human fetuses. But this ambiguity, according to the independent scientists, has been manipulated by the chemical industry. These researchers have claimed that, whereas all 14 of the studies showing that BPA is safe were funded by the chemical industry, of the 204 studies funded through government grants, 93% concluded that BPA is harmful (vom Saal 2008). As such a dichotomy points out, there are numerous problems, both scientific and political, in assessing the effects of BPA or any other endocrine disruptor on human individuals and populations.

Scientific issues

The first set of problems is scientific. One overarching issue, as noted several times in this chapter, is that the effects of BPA and other endocrine disruptors are not anatomically visible and are often not functionally apparent until adult life. This temporal disconnect makes it difficult to be certain that problems observed in adults are the result of prenatal exposure to a given substance.

In virtually all biomedical research, there are issues based on the differences between laboratory animals and humans. This is especially relevant to endocrine disruptor research, because animals differ widely in their responses to hormone disruptors. (This problem was evident in the case of teratogen studies described in Chapter 5, in which laboratory mice and rats were not susceptible to thalidomide, whereas rabbits were sensitive.) Genetic differences among individuals affect sensitivity to these substances (see the box, pp. 216–217). And, whereas laboratory animals are highly inbred, humans are not; thus, among humans one would expect to find greater genetic differences in regulatory and homeostatic mechanisms between individuals, meaning that some humans will be more susceptible than others to different chemicals. This means that the data on humans will probably be epidemiological and not as clear-cut as the evidence from rodents. Such epidemiological data will be difficult to document properly, however, since there is no control group—the entire human population is exposed to these ubiquitous compounds.

Another confounding factor in endocrine disruptor research is the interaction of environmental estrogens. Silva and colleagues (2002) have shown that if cells are exposed to a mixture of environmental estrogens, each of which is at a concentration that induces an estrogen-responsive gene only very weakly, the mixture can induce a response that is much more than the additive responses of the individual components (see the figure). Moreover, Rajapakse and colleagues (2001, 2002) have shown that when normal estrogen is supplemented with amounts of environmental xenoestrogens so small

A mixture of estrogenic compounds at low concentrations activates an estrogen-responsive gene. Plasmids containing an estrogen response element attached to a *lacZ* gene were added to cells, and each of eight environmental estrogens was added at concentrations that only weakly induced the expression of the *lacZ* gene. However, when a mixture of these eight compounds was added, the response was far greater than the additive values of the separate responses. (After Silva et al. 2002.) ▶

that they have inconsequential effects alone, the small amount of exogenous xenoestrogen potentiates the effect of normal estrogen and makes it much greater. This may make the mixture of endocrine disruptors we receive daily even more problematic for us and our offspring. It also makes blaming a particular compound for a specific effect very difficult.

Political issues

In the United States, rigorous scientific proof is needed to claim that the product is harmful. As in a homicide case, the product is assumed to be innocent until proven guilty beyond a reasonable doubt. In past instances of environmental concern (lead, mercury, and asbestos toxicity, tobacco, global warming, and ozone depletion), the affected industries have financed scientific studies that have been able to raise such doubt (see Markowitz and Rossner 2000; Steingraber 2001; Michaels 2005, 2008). In 2008, when researchers at Yale University (Leranth et al. 2008) showed that BPA disrupted brain development in monkeys (at concentrations lower than what the EPA considers safe), the American Chemical Council, an industry trade group, replied that "there is no direct evidence that exposure to bisphenol A adversely effects human reproduction or development" (see Layton 2008). But the "catch-22" exists in that such "direct evidence" (either for or against) may be impossible to obtain, given that we cannot test drugs on human fetuses.

Another series of problems comes from the way that chemicals are patented and licensed. Heavily brominated chemicals, such as polybrominated diphenyl ethers (PBDEs), are a family of chemicals used as flame retardants in fabrics, furniture, and plastics. These substances, which are related to PCBs, were discovered to bioaccumulate in humans and wildlife and to be hazardous to human health. Some of them have been banned in the United States and Europe, some are banned only in Europe, and some are banned in a few particular U.S. states. When one of these compounds is banned, however, the industry can easily manufacture another that is not much different from the original. Most of these alternative retardants still contain bromine and similarly bioaccumulate. Donald Kennedy, editor of the journal *Science,* compared the regulation of PBDEs to the game of "Whack-a-Mole," in which whacking down one problem only leads to another one popping up (Kennedy 2007).

Under current regulations, a new chemical cannot be distributed in the United States until the manufacturer has submitted data to the U.S. Environmental Protection Agency detailing the properties of the chemical and probability of any harmful effects. The EPA then has 90 days to respond. But the required data do not currently include assessment of endocrine disruptor or teratogenic effects, and so far lobbyists have prevented implementation of changes intended to legislate such testing (Michaels 2008). These tests, they note, would be extremely expensive, which would retard innovation and make products more costly to consumers.

(*continued on next page*)

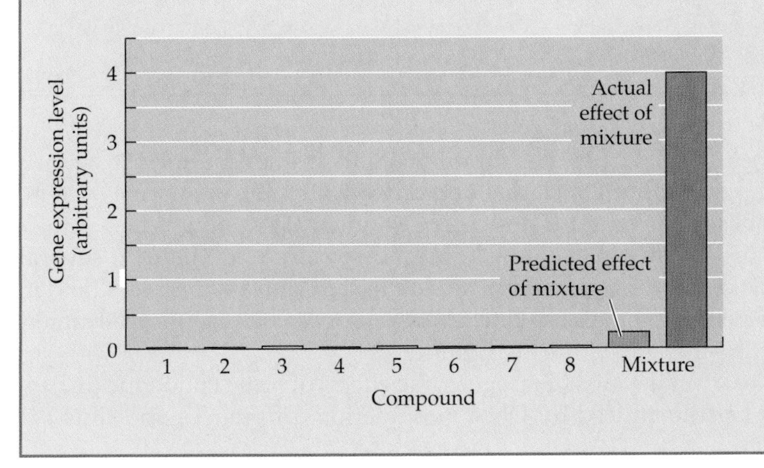

Regulatory and Policy Decisions on BPA and Other Endocrine Disruptors (*continued*)

International perspectives

In contrast to the "harmless unless proven otherwise" approach common in the United States, some Western European nations are apt to invoke the "precautionary principle" when approaching issues of public health and safety (see Harremoes et al. 2002). This principle asserts that when an action or a policy could seriously or irreversibly damage the public welfare or the environment, in the lack of full scientific certainty that harm will not take place, the burden of proof falls on those who advocate the policy or action (Raffensberg and Tickner 1999). This, however, calls for a lot of judgment as to what constitutes irreversible or serious damage, how much scientific data are needed for such a case to be made, and how one balances the risk of bad outcomes versus the potential gains from the innovation (Sunstein 2005).

Currently the European Union (EU) is attempting to develop a new screening approach for chemicals, referred to as REACH (Registration, Evaluation, and Authorization of Chemicals). This program aims to develop tests to assess the toxicity of thousands of chemicals, including endocrine disruptor and reproductive and developmental toxicity assays. The size of the EU market may result in most international chemical companies conforming to the EU guidelines (see Schapiro 2007).

Right now, public relations and consumer confidence seem to be the major forces driving the regulation of potential endocrine disruptors. Canada, for instance, has banned the use of BPA in the manufacture of baby bottles, while the United States, the European Union, and Japan still allow it. Commercial manufacturers, however, have independently moved to discontinue BPA-containing products. Wal-Mart has stopped selling BPA-containing baby bottles in Canada and plans to discontinue these products in the United States in 2009. Similarly, Nalgene (the producer of those brightly colored reusable water bottles) has stated that it will discontinue the use of BPA in the production of these bottles and will recall BPA-containing products already in stores (Layton and Lee 2008).

Because endocrine disruptors recognize no political boundaries and persist through many administrations and regimes, international agencies can be critical. As of May, 2001, The Stockholm Conventions of the United Nations had been signed by delegates from 122 nations (including the United States). This treaty, which seeks to regulate certain industrial pollutants, incorporates the precautionary principle (Kaiser and Enserink 2000; UNEP 2008). However, it has not gone into effect because it has not been ratified by the necessary 50 nations and, as of September 2008, the United States Congress has not endorsed it.

The crucial question, as the authors of *Our Stolen Future* (Colborn et al. 1996) ask, is "Are we threatening our fertility, intelligence, and survival?" And how much is at risk if the answer should turn out to be "yes"?

high doses. Second, they are ubiquitous, rather than confined to a small exposed population. Third, by operating through the endocrine system, they establish an inverted-U dose curve, wherein high amounts may actually have less of an effect than slightly lower amounts. Fourth, several endocrine disruptors are usually present in the same environment (and in the same person), and each endocrine disruptor can be synergistically interactive with other endocrine disruptors.

Moreover, unlike most teratogens, the effects of some endocrine disruptors can be transmitted by DNA methylation differences and thus are transgenerational. Endocrine disruptors can alter DNA methylation in all cells (including the precursors of the germline cells), and if these methylation patterns take place in the germ cells, the change can become stably inherited. As we will see in the next chapter, other types of stress during fetal development, such as low-protein diets, can act very similarly to endocrine disruptors and modify gene expression patterns to generate phenotypes that are not apparent at birth but that show up later in the adult.

References

Adeoya-Osiguwa, S. A., S. Markoulaki, V. Pocock, S. R. Milligan and L. R. Fraser. 2003. 17β Estradiol and environmental estrogens significantly affect mammalian sperm function. *Human Reprod.* 18: 100–107.

Aitken, R. J., P. Koopman and S. E. Lewis. 2004. Seeds of concern. *Nature* 432: 48–52.

AMPA (Arctic Monitoring and Assessment Programme). 2002. *Arctic Pollution 2002.* Oslo, Norway.

Anway, M. D., A. S. Cupp, M. Uzumcu and M. K. Skinner. 2005. Epigenetic transgenerational actions of endocrine disruptors and male fertility. *Science* 308: 1466–1469.

Anway M. D., C. Leathers and M. K. Skinner. 2006. Endocrine disruptor vinclozolin induced epigenetic transgenerational adult onset disease. *Endocrinology* 147: 5515–5523.

Anway, M. D. and M. K. Skinner. 2006. Epigenetic transgenerational actions of endocrine disruptors. *Endocrinology* 147 (suppl.): S43–S49.

Anway M.D., S. S, Rekow, and M. K. Skinner. 2008. Transgenerational epigenetic programming of the embryonic testis transcriptome. *Genomics* 91: 30–40.

Auger, J., J. M. Kunstmann, F. Czyglick and P. Jouannet. 1995. Decline in semen quality among fertile men in Paris during the past 20 years. *New Engl. J. Med.* 332: 281–285.

Bell, S. E. 1986. A new model of medical technology development: A case study of DES. *Sociol. Health Care* 4: 1–32.

Bell, S. E. 1995. Gendered medical science: Producing a drug for women. *Feminist Studies* 21: 469–500.

Benson, G. V., H. Lim, B. C. Paria, I. Satokata, S. K. Dey and R. Maas. 1996. Mechanisms of female infertility in *Hoxa-10* mutant mice: Uterine homeosis versus loss of maternal *Hoxa-10* expression. *Development* 122: 2687–2696.

Block, K., A. Kardanga, P. Igarashi and H. S. Taylor. 2000. In utero DES exposure alters Hox gene expression in the developing Müllerian system. *FASEB J.* 14: 1101–1108.

Blumenstyk, G. 2003. The story of Syngenta and Tyrone Hayes at UC Berkeley: A Berkeley scientist says a corporate sponsor tried to bury his unwelcome findings and then buy his silence. *Chron. Higher Ed.* 50 (10) October. http: //www.mindfully.org/Pesticide/2003/Syngenta-Tyrone-Hayes31oct03.htm.

Bouquet, C. and F. Nothias. 2007. Molecular mechanisms of axonal growth. *Adv. Exp. Med. Biol.* 621: 1–16.

Boutin, E. L. and G. R. Cunha. 1997. Estrogen-induced proliferation and cornification are uncoupled in sinus vaginal epithelium associated with uterine stroma. *Differentiation* 62: 171–178.

Brodkin, M. A., H. Madhoun, M. Rameswaran and I. Vatnick. 2007. Atrazine is an immune disruptor in adult northern leopard frogs (*Rana pipiens*). *Environ. Toxicol. Chem.* 26: 80–84.

Brosens, J. J. and M. G. Parker. 2003. Oestrogen receptor hijacked. *Nature* 423: 487–488.

Brouwer, A. and 7 others. 1999. Characterization of potential endocrine-related health effects at low-dose levels of exposure to PCBs. *Environ. Health Persp.* 107 (suppl. 4): 639–649.

Cabanes, A., M. Wang, S. Olivo, S. DeAssis, J. A. Gustafsson, G. Khan and L. Hilakivi-Clarke. 2004. Prepubertal estradiol and genistein exposures upregulate *BRCA1* mRNA and reduce mammary tumorigenesis. *Carcinogenesis* 25: 741–748.

Calafat, A. M., Z. Kuklenyik, J. A. Reidy, S. P. Caudill, J. Ekong and L. L. Needham. 2005. Urinary concentrations of bisphenol A and 4-nonylphenol in a human reference population. *Env. Health Persp.* 113: 391–395.

Calafat , A. M., X. Ye, L. Y. Wong, J. A. Reidy and L. L. Needham. 2008. Exposure of the U.S. population to bisphenol A and 4-tertiary-octylphenol, 2003–2004. *Env. Health Persp.* 116: 39–44.

Capel, P. and S. Larson. 2001. Effect of scale on the behavior of atrazine in surface waters. *Environ. Sci. Tech.* 35: 648–657.

Carlsen, E., A. Giwercman, N. Keiding and N. Skakkebæk. 1992. Evidence for decreasing quality of semen during past 50 years. *Brit. Med. J.* 305: 609–613.

Carson, R. 1962. *Silent Spring.* Houghton Mifflin, New York.

Carta, L. and D. Sassoon. 2004. Wnt7a is a suppressor of cell death in the female reproductive tract and is required for postnatal and estrogen-mediated growth. *Biol. Reprod.* 71: 444–454.

Cederroth, C. R. and 9 others. 2007. A phytoestrogen-rich diet increases energy expenditure and decreases adiposity in mice. *Env. Health Persp.* 115: 1467–1473.

Chang, H.-S., M. D. Anway, S. S. Rekow and M. K. Skinner. 2006. Transgenerational epigenetic imprinting of the male germline by endocrine disruptor exposure during gonadal sex determination. *Endocrinology* 147: 5524–5541.

Chavarro, J. E., T. L. Toth, S. M. Sadio and R. Hauser. 2008. Soy food and isoflavone intake in relation to semen quality parameters among men from an infertility clinic. *Human Reprod.* DOI: 10.1093/humrep/den243.

Christian, S. and 8 others. 2006. Xenoendocrine pollutants can reduce size of sexual organs in East Greenland polar bears (*Ursus maritimus*). *Environ. Sci. Tech.* 40: 5668–5674.

Christin, M.-S. and 7 others. 2003. Effects of agricultural pesticides on the immune system of *Rana pipiens* and on its resistance to parasitic infection. *Environ. Toxicol. Chem.* 22: 1127–1133.

Colborn, T., F. S. vom Saal and A. M. Soto. 1993. Developmental effects of endocrine-disrupting chemicals in wildlife and humans. *Environ. Health Persp.* 101: 378–384.

Colborn, T., D. Dumanoski and J. P. Myers. 1996. *Our Stolen Future*. Dutton, New York.

Cook, J. D., B. J. Davis, S. L. Cai, J. C. Barrett, C. J. Conti and C. L. Walker. 2005. Interaction between genetic susceptibility and early-life environmental exposure determines tumor-suppressor gene penetrance. *Proc. Natl. Acad. Sci. USA* 102: 8644–8649.

Cooke, A. S. 1973. Shell thinning in avian eggs by environmental pollutants. *Environ. Pollut.* 4: 85–152.

Crain, D. A. and L. L. Guillette, Jr. 1998. Reptiles as models of contaminant-induced endocrine disruption. *Anim. Reprod. Sci.* 53: 77–86.

Crain, D. A., A. Rooney, E. F. Orlando and L. J. Guillette. 1998. Endocrine-disrupting contaminants and hormone dynamics: Lessons from wildlife. In L. L. Guillette, Jr., and D. A. Crain (eds.), *Environmental Endocrine Disruptors: An Evolutionary Perspective*. Taylor & Francis, New York, pp. 1–21.

Davis, D. L., H. L. Bradlow, M. Wolff, T. Woodruff, D. G. Hoel and H. Anton-Culver. 1993. Xenoestrogens as preventable causes of breast cancer. *Environ. Health Persp.* 101: 372–377.

Den Hond, E. and G. Shoeters. 2006. Endocrine disrupters and human puberty. *Internatl. J. Androl.* 29: 264–271.

Dent, E. W., F. Tang and K. Kalil. 2003. Axon guidance by growth cones and branches: Common cytoskeletal and signaling mechanisms. *Neuroscientist* 9: 343–353.

Dolinoy, D. C., D. Huang and R. L. Jirtle. 2007. Maternal nutrient supplementation counteracts bisphenol A–induced DNA hypomethylation in early development. *Proc. Natl. Acad. Sci. USA* 104: 13056–13061.

Domingo, J. L. and A. Bocio. 2007. Levels of PDCC/PCDFs and PCBs in edible marine species and human uptake: A literature review. *Environ. Internatl.* 33: 397–405.

Durando, M., L. Kass, J. Piva, C. Sonnenschein, A. M. Soto, E. H. Luque and M. Muñoz-de-Toro. 2007. Prenatal bisphenol A exposure induces preneoplastic lesions in the mammary gland in Wistar rats. *Environ. Health Persp.* 115: 80–86.

EPA-PCB. 2008. www.epa.gov/pcb/pubs/effects/html.

Facemire, C. F., T. S. Gross and L. J. Guillette, Jr. 1995. Reproductive impairment in the Florida panther: Nature or nurture, *Environ. Health Persp.* 103 (suppl. 4): 79–86.

Fairchild, W. L., E. O. Swansburg, J. T. Arsenault and S. B. Brown. 1999. Does an association between pesticide use and subsequent declines in catch of Atlantic salmon (*Salmo salar*) represent a case of endocrine disruption? *Environ. Health Perspect.* 107: 349–358.

Fan, W. and 10 others. 2007. Atrazine-induced aromatase expression is SF1-dependent: Implications for endocrine disruption in wildlife and reproductive cancers in humans. *Environ. Health Persp.* 115: 720–727.

FDA. 2008a. http://www.fda.gov/oc/opacom/hottopics/bpa.html

FDA. 2008b. http://www.fda.gov/ohrms/dockets/ac/08/briefing/2008-0038b1_01_02_FDA%20BPA%20Draft%20Assessment.pdf

Filipov, N. M., L. M. Pinchuk, B. L. Boyd and P. L. Crittenden. 2005. Immunotoxic effects of short-term atrazine exposure in young male C57BL/6 mice. *Toxicol. Sci.* 86: 324–32.

Fisher, J. S., S. Macpherson, N. Marchetti and R. M. Sharpe. 2003. Human testicular dysgenesis syndrome: A possible model using in utero exposure of the rat to dibutyl phthalate. *Human Reprod.* 18: 1383–1394.

Forson, D. and A. Storfer. 2006a. Effects of atrazine and iridovirus infection on survival and life history traits of the long-toed salamander (*Ambystoma macrodactylum*). *Environ. Toxicol. Chem.* 25: 168–173.

Forson, D. and A. Storfer. 2006b. Atrazine increases Ranavirus susceptibility in the tiger salamander, *Ambystoma tigrinum*. *Ecol. Applic.* 16: 2325–2332.

Gendron, A. and 7 others. 2003. Exposure of leopard frogs to a pesticide mixture affects life history characteristics of the lungworm *Rhabdias ranae*. *Oecologia* 135: 469–476.

George, O., B. K. Bryant, R. Chinnasamy, C. Corona, J. B. Arterbum and C. B. Shuster. 2008. Bisphenol A directly targets tubulin to disrupt spindle organization in embryonic and somatic cells. *ACS Chem. Biol.* 3: 167–179.

Gilbert, S. F. 2006. *Developmental Biology*, 8th Ed. Sinauer Associates, Sunderland, MA.

Grey, L. E., Jr., J. Ostby, R. L. Cooper and W. R. Kelce. 1999. The estrogenic and antiandrogenic pesticide methoxychlor alters the reproductive tract and

behavior without affecting pituitary size or LH and prolactin secretion in male rats. *Toxicol. Indust. Health*. 15: 37–47.
Guillette, L. J., Jr. 2006. Endocrine disrupting contaminants: Beyond the dogma. *Environ. Health Persp.* 114 (suppl 1): 9–12.
Guillette, L. J., Jr., T. M. Edwards and B. C. Moore. 2007. Alligators, contaminants and steroid hormones. *Environ. Sci.* 14: 331–347.
Guillette, L. J., Jr., T. S. Gross, G. R. Masson, J. M. Matter, H. F. Percival and A. R. Woodward. 1994. Developmental abnormalities of the gonad and abnormal sex hormone concentrations in juvenile alligators from contaminated and control lakes in Florida. *Environ. Health Persp.* 102: 680–688.
Guo, Y. L., G. H. Lambert, C. C. Hsu and M. M. Hsu. 2004. Yucheng: Health effects of prenatal exposure to polychlorinated biphenyls and dibenzofurans. *Internatl. Arch. Occup. Environ. Health* 77: 153–158.
Harremoes, P., D. Gee, M. MacGarvin, A. Stirling, J. Keys, B. Wynne and S. G. Vaz (eds.). 2002. *The Precautionary Principle in the 20th Century: Late Lessons from Early Warnings.* Earthscan Publications Ltd., London.
Hayes, T. 2005. Welcome to the revolution. Integrative biology and assessing the impact of endocrine disruptors on environmental and public health. *Integr. Comp. Biol.* 45: 321–329.
Hayes, T. 2007. http://www.atrazinelovers.com/m7.html.
Hayes, T., K. Haston, M. Tsui, A. Hoang, C. Haeffele and A. Vonk. 2003. Atrazine-induced hermaphroditism at 0.1 ppb in American leopard frogs (*Rana pipiens*): Laboratory and field evidence. *Environ. Health Persp.* 111: 568–575.
Hayes, T. and 10 others. 2006. Pesticide mixtures, endocrine disruption and amphibian declines: Are we underestimating the impact? *Environ. Health Persp.* 114 (suppl 1): 40–50.
Heindel, J. 2006. Role of exposure to environmental chemicals in the developmental basis of reproductive disease and dysfunction. *Semin. Reprod. Med.* 24: 168–177.
Heindel, J. 2007. *Endocrine disruption: Status and outlook.* Talk presented at the symposium Endocrine Disruptors: Relevance to Humans, Animals and Ecosystems in Macolin, Switzerland.
Hennion, M., V. Pichon, P. Legeay and M. Cohen. 2004. *A ten-year survey of ground water: High persistence of some pesticide metabolites* (Abstract). Presented at the Ninth Symposium on Chemistry and Fate of Modern Pesticides, 16–19 August 2004, Vail, CO.
Herman-Giddens, M. E., E. J. Slora, R. C. Wasserman, C. J. Bourdony, M. V. Bhapkar, G. G. Koch and C. M. Hasemeir. 1997. Secondary sexual characteristics and menses in young girls seen in office practice: A study from the pediatric research in office settings network. *Pediatrics* 99: 505–512.
Ho, S.-M., W.-Y. Tang, J. Belmonte de Frausto and G. S. Prins. 2006. Developmental exposure to estradiol and bisphenol A increases susceptibility to prostate carcinogenesis and epigenetically regulates phosphodiesterase type 4 variant 4. *Cancer Res.* 66: 5624–5632.
Holm, L., A. Blomqvist, I. Brandt, B. Brunstrom, Y. Ridderstrale and C. Berg. 2006. Embryonic exposure to *o,p_*-DDT causes eggshell thinning and altered shell gland carbonic anhydrase expression in the domestic hen. *Environ. Toxicol. Chem.* 25: 2787–2793.
Howdeshell, K. L., A. K. Hotchkiss, K. A. Thayer, J. G. Vandenbergh and F. S. vom Saal. 1999. Plastic bisphenol A speeds growth and puberty. *Nature* 401: 762–764.
Howdeshell, K. L. and F. S. vom Saal. 2000. Developmental exposure to bisphenol A: Interaction with endogenous estradiol during pregnancy in mice. *Amer. Zool.* 40: 429–437.
Howdeshell, K. L. and 7 others. 2003. Bisphenol A is released from used polycarbonate animal cages into water at room temperature. *Environ. Health Persp.* 111: 1180–1187.
Huang, P. C., P. L. Kuo, Y. Y. Chou, S. J. Lin and C. C. Lee. 2008. Association between prenatal exposure to phthalates and the health of newborns. *Environ. Internatl.* DOI: 10.1016/jenvint.2008.05.012.
Hunt, P. A. and 8 others. 2003. Bisphenol A exposure causes meiotic aneuploidy in the female mouse. *Curr. Biol.* 13: 546–553.
Ikezuki, Y., O. Tsutsumi, Y. Tahai, Y. Kamzi and Y. Taketa. 2002. Determination of bisphenol A concentrations in human biological fluids reveals significant prenatal exposure. *Human Reprod.* 17: 2839–2841.
Irvine, S., E. Cawood, D. Richardson, E. MacDonald and J. Aitken. 1996. Evidence of deteriorating semen quality in the United Kingdom: Birth cohort study in 577 men in Scotland over 11 years. *Brit. Med. J.* 312: 467–471.
Jefferson, W. N., E. Padilla-Banks and R. R. Newbold. 2007. Disruption of the developing female reproductive system by phytoestrogens: Genistein as an example. *Mol. Nutr. Food Res.* 51: 832–844.
Ji, L., R. C. Domanski, R. C. Skirrow and C. C. Helbing. 2007. Genistein prevents thyroid hormone-dependent tail regression of *Rana catesbeiana* tadpoles by targeting protein kinase C and thyroid hormone receptor α. *Devel. Dynam.* 236: 777–790.
Kaiser, J. and M. Enserink. 2000. Treaty takes a POP at the dirty dozen. *Science* 290: 2053.
Karrow, N. A., J. A. McCay, R. D. Brown, D. L. Musgrove, T. L. Guo, D. R. Germolec and K. L. White, Jr. 2005. Oral exposure to atrazine modulates cell-mediated immune function and decreases host resistance to the B16F10 tumor model in female B6C3F1 mice. *Toxicology* 209: 15–28.
Kelce, W. R., C. R. Stone, S. C. Laws, L. E. Gray, J. A. Kemppainen and E. M. Wilson. 1995. Persistent DDT

metabolite *p,p*_-DDE is a potent androgen receptor antagonist. *Nature* 375: 581–585.

Kennedy, D. 2007. Toxic dilemmas. *Science* 318: 1217.

Kettles, M. A., S. R. Browning, T. S. Prince and S. W. Horstman. 1997. Triazine exposure and breast cancer incidence: An ecologic study of Kentucky counties. *Environ. Health Perspect.*105: 1222–1227.

Kiesecker, J. M. 2002. Synergism between trematode infection and pesticide exposure: A link to amphibian limb deformities in nature? *Proc. Natl. Acad. Sci. USA* 99: 9900–9904.

Kim, H. S. and 8 others. 2002. Comparative estrogenic effects of *p*-nonylphenol by 3-day uterotrophic assay and female pubertal onset assay. *Reprod. Toxicol.* 16: 259–268.

Knight, W. M. 1980. Estrogens administered to food-producing animals: Environmental considerations. In J. McLachlan (ed.), *Estrogens in the Environment*. Elsevier, Amsterdam, pp. 391–401.

Kohno, S., D. S. Bermudez, Y. Katsu, T. Iguchi and L. B. Guillette, Jr. 2008. Gene expression patterns in juvenile American alligators (*Alligator mississippiensis*) exposed to environmental contaminants. *Aquatic Toxicol.* 88: 95–101.

Kolpin, D., J. Barbash and R. Gilliom. 1998. Occurrence of pesticides in shallow groundwater of the United States: Initial results from the National Water Quality Assessment Program. *Environ. Sci. Technol.* 32: 558–566.

Krimsky, S. 2000. *Hormonal Chaos: The Scientific and Social Origins of the Environmental Endocrine Hypothesis.* Johns Hopkins University Press, Baltimore.

Krishnan, A. V., P. Starhis, S. F. Perlmuth, I. Tokes and D. Feldman. 1993. Bisphenol A: An estrogenic substance is released from polycarbonate flasks during autoclaving. *Endocrinology* 132: 2279–2286.

Kubo, K., O. Arai, M. Omura, R. Watanabe, R. Ogata and S. Aou. 2003. Low-dose effects of bisphenol A on sexual differentiation of the brain and behavior in rats. *Neurosci. Res.* 45: 345–356.

Kurihara, R., E. Watanabe, Y. Ueda, A. Kakuno, K. Fujiki, F. Shiraishi and S. Hishimoto. 2007. Estrogenic activity in sediments contaminated by nonylphenol in Tokyo Bay (Japan) evaluated by vitellogenin induction in male mummichogs (*Fundulus heteroclitus*). *Mar. Pollut. Bull.* 54: 1315–1320.

Lannoo, M. 2008. *Malformed Frogs: The Collapse of Aquatic Ecosystems.* University of California Press, Berkeley.

Layton, L. 2008. BPA linked to primate health issues. *Washington Post*, September 4, 2008, p. A2.

Layton, L. and C. Lee. 2008. Canada bans BPA from baby bottles. *Washington Post,* April 19, 2008, p. A3.

Lear, L. 1998. *Rachel Carson: Witness for Nature.* Holt, New York.

Leranth, C., T. Hajszan, K. Szigeti-Buck, J. Bober and N. J. Maclusky. 2008. Bisphenol A prevents the synaptogenic response to estradiol in hippocampus and prefrontal cortex of ovariectomized nonhuman primates. *Proc. Natl. Acad. Sci. USA*, in press.

Li, S., S. D. Hursting, B. J. Davis, J. A. McLachlan and J. C. Barrett. 2003. Environmental exposure, DNA methylation, and gene regulation: Lessons from diethylstilbestrol-induced cancers. *Ann. N.Y. Acad. Sci.* 983: 161–169.

Lin, T.-C. E., S.-C. Chien, P.-C. Hsu and L.-A. Li. 2006. Mechanistic study of polychlorinated bisphenyl 126-induced CYP11B1 and CYP11B2 upregulation. *Endocrinology* 147: 1536–1544.

Lundholm, C. D. 1997. DDE-induced eggshell thinning in birds: Effects of p,p_-DDE on the calcium and prostaglandin metabolism of the eggshell gland. *Comp. Biochem. Physiol. C: Pharmacol. Toxicol. Endocrinol.* 118: 113–128.

Ma, L., G. V. Benson, H. Lim, S. K. Dey and R. Maas. 1998. Abdominal B (*AbdB) Hoxa* genes: Regulation in adult uterus by estrogen and progesterone and repression in Mullerian duct by the synthetic estrogen diethylstilbestrol (DES). *Dev. Biol.* 197: 141–154.

MacLennan, P. and 7 others. 2002. Cancer incidence among triazine herbicide manufacturing workers. *J. Occup. Environ. Med.* 44: 1048–1058.

Mahood, I. K., H. M. Scott, R. Brown, N. Hallmark, M. Walker and R. M. Sharpe. 2007. In utero exposure to di(n-butyl)phthalate and testicular dysgenesis: Comparison of fetal and adult endpoints and their dose sensitivity. *Environ. Health Persp.* 115 (Supp 1): 55–61.

Markey, C. M., M. A. Coombs, C. Sonnenschein and A. M. Soto. 2003. Mammalian development in a changing environment: Exposure to endocrine disruptors reveals the developmental plasticity of steroid hormone target organs. *Evol. Dev.* 5: 67–75.

Markey, C. M., E. H. Luque, M. Muñoz-de-Toro, C. Sonnenschein and A. M. Soto. 2001. In utero exposure to bisphenol A alters the development and tissue organization of the mouse mammary gland. *Biol. Reprod.* 65: 1215–1223.

Markowitz, G. and D. Rosner. 2000. "Cater to the children": The role of the lead industry in a public health tragedy, 1900–1955. *Am. J. Publ. Health* 90: 36–46.

Matsushima, A. and 9 others. 2007. Structural evidence for endocrine disruptor bisphenol A binding to human nuclear receptor ERR gamma. *J. Biochem.* 142: 517–524

Matter, J. M. and 8 others. 1998. Effects of endocrine-disrupting contaminants in reptiles: Alligators. In R. J. Kendall, R. L. Dickerson, J. P. Geisy and W. A. Suk (eds.), *Principles and Processes for Evaluating Endocrine Disruptions in Wildlife*. SETAC Press, Pensacola, FL, pp. 267–289.

McCoy, K. A., L. J. Bortnick, C. M. Campbell, H. J. Hamlin, L. J. Guillette, Jr., and C. M. St. Mary. 2008. Agriculture alters gonadal form and function in the toad *Bufo marinus*. *Env. Health Persp.* DOI: 10.1289/ehp11536.

McNabb, A. and 7 others. 1999. Basic physiology. In R. T. Di Giulio and D. E. Tillitt (eds.), *Reproductive and Developmental Effects of Contaminants in Oviparous Vertebrates*. SETAC Press, Pensacola, FL, pp. 113–223.

Michaels, D. 2005. Doubt is their product. *Sci. Am.* 292(6): 96–101.

Michaels, D. 2008. *Doubt Is Their Product: How Industry's Assault on Science Threatens Your Health.* Oxford University Press, New York.

Miller, C., K. Degenhardt and D. A. Sassoon. 1998. Fetal exposure to DES results in de-regulation of *Wnt7a* during uterine morphogenesis. *Nature Genet.* 20: 228–230.

Miller, S. M., C. W. Sweet, J. V. DePinto and K. C. Hornbuckle. 2000. Atrazine and nutrients in precipitation: Results from the Lake Michigan mass balance study. *Environ. Sci. Tech.* 34: 55–61.

Milnes, M. R., T. A. Bryan, Y. Katsu, S. Kohno, B. C. Moore, T. Iguchi and L. J. Guillette, Jr. 2008. Increased post-hatching mortality and loss of sexually dimorphic gene expression in alligators (*Alligator mississippiensis*) from a contaminated environment. *Biol. Reprod.* 78: 932–938.

Miyakoda, H., M. Tabata, S. Onodera and K. Takeda. 1999. Passage of bisphenol A into the fetus of the pregnant rat. *J. Health Sci.* 45: 318–323.

Mol, N. M. and 7 others. 2002. Spermaturia and serum hormone concentrations at the age of puberty in boys prenatally exposed to polychlorinated biphenyls. *Eur. J. Endocrinol.* 146: 357–363.

Monosson, E., W. R. Kelce, C. Lambrigh, J. Ostby and L. E. Grey, Jr. 1999. Peripubertal exposure to the antiandrogenic fungicide, vinclozolin, delays puberty, inhibits the development of androgen-dependent tissues and alters androgen receptor function in the male rat. *Toxicol. Indust. Health* 15: 65–79.

Muñoz-de-Toro, M., C. Markey, P. R. Wadia, E. H. Luque, B. S. Rubin, C. Sonnenschein and A. M. Soto. 2005. Perinatal exposure to bisphenol A alters peripubertal mammary gland development in mice. *Endocrinology* 146: 4138–4147.

Murray, T. J., M. V. Maffini, A. A. Ucci, C. Sonenschein and A. M. Soto. 2007. Induction of mammary gland ductal hyperplasias and carcinoma in situ following bisphenol-A exposure. *Reprod. Toxicol.* 23: 383–390.

Narita, M., K. Miyagawa, K. Mizui, T. Yoshida and T. Suzuki. 2006. Prenatal and neonatal exposure to low-dose of bisphenol A enhance the morphone-induced hyperlocomotion and rewarding effect. *Neurosci. Lett.* 402: 249–252.

NCI (National Cancer Institute). 2007. Letrozole. http://www.cancer.gov/cancertopics/druginfo/letrozole.

Newbold, R. R. 2004. Lessons learned from prenatal exposure to diethylstilbestrol. *Toxicol. Appl. Pharmacol.* 199: 142–150.

Newbold, R. R., R. B. Hanson, W. N. Jefferson, B. C. Bullock, J. Haseman and J. A. McLachlan. 1998. Increased tumors but uncompromised fertility in the female descendants of mice exposed developmentally to diethylstilbestrol. *Carcinogenesis* 19: 1655–1663.

Newbold, R. R., R. B. Hanson, W. N. Jefferson, B. C. Bullock, J. Haseman and J. A. McLachlan. 2000. Proliferative lesions and reproductive tract tumors in male descendants of mice exposed developmentally to diethylstilbestrol. *Carcinogenesis* 21: 1355–1363.

Newbold, R. R., E. Padilla-Banks, B. C. Bullock and W. N. Jefferson. 2001. Uterine adenocarcinoma in mice treated neonatally with genistein. *Cancer Res.* 61: 4325–4328.

Newbold, R. R., E. P. Banks and W. N. Jefferson. 2006. Adverse effects of the model environmental estrogen diethylstilbestrol are transmitted to subsequent generations. *Endocrinology* 147 (suppl): S11–S17.

Newbold, R. R., E. P. Banks, R. J. Snyder and W. N. Jefferson. 2007. Perinatal exposure to environmental estrogens and the development of obesity. *Mol. Nutr. Food Res.* 51: 912–917.

Newbold, R. R., S. Tyrey, A. F. Haney and J. A. MacLachlan. 1983. Developmentally arrested oviduct: A structural and functional defect in mice following prenatal exposure to diethylstilbestrol. *Teratology* 27: 417–426.

NIEHS (National Institute of Environmental Health Sciences). 2008. http://www.niehs.nih.gov/news/media/questions/sya-bpa.cfm#4.

Novartis. 1997. http://www.novartisoncology.com/page/extended_adjuvantbreast_therapy.jsp?usertrack.filter_applied=true&NovaId=2229644963416064359.

NRC (National Research Council). 1999. *Hormonally Active Agents in the Environment.* National Academies Press, Washington, DC.

Orlando, L. 2002. Industry attacks on dissent: From Rachel Carson to Oprah. http://www.dollarsandsense.org/archives/2002/0302orlando.html

Padilla-Banks, E., W. N. Jefferson and R. R. Newbold. 2006. Neonatal exposure to the phytoestrogen genistein alters mammary gland growth and developmental programming of hormone receptor levels. *Endocrinology* 147: 4871–4882.

Pajarinen, J., P. Laippala, A. Penttila and P. J. Karhunen. 1997. Incidence of disorders of spermatogenesis in middle-aged Finnish men 1981–1991: Two necropsy series. *Brit. Med. J.* 314: 13–18.

Palanza, P., K. L. Howdeshell, S. Parmigiani and F. S. vom Saal. 2002. Exposure to a low dose of bisphenol A during fetal life or in adulthood alters maternal behavior in mice. *Environ. Health Persp.* 110 (suppl.): 415–422.

Palmer, J. R. and 10 others. 2006. Prenatal DES exposure and risks of breast cancer. *Cancer Epidem. Biomarkers Prevent.* 15: 1509–1514.

Palmund, I. 1996. Exposure to a xenoestrogen before birth: The diethylstilbestrol experience. *J. Psychosom. Obstet. Gynecol.* 17: 71–84.

Pintér, A., G. Török, M. Börzsönyi, A. Surján, M. Csík, Z. Kelecsényi and Z. Kocsis. 1990. Long-term carcinogenicity bioassay of the herbicide atrazine in F344 rats. *Neoplasma* 37: 533–544.

Powell, C. E., A. M. Soto and C. Sonnenschein. 2001. Identification and characterization of membrane estrogen receptor form MCF7 estrogen-target cells. *J. Steroid Biochem. Mol. Biol.* 77: 97–108.

Raffensberger C. and J. Tickner (eds.). 1999. *Protecting Public Health and the Environment: Implementing the Precautionary Principle*. Island Press, Washington, D.C.

Rajapakse, N., D. Ong and A Kortenkamp. 2001. Defining the impact of weakly estrogenic chemicals on the action of steroidal estrogens. *Toxicol. Sci.* 60: 296–304.

Rajapakse, N., E. Silva and A. Kotenkamp. 2002. Combining xenoestrogens at levels below individual no-observed-effect concentrations dramatically enhances steroid hormone action. *Environ. Health Persp.* 110: 917–921.

Ribas-Fito, N., M. Sala, M. Kogevinas and J. Sunyer. 2001. Polychlorinated biphenyls (PCBs) and neurological development in children: A systematic review. *J. Epid. Comm. Health* 55: 537–546.

Robboy, S. J., R. H. Young and A. L. Herbst. 1982. Female genital tract changes related to prenatal diethylstilbestrol exposure. In A. Blaustein (ed.), *Pathology of the Female Genital Tract*, 2nd Ed. Springer-Verlag, New York, pp. 99–118.

Roberge, M., H. Hakk and G. Larsen. 2004. Atrazine is a competitive inhibitor of phosphodiesterase but does not affect the estrogen receptor. *Toxicol. Lett.* 154: 61–68.

Ross, G. 2004. The public health implications of polychlorinated biphenyls (PCBs) in the environment. *Ecotoxicol. Environ. Safety* 59: 275–291.

Rowe, A., K. M. Brundage. R. Schafer and J. B. Barnett. 2006. Immunomodulatory effects of maternal atrazine exposure on male Balb/c mice. *Toxicol. Appl. Pharmacol.* 214: 69–77.

Royland, J. E. and P. R. S. Kodavanti. 2008. Gene expression profiles following exposure to a developmental neurotoxicant, Aroclor 1254: Pathway analysis for possible mode(s) of action. *Toxicol. Appl. Pharmacol.* DOI: 10.1026/j.taap.2008.04.023.

Royland, J. E., J. Wu, N. H. Zawia and P. R. S. Kodaventi. 2008. Gene expression profiles in the cerebellum and hippocampus following exposure to a neurotoxicant, Aroclor 1254: Developmental effects. *Toxicol. Appl. Pharmacol.* DOI: 10.1016/j.taap.2008.04.022.

Rozman, K. K. and 13 others. 2006. NTP-CERHR expert panel report on the reproductive and developmental toxicity of genistein. *Birth Def. Res.* 77: 485–638.

Rubin, B. S., J. R. Lenkowski, C. M. Schaeberle, L. N. Vandenberg, P. M. Ronscheim and A. M. Soto. 2006. Evidence of altered brain sexual differentiation in mice exposed perinatally to low, environmentally relevant levels of bisphenol A. *Endocrinology* 147: 3681–3691.

Rubin, B. S., M. K. Murray, D. A. Damassa, D. A. King and A. M. Soto. 2001. Perinatal exposure to low doses of bisphenol A affects body weight, patterns of estrous cyclicity and plasma LH levels. *Environ. Health Persp.* 109: 675–680.

Ruden, D. M., L. Xiao, M. D. Garfinkel and X. Lu. 2005. Hsp90 and environmental impacts on epigenetic states: A model for the trans-generational effects of diethylstilbestrol on uterine development and cancer. *Hum. Mol. Genet.* 14 (suppl. 1): R149–R155.

Sanderson, J. T., W. Seinen, J. P. Giesy and M. van den Berg. 2000. 2-Chloro-*s*-triazine herbicides induce aromatase (*CYP19*) activity in H295 human adrenocortical carcinoma cells: A novel mechanism of estrogenicity? *Toxicol. Sci.* 54: 121–127.

Sass, J. 2003. Letter to the editor. *J. Occup. Environ. Med.* 45: 343–344

Schantz, S. L., J. J. Widholm and D. C. Rice. 2003. Effects of PCB exposure on neuropsychological function in children. *Environ. Health Persp.* 111: 357–376.

Schapiro, M. 2007. *Exposed: The Toxic Chemistry of Everyday Products and What's at Stake for American Power.* Chelsea Green Publishing, White River Junction, VT.

Schlumpf, M. and 14 others. 2004. Endocrine activity and developmental toxicity of cosmetic UV filters: An update. *Toxicology* 205: 113–122.

Schönfelder, G., W. Wittfoht, H. Hopp, C. E. Talsness, M. Paul and I. Chahoud. 2002. Parent bisphenol A accumulation in the human maternal-fetal-placental unit. *Environ. Health Persp.* 110: A703–A707.

Sharpe, R. M. 1994. Could environmental oestrogenic chemical be responsible for some disorders of human male reproductive development? *Curr. Opin. Urol.* 4: 295–301.

Sharpe, R. M. and D. S. Irvine. 2004. How strong is the evidence of a link between environmental chemicals and adverse effects on human reproductive health? *Brit. Med. J.* 328: 447–451.

Silva, E., N. Rajapakse and A. Kortenkamp. 2002. Something from "nothing": Eight weak estrogenic chemicals combined at concentration below NOECs produce significant mixture effects. *Environ. Sci. Technol.* 36: 1751–1756.

Skakkebaeck, N. E. 2004. Testicular dysgenesis syndrome: New epidemiological evidence. *Internatl. J. Andrology* 27: 189–191.

Skinner M. K. and M. D. Anway. 2005. Seminiferous cord formation and germ-cell programming: Epigenetic transgenerational actions of endocrine disruptors. *Ann. NY Acad. Sci.* 1061: 18–32.

Smith, M. D. and E. M. Hill. 2006. Profiles of short-chain oligomers in roach (*Rutilus rutilus*) exposed to waterborne polyethoxylated nonulphenols.. *Sci. Total Environ.* 356: 100–111.

Soto, A. M., H. Justicia, J. Wray and C. Sonnenschein. 1991. *p*-Nonylphenol: An estrogenic xenobiotic released from "modified" polystyrene. *Environ. Health Persp.* 92: 167–173.

Soto, A. M., M. U. Maffini and C. Sonnenschein. 2008. Neoplasia as development gone awry: The role of endocrine disruptors. *Internatl. J. Androbgy* 31: 280–293.

Spearow, J. L., P. Doemeny, R. Sera, R. Leffler and M. Barkley. 1999. Genetic variation in susceptibility to endocrine disruption by estrogen in mice. *Science* 285: 1259–1261.

Stoker, T. E., C. L. Robinette and R. L. Cooper. 1999. Maternal exposure to atrazine during lactation suppresses suckling-induced prolactin release and results in prostatitis in the adult offspring. *Toxicol. Sci.* 52: 68–79.

Sunstein, C. 2005. *Laws of Fear: Beyond the Precautionary Principle* (The Seeley Lectures). Cambridge University Press.

Susiarjo, M., T. J. Hassold, E. Freeman and P. Hunt. 2007. Bisphenol A exposure in utero disrupts early oogenesis in the mouse. *PloS Genetics* 3: 63–70.

Swan, S. H. 2006. Semen quality in fertile U.S. men in relation to geographical area and pesticide exposure. *Internatl. J. Androl.* 29: 62–68.

Swan, S. H. et al. and the Study for Future Families Research Group. 2003. Semen quality in relation to biomarkers of pesticide exposure. *Environ. Health Persp.* 111: 1478–1484.

Swan, S. H. and 11 others. 2005. Decrease in anogenital distance among male infants with prenatal phthalate exposure. *Env. Health Persp.* 113: 1056–1063.

Thomas, P., Y. Pang, E. J. Filardo and J. Dong. 2006. Identity of an estrogen membrane receptor coupled to a G-protein in human breast cancer cells. *Endocrinology* 146: 624–632.

Thurman, E. and A. Cromwell. 2000. Atmospheric transport, deposition and fate of triazine herbicides and their metabolites in pristine areas at Isle Royale National Park. *Environ. Sci. Tech.* 34: 3079–3085.

Timms, B. G., K. Howdeshell, L. Barton, S. Bradley, C. A. Richter and F. S. vom Saal. 2005. Estrogenic chemicals in plastic and oral contraceptives disrupt development of the fetal mouse prostate and urethra. *Proc. Natl. Acad. Sci. USA* 102: 7014–7019.

Turusov, V. S., L. S. Trukhanova, Y. D. Parfenov and L. Tomatis. 1992. Occurrence of tumours in the descendants of CBA male mice prenatally treated with diethylstilbestrol. *Internatl. J. Cancer* 50: 131–135.

Tyl, R. W. and 16 others. 2002. Three-generation reproductive toxicity study of dietary bisphenol A in CD Sprague-Dawley rats. *Toxicol. Sci.* 68: 121–146.

Tyl, R. W. and 12 others. 2008. Two-generation reproductive toxicity study of dietary bisphenol A in CD-1 (Swiss) mice. *Toxicol. Sci.* 104: 362–384.

Ueda, M., T. Imai, T. Takizawa, H. Onodera, K. Mitsumori, T. Matsui and M. Hirose. 2005. Possible enhancing effects of atrazine on growth of 7,12-dimethylbenz(*a*) anthracene induced mammary tumors in ovariectomized Sprague-Dawley rats. *Cancer Sci.* 96: 19–25.

U.S.G.S. (United States Geological Survey). http://pubs.usgs.gov/bat/fig1.gif.

Vandenberg, L. N., M. V. Maffini, P. R. Wadia, C. Sonnenschein, B. S. Rubin and A. M. Soto. 2007. Exposure to environmentally relevant doses of the xenoestrogen bisphenol A alters development of the fetal mouse mammary gland. *Endocrinology* 148: 116–127.

Vandenberg, L. N., P. R. Wadia, C. M. Schaeberle, B. S. Rubin, C. Sonnenschein and A. M. Soto. 2006. The mammary gland response to estradiol: Monotonic at the cellular level, non-monotonic at the tissue level of organization? *J. Steroid Biochem. Mol. Biol.* 101: 263–274.

vom Saal, F. S. 2000. Very low doses of bisphenol A and other estrogenic chemicals alter development in mice. Endocrine Disruptors and Pharmaceutical Active Compounds in Drinking Water Workshop.

vom Saal, F 2008. Bisphenol A references. http://endocrinedisruptors.missouri.edu/vomsaal/vomsaal.html.

vom Saal, F. S. and C. Hughes. 2005. An extensive new literature concerning low-dose effects of bisphenol A shows the need for a new risk assessment. *Environ. Health Persp.* 113: 926–933.

vom Saal, F. S. and W. Welshons. 2006. Large effects from small exposures. II. The importance of positive controls in low-dose research on bisphenol A. *Environ. Res.* 100: 50–76.

vom Saal, F. S. and 7 others. 1998. A physiologically based approach to the study of bisphenol A and other estrogenic chemicals on the size of reproductive organs, daily sperm production, and behavior. *Toxicol. Indust. Health* 14: 239–260.

vom Saal, F. S. and 9 others. 1997. Prostate enlargement in mice due to fetal exposure to low doses of estradiol or diethylstilbestrol and opposite effects at high doses. *Proc. Natl. Acad. Sci. USA* 94: 2056–2061.

vom Saal, F. and 37 others. 2007. Chapel Hill bisphenol A expert panel consensus statement: integration of mechanisms, effects in animals and potential to impact human health at current levels of exposure. *Reprod Toxicol.*. 24: 131–138.

Wadia, P. R., L. N. Vandenberg, C. M. Schaeberle, B. S. Rubin, C. Sonnenschein and A. M. Soto. 2007. Perinatal bisphenol A exposure increases estrogen sensitivity of the mammary gland in diverse mouse strains. *Environ. Health Persp.* 115: 592–598.

Walker, B. C. and M. I. Haven. 1997. Intensity of multigenerational carcinogenesis from diethylstilbestrol in mice. *Carcinogenesis* 18: 791–793.

Walker, B. C. and L. A. Kerth. 1995 Multi-generational carcinogenesis from diethylstilbestrol investigated by blastocyst transfers in mice. *Internatl. J. Cancer* 61: 249–252.

Wang, S. L., P. H. Su, S. B. Jong, Y. L. Guo, W. L. Chou and O. Päpke. 2005. In utero exposure to dioxins and polychlorinated biphenyls and its relations to thy-

roid function and growth hormone in newborns. *Environ. Health Persp.* 113: 1645–1650.
Watson, C. S., R. A. Alyea, Y. J. Jeng and M. Y. Kochukov. 2007. Nongenomic actions of low concentration estrogens and xenoestrogens on multiple tissues. *Mol. Cell Endocrinol.* 274: 1–7.
Welshons, W. V., K. A. Thayer, B. M. Judy, J. A. Taylor, E. M. Curran and F. S. vom Saal. 2003. Large effects from small exposures. I. Mechanisms for endocrine disrupting chemicals with estrogenic activity. *Environ. Health Persp.* 111: 994–1006.
Wetherill, Y. B., C. E. Petse, K. R. Monk, A. Puga and K. E. Knudson. 2002. Xenoestrogen bisphenol A induces inappropriate androgen receptor activation and mitogenesis in prostatic adenocarcinoma cells. *Mol. Cancer Ther.* 1: 515–524.
Wiszniewski, A. B., A. Cernetich, J. P. Gearhart and S. L. Klein. 2005. Perinatal exposure to genistein alters reproductive development and aggressive behavior in male mice. *Physiol. Behav.* 84: 327–334.
Wozniak, A., N. Bulayeva and C. S. Watson. 2005. Xenoestrogens at picomolar to nanomolar concentrations trigger membrane estrogen receptor-α-mediated Ca^{2+} fluxes and prolactin release in GH3/B6 pituitary tumor cells. *Environ. Health Persp.*113: 431–439.
Xu, L. C., H. Sun, J. F. Chen, Q. Bian, L. Song and X. R. Wang. 2006. Androgen receptor activities of *p,p′*-DDE, fenvalerate, and phoxim detected by androgen receptor reporter assay. *Toxicol. Lett.* 160: 151–157.
Yoshimura, T. 2003. Yusho in Japan. *Ind. Health* 41: 139–148.

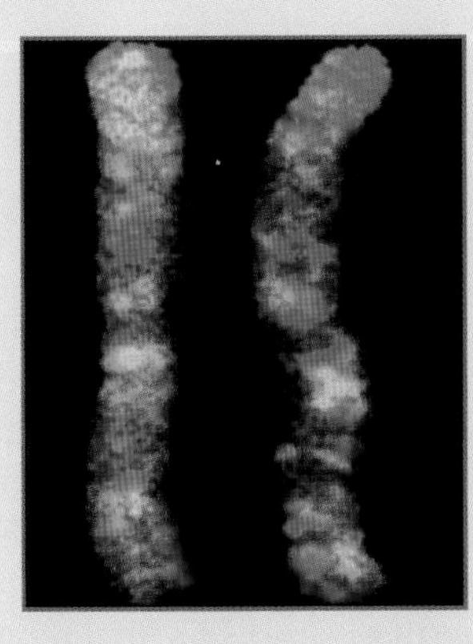

Chapter 7

The Epigenetic Origin of Adult Diseases

Time present and time past
Are both perhaps present in time future
And time future contained in time past.

T. S. Eliot, 1935

Now there are more overweight people in America than average-weight people. So overweight people are now average. Which means you've met your New Year's resolution.

Jay Leno, 2006

Organisms respond to environmental changes on three timescales. First, physiological homeostatic mechanisms can immediately circumvent adverse conditions. On a longer timescale, developmental plasticity can result in adaptive changes in the organism by enabling the emergence of a phenotype appropriate for the environment it is expected to encounter. And, on the longest timescale, as changes in phenotypes interact with the environment, the genetic composition of populations can change as a result of natural selection.

Medicine deals predominantly with the first category, the short-term physiological changes. However, as we document in this chapter, some short-term medical conditions are the result of mismatches generated on the second timescale. Developmental plasticity is key to a strategy that uses cues present in early development to prepare an organism for its future life. The success of such a strategy assumes (1) that the cues are accurate and (2) that the early environment is a good predictor of the adult environment. If either assumption is not met, the survival and/or health of the resulting individual may be compromised.

But developmental plasticity must remain within the boundaries of the grand scale of evolution. Adaptive biological responses evolve to maximize fitness, not necessarily health (see Gluckman and Hanson 2007). Since evolutionary fitness is measured by the relative number of fertile progeny an individual produces, our biological responses have evolved to get us to the point where we can successfully reproduce and rear our offspring. "No sooner than we have a child," writes the melancholy Jean Rostand (1962, p. 9), we become "lateral excrescences" on the tree of life; evolution cares nought for us. However, advances in medicine and pharmacology, along with public health programs and improved sanitation, have enabled much of the population to live well beyond the years of prime reproductive life that marked the extent of average life expectancy throughout most of history. As a result, late-onset diseases such as hypertension, type 2 diabetes, and cancer have become major medical and public health concerns.

One of the most exciting new areas of public health and medicine concerns the epigenetic origins of adult disease. The previous chapter discussed the health implications of endocrine disruption. In this chapter, we will discuss two other areas of major importance for human health. The first is the fetal origin of diseases such as hypertension, obesity, and heart failure. Here, the diet experienced in utero appears to play a role in making the individual more or less susceptible to developing these conditions. The second area involves epigenetic errors that, in adults, may lead to cancers and the syndrome of aging. Both types of epigenetic disease origin may ultimately be based on changes in the patterns of DNA methylation and histone modification.

THE DEVELOPMENTAL ORIGINS OF HEALTH AND DISEASE

Probably the most obvious instance of co-development is mammalian pregnancy. Entire volumes have been written on "maternal-fetal interactions," and "maternal-fetal medicine" is a recognized subspecialty of obstetrics and gynecology. The fetus and its extraembryonic layer (the trophoblast, which forms the portion of the placenta derived from the fetus) induce changes in the structure and physiology of the uterus. The fetus and placenta produce hormones such as progesterone and human chorionic gonadotropin (hCG) that maintain the lining of the uterus. These hormones instruct the uterine stromal cells to proliferate rather than allowing them to be sloughed off during menstruation; they also cause blood vessels to enter into this uterine tissue. (This set of changes to the uterus is called the *decidua reaction*.)

The fetus and its placental tissues also produce substances that block the maternal immune responses against the fetus, a critical action in maintaining the pregnancy. Also critical to the maintenance of pregnancy, the uterus induces proteins on the extraembryonic trophoblast that allow it to adhere to and ingress into the uterus (see Dey et al. 2004). Later, of course, the uterus provides the developing mammal with oxygen, nutrients, and a temperature-regulated residence for the duration of the pregnancy. And, toward the end of pregnancy, trophoblast cells are involved in the production of relaxin, a hormone that softens the maternal cervix and pelvic ligaments in preparation for childbirth.

Almost all of the research on maternal-fetal interaction concerns these permissive aspects of development. The presence of a fetus induces the decidua reaction in the mother, but this reaction has nothing to do with specifying the phenotype of the fetus. Rather, it allows the fetus to survive. Similarly, the transport of nutrients and gases in and out of the fetus is also permissive. It is critical for fetal growth and survival, but it is not usually discussed in terms of providing information to the fetus on how to develop its phenotype. Here we will explore evidence that the mother is indeed providing the fetus with instructions as to particular aspects of its phenotype.

Instructing the Fetus

Mammals developing within the uterus can fine-tune their phenotypes to suit an expected future environment. The mother's diet and hormonal milieu provide information about the environment the offspring will be born into. However, when environmental conditions change greatly between conception and adulthood (as has been the case for many modern human populations), the potential exists that a phenotype developed for one situation may not be particularly healthy in the new situation.

Epidemiological, physiological, and molecular evidence are now converging to show that numerous human diseases—diabetes, obesity, and hypertension among them—can be attributed to the mismatch between (1) the phenotype developed from nutritional cues sensed by the fetus in utero, and (2) the nutritional environment in which that individual finds itself as an adult (Gluckman and Hanson 2006a,b, 2007). The terms "fetal programming" or the less deterministic "fetal plasticity" refer to the processes by which a stimulus given during an intrauterine period of development can have a lasting or lifelong effect (Lucas 1991; Barker 1998).*

*Mammals are not the only animals for whom signals received during embryogenesis can lead to disease or other phenotypic changes in adult life. In the marine environment, "orphan embryos" (see Chapter 4) that experience short-term starvation, salinity stress, or exposure to sublethal concentrations of pollutants may, after metamorphosis, experience altered growth and fecundity rates and heightened mortality rates. The mechanisms of these "latent effects" remain largely unexplored (Pechenik 2006).

Maternal–fetal co-development

The concept that adult human phenotypes can be cued during early life is not new, nor is the concept that developmental decisions made early in life may be detrimental when the environment changes. During World War II, the invasion of the Japanese military into southeast Asia placed its soldiers into environments that were new to them. Some soldiers got heatstroke due to their inability to sweat copiously, whereas other soldiers adapted more readily to the hotter climate. Most of the soldiers who got heatstroke came from the northern provinces of Japan. Their propensity to heatstroke was found to be an inability to sweat efficiently, and this inability was not genetic. Rather, it was due to a developmental decision made during the early years of life (see Diamond 1991; Barker 1998). The researchers found that each person is born with nearly the same number of sweat glands, but that initially none of these sweat glands function. The glands are activated during the first 3 years after birth, and the percentage of sweat glands that becomes functional depends on the temperatures the child experiences during those years: the hotter the conditions, the greater the percentage of sweat glands that function. This is due to the innervation of the sweat glands by the sympathetic nervous system. A sweat gland does not become functional until the immature gland has interacted with axons of the sympathetic nervous system (Stevens and Landis 1988).

This high degree of plasticity allows the adjustment of human sweat glands to new thermal environments within a single generation. Indeed, Gluckman and Hanson (2005) suggest that this ability to adapt to new climates may have been crucial in allowing prehistoric human migrations to occur as rapidly as they did. So we see in this example that an adult human phenotype can develop in a manner consistent with the environment experienced early in life. However, once the phenotype is "programmed," it remains set for the remainder of the person's life; for sweat gland function, the period of plasticity ends after 3 years. Should such persons find themselves in a new environment, the predictive response may not have been appropriate, and disease (in this case, heatstroke) may ensue.

It is also known that the maternal environment contains cues that can override genetic information. One place where this phenomenon is readily seen is in the size of newborn mammals. Among healthy humans, an infant's birth weight is determined much more by its mother than by its father. When the same woman has borne children with different men, the birth weights of the babies are usually similar. However, when the same man fathers children with different women, the birth weights are often very different (Morton 1955; Robson 1955; Gluckman and Hanson 2005). This effect is probably what historically has allowed small women to bear children fathered by large men. The baby must be able to remain inside the uterus before birth and must be small enough to pass through the mother's birth canal (Hanson and Godfrey 2008). Thus, genes alone do not determine the size of the newborn. Maternal constraints play a major role.

In more controlled studies, researchers crossed large Shire horses with small Shetland ponies (Figure 7.1). If a Shire stallion was crossed with a Shetland mare (by artificial insemination, for obvious reasons), the size of the resulting foal at birth was closer to the size of a typical Shetland foal. If,

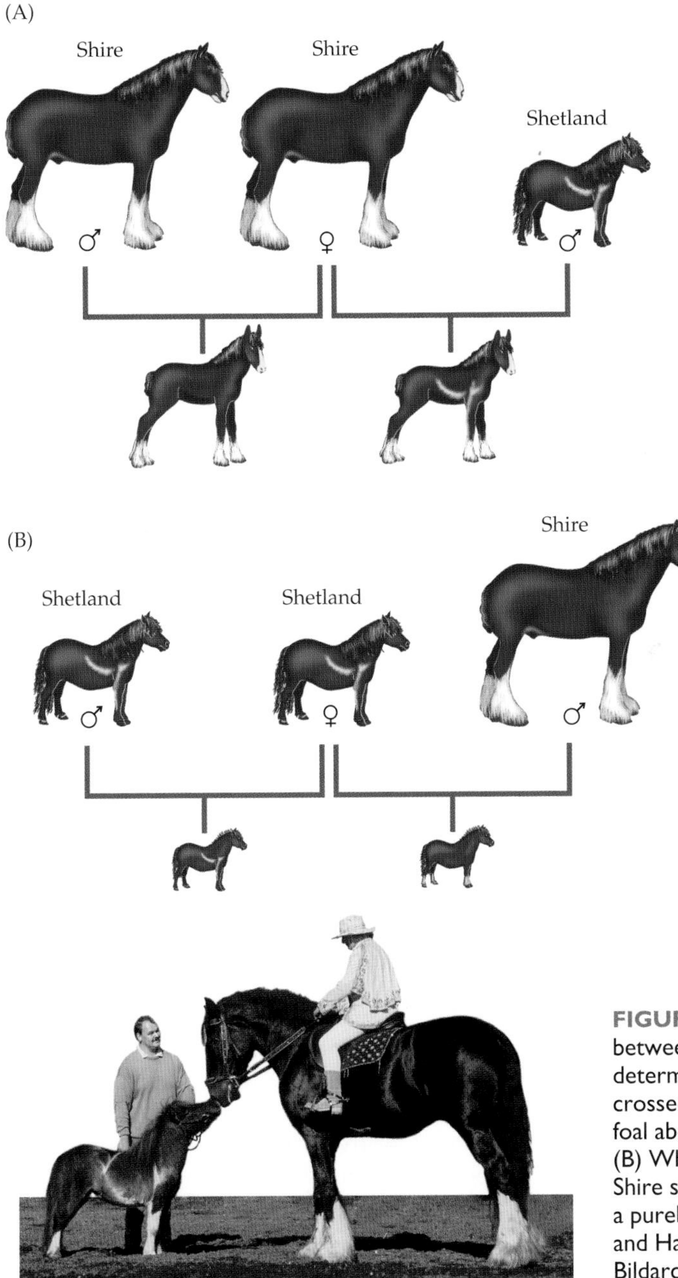

FIGURE 7.1 Foal size in outcrosses between Shire and Shetland horse breeds is determined by the mother. (A) A Shire mare crossed with a Shetland stallion produces a foal about the size of a purebred Shire. (B) When a Shetland mare is crossed with a Shire stallion, the foal is closer to the size of a purebred Shetland pony. (After Gluckman and Hanson 2005; photograph © Juniors Bildarchiv/Alamy.)

however, a Shire mare was crossed with a Shetland stallion, the newborn foal was much larger, closer in size to that of a typical Shire foal. Even though in both cases the foal genomes were 50% Shetland and 50% Shire, the size of the newborn was typically that of the mother's strain. The Shetland mare in some manner limited the growth of a fetus whose genotype, if fully expressed, would make it too large to pass through her birth canal. Uninhibited by such constraints, the same genotype could grow to be much larger, as shown by its growth in the Shire mother. While we don't know the exact mechanism regulating this growth constraint, it probably has to do with the nutrient supply that can be delivered by the smaller placenta. Moreover, the offspring of the mothers with smaller uteruses and placentas developed into smaller adults than did the analogous hybrids whose mothers were of the larger breed (Walton and Hammond 1938; Tischner 1987; Allen et al. 2004).

Fetal plasticity in humans

D. J. P. Barker and colleagues cite evidence that, in humans, certain adult-onset diseases may result in part from conditions in the uterus prior to birth (Barker 1994; Barker and Osmond 1986). Based on epidemiological evidence, they hypothesized that there are critical periods of development during which insults or stimuli cause specific changes in the body's physiology. The "Barker hypothesis" postulates that certain anatomical and physiological parameters get "programmed" during embryonic and fetal development, and that changes in nutrition during this time can produce permanent changes in the pattern of metabolic activity. These changes can predispose the adult to particular diseases.

Specifically, the Barker group showed that individuals whose mothers had experienced nutritional deprivation during certain months of pregnancy (due to war, famine, or migrations) were at high risk for having certain diseases as adults (Figure 7.2). Undernutrition during the first trimester leads to a relatively high risk of having hypertension and stroke as an adult, while those fetuses experiencing undernutrition during the second trimester had a relatively high risk of developing heart disease and diabetes as adults. Fetuses experiencing undernutrition during the third trimester were prone to blood-clotting defects as adults.

Recent studies have tried to determine whether there are molecular or anatomical reasons for these correlations (Gluckman and Hanson 2004, 2005; Lau and Rogers 2004). Anatomically, undernutrition can change the number of cells produced during critical periods of organ formation. When pregnant rats are fed low-protein diets at certain times during pregnancy, the resulting offspring are at high risk for hypertension (high blood pressure) as adults. The poor diet appears to cause low nephron numbers in the adult kidney (see Moritz et al. 2003). Nephrons are the filtering units of the

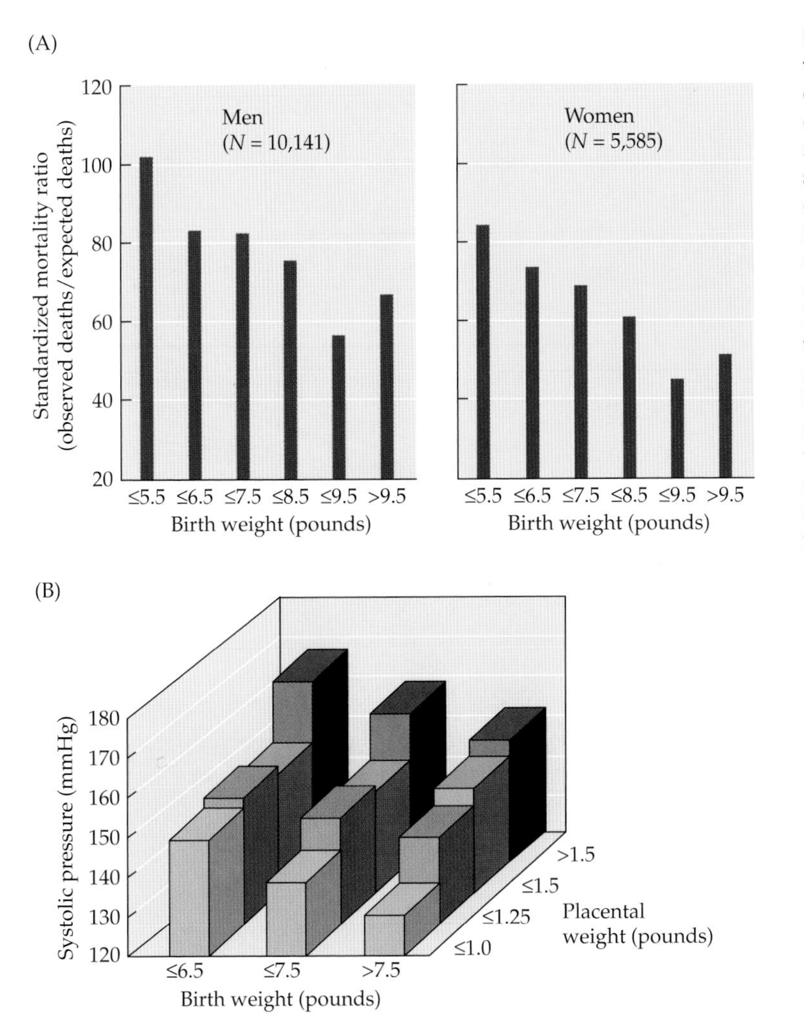

FIGURE 7.2 Low birth weight and adult-onset coronary disease. (A) Epidemiological data demonstrating the association of adult-onset coronary disease with low birth weight in both men and women. (B) The association of mean systolic blood pressure (mm Hg) of 50-year-old men and women is associated with both low birth weight and a large placenta. Those especially at risk would be those whose low birth weight could not be explained by a small placenta. (After Godfrey 2006.)

kidneys, and they synthesize proteins that regulate blood pressure. In humans, men with hypertension were shown to have only about half the number of nephrons found in men without hypertension (Figure 7.3; Keller et al. 2003; Hoppe et al. 2007). It is possible that malnutrition activates a glucocorticoid response, which causes cell death among the developing kidney cells (Moritz et al. 2003; Welham et al. 2005).

The ability to sacrifice nephrons so that the limited nutrition available can go to the brain, heart, and other essential places seems like good evolutionary sense. One can survive well and reproduce with only one kidney. Therefore, when the life expectancy of the average human was less than 40 years, one could function efficiently with half the number of nephrons throughout one's life. The accumulated effects leading to hypertension usually are seen only after one's fortieth birthday, so sacrificing nephron number would not have been problematic during most of human history. It is

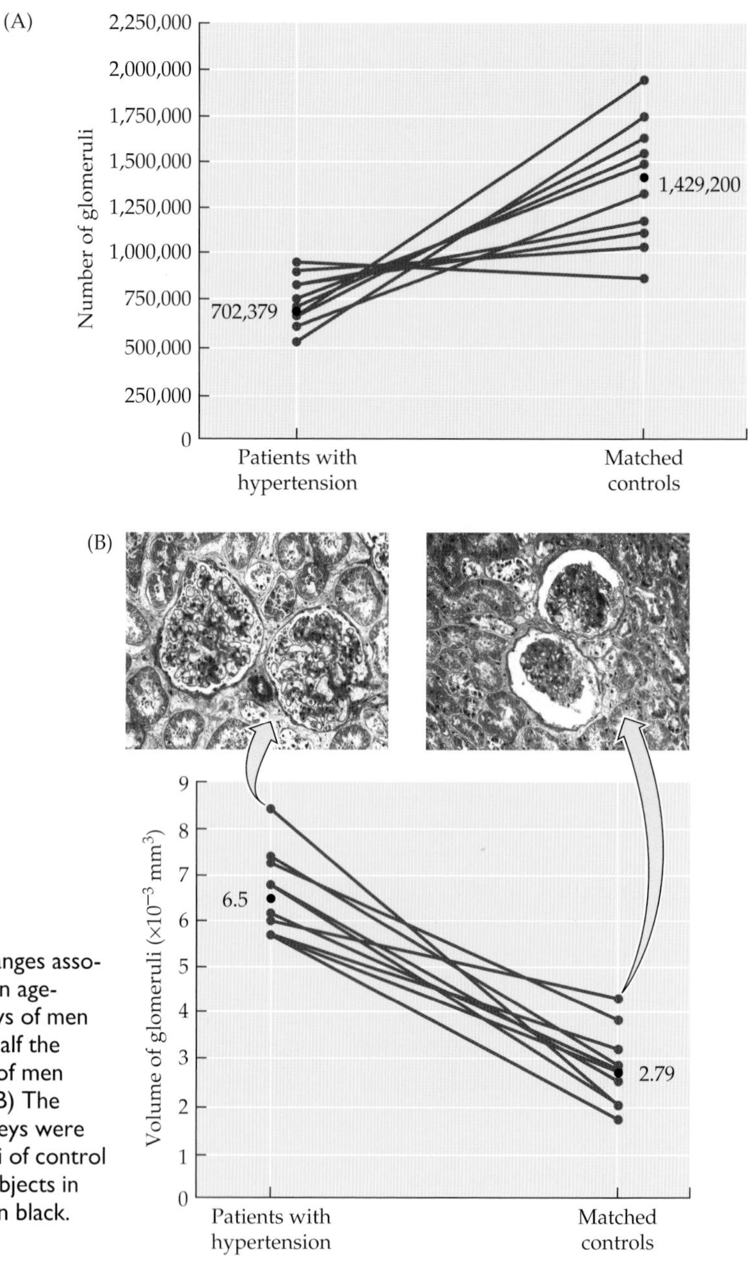

FIGURE 7.3 Anatomical changes associated with hypertension. (A) In age-matched individuals, the kidneys of men with hypertension had about half the number of nephrons as those of men with normal blood pressure. (B) The glomeruli of hypertensive kidneys were much larger than the glomeruli of control subjects. Average values for subjects in both categories are indicated in black. (After Keller et al. 2003.)

only since life expectancy in the industrialized world has reached the seventh decade that the detrimental effects of nephron deficiency have become apparent.

Developmentally plastic changes in anatomy have also been reported for the pancreas and liver. Poor nutrition during fetal development reduces the number of insulin-secreting cells in the pancreas, causing the person to synthesize smaller amounts of insulin (Hales et al. 1991; Hales and Barker 1992; Petrik et al. 1999). This insulin deficiency predisposes these individuals to type 2 diabetes and metabolic syndrome. In rats, undernutrition before (but not after) birth changes the histological architecture in the liver. Liver cells (hepatocytes) are roughly divided into two types, depending upon their location in the liver lobule: an upstream, "periportal" population that maintains gluconeogenesis (glucose production), and a downstream, "perivenous" population that is involved in glycolysis. A low-protein diet during gestation appears to increase the number of periportal cells that produce the glucose-synthesizing enzyme phosphoenolpyruvate carboxykinase, and to decrease the number of perivenous cells that synthesize the glucose-degrading enzyme glucokinase (Desai et al. 1995). As a result, more glucose is made and less is degraded. As in the case of kidney nephrons, it is thought that these changes may be coordinated by glucocorticoid hormones that are stimulated by malnutrition and that act to conserve resources, even though such actions might make the person prone to disease later in life (see Fowden and Forhead 2004).

The opposite side of this coin—that offspring who experience high-protein diets in utero perform less well in calorie-impoverished environments than do their siblings who experience low-protein diets in utero—has not been as well studied. However, during famines in Ethiopia, low-birth weight children appear to have had less chance of getting rickets than those born with high birth weights (Chali et al. 1998), and Bateson (2007) relates anecdotal evidence that in World War II concentration camps, large individuals appeared to die more readily than smaller inmates.

Gene methylation and the fetal phenotype

How can environmental conditions experienced only while the fetus is in the uterus result in anatomical and biochemical states that are maintained throughout the individual's adulthood? One place to look for the answer to this question is DNA methylation. As we saw in Chapter 2, the chromatin we inherit can be modified by enzymes that add or delete methyl groups (CH_3) to DNA or to the histones that compress DNA and prevent its transcription. A gene is activated when the enzyme RNA polymerase binds to a DNA sequence (the promoter) in front of the actual gene sequence. Once bound to the promoter, RNA polymerase can transcribe the gene sequence into messenger RNA, which in turn enters the cell cytoplasm, where it is translated into a specific sequence of amino acids that forms a protein. RNA

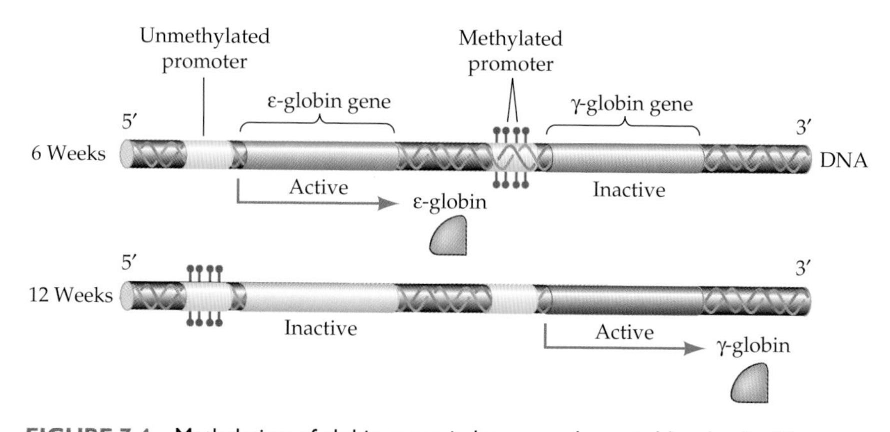

FIGURE 7.4 Methylation of globin genes in human embryonic blood cells. The activity of these human globin genes correlates inversely with the methylation of their promoters. (After Gilbert 2006.)

polymerase is stabilized on the promoter by other DNA sequences called enhancers (see Figure 2.3). The addition of methyl groups to the DNA of enhancer or promoter regions usually prevents the RNA polymerase from binding to the promoter and, as a result, blocks gene activity, whereas demethylation (removing existing methyl groups) from these regions activates gene expression (Figure 7.4; see also Figure 2.4).

Studies have shown that rats born of mothers given a low-protein diet during gestation have a different pattern of liver gene methylation than do the offspring of mothers fed a diet with a normal amount of protein, and that these differences in methylation change the metabolic profile of a rat's liver. For instance, the methylation of the promoter region of the *PPARα* gene (a gene whose product is critical in the regulation of carbohydrate and lipid metabolism) is 20% lower in rats fed protein-restricted diets, and consequently its activity is tenfold greater (Figure 7.5; Lillycrop et al. 2005). In addition, the promoter region of the glucocorticoid receptor gene (critical in blood pressure regulation) was 22% less methylated and three times more transcriptionally active in the pups born of mothers given a protein-restricted diet (Burdge et al. 2006). These methylation patterns persisted after the dietary restrictions ceased, thereby showing stable modifications in gene expression due to nutritional influences (Lillycrop et al. 2008). DNA methylation thereby provides a mechanism for fetal alterations to persist throughout life.

Interestingly, the difference in methylation patterns could be abolished by including folate in the protein-restricted diet. Folate is a methyl-group donor. Thus, the difference in methylation is a result of changes in folate

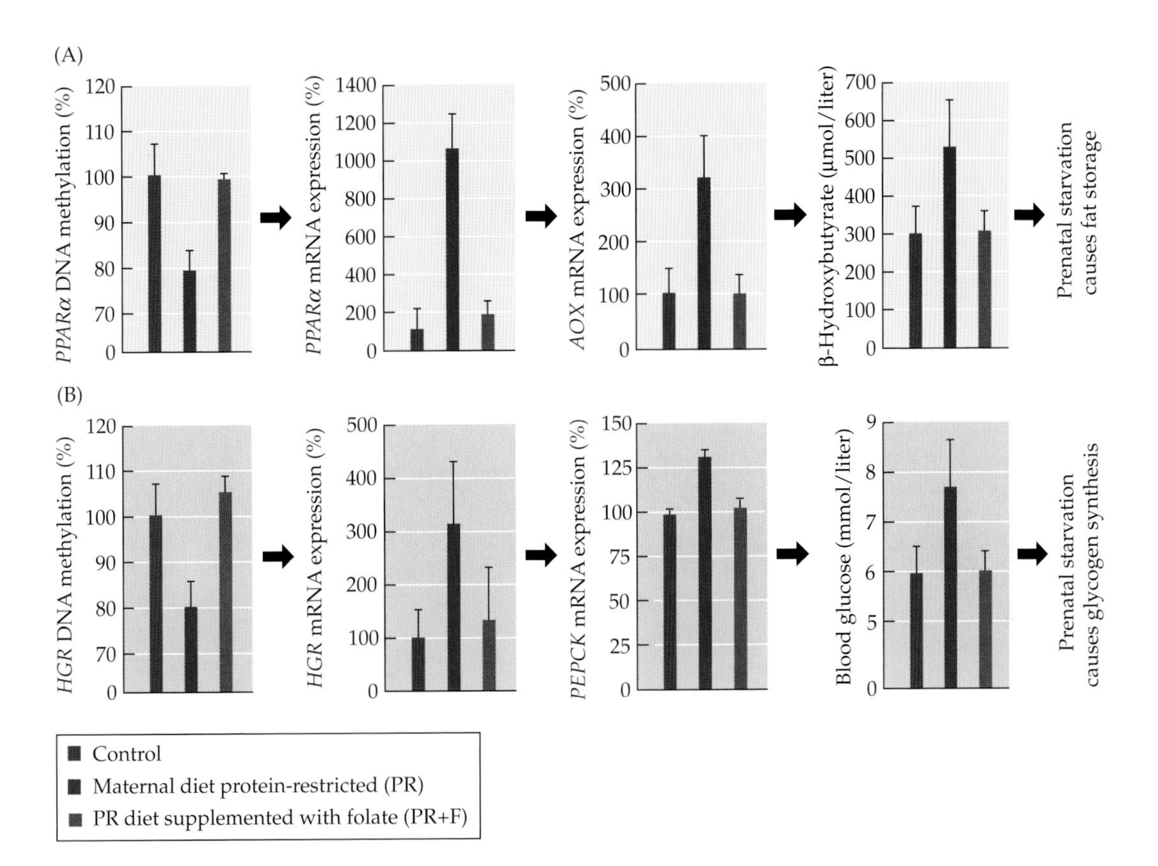

FIGURE 7.5 The offspring of rats fed a protein-restricted diet (PR) showed higher fat synthesis and breakdown of carbohydrates into glucose. The mechanism for this may involve reduced promoter methylation and increased mRNA expression of at least two major genes. (A) Reduced promoter methylation and increased expression of the gene for the transcription factor peroxisome proliferator-activated receptor α (PPARα) in the liver, as compared with control rats fed a normal diet. Increased *PPARα* expression is associated with increased expression of acyl-CoA oxidase (AOX), a key enzyme in fatty acid oxidation, and increased circulating concentrations of the ketone β-hydroxybutyrate. These metabolic effects can be prevented by supplementing the protein-restrictive maternal diet with folate (PR+F). (B) Reduced promoter methylation and increased gene expression for hepatic glucocorticoid receptor (HGR) in rats whose mothers were protein-restricted during pregnancy. Increased *HGR* expression is associated with increased expression of the gene for gluconeogenic enzyme phosphoenolpyruvate carboxykinase (PEPCK) and consequent increased blood glucose levels. Again, the effects are not seen in offspring of PR rats whose diets were supplemented with folate; *p* values for comparisons of the PR+F group with the control group were >0.05 in all cases, indicating no significant difference between these two groups. Data are means; T-bars indicate standard error. (After Gluckman et al. 2008.)

metabolism caused by the limited amount of protein available to the fetus. Folate (folic acid; vitamin B9) is often added as a supplement to cereal grain-based foods because of its importance for reducing neural-tube closure birth defects. However, increased folate may activate other genes.

Predictive Adaptive Responses

Data such as those shown in Figure 7.5 have important consequences. Ever wonder why some people can eat cheesecake and remain thin, while others seem to gain pounds even when they graze on salad? It may have to do with an interaction between genes and environment that played out while they were in the uterus.

Hales and Barker (1992, 2001) proposed the **thrifty phenotype hypothesis**, which holds that malnourished fetuses are "programmed" during their plastic stages to expect an energy-deficient environment and, in response, set their biochemical parameters to conserve energy and store fat. When these individuals develop into adults, if they do indeed find themselves in the expected nutrient-poor environment, they are ready for it and can survive better than if their metabolism had been "set" to utilize energy less efficiently. However, if they find themselves living in a calorie- and protein-rich environment, their energy-efficient metabolism means that their cells simply store the abundance as more and more fat. In addition, because their hearts and kidneys developed to face more stringent conditions, their altered phenotype puts such persons at risk for several diseases (Figure 7.6).

The thrifty phenotype hypothesis has since expanded to link prenatal influences with many other types of adult illness, including lung disease, cognitive development abnormalities, breast cancer, prostate cancer, and leukemia (Godfrey and Barker 2000; De Boo and Harding 2006). The broadening of Barker's hypothesis to encompass a large range of prenatally influenced illnesses has been referred to as the **fetal origins of adult disease** (**FOAD**) hypothesis. This, in turn, has been expanded to become the **developmental origin of health and disease** (**DOHD**) hypothesis. The shift to the DOHD nomenclature was proposed for two reasons. First, research has demonstrated the plasticity of development in both the prenatal and early postnatal periods (as in the case of the sweat gland activation mentioned earlier). Second, the change in terminology emphasizes not only the causes of disease, but also disease prevention and adult health promotion (Gluckman and Hanson 2006a,b). The broader terminology reduces temptation to adopt the fatalistic idea that prenatal events can completely determine adult health. *Both* pre- and postnatal environments are influential in development.

Why is it advantageous for fetal development to pivot on environmental cues? The benefits of fetal developmental plasticity can be realized in

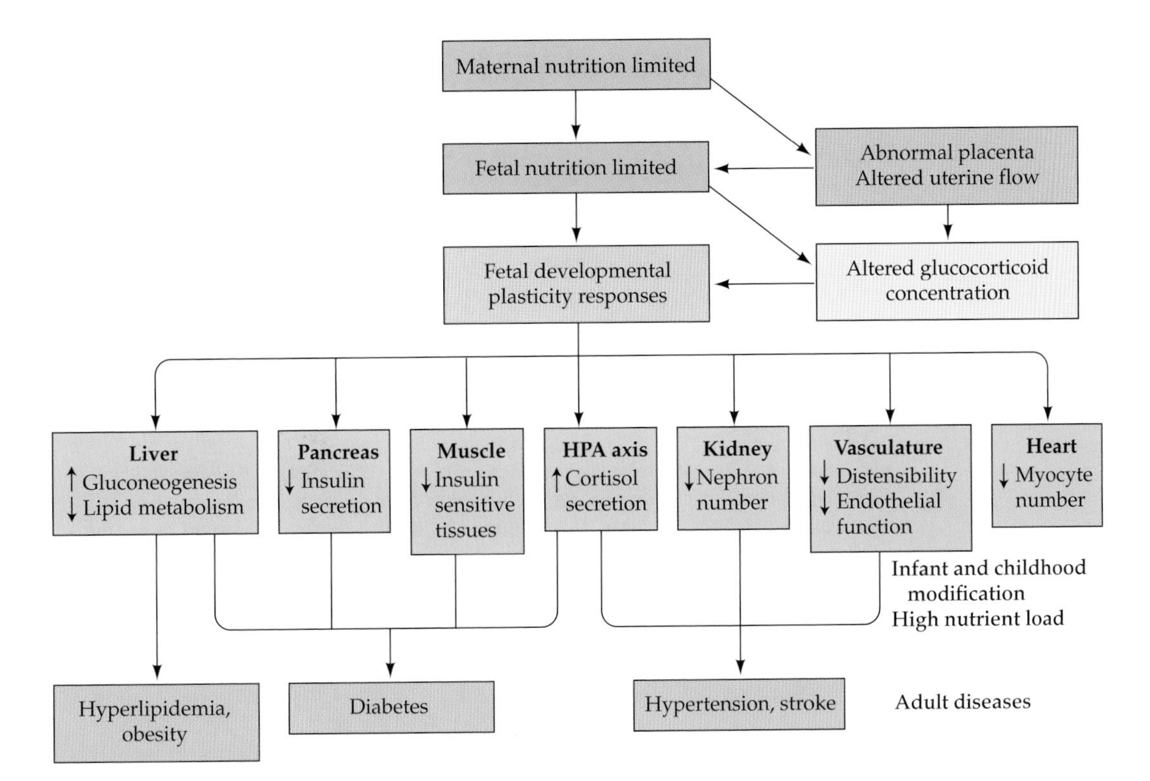

FIGURE 7.6 The thrifty phenotype hypothesis. Altered maternal nutrition leads to a cascade of consequences mediated by the phenotypic plasticity of the fetal organs. The combination of biochemical and anatomical changes in these organs produces a phenotype that stores excess energy as fats and sacrifices kidney growth for brain growth. This phenotype is beneficial under environments of low nutrition, but high nutrition during infancy and/or childhood can result in adult obesity, diabetes, and hypertension. (After De Boo and Harding 2006.)

two ways (Figure 7.7). First, there can be **immediate adaptive responses** that alter development in response to a particular crisis that is already occurring in the embryo. Thus, malnutrition or oxygen deprivation can cause changes in blood flow so that the brain continues to develop at the expense of less critical tissues (Bateson et al. 2004).

A second type of plasticity is the **predictive adaptive response**, or **PAR**. The PAR model states that early environmental cues can shift the developmental pathway to modify the phenotype in expectation of the later environment (Gluckman et al. 2005; Gluckman and Hanson 2005). As opposed to immediate adaptive responses, PARs benefit the organism later in life—

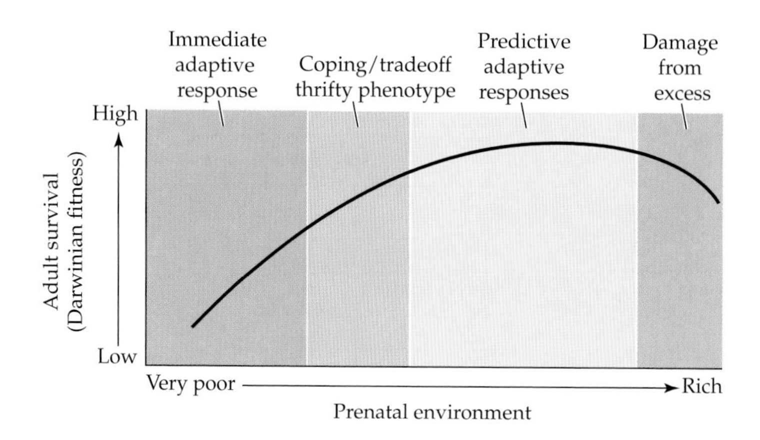

FIGURE 7.7 Developmental responses and fitness. Mutations or severe environmental disturbances can disrupt development and cause *immediate responses* to the injury or defect. Less extreme disturbances can lead to *predictive adaptive responses* (PARs), including the induction of thrifty phenotypes. Within the normal range of variation, an individual's developmental trajectory will be conferred by the actions of PARs. (After Gluckman and Hanson 2006b.)

as long as the predicted and actual postnatal environments match. The properties of PARs include the following characteristics:

- They are induced by early environmental factors and act via developmental plasticity to modify the phenotype to match an anticipated environment.
- They permanently change the physiology and/or anatomy of the organism.
- They respond to a range of developmental environments, not just to extreme conditions.

PARs can give the organism a survival advantage if the response matches the predicted environment; however, if the environment is not the one predicted by the PAR, the mismatched condition can lead to disease.

The essential elements of the PAR model are not new to biologists. Indeed, the plasticity seen in the wing color polyphenisms in butterflies (see Figure 1.5) is essentially a predictive adaptive response, since the color made within the imaginal discs of the larva will be manifest only in the adult. Similarly, many animals have a cyclic development of their fur. In some Arctic animals whose coats are brown in summer but white in winter (Figure 7.8), the decreased sunlight as summer ends initiates increased

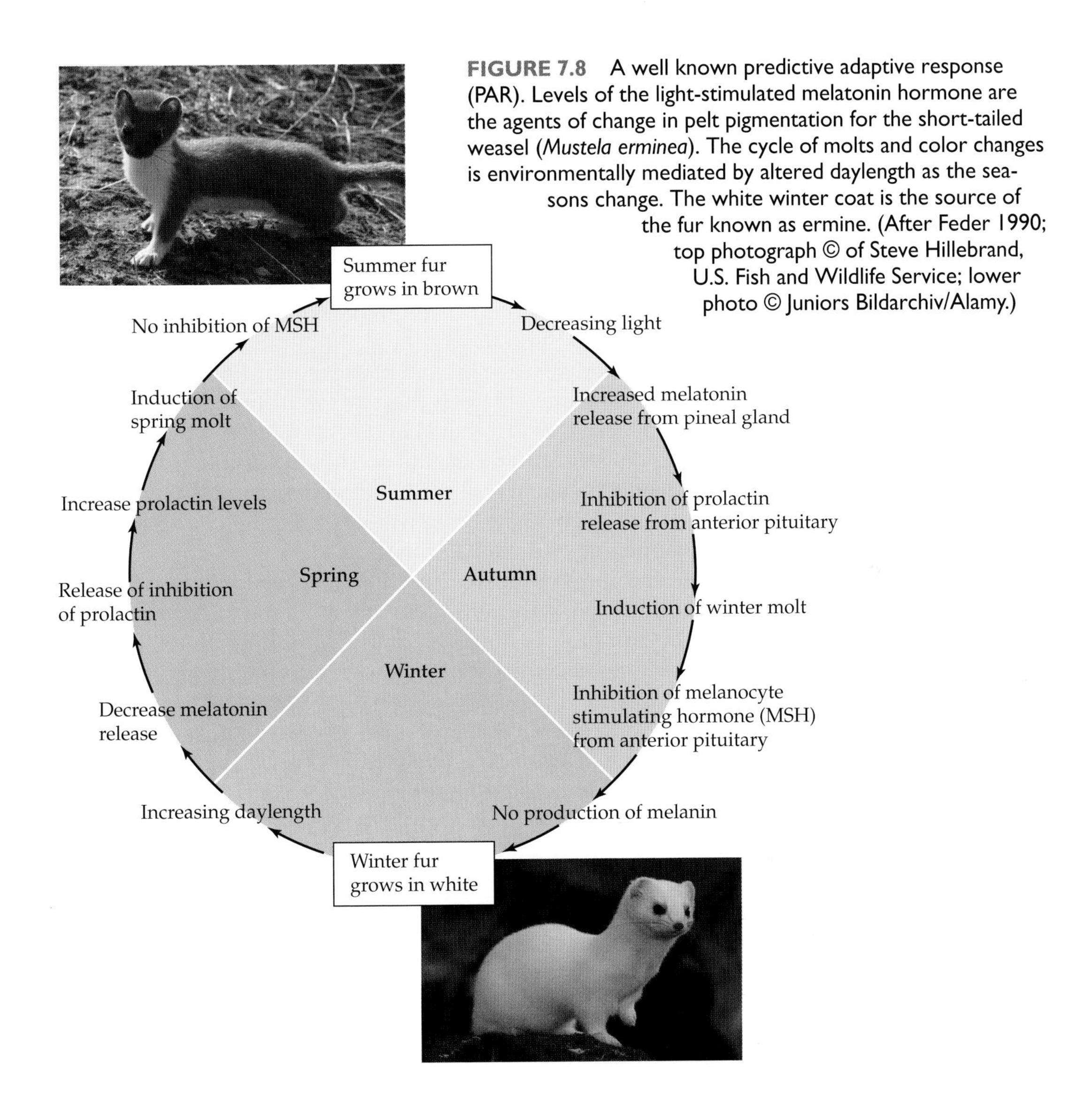

FIGURE 7.8 A well known predictive adaptive response (PAR). Levels of the light-stimulated melatonin hormone are the agents of change in pelt pigmentation for the short-tailed weasel (*Mustela erminea*). The cycle of molts and color changes is environmentally mediated by altered daylength as the seasons change. The white winter coat is the source of the fur known as ermine. (After Feder 1990; top photograph © of Steve Hillebrand, U.S. Fish and Wildlife Service; lower photo © Juniors Bildarchiv/Alamy.)

melatonin secretion from the pineal gland. Melatonin secretion triggers a cascade of hormones that ultimately induces a molt followed by the regrowth of hair, this time in the absence of melanin production (i.e., the fur grows in white; Feder 1990). Again, the processes were initiated by a signal (decreasing photoperiod) long before the first snows of winter fell.

The environmental mismatch hypothesis

The PAR model can be seen as extending the thrifty phenotype hypothesis from the clinical and epidemiological realm into evolutionary biology. The evolutionary advantages of the PAR strategy depend on the accuracy of the cues, the intrinsic costs of plasticity, and the frequency of mismatches (Gluckman et al. 2005). If the predicted and future environments do not match, health risks or other disadvantages may arise. Certain ideas concerning the improper predictions of PARs have come to be known as the **environmental mismatch hypothesis**.

Earlier we mentioned some of the consequences of PARs making the "wrong" decisions. One example was thermal environment triggering a minimal number of active sweat glands in human infants, which became maladaptive when in later life the adults needed to survive in a hot and humid environment (Gluckman et al. 2005). Another example would be the white pelt of an Arctic fox or short-tailed weasel appearing in a winter with no snow (as is happening with global climate change).

One of the most important mismatches involves fetal malnutrition followed by excess nutrition in the adult. This combination of circumstances has become more common as nutrition improves among people born in impoverished societies, as when some of these individuals migrate from regions of poor nutrition into regions of relative affluence. For instance, most of the Third World lives in a state of relatively poor nutrition, especially for pregnant women. Developing fetuses are therefore likely to sense a low-protein condition and become "programmed" to survive in an environment where food is scarce. In other words, the fetus predicts that getting adequate food will be difficult and sets its metabolism for a "thrifty phenotype" that would use every calorie efficiently and conserve any excess food as fat to guard against times of famine. Until recently, such a prediction was likely to be a successful one, enabling the individual to survive to reproductive age while staying small and thin. But rising affluence in some parts of the Third World has led to increasingly plentiful and available food supplies within a single generation. The fetus makes the same "thrifty" PAR (based either on maternal size constraints or on maternal diet), but the adult suddenly lives in a world full of food. The embryo develops resistance to insulin, fewer blood vessels feeding the tissues, and enzyme levels set to convert any unused food into fat. With more than enough food, an adult with this phenotype tends to be obese and diabetic, and to have a high risk of heart attack. Such a syndrome is becoming commonplace throughout the developing world (Gluckman and Hanson 2007).

The problematic effects of an inaccurate predictive response were seen on a smaller scale following the Dutch Hunger Winter of 1944–1945. In reprisal for the Dutch resistance in 1944, Nazi authorities imposed severe food rationing in the western Netherlands. Almost immediately, the typi-

FIGURE 7.9 Near-starvation conditions were imposed on portions of the Dutch population during the Nazi occupation. Those fetuses in utero during the winter of 1944–1945 experienced harsh protein deprivation. (Photograph courtesy of Nationaal Bevrijdingsmuseum Groesbeek/Beeldbank WO2, Netherlands.)

cal 1800 calorie per day diet was reduced to about 600 calories per day (Figure 7.9; Stein 1975; Hart 1993). This food restriction lasted 7 months (until Allied forces liberated the Netherlands), after which food intake returned immediately to the original level. Amazingly, despite the horrors of the war, doctors and midwives continued to deliver babies and to keep careful records of the health of these newborns. When analyzed decades later, these records showed that those adults that were undernourished in utero tended to develop insulin resistance and obesity.

Humans come in all shapes and sizes, and these different shapes and sizes mean different risks for diabetes and heart problems. The incidence of diabetes and cardiovascular disease is much higher in those people who are born small and thin and later become fat. Data from one study of a Finnish population (Eriksson et al. 2001) showed that the risk of coronary heart disease is increased by the presence of two factors: small birth size and then becoming relatively fat in childhood (Figure 7.10). What is important from these data is that high risk does not have to be associated with the extremes of malnutrition (as in the Dutch Hunger Winter.) Even normal variations in nutrition can result in relatively high risks for heart disease and diabetes. It appears the fetus is making developmental decisions in response to low nutrient intake that anticipate its living after birth in a nutritionally deprived environment. Accordingly, it "expects" to stay small and thin and programs its development accordingly. However, if it is born into a nutritionally abundant environment, the thrifty phenotype that the fetus select-

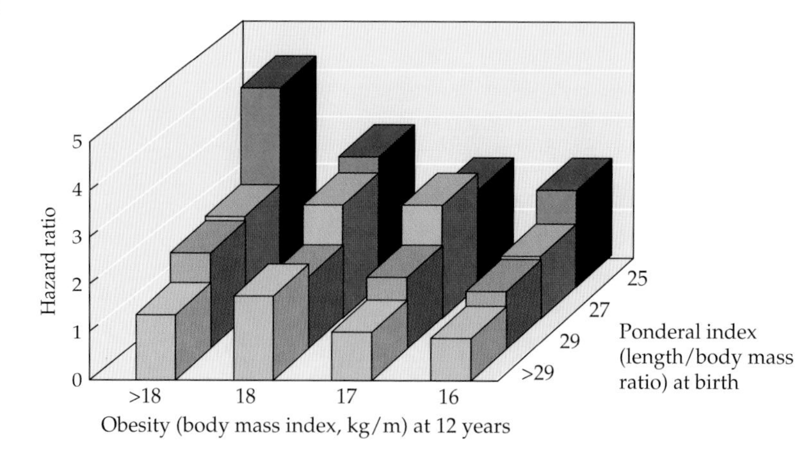

FIGURE 7.10 Data showing that the risk of coronary heart disease is increased by both small birth size and greater fatness in childhood. The graph shows that these conditions interact such that the highest risk of heart disease is in those people who were born at low birth weight (i.e., in conditions that may have caused a thrifty phenotype) but subsequently experienced ample nutrition. (After Eriksson et al. 2001.)

ed stores the extra calories and causes the person to become obese and at risk for cardiovascular problems.

Animal experiments have also provided data consistent with the environmental mismatch hypothesis (see Gluckman and Hanson 2005; Taylor and Poston 2007). As we have mentioned, poor maternal nutrition can lead to low birth weight, which has been correlated (depending on various factors, including the nature of the nutritional deprivation) with heightened risk of heart attack, obesity, and type 2 diabetes. For instance, rat fetuses that are undernourished in utero have a higher probability of having hypertension and insulin resistance as adults (Figure 7.11; Vickers et al. 2000). If, after weaning, these rats are exposed to a high-calorie diet, the level of hypertension and insulin resistance is even greater. As Gluckman and Hanson (2005) concluded:

> [T]he full spectrum of the "couch potato" syndrome could be explained by a combination of an antenatal event coupled with postnatal amplification. (The only part of the syndrome that escaped us was that we could not demonstrate that the rats had a preference for a TV remote control.)

The likelihood that these diseases are caused by mismatched PARs was suggested by experiments in which the newborn rats were given injections of leptin. When rats are injected with leptin, it tells their bodies that they are fat. When newborn rats from protein-starved mothers were given leptin, they set the metabolic phenotype of their livers to levels appropriate to a high-nutrition diet (Gluckman and Hanson 2007; Gluckman et al. 2007).

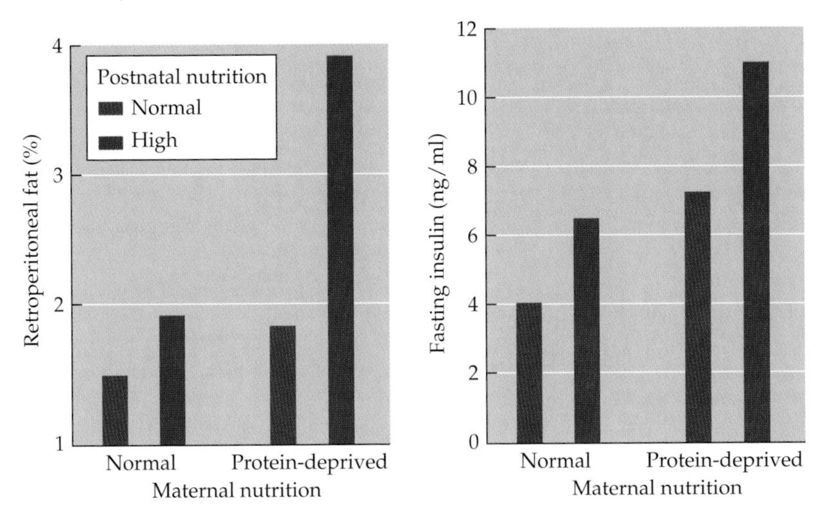

FIGURE 7.11 Experimental evidence for the fetal origin of adult disease in rats. Rat fetuses were exposed to either normal or protein-deficient maternal diets. After birth, the pups were given either normal or high nutrition. They were then measured for (A) obesity (percent fatty tissue in the abdominal area between the peritoneal cavity and the skin) and (B) insulin resistance (as measured by the amount of insulin in their blood after fasting). Obesity and insulin resistance were greater in the pups whose mothers were fed a protein-deficient diet. High nutrition after birth heightened these differences. (After Vickers et al. 2000.)

Environment-genotype interactions in diabetes

As we have seen throughout these chapters, genes do not work in isolation from their environment. Moreover, gene-environment (G × E) interactions early in development induce the predictive adaptive responses that determine postnatal G × E interactions. One of the most interesting of the early G × E interactions is one that alters the individual's sensitivity to the prenatal environment, thus influencing the PAR. This interaction involves one of the "thrifty genes," *PPARγ2*. This gene encodes a transcription factor protein that activates particular genes after it combines with the glucose receptor RXR. The PPARγ2 protein promotes the formation and storage of fat, induces the formation of more fat cells, increases insulin sensitivity, and reduces hypertension (by suppressing angiotensin II). As such, this gene can be critical in susceptibility to diabetes. Humans have several versions of this gene. One allele of this gene, which encodes alanine as its twelfth amino acid, is found in about 10% of the Caucasian and Chinese popula-

Paternal Epigenetic Effects

Maternal effects in mammals can be readily accounted for by the interactions between the mother and the fetus. Paternal effects, however, have not been nearly as well studied (see Friedler 1985, 1996; Hansen 2008). It was assumed that teratogenic effects would be passed on through the mother's exposure to chemicals, and that the father's germline, since it didn't transmit much but a genome, was relatively secure from environmental effects. For example, in 1995, a report from the Institute of Medicine (1995, p. 44), claimed that exposure of human males to environmental chemicals "is generally unrelated to developmental endpoints such as miscarriage, birth defects, growth retardation, and cancer." However, we now know that environmental agents *can* effect epigenetic changes in the male germline that can be passed from one generation to the next. Could epigenetic effects produced by diet and other environmental agents likewise be transmitted?

Epidemiological studies by Pembrey and colleagues (2006) suggest that the environment of prepubescent boys can influence the phenotypes of their sons and grandsons. Using data from a long-term study in Britain, they showed that early (i.e., before teenage years) paternal smoking in males was correlated to a greater body mass index—an indicator of obesity—in their preteen sons (but not their daughters). Furthermore, analysis of medical records from an isolated community in northern Sweden showed that the paternal grandfather's food supply during middle childhood was linked to a greater risk of early deaths in grandsons (but not granddaughters). Conversely, the *paternal grandmother's* food supply during middle childhood was reflected in the higher mortality risk of her granddaughters (and not her grandsons; see figure on facing page). This effect was not observed in the maternal grandmother or maternal grandfather, indicating that this influence was coming from the childrens' father. The report argues that exposure of a male to environmental influences can affect the development and the health of his sons and grandsons.

The molecular basis for this presumably non-genetic hereditary effect is not yet known. As seen in Chapter 6, the effects of endocrine disruptors such as BPA or vinclozolin can be transmitted through the male germline through DNA methylation; that remains a possibility here. Whatever the mechanism of transmission, as Whitelaw (2006) notes in a commentary on the paper by Pembrey and colleagues, "these findings may go some way towards shifting the balance of responsibility for the unborn away from the mother. Fathers-to-be take note!"

The effects of varying food supply of paternal grandparents on the mortality rates of their grandchildren. "Age of paternal grandparent" refers to the beginning of a duration of a year or more of exposure to a poor year in a "good" food supply, or to a "good" duration in a poor food supply. (A) Mean mortality risk ratio of grandsons plotted against paternal grandparents who had a good food supply versus those whose food supply was poor during the same stage of their childhoods shows increased mortality risk in grandsons of those whose "good" food supply was interrupted during a crucial growth period. (B) A similar graph for granddaughters and paternal grandmothers. (C) A graph of grandsons' mortality ratio against paternal grandmothers (or granddaughters' mortality risk against paternal grandfathers, not shown here) shows no such correlations. (After Pembrey et al. 2006.) ▶

tions and is less functional than the more common allele, which has proline at that position.

Research by Lindi and colleagues (2002) showed that the alanine-containing allele was associated with a predisposition to diabetes, especially in those having low birth weight (Figure 7.12). However, the association of low birth weight and adult-onset diabetes was not seen in people having

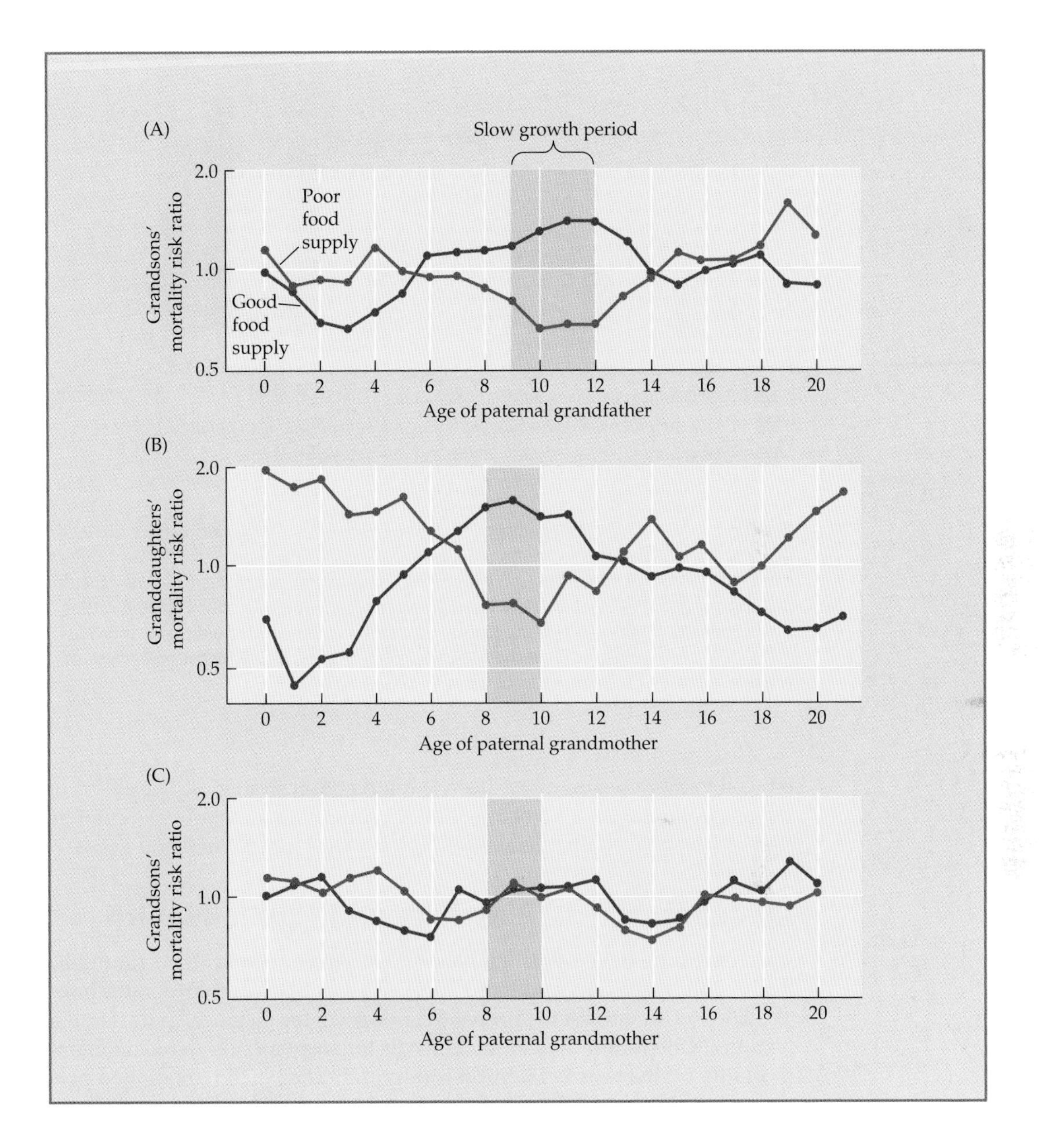

the more predominant form of PPARγ2. In the presence of a genetic situation that made insulin resistance more likely, the low-birth-weight–diabetes connection was apparent. If the genes were such that insulin resistance was not likely, then the effects of fetal programming had little effect later in life. Moreover, in both a Finnish study (Lindi et al. 2002) and a German study (Rittig et al. 2007), the people with the alanine variant of PPARγ2 respond-

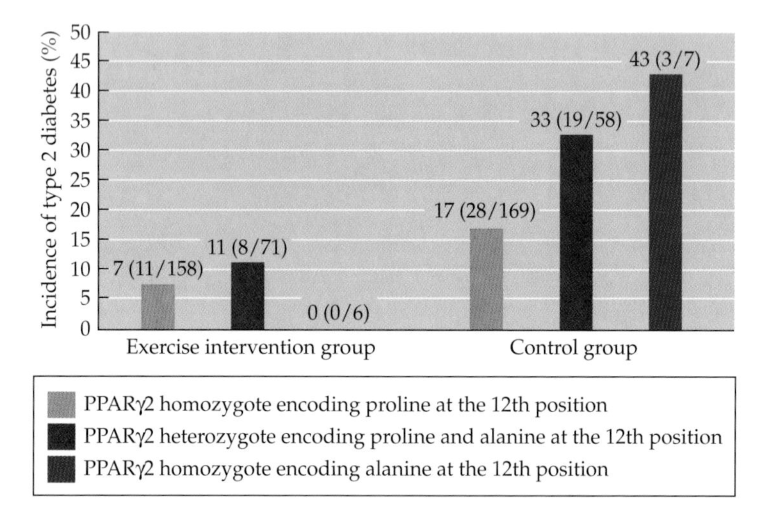

FIGURE 7.12 Genetics, predisposition, and behavioral intervention affect the incidence of type 2 diabetes in adults. Three-year weight change in control groups differs from that in exercise intervention groups with polymorphism of the PPARγ2 gene. In the control group, those with the less common allele (which encodes alanine rather than proline at position 12 of the gene) had a high incidence of the disease. Individuals in an exercise program had the lowest incidence of type 2 diabetes regardless of which allele they carried. (After Lindi et al. 2002.)

ed well to exercise and other lifestyle intervention strategies. The effects of the environment have to be mediated through the metabolic parameters that are established by interactions between the environment and genes.

Developmental Plasticity and Public Health

Understanding developmental plasticity has major implications for public health policy (see Law and Baird 2006). Most public health programs have at their core an interest in survival. This is seen, for instance, in the United Nations Millennium Development Goals for substantially reducing infant mortality by the year 2015. But it is important that such remediation programs not shift the burden of illness from the early years of life to the later ones. We have seen that developmental plasticity needs to become a central part of developmental biology and conservation biology. It also needs to become a central part of public health policy making.

First, interventions designed to improve adult health need to begin early in life or even in utero. Lifestyle interventions aimed at obesity-related problems in adults, for example, may be less effective if the person's gene expression pattern is already "set" to a phenotype that efficiently conserves

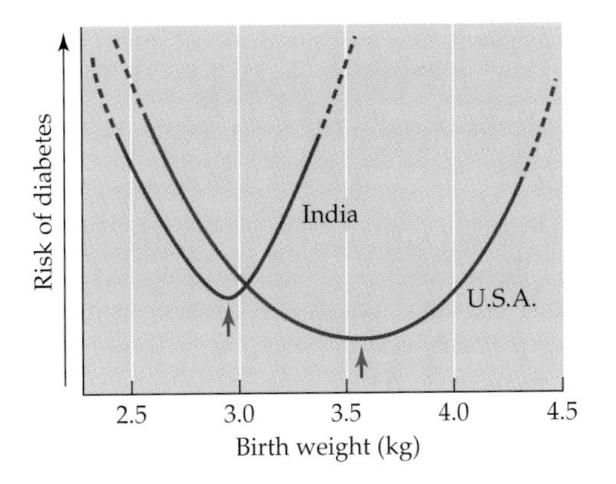

FIGURE 7.13 The risk of adult-onset diabetes is related to birth weight in both India and the United States. The U-shaped curve indicates that the risk for this disease increases at both very high and very low birth weights. However, the optimum birth weight is different for the two populations (arrows), suggesting that developmental plasticity may have resulted in adaptation to distinct adult nutritional milieus. (After Gluckman and Hansen 2005.)

energy and stores it as fat (Wells 2007). Second, interventions designed for infant survival must take into account the effects of such changes. For instance, babies born weighing 2.5 kilograms in rural India may be well adapted both to their maternal milieu and to the nutritional environment where they will live as adults (Figure 7.13). Food supplementation programs aimed at increasing the birth weight of such babies or providing food in schools may increase the chance of those individuals' getting diabetes later in life (Yajnik 2005; Bhargava 2004). Third, it must be recognized that different evolutionary forces have acted on different populations. A "universal standard" of human growth erroneously assumes that life history strategies are the same worldwide and does not take into account the plasticity that may have evolved in different locations (Wells 2007; Adams and White 2005). This new area demands research from many disciplines.

AGING AND CANCER AS DISEASES OF EPIGENESIS

Epigenetic Methylation, Disease, and Aging

Given that appropriate methylation is essential for normal development, one can immediately see that diseases would result as a consequence of inappropriate epigenetic methylation. First proposed by Boris Vanyushin in 1973, recent studies have confirmed that inappropriate methylation not only causes metabolic diseases, but also may be the critical factor in aging and cancers.

Evidence from identical twins

Some of the first evidence for the roles of epigenetic methylation in aging and diseases came from studies of identical twins. Human monozygotic ("identical") twins account for 1 in every 250 live births. Monozygotic twins, as the name indicates, arise from a single zygote and therefore have exactly the same DNA. However, such twins can have a relatively high rate of discordance in many characteristics,* including disease susceptibility—that is, identical twins often develop different diseases. These diseases include conditions that are seen early, such as juvenile diabetes and autism, as well as conditions that manifest later in life, such as ulcerative colitis and various cancers. There is no correlation between the age of a disease's onset and its concordance between twins (Figure 7.14; Petronis 2006). The cause of twin-pair discordance is not known, but recent evidence suggests that differences in DNA methylation may be involved: DNA methylation patterns can differ between twins, even though their DNA is identical.

If a gene becomes methylated when it should not be, it will usually become inactive, just as if the DNA had been mutated. The gene's function is lost either way and, in fact, methylation is a much easier way to lose function. Similarly, anomalies arise if a gene becomes unmethylated and thus activated in the "wrong" cells. Such methylation differences appear to be at the root of one case in which one of a pair of monozygotic twin girls had a severe anomaly—a duplicated portion of the spine in the posterior portion of her body. Her phenotype reminded clinicians of a similar phenotype in mice where the *Axin1* gene is mutated. Axin protein is an inhibitor of the Wnt pathway in development and, in mice, mutations of the *Axin1* gene cause duplications of the caudal axis, resulting in extra spines and bifurcated tails. Research revealed that methylation of the *Axin1* promoter repressed *Axin1*, preventing it from functioning and giving an abnormal tail phenotype like that produced by the mutation. Blood samples from the twin girls showed that, although both girls had the same allele for *AXIN1* (the human analogue of the mouse gene), there was significantly more methylation at that locus in the affected twin than in the unaffected twin,

*Stephen Jay Gould liked to point out that conjoined twins Eng and Chang Bunker (who inspired the term "Siamese twins") were two very different people (see Gould 1997). Eng was a quiet, content vegetarian who didn't touch alcoholic beverages, while Chang was a morose and aggressive fellow who liked strong drink. This, despite the fact that the twins shared the same genome *and* were subject to exactly the same environments throughout their lives. So something else must contribute to their phenotypes. Random epigenetic effects may provide one answer. (Eng and Chang, by the way, worked out a scheme whereby they adhered to the lifestyle of the twin in whose house they were living that week.)

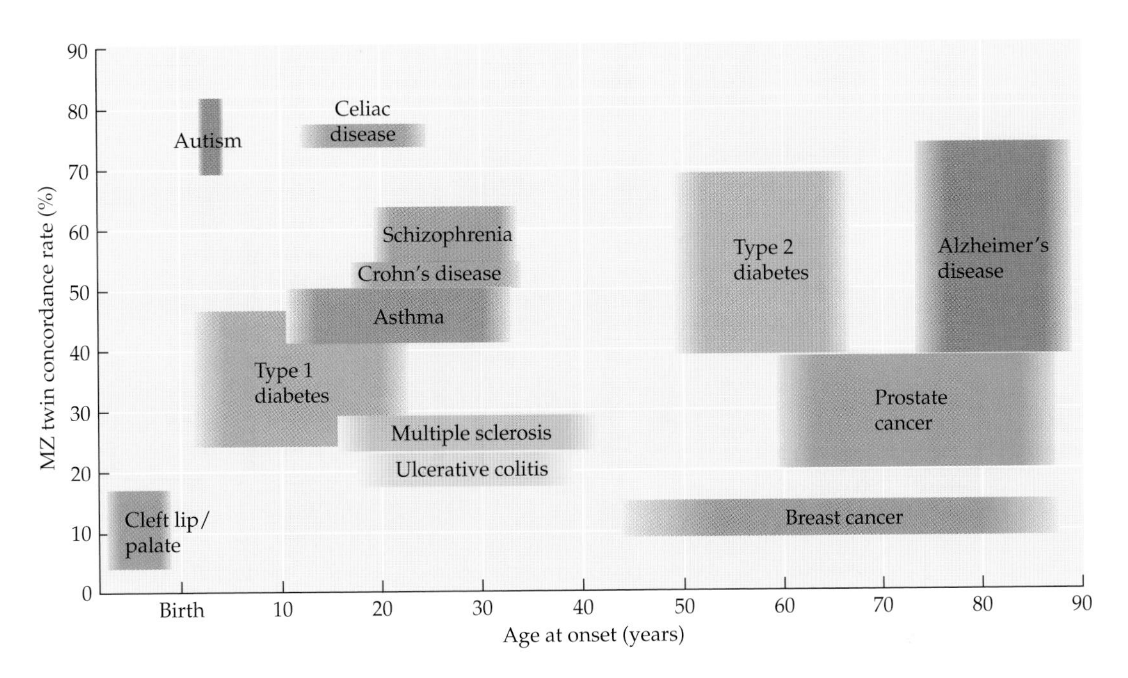

FIGURE 7.14 Age of disease onset and concordance of occurrence among monozygotic (MZ) twins for a number of pathologies. The concordance rate reflects the probability that both twins will be affected; a low rate indicates that often one twin has the disease and the other does not. Note that there is no correlation between age of onset and the degree of concordance. (After Petronis 2006.)

or in either parent (Oates et al. 2006). The regions around the gene showed no significant differences in methylation.

Most "identical" twins start off life with very few differences in appearance or behaviors, but these differences accumulate with age. Experience counts, and both random events and lifestyles may be reflected in phenotypes. Mario Fraga and colleagues (2005) in Manel Esteller's laboratory in Madrid found that the gene methylation patterns of twin pairs were nearly indistinguishable when the pair was young, but older monozygous twin pairs exhibited very divergent patterns of methylation. Methylation patterns affected their gene expression patterns, such that older twins had different patterns of DNA expression, while younger twin pairs had very similar expression patterns. Figure 7.15 shows that monozygotic twins start off with identical amounts of methylated DNA, histone H4 acetylation, and histone H3 acetylation—three important epigenetic markers. As the twins age, methylation increases in both, but to different extents. Acetylation differences also increase.

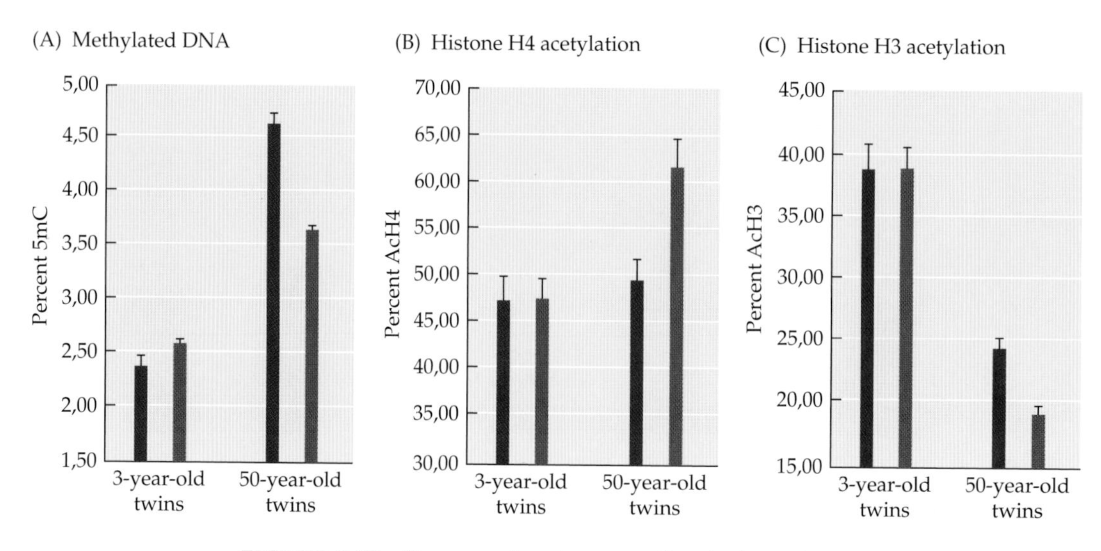

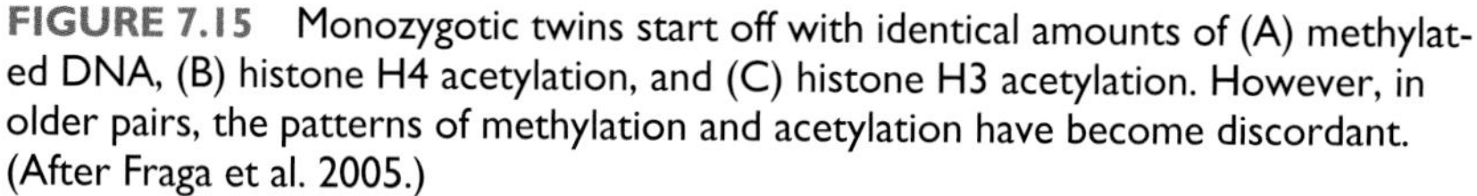
FIGURE 7.15 Monozygotic twins start off with identical amounts of (A) methylated DNA, (B) histone H4 acetylation, and (C) histone H3 acetylation. However, in older pairs, the patterns of methylation and acetylation have become discordant. (After Fraga et al. 2005.)

Not only do the amounts of methylation and acetylation change as twins age, but so does the pattern. The differences can be seen at specific genetic sequences when one cuts the DNA with enzymes sensitive to methyl groups on the cytosines. Thus, there are some enzymes that will cleave DNA at a sequence containing a cytosine residue (C) but will not cut that sequence if the cytosine is methylated. Figures 7.16 show that, as a twin pair ages, there is an increase in the discrepancies between their DNAs. Indeed, more recent studies that look at methylation differences in the same individuals at different times in their respective lives show that DNA methylation patterns change as a person ages. Moreover, these same studies suggested that methylation maintenance might be genetically controlled, since different families showed different patterns of methylation changes (Bjornsson et al. 2008). It certainly seems likely that methylation differences will turn out to be critical in explaining both phenotypic divergence and discordant susceptibility for different diseases.

Aging and random epigenetic drift

The idea that **random epigenetic drift** inactivates important genes without any particular environmental cue gives rise to an entirely new hypothesis of aging. Instead of randomly accumulated mutations—which might be

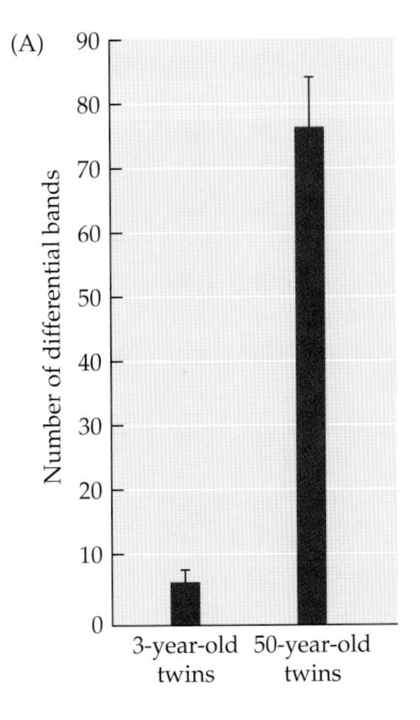

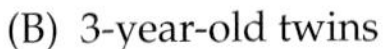

FIGURE 7.16 Differential DNA methylation patterns in aging twins. (A) In bisulfite sequence mapping, regions of DNA that are unmethylated will be cut by restriction enzymes (because bisulfite converts unmethylated cytosine to uracil) but methylated sites will not. The histogram summarizes the number of differences in the resulting restriction maps of 3-year-old twins and 50-year-old twins. (B) A more recent technique of revealing methylation differences and similarities between twins is to mark the DNA from one twin with a red dye and that from the other with a green dye. One can then collect only the nonmethylated DNA and bind it to metaphase chromosomes. If the bands are red or green, it means that the DNA from one twin bound but the DNA from the other twin did not. If the region is yellow, it means that the red and green DNAs bound equally. (After Fraga et al. 2005; photographs courtesy of M. Esteller.)

(B) 3-year-old twins

50-year-old twins

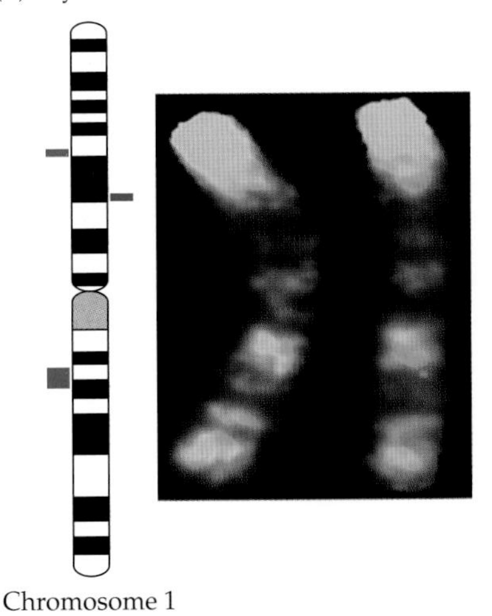

Chromosome 1

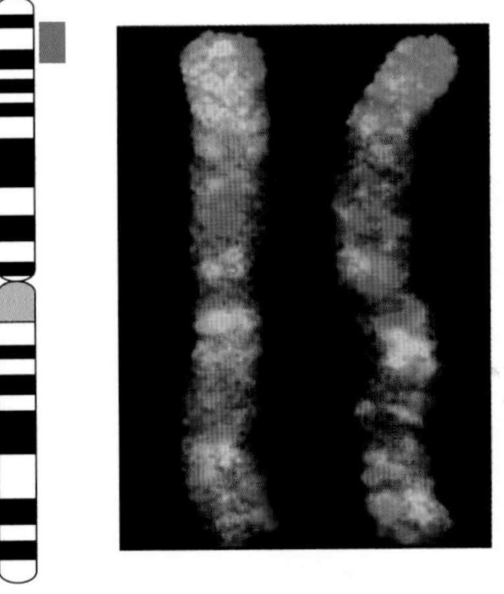

Chromosome 1

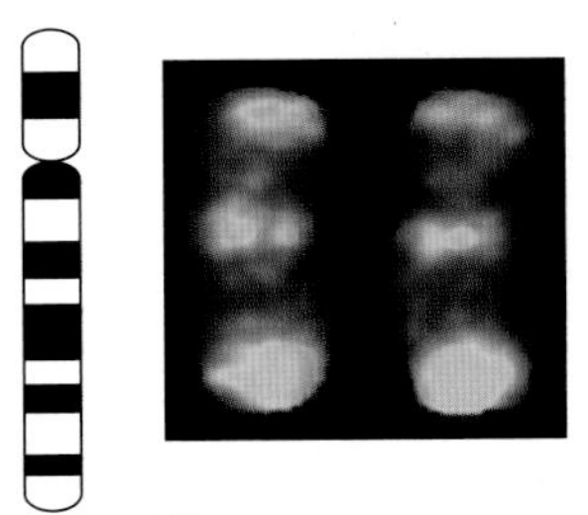

Chromosome 12

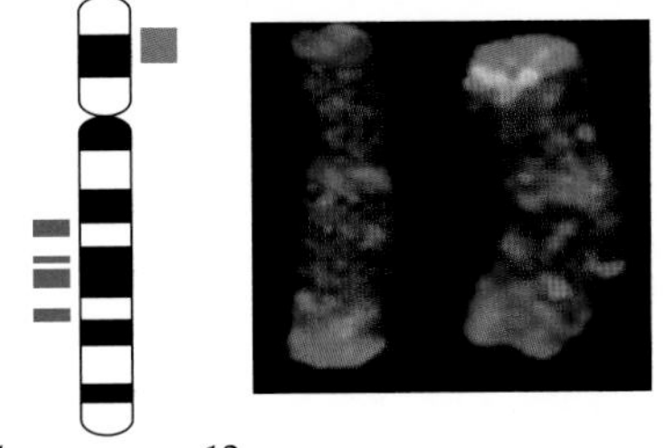

Chromosome 12

due to specific mutagens—we are at the mercy of chance accumulations of errors made by the DNA methylating and demethylating enzymes. Indeed, unlike the DNA polymerases, our DNA-methylating enzymes are prone to errors. At each round of DNA replication, DNA methyltransferases must methylate the appropriate Cs and leave the other Cs unmethylated, and they are not the most fastidious of enzymes, making errors at the rate of about 4% (see Appendix B). Within certain genetic parameters (which may affect the speed at which methylation changes occur and which may differ between species and between individuals), our cells may be accumulating errors of gene expression throughout our lives.

Random epigenetic drift may have profound effects on our physiology. For instance, methylation of the promoter regions of the α and β estrogen receptors is known to increase linearly with age (Figure 7.17; Issa et al. 1994), and such methylation is thought to bring about the inactivation of these genes in the smooth muscle cells of the blood vessels. This decline in estrogen receptors would prevent estrogen from maintaining the elasticity of these muscles and would thereby lead to "hardening of the arteries." Increased methylation of the estrogen receptor genes is even more prominent in the atherosclerotic plaques that occlude the blood vessels (Figure

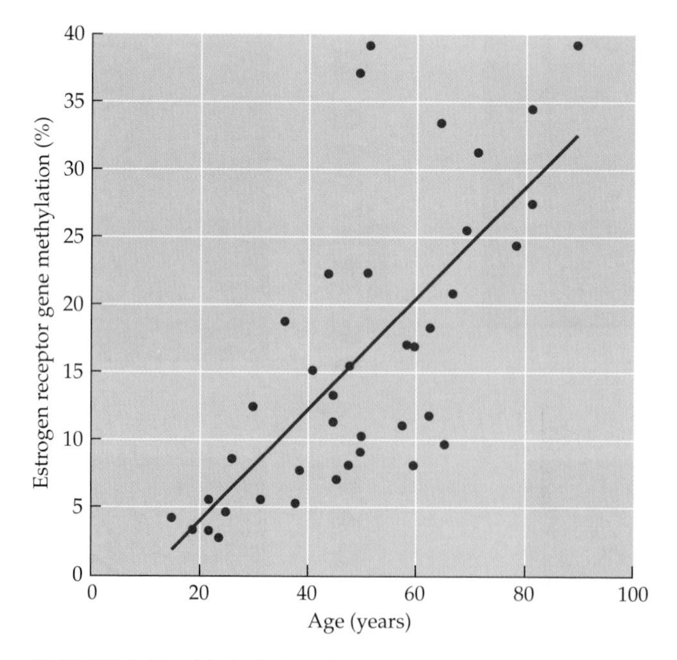

FIGURE 7.17 Methylation of an estrogen receptor gene occurs as a function of normal physiological aging. (After Issa et al. 1994.)

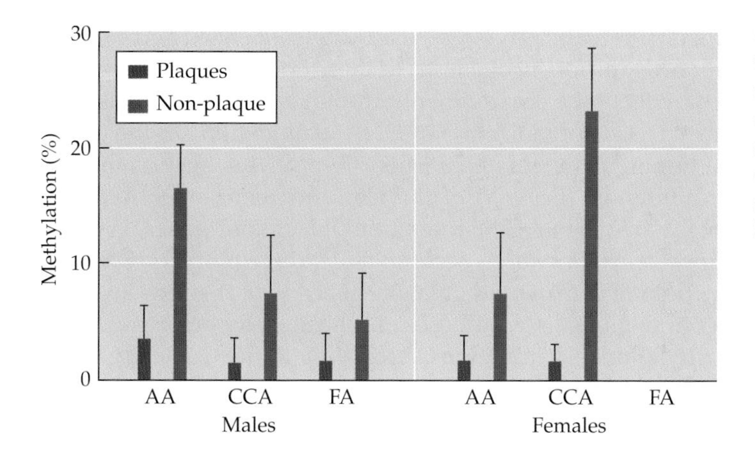

FIGURE 7.18 Methylation of the estrogen receptor-β gene in atherosclerotic plaques and adjacent non-plaque blood vessel tissue in the ascending aorta (AA), common carotid artery (CCA), and femoral artery (FA). (After Kim et al. 2007.)

7.18). Atherosclerotic plaques show more methylation of estrogen receptor genes than do the surrounding tissues (Post et al. 1999; Kim et al. 2007). Thus, methylation-associated inactivation of the estrogen receptor genes in these cells may play a role in the age-related deterioration of the vascular system. This potentially reversible defect may provide a new target for intervention in heart disease.

So in this new hypothesis for aging, there appears to be random epigenetic drift that is not determined by the type of allele or any specific environmental factor. Random epigenetic drift may be the cause of the various phenotypes associated with aging as different genes randomly get repressed or ectopically activated. Mistakes in the DNA methylation process accumulate with age and may be responsible for the deterioration of our physiology and anatomy. If this is so, some genes may be more important targets than others. (The above-mentioned estrogen receptors, for instance, are critical in not only in the vascular system but also for skeletal and muscular health.)

Epigenetic Origins of Cancer

A widely accepted model of cancer formation (oncogenesis) claims that an accumulation of mutations in a somatic cell (that is, a non-germline cell) gives that cell a growth advantage over other somatic cells, allowing it to proliferate and become a tumor. This **somatic mutation hypothesis** holds that cancers are caused by mutations in the DNA (see Weinberg 1998; Wood et al. 2007). This model for tumor development is supported by the observations that mutations in certain genes make cells susceptible to tumorous growth, that some cancers (about 5%) are transferred through the germline, and that normally growing cells can be converted into fast-growing cancer cells by mutation of certain genes.

Today the somatic mutation hypothesis is being challenged and expanded by epigenetic accounts of oncogenesis. In these epigenetic hypotheses, a cell is regulated by its tissue environment through epigenetic signals. Any agent that disrupts tissue organization or DNA methylation has the potential to initiate a tumor. In one set of epigenetic hypotheses, such as the **epigenetic progenitor model** (Feinberg et al. 2006), alterations of DNA methylation can prevent the normal functioning of DNA repair genes. Without the DNA repair proteins, mutations accumulate. If these mutations prevent the normal functioning of growth regulatory genes, cancers may develop.

A second set of epigenetic hypotheses emphasizes tissue organization rather than methylation and mutations. According to the **tissue organization field** (TOF) hypothesis (Sonnenschein and Soto 1999), cancer arises from disruptions in tissue organization. Mutations are not required for a cell to become cancerous, and a cell that remains normal in one tissue context may become cancerous in another. Moreover, such cancer cells can revert back to normal growth patterns if placed back in their original environment, or if they are instructed to differentiate.

Cancer as caused by altered epigenetic methylation

In the discussion of aging, we mentioned that altered methylation of physiologically important genes may be important for generating the phenotypes of senescence. There are also genes that are critical for regulating cell division, and the aberrant expression of these genes can give rise to cancers. The somatic mutation hypothesis of oncogenesis recognizes two types of genes involved in cancer production:

- **Oncogenes** promote tumor formation and metastasis (the spread of cancer cells throughout the body). They are the genes that promote cell division, reduce cell adhesion, and prevent apoptosis (cell death).
- **Tumor-suppressor genes** are genes that usually put brakes on cell division and increase cell-cell adhesion. Activation of these genes can lead to the apoptosis of rapidly dividing cells.

The expression of oncogenes and tumor-suppressor genes has to be very finely regulated. In the stem cells of the skin, gut, and blood, for instance, there has to be a very carefully controlled rate of cell division. Either too much or too little cell proliferation is potentially lethal. Thus, it seems logical that cancers might easily arise if tumor-suppressor genes were to become inappropriately methylated, or if oncogenes were to become inappropriately demethylated (Figure 7.19).

This situation has indeed been found to be the case in human tumors (see Esteller 2008). In breast cancers, for instance, the tumor suppressor genes encoding BRCA1 (which restricts cell division) and E-cadherin (which tightens the junctions between cells) become heavily methylated,

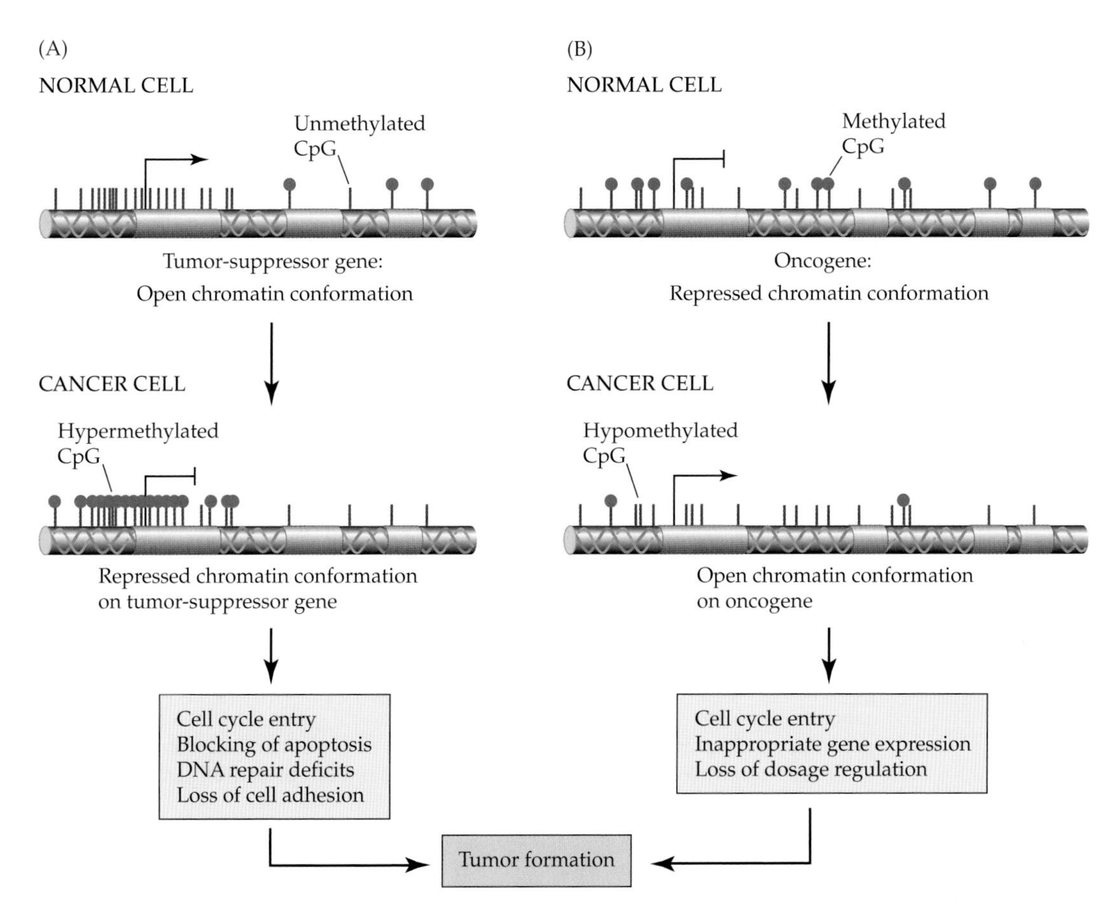

FIGURE 7.19 Cancer can arise (A) if tumor-suppressor genes are inappropriately methylated (and thereby inactivated) or (B) if oncogenes are inappropriately demethylated (and thus activated). (After Esteller 2007.)

while the genes for growth promoting factors such as cyclooxygenase-2 are demethylated (Ma et al. 2004). In gliomas (brain tumors), the gene encoding the DNA repair enzyme MGMT is hypermethylated, as is the gene encoding thrombospondin, a tumor-suppressor protein needed for cell-cell and cell-matrix adhesion.

The methylation of tumor-suppressor genes may explain the increased prevalence of sporadic tumors with age (Fraga and Esteller 2007). In the colon, where estrogen receptor genes function as tumor suppressors, there is a linear increase of gene methylation with age (see Figure 7.16). In addition, Issa and colleagues (1994) showed extremely high levels of DNA methylation in the estrogen receptor genes of colon cancer cells, even the

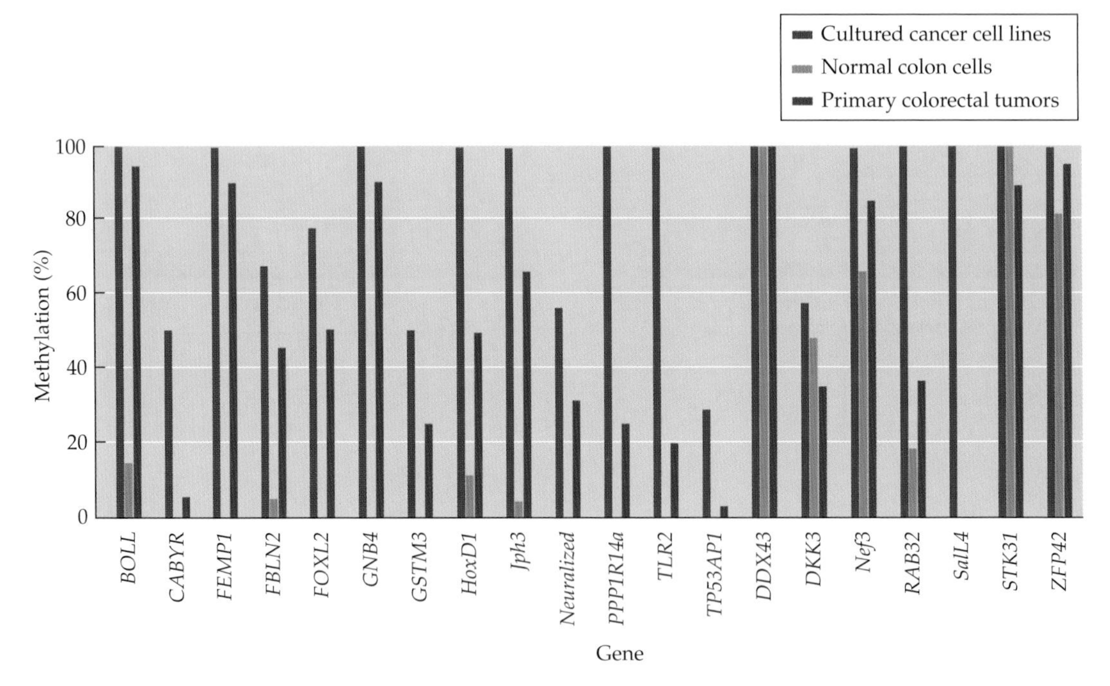

FIGURE 7.20 Gene methylation is increased in cancerous colorectal cells. The genes were selected from a list of genes analyzed by microarrays. Gold bars represent samples from normal colon cells; note that several genes show essentially no methylation in normal tissue. The red bars represent genes sampled from primary colon tumors, while the blue bars represent samples from colorectal cancer cell cultures. (After Schuebel et al. 2007.)

smallest of which had nearly 100% methylation of the estrogen receptor promoter region. Moreover, microarrays (which detect differences in mRNA populations) found several genes whose methylation patterns changed dramatically in colon cancer cells (Figure 7.20). In normal human breast tissue, there is an age-dependent increase in the promoter methylation of another tumor-suppressor gene, *RASSF1A*, and this methylation is highly correlated with breast cancer risk (Euhus et al. 2008).

Epigenetic changes help explain the causes of tumor cell formation and malignancy. In Chapter 3, we mentioned that the bacteria *Helicobacter pylori* may cause cancer by altering the methylation patterns of certain genes in the stomach epithelium. The promoter of the tumor-suppressor gene encoding E-cadherin is hypermethylated by secretions of *H. pylori* (Chan et al. 2003, 2006; Miyazaki et al. 2007), while the gene encoding the growth-promoting enzyme cyclooxygenase-2 has reduced methylation (Akhtar et al. 2001). In lung cancer, the malignancy of a given tumor has been correlated with the promoter methylation of four tumor-suppressor genes, including a gene that regulates cell division (*p16*), one that increases cell adhesion

(*H-cadherin*), one that causes the death of dividing cells (*APC*), and one that is necessary for stabilizing epithelial cell structure (*RASSF1A*). These finding not only help us understand the mechanisms of malignancy, they also may lead to better identification of metastatic cells and better risk assessment and treatment (Brock et al. 2008)

DNA methylation can also explain several lifestyle-related tumors. For instance, tobacco smoke is known to initiate lung cancers even though it is not a very powerful mutagen. Russo and colleagues (2005) and Liu and colleagues (2006) found that in lung cancers attributed to cigarette smoke, several tumor-suppressor genes in the early cancerous cells were heavily methylated. These included genes encoding cell adhesion proteins, apoptosis accelerators, and mitosis inhibitors. In addition, Liu and colleagues (2007) showed that the gene for synuclein-γ, which is not normally expressed in lung tissues, is activated by cigarette smoke. This gene appears to become demethylated, and its subsequent activation promotes the spread of the tumor to other parts of the body. The mechanism by which the synuclein-γ gene is demethylated gives us an important clue as to how tumors can occur: the demethylation mechanism appears to be the *down-regulation* of the gene encoding the methyltransferase DNMT3B, an enzyme that usually adds methyl groups to DNA.

In summary, for the past half-century, the model of tumor formation has been a gene-centered one, wherein a particular cell becomes tumorous and malignant by accumulating numerous new mutations. However, we now see that changes in DNA methylation patterns can induce the formation of tumors, and that a cell can become malignant through the repression of gene expression rather than by the mutation of particular genes. Moreover, when mutated genes are found in the tumors, they are often secondary to epigenetic changes in the genes encoding DNA repair enzymes or methyltransferases.

The reciprocity of epigenetic and genetic causation in cancer

The epigenetic model of cancer does not exclude genetic causes. Indeed, several studies indicate that the two mechanisms augment each other. Numerous mutations occur in each cancer cell, and recent evidence suggests that as many as fourteen significant tumor-promoting mutations are found in each cancer cell (Sjöblom et al. 2006). Jacinto and Esteller (2007) have shown that the large number of mutations that accumulate in cancer cells may have an epigenetic cause.

DNA has several means of protecting itself from mutations. One is the editing subunits on DNA polymerase, which "proofread" duplicated DNA to remove mismatched bases and insert the correct ones. Another mechanism is the set of DNA repair enzymes. These enzymes repair DNA that has been damaged by light or by cellular compounds that are products of metabolism. In cancer cells, the genes encoding these DNA repair enzymes appear to be susceptible to inactivation by methylation. Once these repair enzymes have been downregulated, the number of mutations increases.

It seems, then, that aging and cancer may be linked in having the common denominator of aberrant DNA methylation. If metabolically or structurally important genes (such as the estrogen receptors) become heavily methylated, they don't produce enough receptor proteins, and our bodies function more poorly. If tumor-suppressor genes or the genes encoding DNA repair enzymes are heavily methylated, tumors can arise.

However, even as one finds that epigenetic mechanisms can cause mutations, one also finds that mutations can be the cause of epigenetic silencing. For instance, as many as 5% of all the genes in tumor cells may be hypermethylated (Schuebel et al. 2007). One of the most important oncogenes encodes a small G-protein known as Ras. This oncogene product is known for its ability to initiate signaling transduction cascades that phosphorylate certain transcription factors and promote tumorigenesis. However, Ras can also act via a second, epigenetic, cascade (Gazin et al. 2007; Cheng 2008). Through a pathway composed of at least 28 proteins (including DNMT1 methyltransferase), Ras protein is able to silence the *Fas* gene that enables the rapidly dividing cells to commit apoptosis. The active Ras protein recruits DNMT1 to the promoter of the *Fas* gene, where it hypermethylates the promoter and stops its tumor-suppression function.

The tissue organization field hypothesis

In the above descriptions of oncogenesis, the cell becomes tumorigenic through the accumulation of mutations. The initiating cause may be genetic (mutation) or epigenetic (methylation), but the end result is that DNA repair breaks down and mutations occur that predispose the cell to proliferate. However, a great deal of data indicate that many cancers have normal genomes (see reviews by Sonnenschein and Soto 1999, 2005). Indeed, in Chapter 6 we saw that prenatal exposure of female rats to certain endocrine disruptors (which do not cause mutations) induced in situ carcinomas in their mammary gland ducts (see Figure 6.16). Some cancer cells can actually revert back to being normal cells when placed into appropriate environments, or when given chemicals that cause them to differentiate. For instance, the stem cells of teratocarcinomas resemble embryonic stem cells. They are pluripotent and can generate cells of all three germ layers (Figure 7.21). Stewart and Mintz (1981) not only showed that these cancerous teratocarcinoma cells had no obvious mutations, they also showed that when these tumorous cells were inserted into the inner cell mass of a normal embryo, they integrated into that normal embryo. This created chimeric mice, some cells of which were derived from the original inner cell mass and some cells of which derived from the tumor cell. Moreover, when the adult chimeric mice were

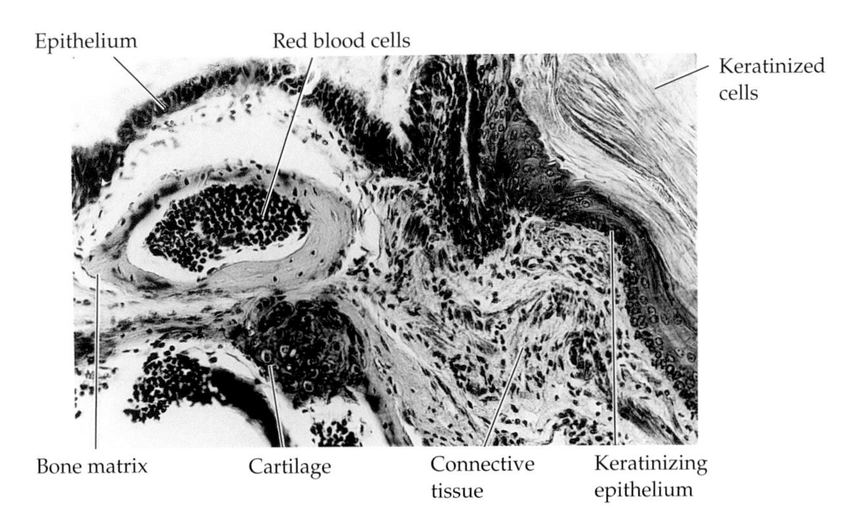

FIGURE 7.21 Micrograph of a section through a teratocarcinoma shows numerous differentiated cell types. (From Gardner 1982; photograph by C. Graham, courtesy of R. L. Gardner.)

mated with one another, some of their offspring were derived solely from the tumor cells (Figure 7.22). Thus, despite being malignant, these cells had a genome so normal that it could generate an entire mouse.

The tissue organization field hypothesis holds that normal interactions between cells inhibit the normal propensity of a cell to proliferate, and that cancers arises from the disruption of these repressive interactions (Soto and Sonnenschein 2005). Numerous studies have provided evidence for this hypothesis. Mina Bissell and her colleagues have found that when the mammary cells are taken out of their normal breast tissue environment, they have a propensity to become malignant. Moreover, certain human breast cancer cells can revert to a normal phenotype when placed into an appropriate environment (Petersen et al. 1992; Weaver et al. 1997; Bissell et al. 2002). The extracellular matrix surrounding the mammary epithelium is critical in activating the genes that are involved in mammary gland differentiation (lactoferrin, casein) and in repressing those genes (*c-myc* and *cyclin D1*) involved in promoting cell division (Figure 7.23). It appears that the extracellular matrix constrains the actin cytoskeleton of the cell; but when the actin cytoskeleton is perturbed, the acetylation patterns of histones H3 and H4 are dramatically altered (Le Beyec et al. 2007). Thus, by perturbing tissue organization, the mammary cell (probably a mammary stem cell whose division is tightly controlled) changes its epigenetic state into that of a proliferating cell. So these cancer cells have been thought to be epigenetically rather than genetically altered.

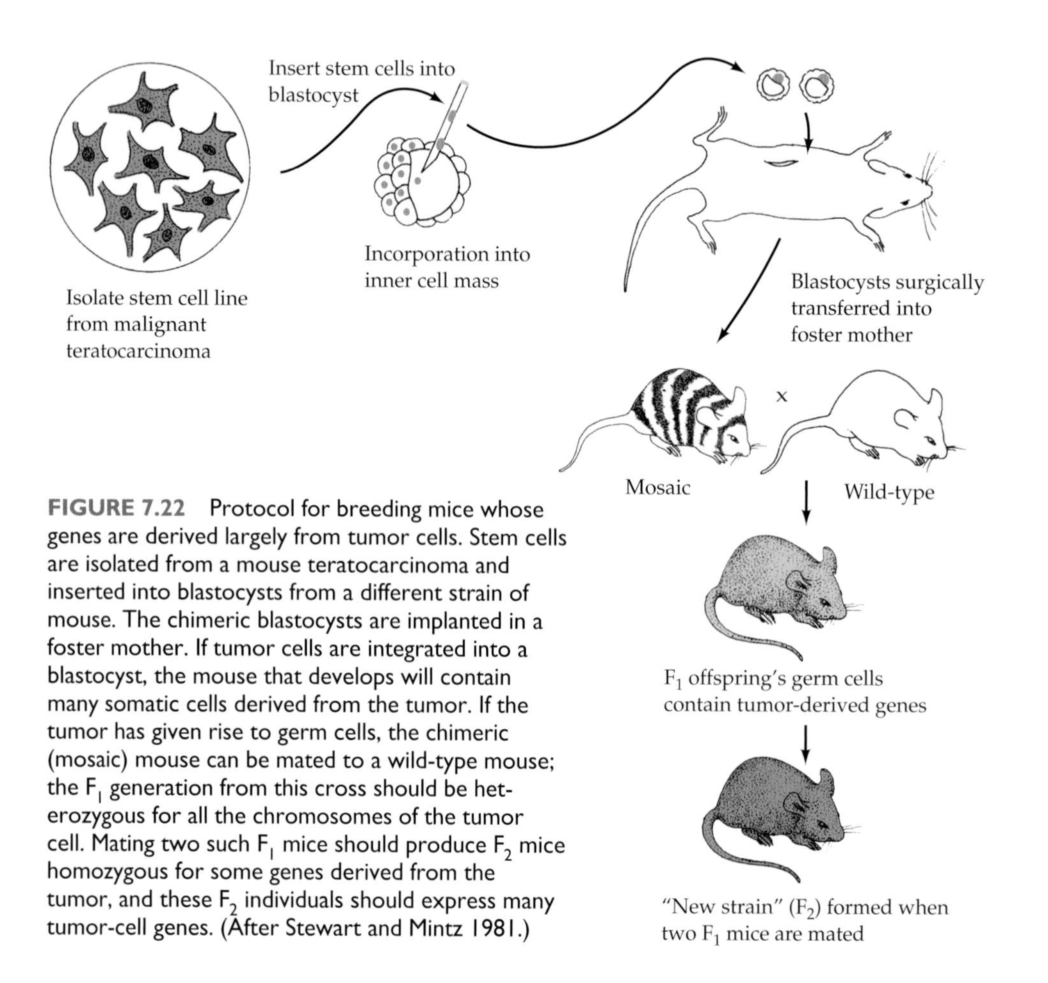

FIGURE 7.22 Protocol for breeding mice whose genes are derived largely from tumor cells. Stem cells are isolated from a mouse teratocarcinoma and inserted into blastocysts from a different strain of mouse. The chimeric blastocysts are implanted in a foster mother. If tumor cells are integrated into a blastocyst, the mouse that develops will contain many somatic cells derived from the tumor. If the tumor has given rise to germ cells, the chimeric (mosaic) mouse can be mated to a wild-type mouse; the F_1 generation from this cross should be heterozygous for all the chromosomes of the tumor cell. Mating two such F_1 mice should produce F_2 mice homozygous for some genes derived from the tumor, and these F_2 individuals should express many tumor-cell genes. (After Stewart and Mintz 1981.)

Indeed, while most tumors are of epithelial origin, the development of epigenetic-origin tumors appears to be regulated by the stromal cells (fibroblasts, myoepithelial cells, etc.) surrounding them. These stromal cells make the extracellular matrix that stabilizes the epithelial cell cytoskeleton and provides the paracrine factors that inhibit cell proliferation (Bhowmick et al. 2004; Tlsty and Coussens 2006; Hu and Polyak 2008). The growth of mammary epithelial cells appears to be kept in place not only by the extracellular matrix, but also by the paracrine factor transforming growth factor-β (TGF-β) secreted by the stromal cells (Figure 7.24; Cheng et al. 2005). TGF-β is important (1) for preventing the division of the neighboring epithelial cells and (2) for preventing the fibroblast from secreting positive growth factors that would stimulate epithelial cell division.

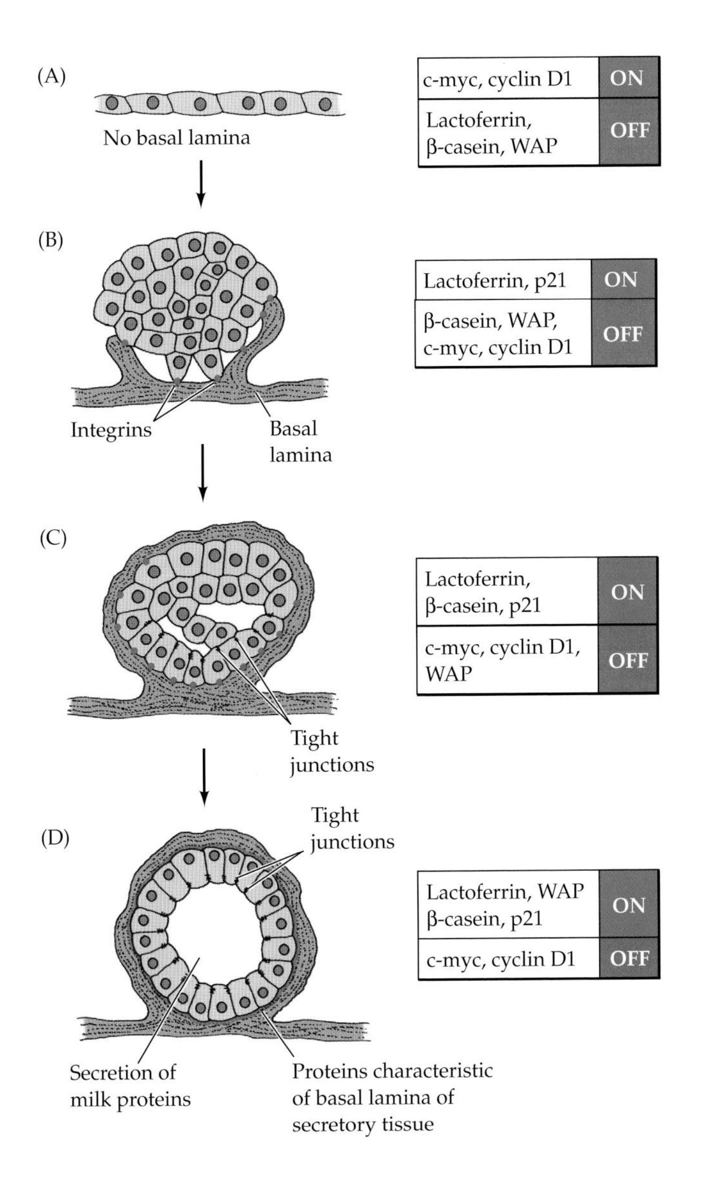

FIGURE 7.23 Basement membrane-directed gene expression in mammary gland tissue. (A) Mouse mammary gland tissue divides when placed on tissue culture plastic. The genes encoding cell division proteins are on, and the genes capable of synthesizing the differentiated products of the mammary gland (lactoferrin, casein, and whey acidic protein WAP) are off. (B) When these cells are placed on a basement membrane that contains laminin (basal lamina), the genes for the cell division proteins are turned off, while the genes encoding inhibitors of cell division (such as p21) and the gene for lactoferrin are turned on. (C,D) Mammary gland cells wrap the basal lamina around them, forming a secretory epithelium. The genes for casein and WAP are sequentially activated. (After Bissell et al. 2002.)

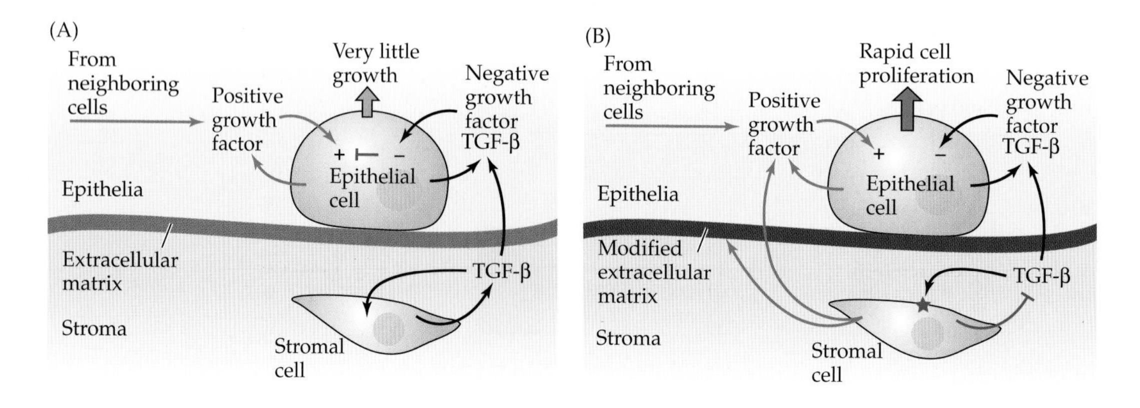

FIGURE 7.24 Regulation of the proliferation and differentiation of both the stroma and epithelia by the stromal cells of an organ. (A) Stromal cells secrete factors such as TGF-β that signal both stromal and epithelial cells to stop proliferation. Each epithelial cell, however, secretes and receives (from itself and its neighbors) positive growth factors that promote proliferation (green arrows). But as long as the stromal cells counter these positive signals, proliferation is tightly regulated. (B) Epigenetic or mutational changes in stromal cells block their inhibitory signaling (red starburst). The stroma can then start secreting paracrine factors (for example, hepatocyte growth factor; green arrows) that modify the extracellular matrix and promote the proliferation and transformation of the epithelial cells. (After Bhowmick et al. 2004.)

One study (Maffini et al. 2004) exposed mammary stroma and mammary epithelial cells, in separate compartments, to cancer-causing drugs. When the researchers recombined the compartments, the epithelial cells were able to initiate tumors only when combined with the chemically treated stroma. The carcinogen-exposed mammary epithelium was not able to form tumors if placed into normal stroma. Here, the stroma appears to be the critical element of cancer formation, and to be necessary and sufficient to induce cancer in the neighboring epithelial cells. If this type of study can be confirmed, it would be of great significance for cancer chemotherapies. As Finak and colleagues (2008) conclude:

> Under normal physiological conditions, stroma serves as an important barrier to epithelial transformation [into cancers]; the interplay between epithelial cells and the microenvironment maintains epithelial polarity and modulates growth inhibition.

These studies show that, in human breast cancer, stromal gene expression patterns predict clinical outcome better than epithelial gene expression patterns. Thus, cancer is not only a disease of cell proliferation, it is also a disease of tissue organization and epigenetic cross-talk between cells.

Tumors can be generated by a combination of genetic and epigenetic means. Changes in DNA methylation can activate oncogenes and repress tumor-suppressor genes, thereby initiating tumor formation. Oncogenes can cause the methylation of tumor-suppressor genes, which also aids in tumorigenesis. The tissue environment of the cell may be critical in regulating these processes. The complexities of tumors, including their multiple somatic mutations and their resistance to agents that induce apoptotic cell death, may better be explained by a combination of genetic and epigenetic factors than by mutations alone. Moreover, it is possible that knowledge of the epigenetic causes of cancer can provide the basis for new methods of cancer therapy.

Summary

Although the epigenetic origin of human disease remains a hypothesis, we have seen that many of the most critical diseases of adult humans—diabetes, hypertension, heart attack, obesity, cancer, and even the aging syndrome itself—have major epigenetic components. The susceptibility to such diseases, whether by maternal diet, from the environment, or through lifestyle, is regulated both by genetic alleles and by epigenetic changes in the genome. These new insights can lead us to better preventive measures in public health and better treatments through drug design. While the epigenetic origin of adult diseases is a relatively new hypothesis and not as well established as others (such as the germ theory of infectious disease), it is providing explanations for many medical phenomena that had until now resisted our understanding. The evidence certainly provides reasons to believe that more research in this area is likely to illuminate important new ways to explain and possibly prevent certain human illnesses.

References

Adams, J. and M. White. 2005. When the population approach to prevention puts the health of individuals at risk. *Internatl. J. Epidem.* 34: 40–43.

Akhtar, M., Y. Cheng, R. M. Magno, H .Ashktorab, D. T. Smoot, S. J. Melter and K. T. Wilson. 2001. Promoter methylation regulates *Helicobacter pylori*-stimulated cyclooxygenase-2 expression in gastric epithelial cells. *Cancer Res* 61: 2399–2403.

Allen, W. R., S. Wilsher, C. Tiplady, and R. M. Butterfield. 2004. The influence of maternal size on pre- and postnatal growth in the horse. III. Postnatal growth. *Reproduction* 127: 67–77.

Bhargava, S. K. 2004. Relation of serial changes in childhood body mass index to impaired glucose tolerance in young adulthood. *New Engl. J. Med.* 350: 865–875.

Barker, D. J. P. 1994. *Mothers, Babies and Disease in Later Life.* Churchill Livingstone, London.

Barker, D. J. P. 1998. In utero programming of chronic disease. *Clin. Sci.* 95: 115–128.

Barker, D. J. P. and C. Osmond. 1986. Infant mortality, childhood nutrition, and ischaemic heart disease in England and Wales. *Lancet* 1: 1077–1081.

Bateson, P. 2007. Developmental plasticity and evolutionary biology. *J. Nutrit.* 137: 1060–1062.

Bateson, P. and 14 others. 2004. Developmental plasticity and human health. *Nature* 430: 419–421.

Bhargava, S. K. 2004. Relation of serial changes in childhood body mass index to impaired glucose tolerance in young adulthood. *New Engl. J. Med.* 350: 865–875.

Bhowmick, N. A., E. G. Neilson and H. L. Moses. 2004. Stromal fibroblasts in cancer initiation and progression. *Nature* 432: 332–337.

Bjornsson, H. T. and 15 others. 2008. Intraindividual change over time in DNA methylation with familial clustering. *J. Amer. Med. Assoc.* 299: 2877–2883.

Bissell, M. J., D. C. Radisky, A. Rizki, V. M. Weaver and O. W. Petersen. 2002. The organizing principle: Microenvironmental influences in the normal and malignant breast. *Differentiation* 70: 537–546.

Brock, M. V. and 14 others. 2008. DNA methylation markers and early recurrence in stage 1 lung cancer. *New Engl. J. Med.* 358: 1118–1128.

Burdge, G. C., J. Slater-Jefferies, C. Torrens, E. S. Phillips, M. A. Hanson and K. A. Lillycrop. 2006. Dietary protein restriction of pregnant rats in the F_0 generation induces altered methylation of hepatic gene promoters in the adult male offspring in the F_1 and F_2 generations. *Br. J. Nutr.* 97: 435–439.

Chali, D., E. Enquselassie and M. Gesese. 1998. A case-control study of determinants of rickets. *Ethiopian Med. J.* 36: 227–234. (Quoted in Bateson 2007.)

Chan, A. O. and 8 others. 2003. Promoter methylation of E-cadherin gene in gastric mucosa associated with *Helicobacter pylori* infection and in gastric cancer. *Gut* 52: 502–506.

Chan, A. O. and 9 others. 2006. Eradication of *Helicobacter pylori* infection reverses E-cadherin promoter hypermethylation. *Gut* 55: 463–468.

Cheng, N. and 9 others. 2005. Loss of TGFβ type II receptor in fibroblasts promotes mammary carcinoma growth and invasion through upregulation of TGFβ-, MSP-, and HGF-mediated signaling networks. *Oncogene* 24: 5053–5068.

Cheng, X. 2008. Silent assassin: Oncogeneic Ras directs epigenetic inactivation of target genes. *Science Signal* 1(13), pe 14. (DOI: 10.1126/stke.113pe14.)

De Boo, H. A. and J. E. Harding. 2006. The developmental origins of adult disease (Barker) hypothesis. *Austral. NZ J. Obst. Gyn.* 46: 4–14.

Desai, M., N. Crowther, A. Lucas and C. M. Hales. 1995. Adult glucose and lipid metabolism may be programmed during fetal life. *Biochem. Soc. Transact.* 23: 331–335.

Dey, S. K., H. Lim, S. K. Das, J. Reese, B. C. Paria, T. Daikoku and H. Wang. 2004. Molecular cues to implantation. *Endocr. Rev.* 25: 341–373.

Diamond, J. 1991. Pearl Harbor and the Emperor's physiologists. *Natural History* 1991(12): 2–7.

Eriksson, J. G., T. Forsen, J. Tuomilehto, C. Osmond and D. J. Barker. 2001. Early growth and coronary heart disease in later life: Longitudinal study. *Brit. Med. J.* 322: 949–953.

Esteller, M. 2007. Cancer epigenomics: DNA methylomes and histone modification maps. *Nature Rev. Genet.* 8: 286–298.

Esteller, M. 2008. Epigenetics in cancer. *New Engl. J. Med.* 358: 1148–1159.

Euhus, D. M., D. Bu, S. Milchgrub, X.-J. Xie, A. Bian, A. M. Leitch and C. M. Lewis. 2008. DNA methylation in benign breast epithelium in relation to age and breast cancer risk. *Canc. Epidemiol. Biomarkers Prev.* 17: 1051–1059.

Feder, S. 1990. Environmental determinants of seasonal coat color change in weasels (*Mustela erminea*) from two populations. M.S. Thesis, University of Alaska, Fairbanks.

Feinberg A. P., R. Ohlsson and S. Henikoff. 2006. The epigenetic progenitor origin of human cancer. *Nature Rev. Genet.* 7: 21–33.

Finak, G. and 11 others. 2008. Stromal gene expression predicts clinical outcome in breast cancer. *Nature Medicine* 14: 518–527.

Fowden, A. L. and A. J. Forhead. 2004. Endocrine mechanisms of intrauterine programming. *Reproduction* 127: 515–526.

Fraga, M. F. and M. Esteller. 2007. Epigenetics and aging: The targets and the marks. *Trends Genet.* 23: 413–418.

Fraga, M. F. and 20 others. 2005. Epigenetic differences arise during the lifetime of monozygotic twins. *Proc. Nat. Acad. Sci USA* 102: 10604–10609.

Friedler, G. 1985. Effects of limited paternal exposire to xenobiotic agents on the development of progeny. *Neurobehav. Toxicol. Teratol.* 7: 739–743.

Friedler, G. 1996. Paternal exposures: Impact on reproductive and developmental outcome—An overview. *Phamacol. Biochem. Behav.* 55: 691–700.

Gardner, R. L. 1982. Manipulation of development. In C. R. Austin and R. V. Short (eds.), *Embryonic and Fetal Development*. Cambridge University Press, Cambridge, pp. 159–180.

Gazin, C., N. Wajapeyee, S. Gobeil, C.-M. Virbasius and M. R. Green. 2007. An elaborate pathway required for Ras-mediated epigenetic silencing. *Nature* 449: 1073–1077.

Gilbert, S. F. 2006. *Developmental Biology*, 8th Ed. Sinauer Associates, Sunderland, MA, pp. 677–681.

Gluckman, P. D. and M. A. Hanson. 2004. Living with the past: Evolution, development, and patterns of disease. *Science* 305: 1733–1736.

Gluckman, P. D. and M. A. Hanson. 2005. *The Fetal Matrix: Evolution, Development, and Disease.* Cambridge University Press, Cambridge.

Gluckman, P. D. and M. A. Hanson. 2006a. The developmental origins of health and disease: An overview. In P. D. Gluckman and M. A. Hanson (eds.), *Developmental Origins of Health and Disease*. Cambridge University Press, Cambridge, pp. 1–5.

Gluckman, P. D. and M. A. Hanson. 2006b. The conceptual basis for the developmental origins of health and disease. In P. D. Gluckman and M. A. Hanson (eds.), *Developmental Origins of Health and Disease*. Cambridge University Press, Cambridge, pp. 33–50.

Gluckman, P. D. and M A. Hanson. 2007. *Mismatch: Why Our World No Longer Fits Our Bodies*. Oxford University Press, Oxford.

Gluckman. P. D., M. A. Hanson and H. G. Spencer. 2005. Predictive adaptive responses and human evolution. *Trends Ecol. Evol.* 20: 527–533.

Gluckman, P. D., A. S. Beedle, M. A. Hanson and M. H. Vickers. 2007. Leptin reversal of the metabolic phenotype: Evidence for the role of developmental plasticity in the development of the metabolic syndrome. *Hormone Res.* 67 (suppl. 1): 115–120.

Gluckman, P. D. and 7 others. 2007. Metabolic plasticity during mammalian development is directionally dependent on early nutritional status. *Proc. Natl. Acad. Sci. USA* 104: 12796–12800.

Gluckman, P. D., M. A. Hanson, C. Cooper and K. L. Thornburg. 2008. Effect of the in utero and early-life conditions on adult health and disease. *New Engl. J. Med.* 359: 61–73.

Godfrey, K. M. 2006. The "developmental origins" hypothesis. In P. D. Gluckman and M. A. Hanson (eds.), *Developmental Origins of Health and Disease* Cambridge University Press, Cambridge, pp. 6–32.

Godfrey, K. M. and D. Barker. 2000. Fetal nutrition and adult diseases. *Am. J. Clin. Nutr.* 71 (suppl.): S1344–S1355.

Gould, S. J. 1997. Dolly's fashion, Louis' passion. *Natural History* 106(5): 18–23.

Hales, C. N. and D. J. Barker. 1992. Type 2 (non-insulin-dependent) diabetes mellitus: The thrifty phenotype hypothesis. *Diabetologia* 35: 595–601.

Hales, C. N. and D. J. Barker. 2001. The thrifty phenotype hypothesis. *Brit. Med. Bull.* 60: 5–20.

Hales, C. N. and 6 others. 1991. Fetal and infant growth and impaired glucose tolerance at age 64. *Brit. Med. J.* 303: 1019–1022.

Hansen, D. A. 2008. Paternal environmental exposures and gene expression during spermatogenesis: Research review to research framework. *Birth Def. Res. C: Embryo Today* 84: 155–163.

Hanson, M. A. and K. M. Godfrey. 2008. Commentary: Maternal constraint is a pre-eminent regulator of fetal growth. *Internatl. J. Epidem.* 37: 252–254.

Hart, N. 1993. Famine, maternal nutrition and infant mortality: A re-examination of the Dutch Hunger Winter. *Population Studies* 47: 27–46.

Hoppe, C. C., R. G. Evans, J. F. Bertram and K. M. Moritz. 2007. Effects of dietary protein restriction on nephron number in the mouse. *Am. J. Regul. Integr Comp. Physiol* 292: R1768–1774.

Hu, M. and K. Polyak. 2008. Microenvironmental regulation of cancer development. *Curr. Opin. Genet. Devel.* 18: 27–34.

Institute of Medicine. 1995. *Adverse Reproductive Outcomes in Families of Atomic Veterans: The Feasibility of Epidemiologic Studies.* National Academy Press, Washington, D.C.

Issa, J.-P., Y. L. Ottaviano, P. Celano, S. R. Hamilton, N. E. Davidson and S. B. Baylin. 1994. Methylation of the oestrogen receptor CpG island links aging and neoplasia in human colon. *Nature Genetics* 7: 536–540.

Jacinto, F. V. and M. Esteller. 2007. Mutator pathways unleashed by epigenetic silencing in human cancer. *Mutagenesis* 22: 247–253.

Keller, G., G. Zimmer, G. Mall, E. Ritz, K. Amann. 2003. Nephron number in patients with primary hypertension. *N. Engl J. Med.* 348: 101–108.

Kim, J. and 8 others. 2007. Epigenetic changes in estrogen receptor beta gene in atherosclerotic cardiovascular tissues and in vitro vascular senescence. *Biochim. Biophys. Acta* 1772: 72–80.

Lau, C. and J. M. Rogers. 2004. Embryonic and fetal programming of physiological disorders in adulthood. *Birth Defects Res. C: Embryo Today* 72: 300–312.

Law, C. and J. Baird. 2006. Developmental origins of health and disease: Public-health perspectives. In P. D. Gluckman and M. A. Hanson (eds.), *Developmental Origins of Health and Disease*. Cambridge University Press, Cambridge, pp. 446–455.

Le Beyec, J., R. Xu, S. Y. Lee, C. M. Nelson, A. Rizki, J. Alcaraz and M. J. Bissell. 2007. Cell shape regulates global histone acetylation in human mammary epithelial cells. *Exp. Cell Res.* 313: 3066–3075.

Lillycrop, K. A., E. S. Phillips, A. A. Jackson, M. A. Hanson and G. C. Burdge. 2005. Dietary protein restriction of pregnant rats induces and folic acid supplementation prevents epigenetic modification of hepatic gene expression in the offspring. *J. Nutrit.* 135: 1382–1386.

Lillycrop, K. A., E. S. Phillips, C. Torrens, M. A. Hanson, A. A. Jackson and G. C. Burdge. 2008. Feeding pregnant rats a protein-restricted diet persistently alters the methylation of specific cytosines in the hepatic PPARα promoter of the offspring. *Brit. J. Nutrit.* 100: 278–282.

Lindi, V. I. and 11 others. 2002. Association of the Pro12Ala polymorphism in the PPAR-γ2 gene with 3-year incidence of type 2 diabetes and body weight change in the Finnish diabetes prevention study. *Diabetes* 51: 2581–2586.

Liu, H., Q. Lan, J. M. Siegfried, J. D. Luketich and P. Keohavong. 2006. Aberrant promoter methylation of *p16* and *MGMT* genes in lung tumors from smoking and never-smoking patients. *Neoplasia* 8: 46–51.

Liu, H., Y. Zhou, S. E. Boggs, S. A. Belinsky and J. Liu. 2007. Cigarette smoke induces demethylation of prometastatic oncogene synuclein-γ in lung cancer cells by downregulation of DNMT3B. *Oncogene* 26: 5900–5910.

Lucas, A. 1991. Programming by early nutrition in man. In G. R. Bock and J. Whelan (eds.), *The Childhood Environment and Adult Disease*. Wiley, Chichester, UK, pp. 38–55.

Ma, X., Q. Yang, K. T. Wilson, N. Kundu, S. J. Meltzer and A. M. Fulton. 2004. Promoter methylation regulates cyclooxygenase expression in breast cancer. *Breast Canc. Res.* 6: R316–R321.

Maffini, M. V., A. M. Soto, J. M.Calabro, A. A. Ucci and C. Sonenschein. 2004. The stroma as a crucial target in rat mammary gland carcinogenesis. *J. Cell Sci.* 117: 1495–1502.

Miyazaki, T. and 8 others. 2007. E-cadherin promoter methylation in *H. pylori*-induced enlarged fold gastritis. *Helicobacter* 12: 523–531.

Moritz, K. M., M. Dodi, and E. M. Wintour. 2003. Kidney development and the fetal programming of adult disease. *BioEssays* 25: 212–220.

Morton, N. E. 1955. The inheritance of human birth weight. *Ann. Hum. Genet. 20*: 125–134.

Oates, N. A. and 9 others. 2006. Increased DNA methylation at the *AXIN1* gene in a monozygotic twin from a pair discordant for a caudal duplication anomaly. *Am. J. Hum. Genet*. 79: 155–162.

Pechenik, J. A. 2006. Larval experience and latent effects: Metamorphosis is not a new beginning. *Int. Comp. Biol.* 46: 323–333.

Pembrey, M. E., L. O. Bygren, G. Kaati, S. Edvinsson, K. Northstone, M. Sjöström, J. Golding and the ALSPAC team. 2006. Sex-specific, male-line transgenerational responses in humans. *Eur. J. Human Genet.* 14: 159–166.

Petersen, O. W., L. Rønnov-Jessen, A. R. Howlett and M. J. Bissell. 1992. Interaction with basement membrane serves to rapidly distinguish growth and differentiation pattern of normal and malignant human breast epithelial cells. *Proc. Natl. Acad. Sci. USA* 89: 9064–9068.

Petrik, J., B. Reusens, E. Arany, C. Remacle, J. J. Hoet and D. J. Hill. 1999. A low protein diet alters the balance of islet cell replication and apoptosis in the fetal and neonatal rat and is associated with reduced pancreatic expression of insulin-like growth factor II. *Endocrinology* 140: 4861–4873.

Petronis, A. 2006. Epigenetics and twins: Three variations on a theme. *Trends Genet*. 22: 347–350.

Post, W. S. and 8 others. 1999. Methylation of the estrogen receptor gene is associated with aging and atherosclerosis in the cardiovascular system. *Cardiovasc. Res.* 43: 985–991.

Rittig, S. and 7 others. 2007. The Pro12Ala polymorphism in PPARγ2 increases the effectiveness of primary prevention of cardiovascular disease by a lifestyle intervention. *Diabetologia* 50: 1345–1347.

Robson, E. B. 1955. Birth weight in cousins. *Ann. Hum. Genet.* 19: 262–268.

Rostand, J. 1962. *The Substance of Man*, translated by I. Brandeis. Doubleday, New York.

Russo, A. L. and 9 others. 2005. Differential DNA hypermethylation of critical genes mediates the stage-specific tobacco smoke–induced neoplastic progression of lung cancer. *Clinical Cancer Res.* 11: 2466–2470.

Schuebel, K. E. and 18 others. 2007. Comparing the DNA hypermethylome with gene mutations in human colonorectal cancer. *PLoS Genet.* 3(9): e157. doi:10.1371/journal.pgen.0030157.

Sjöblom, T. and 28 others. 2006. The consensus coding sequences of human breast and colorectal cancers. *Science* 314: 268–274.

Sonnenschein, C. and A. Soto. 1999. *A Society of Cells: Cancer and Control of Cell Proliferation.* Oxford University Press, Oxford.

Soto, A. and C. Sonnenschein. 2005. Emergentism as a default: Cancer as a problem of tissue organization. *J. Biosci.* 30: 103–118.

Stein, Z. 1975. *Famine and Human Development: The Dutch Hunger Winter of 1944–1945.* Oxford University Press, New York.

Stevens, L. M. and S. C. Landis. 1988. Developmental interactions between sweat glands and the sympathetic neurons which innervate them: Effects of delayed innervation on neurotransmitter plasticity and gland maturation. *Dev. Biol.* 130: 703–720.

Stewart, T. A. and B. Mintz. 1981. Successive generations of mice produced from an established culture line of euploid teratocarcinoma cells. *Proc. Natl. Acad. Sci. USA* 78: 6314–6318.

Taylor, P. D. and L. Poston. 2007. Developmental programming of obesity in mammals. *Exper. Physiol.* 92: 287–298.

Tischner, M. 1987. Development of Polish-pony foals born after embryo transfer to large mares. *J. Reprod. Fertil.* 35 (suppl): 705–709.

Tlsty, T. D. and L. M. Coussens. 2006. Tumor stroma and regulation of cancer development. *Annu. Rev. Pathol. Mech. Dis.* 1: 119–150.

Vanyushin, B. L., L. E. Nemirovsky, V. V. Klimenko, V. K. Vasilev and A. N. Belozersky. 1973. The 5-methylcytosine in DNA of rats: Tissue and age specificity and the changes induced by hydrocortisone and other agents. *Gerontologia* 19: 138–152.

Vickers, M. H., B. H. Breier, W. S. Cutfield, P. I. Hofmann and P. D. Gluckman. 2000. Fetal origins of hyperphagia, obesity, and hypertension and its postnatal amplification by caloric nutrition. *Am. J. Physiol.* 279: E83–87.

Walton, A. J. and J. Hammond. 1938. The maternal effects on growth and conformation in Shire horses–Shetland pony crosses. *Proc. Roy. Soc. B* 125: 311–335.

Weaver V. M., O. W. Petersen, F. Wang, C. A. Larabell, P. Briand, C. Damsky and M. J. Bissell. 1997. Reversion of the malignant phenotype of human breast cells in three-dimensional culture and in vivo using integrin blocking antibodies. *J. Cell Biol.* 137: 231–246.

Weinberg, R. A. 1998. *One Renegade Cell: How Cancer Begins*. New York: Basic Books.

Welham, S. J. M., P. R. Riley, A. Wade, M. Hubank and A. S. Woolf. 2005. Maternal diet programs embryonic kidney gene expression. *Physiol. Genet*. 22: 48–56.

Wells, J. C. K. 2007. Commentary: Why are South Asians susceptible to central obesity? The El Niño hypothesis. *Internatl. J. Epidemiol*. 36: 226–227.

Whitelaw, E. 2006. Sins of the fathers and their fathers. *Eur. J. Hum. Genet*. 14: 131–132.

Wood, L. D. and 41 others. 2007. The genomic landscapes of human breast and colorectal cancers. *Science* 318: 1108–1113.

Yajnik, C. S. 2005. Size and body composition at birth and risk of type-2 diabetes: A critical evaluation of "fetal origins" hypothesis for developing countries. In G. Hornstra, R. Uauy and X. Yang (eds.), *The Impact of Maternal Nutrition on the Offspring*. Karger AG, Basel, pp. 169–183.

PART 3

Toward a Developmental Evolutionary Synthesis

Chapter 8

The Modern Synthesis

Natural Selection of Allelic Variation

Biology points out the individuality of every being, and at the same time reminds us of the brotherhood of all.

Jean Rostand, 1962

The current instar of the evolutionary theory may be defined by such books as those of Huxley, Simpson, Dobzhansky, Mayr, and Stebbins. We are certainly not ready for a new moult, but signs of new organs are perhaps visible.

J. B. S. Haldane, 1953

During the past decade, a new interpretation of evolutionary biology has matured. This approach to evolution, called evolutionary developmental biology or "evo-devo," sees evolutionary change as hereditable alterations in development. If, as ecological developmental biology claims, development encompasses the use of environmental signals in producing phenotypes, then this has profound implications for evolutionary biology. This section of the book looks at how developmental gene regulation and developmental plasticity may account for inherited and selectable variation.

Evolutionary biology is the scientific attempt to explain the origins and patterns of biodiversity. The field has itself evolved through a series syntheses. The idea that one form of life could give rise to another form had been mooted by biologists such as Jean-Baptiste Lamarck, Robert Chambers, as well as by Charles Darwin's grandfather, Erasmus Darwin. These schemes were often referred to as "transformationist theories." However, it was

Charles Darwin's *The Origin of Species* (1859) that synthesized evidence from embryology, biogeography, paleontology, agricultural breeding, and anatomy to show how species originated from other species and that life's diversity was not the result of myriad independent acts of creation. Moreover, Darwin proposed a mechanism for evolution—natural selection—that could be demonstrated on a smaller scale in the changing of the phenotype that characterized a species.*

Beginning in the 1920s, a synthesis emerged that fused the science of genetics to Darwin's theory of evolution. This **Modern Synthesis** (sometimes called **neo-Darwinism**) emphasized gene mutations and rearrangements as the mechanisms producing the variation that is the grist for natural selection. The Modern Synthesis has been extremely successful in showing how environmental pressures favor the survival and propagation of a subset of variations ("the fittest") within a species. We believe that we are currently seeing the third evolutionary synthesis, the "Developmental Synthesis," which emphasizes the idea that evolution consists of inherited changes in the patterns of development and focuses on the mechanisms whereby such changes are effected. This third synthesis emphasizes the embryonic origin of adult variations, and it attempts to explain evolution above the species level through alterations in the use of regulatory genes.

This chapter will discuss both Darwin's ideas and what Haldane (1953) called "the current instar of evolutionary theory," the Modern Synthesis. It will concern evolution within species and show that such evolution can be explained by natural selection (differential reproductive success) and that natural selection works by changing the frequency of alleles† within the population. The subsequent two chapters will document the development of Haldane's "new moult," an expanded evolutionary theory that brings development and developmental plasticity into the synthesis. These latter chapters will concern evolution both within and above the species level and will discuss the view that evolution can involve changes in the time, place, and amount of gene expression as well as in the changes of the type of gene being expressed.

*One of the reasons we recognize Darwin as the major proponent of evolutionary biology rather than his contemporary, A. R. Wallace, is that *The Origin of Species* provided a synthesis that encompassed nearly all of the biology of its time. Wallace's original paper (read, as was Darwin's abstract, at an 1858 meeting of the Linnean Society in London) shared Darwin's views of natural selection and biogeography, but it was Darwin's volume that showed how descent with modification would explain many of the biological enigmas that confounded scientists in the late nineteenth century. For other reasons why Darwin gets the lion's share of the credit for the theory of evolution, see the excellent short essay by Berry and Browne (2008).

†A gene *locus* is a place. Thus, one can talk about the human β-globin gene as a locus on chromosome 11. An *allele* is the specific DNA sequence at that locus—for instance, the sickle-cell allele of the β-globin gene. Thus, the locus can be viewed as a parking spot, the gene is the car in that spot, and the allele might be a Toyota, Ford, or BMW.

Charles Darwin's Synthesis

In the nineteenth century, debates over the classification of species pitted two ways of viewing nature against each other. One view, championed by Georges Cuvier and Charles Bell, focused on the *differences* that allowed each distinct species to adapt to its environment. Thus, the hand of the human, the flipper of the seal, and the wing of the bird were seen as marvelous contrivances, each fashioned by the Creator to allow these animals to adapt to their "conditions of existence." The other view, championed by Étienne Geoffroy Saint-Hilaire and Richard Owen, was that the "unity of type" (the similarities among organisms, which Owen called "homologies") was critical. The human hand, the seal's flipper, and the bird's wing were all modifications of the same basic plan (Figure 8.1). In discovering that plan, one could find the form upon which the Creator designed these animals; the distinctive adaptations were secondary.

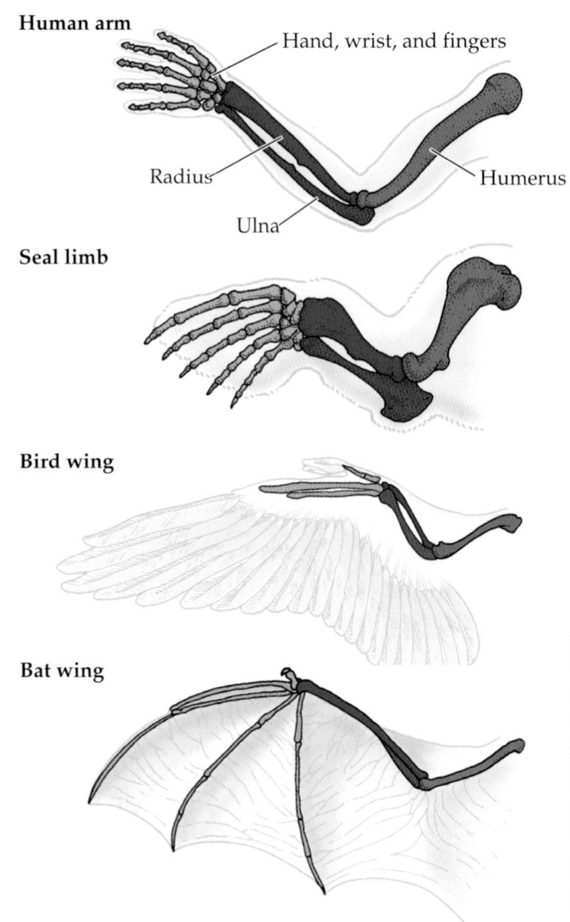

FIGURE 8.1 Homologies of structure among a human arm, a seal forelimb, a bird wing, and a bat wing. (Homologous structures are shown in the same color.) All four are homologous as forelimbs because they derive from a common tetrapod ancestor. The adaptations of bird and bat forelimbs for flight, however, evolved independently of each other, after the two lineages diverged from a common ancestor. Therefore they are homologous as forelimbs but analogous as wings.

Darwin acknowledged his debt to both sides of this debate when he wrote in the *Origin*, "It is generally acknowledged that all organic beings have been formed on two great laws—Unity of Type, and Conditions of Existence." His theory, he stated, would explain "unity of type" as descent from a common ancestor. The changes creating the marvelous adaptations to the "conditions of existence" would be explained by natural selection. Darwin called this concept **descent with modification**.

Darwin noted that the homologies between the embryonic and larval structures of different phyla provided excellent evidence for descent with modification. He also argued that adaptations that depart from the "type" and allow an organism to survive in the "conditions" of its particular environment develop later in the embryo. Thus, one could emphasize *common descent* in the embryonic homologies between two or more groups of animals, or one could emphasize the *modifications* by showing how development was altered to produce diverse adaptive structures (Gilbert 2003).

Classical Darwinism: Natural selection

Classical Darwinian emphasis on natural selection can be summarized in a few sentences:

1. There is variation among the individual organisms that make up a population of a species.
2. There is an enormous amount of death, and most individuals will not survive to reproduce.
3. Death is selective. Those individuals that best fit into the environment they encounter are more likely to survive; those that do not fit the environment well are usually eliminated.
4. When those individuals that survive reproduce, their progeny have a high likelihood of inheriting the variations that allowed their parents to survive. If individuals who carry those variations continue to be favored (selected), over time this natural selection will alter the overall characteristics of the population.
5. When populations of a species become reproductively isolated (i.e., separated in such a way that members of one population cannot mate with members of another*), each population can randomly acquire a distinct and separate suite of variations. If the environmental conditions faced by the isolated populations are different, different variations will be selected. Anatomical and physiological

*While geographic isolation, as Darwin believed happened to finch and turtle populations on the Galápagos Islands, could explain reproductive isolation, Darwin believed that behavioral and ecological differences could also cause such isolation (Darwin 1844, 1859 p. 339). In his second edition of the *Origin*, Darwin faced a major battle with Moritz Wagner, who thought geographic isolation the critical feature in species production.

differences can result from the accumulation of even small distinctions, eventually causing the two populations to be recognized as different species.

There are many contemporary examples of classical Darwinism at work, some of the best known of which come from research done in a place that inspired Darwin himself. Peter and Rosemary Grant headed a team that observed evolution in action on the finches of the Galápagos Islands (see Grant 1986). In one of their several studies, drought conditions in 1976–1977 caused a dearth of nuts (especially the softer-shelled varieties), the main food for the finch *Geospiza fortis*. As the *G. fortis* population declined (Figure 8.2A), the individuals that survived were those with larger bodies (Figure 8.2B), whose stronger beaks could crack large, hard-shelled nuts, thus allowing them access to food that was unavailable to their smaller conspecifics.

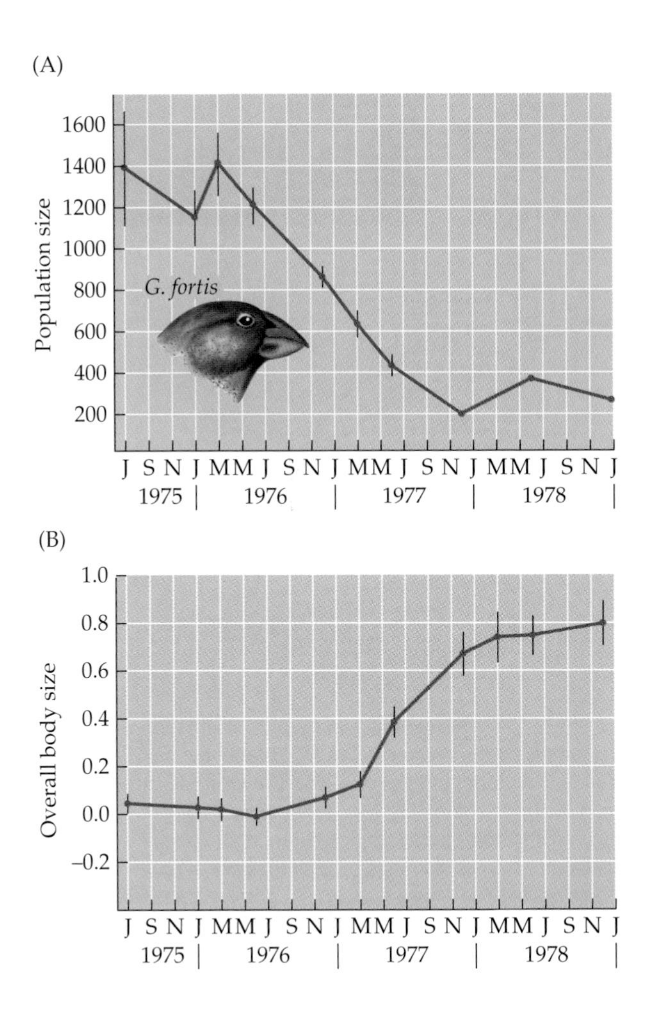

FIGURE 8.2 A contemporary example of Darwinian natural selection. (A) In 1976–1977, during a severe drought in the Galápagos Islands, the ground finch *Geospiza fortis* populations underwent a large drop in numbers. (B) Those that survived were the larger birds that could crack the larger, hard-shelled seeds—the only food available during the drought. (After Grant 1986.)

HUMAN BEINGS AND SELECTION PRESSURE Human beings and their technology have significantly altered the natural world and in so doing have introduced selection pressures on a wide range of organisms. One of the first such cased to be documented was that of the peppered moth (*Biston betularia*) in the United Kingdom. During the nineteenth and early twentieth centuries, there was a dramatic rise in the frequency of an allele (the *carbonaria* allele) that produced a black (melanic) phenotype and a corresponding drop in the wild-type allele (the *typia* allele) that cause the lightly peppered phenotype (Figure 8.3A).

FIGURE 8.3 Industrial melanism. (A) Melanic (left) and typical (right) forms of the moth *Biston betularia.* (B) Frequency of the melanic form from 1959 to 1995 in industrialized areas in England (Caldy Common; blue circles) and the United States (George Reserve, Michigan; red circles). Sampling was continuous in Britain, while the data for the America region only compares samples taken at the beginning and end of the time period. The green plot represents mathematical predictions based on a constant selection coefficient against the melanic allele. (A, photographs © Steve McWilliam and Sergey Chushkin/ShutterStock; B after Grant et al. 1996.)

The first melanic *Biston* specimens appear in collections from the mid-1800s. By the 1890s, the proportion of melanic moths reached 90% in several urban centers, and was a staggering 98% in Manchester (Grant et al. 1996; Grant and Wiseman 2002.) By contrast, the black phenotype remained rare in the countryside. This "industrial melanism" has been well correlated with the darkening of trees and buildings by industrially generated soot, and the corresponding conspicuousness of the wild-type form of the moth as opposed to the melanic form. It is widely believed that the change in allele frequency was the result of selection pressure from bird predation, wherein the more conspicuous moths were preferentially eaten (Kettlewell 1973; Majerus 1998; Grant 1999).

But the story doesn't end there. If the prevalence of the dark phenotype was caused by airborne soot, one might expect (if the same predators were present) that selection would revert to favor the *typia* allele once air quality improved. In fact, this is precisely what has happened. In both Britain and America, the frequency of melanic moths dropped precipitously as clean air legislation lessened the amount of soot that darkened urban areas (Figure 8.3B; Grant et al. 1996; Grant and Wiseman 2002).

Other human innovations have had dramatic effects on the evolution of certain populations. When insecticides were first used to kill malaria-carrying mosquitoes, most of the insects died. However, some few individuals survived because they carried preexisting mutations that either prevented their systems from absorbing the pesticidal chemicals, or else enabled them to degrade those chemicals (Hemingway and Karunaratne 1998; Martinez-Torres et al. 1998). Some of the surviving individuals, for example, had duplications of the esterase gene whose product allows the cells to destroy DDT. In other individuals, a mutation in the potassium ion channel prevented certain insecticides from binding to their targets. Because only individuals with such mutations survived to mate with one another, the resistant genotypes quickly became the norm in many insect populations, with the result that it takes much higher concentrations of insecticides to kill them; indeed, some strains have become completely immune to the effects of many pesticides. A similar trajectory of human intervention explains the evolution of antibiotic resistance in many strains of bacteria.

Several recent studies have shown that human predation practices have caused the evolution of different growth rates among prey species. Intensive selection pressure from human fishing has caused Norwegian graylings and North Atlantic cod to reach sexual maturity at smaller sizes and, in some cases, a year earlier, than they did a century ago (Haugen and Vøllestad 2001; Andersen et al. 2007). This is probably because fish that reproduce at younger ages have a better chance of leaving offspring. Sport hunting of bighorn trophy rams has caused phenotype-selective mortality, wherein those rams with big horns are disproportionately killed. This practice inadvertently selects for rams with smaller horns, and indeed today's

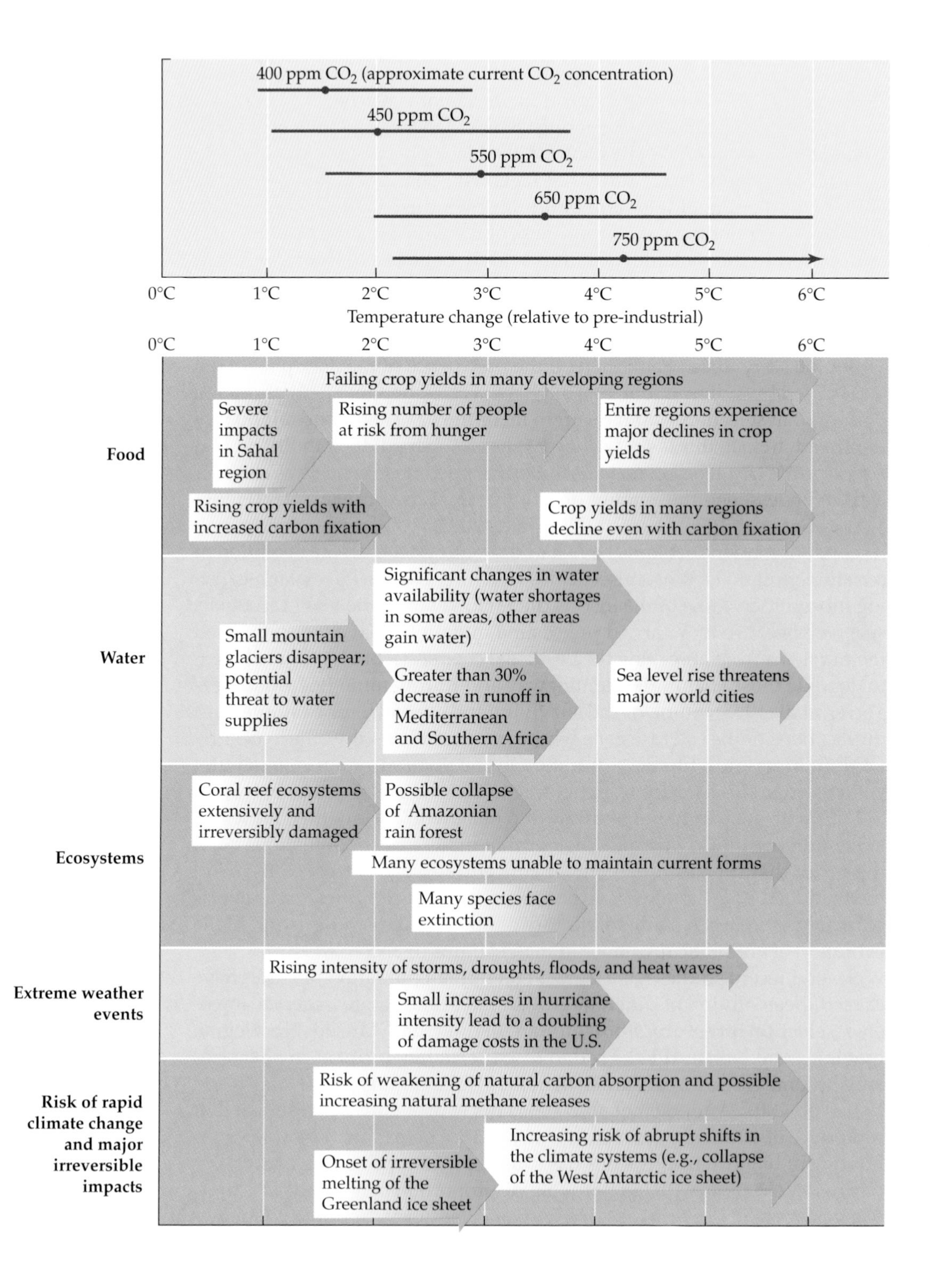

400 ppm CO_2 (approximate current CO_2 concentration)
450 ppm CO_2
550 ppm CO_2
650 ppm CO_2
750 ppm CO_2
0°C
1°C
2°C
3°C
4°C
5°C
6°C
Temperature change (relative to pre-industrial)
Food
Failing crop yields in many developing regions
Severe impacts in Sahal region
Rising number of people at risk from hunger
Entire regions experience major declines in crop yields
Rising crop yields with increased carbon fixation
Crop yields in many regions decline even with carbon fixation
Water
Significant changes in water availability (water shortages in some areas, other areas gain water)
Small mountain glaciers disappear; potential threat to water supplies
Greater than 30% decrease in runoff in Mediterranean and Southern Africa
Sea level rise threatens major world cities
Ecosystems
Coral reef ecosystems extensively and irreversibly damaged
Possible collapse of Amazonian rain forest
Many ecosystems unable to maintain current forms
Many species face extinction
Extreme weather events
Rising intensity of storms, droughts, floods, and heat waves
Small increases in hurricane intensity lead to a doubling of damage costs in the U.S.
Risk of rapid climate change and major irreversible impacts
Risk of weakening of natural carbon absorption and possible increasing natural methane releases
Increasing risk of abrupt shifts in the climate systems (e.g., collapse of the West Antarctic ice sheet)
Onset of irreversible melting of the Greenland ice sheet

◀ **FIGURE 8.4** Summary of global changes that could be brought about by increased levels of carbon dioxide (CO_2, the major anthropogenic greenhouse gas) and the elevated temperatures these CO_2 levels would be expected to generate. A total environmental overhaul is to be expected if the average global temperature increases by 4°C, which is certainly a possible scenario. We are currently seeing the effects of an increase of about 0.8°C, and the amount of CO_2 already in the atmosphere may have irreversibly committed us to a 2°C increase. (After Stern 2006.)

male bighorn sheep display a slower rate of horn growth than they did 30 years ago* (Coltman et al. 2003).

GLOBAL CLIMATE CHANGE AS A SELECTIVE AGENT Today we are seeing Darwinian selection in action as a result of the ubiquitous environmental stress of global climate change (see Stern 2006). Global climate change (sometimes called "global warming" or "global planetary disruption") threatens to change our planet into a different place, one that will almost certainly not be favorable to the diversity of life as we currently know it (Figure 8.4). Both the abundance of species at different locations and the ranges over which particular species can survive have been shifting. In the Northern Hemisphere, for instance, the ranges where certain trees and other plants grow and over which certain insects are found has extended north as winter temperatures get milder and spring temperatures occur earlier. Conversely, the range of hundreds of Arctic species is narrowing (Parmesan and Yohe 2003). Furthermore, the rate of these changes became significant in the mid-1970s—at just the same time the rise in global temperatures became steep and significant.

We can already see natural selection at work, especially in those animals (such as insects) that produce numerous progeny and have short generation times. These are the organisms most likely to evolve via natural selection rather than to face extinction due to temperature change. Sometimes selection can be visually obvious. The two-spotted ladybug beetle (*Adalia bipunctata*) has two major forms (Figure 8.5). The black-spotted red form is found predominantly in warmer climes, while a red-spotted black form is more common in the colder regions. The black (melanic) color absorbs solar heat and is an advantage (especially for reproductive activity) where temperatures are lower (de Jong et al. 1996). In the past 20 years, temperatures in the northern Netherlands have warmed, and the ratio of red ladybugs to black ladybugs in that region has begun to resemble that traditionally found in the southern half of the nation (de Jong and Brakefield 1998).

We are also able to observe shifts in gene and allele frequencies in insects. The genetic compositions of certain insect species (especially fruit

*Olivia Judson (2008) has noted that all of these growth changes were predicted by evolutionary theory.

FIGURE 8.5 Two genetically variant polymorphisms—black-on-red and red-on-black—of the spotted ladybug *Adalia bipunctata*. The frequency of these polymorphisms at any given location depends on temperature, with the red-on-black morph (which, being darker, absorbs sunlight better) found in colder latitudes. (Photograph © Papilio/Alamy.)

flies of the genus *Drosophila*) have been studied by scientists throughout the world for over 50 years, giving us an extensive database with which to compare more recent surveys. For instance, *Drosophila subobscura* is known to have particular inherited chromosomal rearrangements (inversions) that correlate with temperature. Some inversions are seen more frequently at low temperatures, while others appear to confer greater fitness at higher temperatures. Balanyá and colleagues (2006) found that in North America, South America, and Europe, the inversion frequencies have shifted away from the Equator by about 1° of latitude from their historical norm. In other words, the inversion frequencies once found around Naples are now being seen as far north as Rome.

In Australia, the shift may be much greater. The climate along the eastern coast of Australia is becoming warmer and drier, and recent evidence demonstrates that in the past 50 years, the average temperature has increased dramatically as a result of human activities (see Williams 2008). The variants of the alcohol dehydrogenase gene in *Drosophila* correlate well with temperature, and the allele frequencies of this gene show a shift of 4° of latitude from the period 1978–1981 to the period 2000–2003 (Umina et al. 2005). Laboratory studies suggest that temperature (and, to some extent, humidity) selects for particular alleles and that this shift in alleles is probably due to natural selection rather than migration.

In many species, the timing of developmental events such as flowering and mating is also changing because of phenotypic plasticity. In Mediterranean ecosystems, the leaves of most deciduous plants emerge 2 weeks earlier and fall about 2 weeks later than they did 50 years ago, and in western Canada, the quaking aspen tree (*Populus tremuloides*) blooms almost a full month earlier than it did during the 1950s (Peñuelas and Filella 2001). Butterflies in northern Spain are eclosing 11 days earlier than they did in 1952, and the mating calls of some frog species in upstate New York are

Selection for Heritable Plasticity

Many of the phenotypic changes seen in response to global warming involve phenotypic plasticity. Bradshaw and Holzapfel (2001) demonstrated that over the past 30 years, the pitcher plant mosquito (*Wyeomyia smithii*) has evolved an adaptive response to global warming by changing its photoperiodic response. With longer growing seasons now the norm in the northern part of their range, mosquitoes there have shifted toward using shorter day length cues to initiate larval dormancy. This is because winter is arriving later in the northern latitudes, and conditions remain favorable for reproductive activity longer. The northerly mosquitoes now use a photoperiod cue similar to that used by the conspecifics to the south; so what used to be a southerly behavior is now prevalent in the north as well. This change in behavior is the result of genetic alterations in the northerly populations.

The great tit (*Parus major*) has experienced selection for phenotypic plasticity in the timing of its reproduction. These birds feed caterpillars to their young. However, as spring has come progressively earlier to the Northern Hemisphere, the caterpillars have been maturing earlier, to the point that they now often pupate before the young birds hatch. The disappearance of their food source has led to a decline in the tit population. In both Great Britain and the Netherlands, researchers have documented that phenotypic plasticity has enabled some *Parus* individuals to adjust their egg-laying behavior to this changing environmental state of affairs.

There is genetic variation among individual females as to how early they can lay eggs, and those birds who can lay their eggs the earliest have the best chance of having caterpillars available to their young. The birds that survive are the offspring of parents who are most plastic in their egg-laying response (i.e., those with the broadest norm of reaction) and whose egg-laying date thus can be advanced the most. Using a 47-year-old database, Charmantier and colleagues (2008) were able to show that the egg-laying date of female tits in Great Britain has advanced by 14 days over the past half century. In both this and the Dutch study (Nussey et al. 2005), selection has altered the boundaries of phenotypic plasticity for reproductive timing, allowing it to change with the ecological conditions. The broadest reaction norm has been selected such that the birds have a better chance of having their offspring survive.

now heard an average of 10 days earlier than was common a century ago. All of these traits are developmentally plastic over a norm of reaction (see Chapter 1), and the populations mentioned above have not shown changes in gene frequency. However, other plants and animals that are not so developmentally plastic are perishing. Since Henry David Thoreau's detailed observations of the Walden Pond ecosystem in the 1850s, the average temperature there has risen 2°C. About 20% of the more than 400 plant species observed by Thoreau can no longer be found there, and other species that have not shifted their flowering times (especially irises, orchids, bladderworts, and lilies) are declining (Miller-Rushing and Primack 2008).

Embryology and Darwin's synthesis

Darwin recognized that embryonic resemblances were a very strong argument in favor of the common ancestry of different animal groups (see Oppenheimer 1959; Ospovat 1981). "Community of embryonic structure

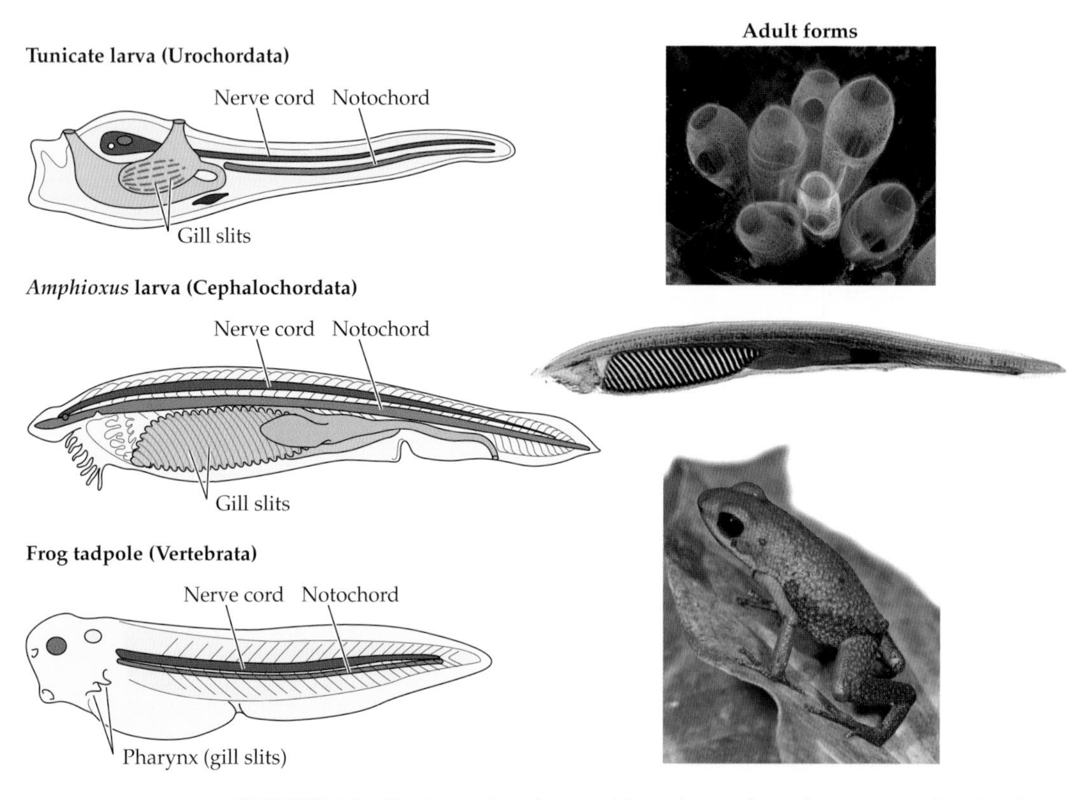

FIGURE 8.6 Tunicates, lancelets, and frogs have a homologous notochord and look similar as larvae but have very dissimilar adult morphologies. Photographs © Norbert Wu/Visuals Unlimited (top), J. D. Cunningham/Visuals Unlimited (center), and © Luís Louro/ Shutterstock (bottom).

reveals community of descent," Darwin said in *Origin of Species*. Thus, he looked to embryonic and larval stages for homologies that would be obscured in the adult. In the *Origin*, Darwin celebrated the case of the barnacle, whose larvae showed it was a shrimplike crustacean rather than a clamlike mollusc, and in the *Descent of Man* (1874), he gloried in Alexander Kowalevsky's (1866) discovery that the tunicate—hitherto also classified as a mollusc—was actually a chordate (Figure 8.6). Tunicate larvae have a notochord and pharyngeal slits arising from the same cell layers that give rise to those structures in fishes and chicks. Thus, the "great divide" between the invertebrates and the vertebrates was bridged by the discovery of larval homologies.

Comparative embryology became evolutionary morphology as new information about the homologies of the germ layers in various animals became paramount in answering questions about phylogeny (e.g.,

Lankester 1877; Balfour 1880–1881; Oppenheimer 1940). Müller (1869) united the Crustacea (including in this phylum not only the barnacles but a whole range of parasites) through their possession of a larval type called the nauplius, and E. B. Wilson (1898) proposed a superphyletic taxon that included all those animals with spiral cleavage and possessing a particular mesoderm-forming cell. A century later, Wilson's proposed taxon was recognized by molecular and cladistic techniques and named the Lophotrochozoa (see Halanych et al. 1995; Halanych 2004).

To many in Darwin's generation, the embryo's development, or unfolding, could be seen as the motor driving evolution. Ernst Haeckel (1866) and others envisioned development as producing new structures that were added onto earlier stages of development. Individual development (ontogeny) would recapitulate (follow) the stages of previous generations of ancestors and then add something new. Haeckel proposed a causal parallelism between an animal's embryological development (ontogeny) and its evolutionary history (phylogeny). His so-called biogenetic law—that "ontogeny recapitulates phylogeny"—was based on the idea that the successive (and, to Haeckel, progressive) origin of new species was based on the same laws as the successive and progressive origin of new embryonic structures. Natural selection would prune the tree of life by eliminating the earlier (and hence less fit) species. To Haeckel, development produced novelties, and natural selection would get rid of entire species that had become obsolete. While this view of development and evolution was not tenable, the next generation of embryologists worked with and modified Haeckel's notions and showed how new morphological structures can arise via the hereditary modification of embryonic development.

One example of such modification, studied by Frank Lillie in 1898, is brought about by an alteration in the typical pattern of spiral cleavage in unionid clams. Unlike most clams, *Unio* and its relatives live in swiftly flowing streams. Streams create a problem for the dispersal of free-swimming larvae because such larvae are always carried downstream by the current. *Unio*, however, adapted to the fast-flowing environment by modifying its development. This modification begins with an alteration in embryonic cleavage. In typical molluscan cleavage, either all the macromeres (larger set of blastomeres) are equal in size or the 2D blastomere is the largest cell. However, cell division in *Unio* is such that the 2d blastomere gets the largest amount of cytoplasm (Figure 8.7). The 2d blastomere then divides to produce most of the larval structures, including a gland capable of producing a large shell on the resulting larva, which is called a glochidium. The glochidium resembles a tiny bear trap; sensitive hairs cause the valves of the shell to snap shut when the larva encounters the gills or fins of a fish. The larva thus attaches itself to the fish and "hitchhikes" until it is ready to drop onto the substrate and metamorphose into an adult clam. In this manner, clam larvae can spread upstream as well as

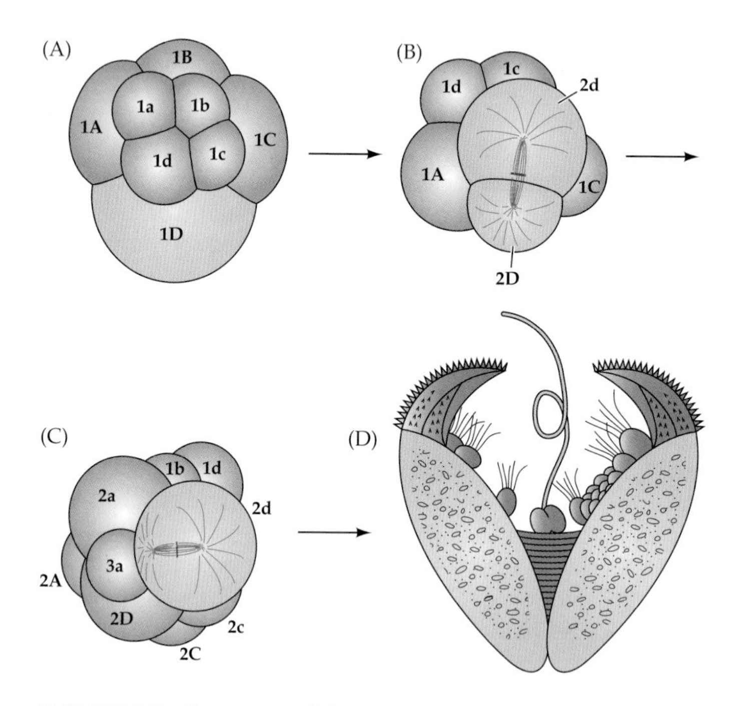

FIGURE 8.7 Formation of the glochidium ("bear trap") larva by modification of spiral cleavage. After the 8-cell embryo is formed (A), the placement of the mitotic spindle causes most of the D cytoplasm to enter the 2d (B). This enlarged 2d blastomere divides (C), eventually giving rise to the "bear trap" shell of the larva (D). This adaptation allows the larva to attach to the gills of passing fish. (After Raff and Kaufman 1983.)

downstream and are less likely to be carried outside their favorable habitat. The marine biologist Walter Garstang used such evidence (and his own studies of snail larvae) to show that the evolution of new features was based on changes in developmental stages, not in adult stages. Garstang reversed Haeckel's relationship between ontogeny and phylogeny. In his address to the Linnean Society, Garstang (1922, p. 724) remarked that "ontogeny does not recapitulate phylogeny: it creates it."

The unionid clams continue to provide excellent examples of evolution through the action of natural selection on changes in development. In some unionid species, glochidia are released from the female's brood pouch and then wait passively for a fish to swim by. Some other species, such as *Lampsilis altilis*, have increased the chances of their larvae finding a fish by yet another series of developmental modifications (Welsh 1969). Many clams develop a thin mantle that flaps around the shell and surrounds the brood pouch. In some unionids, the shape of the brood pouch (marsupium) and the undulations of the mantle mimic the shape and swimming behavior of

FIGURE 8.8 Phony fish atop the unionid clam *Lampsilis altilis*. The "fish" is actually the brood pouch and mantle of the clam. The "eyes" and flaring "tail" attract predatory fish, and the glochidium larvae attach to the fish's gills. (Photograph courtesy of Wendell R. Haag/USDA Forest Service.)

a minnow. To make the deception even better, the clams develop a black "eyespot" on one end and a flaring "tail" on the other. The "fish" in Figure 8.8 is not a fish at all, but the brood pouch and mantle of the female clam beneath it. When a predatory fish is lured within range of this "prey," the clam discharges the glochidia from the brood pouch. Thus, the modification of existing developmental patterns to produce new phenotypes has permitted unionid clams to survive in challenging environments.

The failure of developmental morphology to explain evolution

While embryology played an important role in providing evidence for evolution, there were no mechanisms of development that allowed embryology to be cited as a causal agent for evolution. In the early twentieth century, embryology was predominantly a descriptive science, and any experimentation tended to be concerned with how embryonic development might be keyed to its environment (more about that in the next chapter). There were arguments over which structures were homologous (similar because of common ancestry, as in the chicken wing and the human hand), which were analogous (achieving the same function but arising from independent sources, as do the limbs of insects and vertebrates), and which were parallel (arising from two related but independent sources that diverged in the same way, as in the curved beaks of the diverse species of nectar-feeding birds, which evolved independently from different straight-beaked ancestral birds).

For example, to pose a broad question, are the segmented body plans that are seen in arthropods, annelid worms, and vertebrates (but not in other animals) homologous (arising from the same segmented ancestor), analogous (arising separately and independently because segmentation is a strongly favored adaptation that enhances movement), or parallel (arising not from a single common ancestor, but from several separate but related ancestors that all possessed the genetic basis for achieving segmentation)? No one knew. Fin de siècle evolutionary embryology had no independent method for ascertaining homology, no independent method of classification, and no knowledge of the cellular interactions that constituted embryogenesis. As such, the field became mired in speculation (see Bowler 1996; Gilbert 1998).

Wilhelm Roux, a student of Haeckel, suggested that embryological ideas about evolution were premature and would remain so until we learned something about the specific ways in which animal bodies were constructed. According to Roux (1894), embryology had to leave the seashore and forest and go into the laboratory. Rather than being handmaiden to evolution, embryology should emulate physiology and look for the proximate causes of alterations in development. However, Roux predicted that embryology would someday return to prominence in evolutionary biology, bringing with it new knowledge of how animals were generated and how evolutionary changes might occur. But in the twentieth century, embryology "turned inward," away from evolution, and focused on the fascinating questions of how a fertilized egg gives rise to all the cells, tissues, and organs of the body.

The Modern Synthesis

The position vacated by embryology was soon filled by genetics. The fusion of Darwinian ideas of natural selection with the revelation of the gene as the agent of heredity resulted in the **Modern Synthesis**, sometimes called **neo-Darwinism**. This crucial evolutionary synthesis provided the theoretical and mathematical framework for evolutionary biology that embryology could not provide.

THE ORIGINS OF THE MODERN SYNTHESIS The existence of genes and their role as the source of hereditary information were not realized until the last years of the nineteenth century. Until that time, genetics and embryology were effectively joined in the science of "heredity," and genetics began largely as a subsection of that field (see Gilbert 1998). The work of the renowned geneticist Thomas Hunt Morgan in the early twentieth century was in large part responsible for separating genetics from embryology and establishing the former as a strong and independent field. The mathematical models of population genetics pioneered by Morgan and elaborated by his successors offered sound and demonstrable mechanical explanations for natural selection. And because population genetics can only study interbreeding populations and the barriers that prevent interbreeding, the *explandum* (that which is to be explained) of evolutionary biology changed from the determination of phylogeny among different species (which was the goal of evolutionary morphology) to the quantification of natural selection within the same species.*

But if there was a "modern" synthesis, then there had to have been some "unmodern" synthesis; to the geneticists, that outdated synthesis was the one in which embryology had been united with evolution. Two embryologists-turned-geneticists were particularly important in making a case that evolutionary morphology had come to an end and that genetics was to take over. The first of these men, William Bateson, stated that "the embryological method has failed" when it came to determining the mechanisms of evolution (Bateson 1894). Bateson had been an evolutionary morphologist looking at developmental changes that could produce variation without selection. However, in 1900, he confessed that despite all that embryology and comparative anatomy had done, we were ignorant of the mechanism by which transmitted traits produce new species:

> Let us recognize at the outset that as to the essential nature of these phenomena we still know absolutely nothing. We have no glimmering of an idea as to what constitutes the essential process by which the likeness of the parent is transmitted to the offspring. We can study the process of fertilization and development in the finest detail which the microscope manifests to us, and we may fairly say that we now have a thorough grasp of the visible phenomena; but of the physical basis of heredity we have no conception at all. ... Not only is our ignorance complete, but no one has the remotest idea how to set to work on that part of the problem.

Bateson admitted to a "crisis of faith" and was at a loss to provide a mechanism for evolution. However, he tells how he was then "converted" to Mendelism and how he came to believe the work of Mendel could explain all the things that embryology couldn't. In his 1922 essay "Evolutionary Faith and Modern Doubts," Bateson announced the birth of a new science out of the decay of the old, proclaiming that "the geneticist is the successor of the morphologist."

The second great champion of genetics, Thomas Hunt Morgan, renounced embryology and wrote an article entitled "The Rise of Genetics" (1932a) and a book entitled *The Scientific Basis of Evolution* (1932b) in which he contrasted the new genetics with the old-fashioned embryology. Genet-

*To be sure, this is an oversimplification. The early evolutionary biologists came from different subdisciplines within biology, and some of them professed more than one opinion during the course of their careers. Still, the ascendancy of the population geneticists can be seen in the frustration of those who felt that there was more to evolution than the mathematics of allele frequencies and "bean bag genetics" (Mayr 1959; Smocovitis 1996).

Sturtevant on Snails

During the early days of the evolutionary synthesis, there was a huge debate between geneticists and embryologists over whether genetics could explain development or evolution. One thing embryologists could argue was that there were no mutations that were known to affect early development, and that the inherited traits that were known to affect early development did not seem to be inherited in the expected Mendelian manner. One of these traits was the sinistral (left-handed) shell-coiling variant in the snail *Limnaea*.

Lillie, Wilson, Garstang, and others (including Piaget, as we will see in Chapter 10) used snail embryos to link embryology to evolution, and molluscan development was well characterized. This coiling variant was widely discussed in embryology texts, but when Boycott and Diver (1923) did a study of its inheritance, they could not explain their observed ratios of right- to left-coiling offspring by Mendelian genetics.

It was Morgan's student Alfred Sturtevant who in 1923 explained the phenomenon, in a paper that covered only two columns in the journal *Science*. But in these two columns, Sturtevant told Boycott and Diver that their data were wrong, that their ratios were merely "fortuitous," and that they would have gotten different results if they had done the experiments properly. He then told them what the results *should* have been and explained why (see the figure). Sturtevant hypothesized that the case is a simple

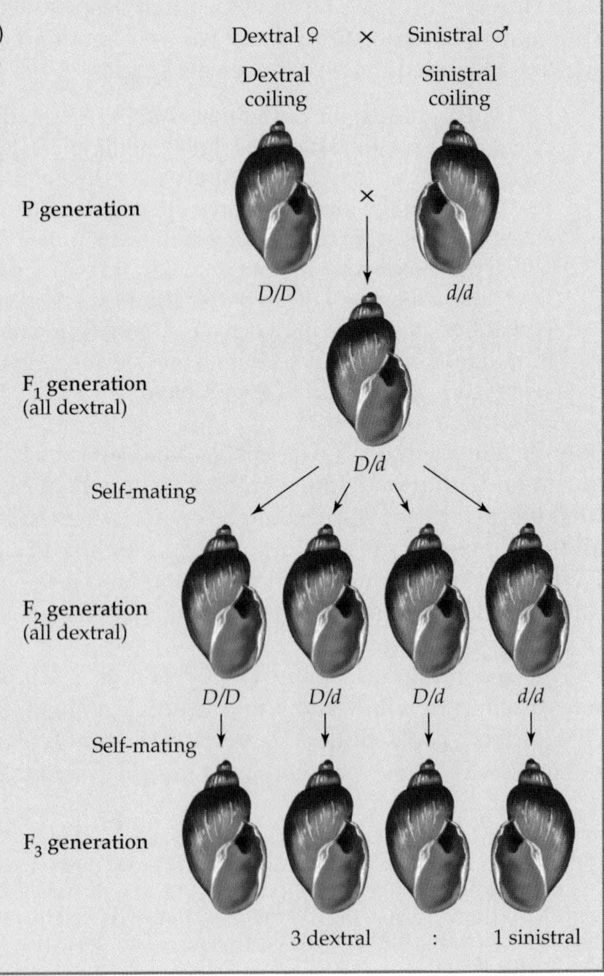

Coiling genetics in the snail *Limnaea*. The factor encoded by the *D* gene is needed in the egg for directing the embryo to coil in a dextral (rightward) direction. Without the factor encoded by *D*, the embryo coils in a sinestral (leftward) direction. Therefore, whether the snail coils to the right or to the left depends on the mother's genotype (i.e., whether or not she put the D factor into her eggs) and not on the genotype of the embryo. In the first cross, since the mother expresses *D*, all her offspring coil to the right, and all are heterozygous. However, when the male parent expresses the D factor and the female does not, all the offspring are similarly heterozygous, but they all coil to the left (since the mother lacked the D factor). When the F_1 offspring are bred to one another, *dd* females are produced. These homozygous females will themselves coil to the right if their mother produced factor D; however, all the *progeny* of these *dd* females will coil to the left. (After Morgan 1927.)

Mendelian one, with the dextral character dominant, but that the phenotype of the individual is determined not by its own genetic constitution, but by that of its mother. He concluded, "It seems likely that we shall have a model case of the Mendelian inheritance of an extremely 'fundamental' character, and a character that is impressed on the egg by the mother." We now call such inheritance (which includes the famous cases of the *Bicoid* and *Nanos* genes that specify the anterior-posterior axis in *Drosophila* embryos) "maternal effects."

Eventually, Boycott, working with senior embryologist S. L. Garstang, got exactly the results that Sturtevant had predicted he would. Boycott and colleagues (1930) called Sturtevant's analysis "his inspired guess." However, it was not so much an inspired guess as a beautiful example of predictive theory. Sturtevant did not have to do the laborious crosses to "know" the result. The geneticists claimed to have superseded the embryologists, and this round ended in their favor.

ics, and only genetics, he wrote, constituted the "scientific basis of evolution." Embryology was one of the "metaphysical" and unscientific approaches that led nowhere. Moreover, Morgan asserted that the proper unit for study for evolutionary biology should be the population (the unit that evolves), not the individual (the unit that natural selection acts on).

THE FISHER-WRIGHT SYNTHESIS It should be clear from the above discussion that the origins of the Modern Synthesis took place in a very anti-embryological environment, and it must be emphasized how complete the replacement of embryology by genetics has been. Darwin had shown that evolution occurred, and he used embryology to support such a view. However, whereas continental evolutionary biologists (notably in Germany) had seen in embryology possible mechanisms for evolutionary change, the British and American evolutionists sought this mechanism in Darwin's own idea of natural selection.

Many embryologists (beginning with Thomas Huxley; see Huxley 1893) accepted natural selection but did not think natural selection alone was sufficient to generate new types of organisms (see Lyons 1999). However, many of the early geneticists, starting with Morgan, did. But the early geneticists did not have the mathematical reasoning to show that this was the case; that task fell to three remarkable men, R. A. Fisher, J. B. S. Haldane, and Sewall Wright. In particular, Fisher's book *The Genetical Theory of Natural Selection* (1930) showed how the interplay of selection and mutation could bring about changes in the populations. Fisher's extensive equations and analyses were expanded, challenged, and restructured by Haldane and Wright (see Provine 1986; Edwards 2001) in what Mayr (2004) called the "Fisherian synthesis." This early version of the Modern Synthesis added a new genetic principle to evolutionary biology: that the inherited factors ("units of heredity') that would make an organism more or less fit for its environment were in fact the genes it received from its parents.

Fisher not only added genetics to evolutionary theory, he also extended the economic metaphor that Darwin (and Wallace) had originally obtained from Malthus's economic theories about competition for scarce resources. For instance, Fisher explained the 1:1 sex ratio in terms of the "biological capital" that each offspring inherits from its parents (Fisher 1930, p. 141). If it costs parents equally to produce male and female offspring, then a 1:1 sex ratio would be the most stable. If male births were less common than female births, then a newborn male would produce more offspring in the next generation than a newborn female. (Think, for example, of one male for eight females, or a 1:8 sex ratio.) In such a case, parents who were genetically predisposed to have male offspring would have more grandoffspring (that is to say, their sons would reproduce proportionally more often). In this way, the male-producing genes would spread through the population, male births would become more common, a 1:1 sex ratio would be approached, and the selective advantage originally associated with having male offspring would then diminish. In modern parlance, the 1:1 sex ratio (often called the "Fisherian ratio") is considered an **evolutionarily stable strategy**,* or **ESS**, and it was modeled mathematically by Shaw and Mohler (1953).

Another important concept for the Modern Synthesis was Sewall Wright's coefficients of relatedness and kinship (Wright 1922). The coefficient of kinship is the numerical probability that the alleles present at a particular locus chosen at random from two related individuals are identical by descent; the **coefficient of relatedness** (*r*) is twice that value. In diploid species, the *r* between parent and offspring is 0.5 (1/2), since half the offspring's genome comes from each parent. Full siblings also have an *r* coefficient of 0.5, indicating that half their genes are expected to be the same. Uncles and aunts have an *r* of 0.25 in relation to their nieces and nephews; half-sibs also have an *r* of 0.25. First cousins have an *r* of 0.125 (1/8).

The kinship coefficient became a very important idea in the Modern Synthesis, leading to the concepts of **inclusive fitness** and **kin selection** (Hamilton 1967; Axelrod and Hamilton 1981), wherein individuals would be expected to favor the reproductive success of those relatives who carried the greatest proportion of the same genes. Thus, when asked if he would risk his life to save his brother, Haldane is said to have replied, "No, but I would do so to save two brothers—or eight cousins" (Figure 8.9; McElreath and Boyd 2007).

THE DOBZHANSKY-MAYR SYNTHESIS Evolutionary syntheses can appear to encapsulate one another like Russian dolls, each synthesis enfolding the

*In this argument, one sees the germs of ideas that will flourish in later versions of the Modern Synthesis. These ideas include the notion of evolutionarily stable strategies (ESS) that are impervious to mutations (Maynard Smith and Price 1973), parental investment (Trivers 1972), and evolutionary game theory (MacArthur 1965; Hamilton 1967).

FIGURE 8.9 Wendy Darling (Peter Pan's sometime girlfriend) shields her two brothers from Captain Hook, thereby illustrating Haldane's take on Wright's coefficient of relatedness, *r*. In this view, sacrificing oneself for two brothers (or eight first cousins) would offset the genes lost in one's own demise.

previous one while explaining further phenomena. Even with Wright's embellishments, the Fisherian synthesis was incomplete. It was based predominantly on research done on experimental organisms in the laboratory and thus seemed more relevant to artificial selection than to natural selection. Moreover, the mathematical underpinnings of this synthesis were not accessible to most biologists.

To show that natural selection actually worked in nature, Sergei Chetverikov demonstrated natural selection in natural populations of *Drosophila*. Chetverikov's program was expanded by Ukrainian-born Theodosius Dobzhansky, who brought this idea to New York City, where he worked with Thomas Hunt Morgan. In 1937, Dobzhansky wrote *Genetics and the Origin of Species*, which in many ways translated Wright and Fisher into accessible English and added insights gained from studying natural populations, thus allowing the framework of population genetics to be expanded into the rest of biology. In his book, Dobzhansky redefined evolution as changes in allelic frequency.

But Dobzhansky's synthesis was also incomplete. For all the hopes that mathematical models could explain the origin of new species, these analyses really showed only how mutation and recombination could change the predominant phenotype of a population from one form to another. It did not show how speciation occurred. To approach this latter question, Ernst Mayr (1942) provided conceptual tools that defined a species as a real or potentially interbreeding population and showed that speciation could take place if this population were geographically divided or otherwise prevented from interbreeding.

By the 1940s, the major tenets of the Modern Synthesis had been established. These included the premises that all evolutionary phenomena could be explained by the known genetic mechanisms; that natural selection was the overwhelmingly important mechanism of evolutionary change; and that the large anatomical changes seen over the course of evolution were the result of accumulations of smaller changes. George Gaylord Simpson had shown that neo-Darwinism was compatible with paleontology (a premise that many other paleontologists, especially those in Germany, did not agree with), and G. Ledyard Stebbins had brought plant biology into the synthesis. By the time the third edition of Dobzhansky's book was published in 1951, he could state that "the study of mechanisms of evolution falls within the province of population genetics." Macroevolution (evolution at the level of the species or higher taxa) could be explained by microevolution (the evolution of populations within a species). Population genetics had become the core of evolutionary biology, and genetics was seen as "Darwin's missing evidence" (Kettlewell 1959).

During the 1970s, studies showed that sexual selection and polymorphic inheritance (that is, when different environmental conditions within a species' range select for different phenotypes of that species) also can be explained by the mathematics of the Modern Synthesis. Research on kin selection then brought a mathematical analysis of behavioral biology into the fold as well. Evolutionary biology went from being a phenotype-based science of genealogical relationships to being the epiphenomenon of the gene pools of populations. The Modern Synthesis truly encapsulated all of biology, and it fully justified Dobzhansky's 1964 declaration that "nothing in biology makes sense except in the light of evolution."

The Triumph of the Modern Synthesis: The Globin Paradigm

But would the Modern Synthesis withstand a new challenge from molecular biology? In the 1960s, groundbreaking discoveries in the field of molecular biology began revolutionizing biology in much the way evolutionary biology had done a century earlier, this time from the "bottom up" rather than the "top down." Like evolution, the field of molecular biology was based on heredity; like evolution, it promised a common language for all biological disciplines; and like evolution, it claimed to be the explanation for life. But while evolution was primarily about ultimate causes (the "why?" questions), molecular biology was about proximate causes (the "how?" questions). Would the two approaches be compatible, or would molecular biology supersede evolution as an explanatory mechanism?

As it turned out, molecular biology was able to integrate the "how" into the "why," and the population genetic theory of allele frequencies was able to integrate the findings of molecular biology. In this reciprocal confirmation, molecular biology provided remarkable evidence for the Modern Synthesis, and evolution provided a new framework for molecular biology.* One of the best examples of this integration was the elucidation of the evolution of malarial resistance in human populations. It is somewhat ironic that, in the modern analysis, some of the best evidence for evolution by natural selection has been found in *Homo sapiens*—a species that fascinated Darwin, but the one he tried to ignore in the *Origin*. However, these molecular studies became paradigmatic examples showing how genetic variation and natural selection produce phenotypes adaptive for a particular environment.

In the following examples, the agent of natural selection is malaria. The eukaryotic malarial parasites of the genus *Plasmodium* are some of the most successful predators of human populations (see the next page).

Hemoglobin S and sickle-cell disease

The course of malarial infection has resulted in natural selection favoring certain people whose red blood cells are better able to resist the entry of *Plasmodium*, or that are able to block the parasite's reproductive cycle inside those cells. One of the most important genetic mechanisms for combating *Plasmodium* involves a mutation of the gene for the β-globin chain of human hemoglobin. The prevalence of this mutation in certain populations was a puzzle at first, because it can result in mutant "hemoglobin S," deformed red blood cells, and sickle-cell disease, a condition that is more deadly than malaria.

THE GENETIC BASIS OF SICKLE-CELL DISEASE The story of the human globins, hemoglobin S, malaria, and sickle-cell disease unites evolutionary biology,

*One of the areas where the Modern Synthesis was *not* confirmed by molecular biology was the idea that all genetic variation was selected. Molecular biology revealed that individuals and populations harbor an enormous amount of genetic variation that does not appear to be acted on by selection (Lewontin and Hubby 1966; Kimura 1983; Kreitman 1983). At first, neo-Darwinians fought this idea, but the neutral mutation theory became a major part of the evolutionary canon (see Nei 2005) and has been important in documenting phylogenetic relationships and the timing of evolutionary changes (Bromham and Penny 2003). In fact, Darwin had written in the *Origin* that some variations would not be affected by natural selection.

Malaria and Evolution

Allelic variation is critical for providing the variation needed for natural selection, and some of the best-studied examples (outside of the extensive laboratory studies of *Drosophila*, *C. elegans*, and rodents) come from human populations. Human health considerations are the primary impetus for these studies, and the widespread tropical disease malaria provides several examples of natural selection by environmental agents. Malaria is endemic in over 91 countries and kills some 2 million people each year. It remains a powerful selective force on many human populations.*

Malaria in humans is the result of infection by a eukaryotic, parasitic protist of the genus *Plasmodium* (notably *P. falciparum*). Plasmodial parasites have a complex life cycle, part of which is spent in the body of an insect vector (i.e., intermediate host), usually a mosquito of the genus *Anopheles*. As seen in the figure, *Plasmodium* invades the mosquito's salivary glands and can enter the human bloodstream via the mosquito's bite, where they develop into parasites that infect and destroy red blood cells. Eventually some of the parasites in the bloodstream form gametocytes, which the mosquitoes ingest when they bite that person, and the cycle starts again.

Malaria has not affected only the human gene pool, but it has also affected the evolution of the vector (i.e., the mosquito) and the plasmodial parasite. Humans have used vast quantities of pesticides, including DDT (see Chapter 6), in our attempt to rid the world of malaria-infested mosquitoes. (Indeed, today DDT use is legal only in those tropical countries where the public health threat of malaria is greater and more immediate than the threat of the pesticide's toxicity.) But in doing so, we have applied a selective pressure that has resulted in pesticide-resistant mosquito populations. In addition, the drug chloroquine has been effective in destroying *Plasmodium*, but many parasite populations are now resistant to this treatment. It seems that in these populations, mutations have been selected for in a protein that transports the drug (Tran and Saier 2004; Yang et al. 2007).

*Malaria has probably been a major factor in the constitution of the southern European gene pool. In 410, the Goths captured Rome, but their victory was thwarted when the conquering army was ravaged by malaria. In 452, when Attila and his Huns pillaged their way to Rome, they were met by Pope Leo I and malaria. While Church historians assure us that the Pope's eloquence convinced Attila not to sack the city, other historians note that malaria may have presented a more convincing argument to the Scourge of God. The Romans paid a high price for this protection. Analysis of bones and DNA from a children's cemetery dating from about 450 shows that falciparum malaria was rampant, and some historians suggest that malaria may have been the most critical factor in why the Roman army there at that time was unable to repel the barbarian invaders (see Sallares 2002).

classical genetics, population genetics, medical genetics, and molecular biology. Sickle-cell disease is characterized by the sickling and loss of flexibility of the normally round and supple red blood cells. Their sickled shape and diminished flexibility cause these cells to obstruct capillaries, preventing blood flow to the areas supplied by the capillaries (Figure 8.10). This causes severe pain to the person and damage to organs. The spleen and the brain, with their narrow capillaries and large flow of blood, are among the organs most severely affected, and the condition is eventually lethal.

Sickle-cell disease was known to be an autosomal recessive trait, so only a person with two copies of the mutant gene (i.e., a homozygote) expresses the disease. In 1949, Linus Pauling discovered that the hemoglobin of

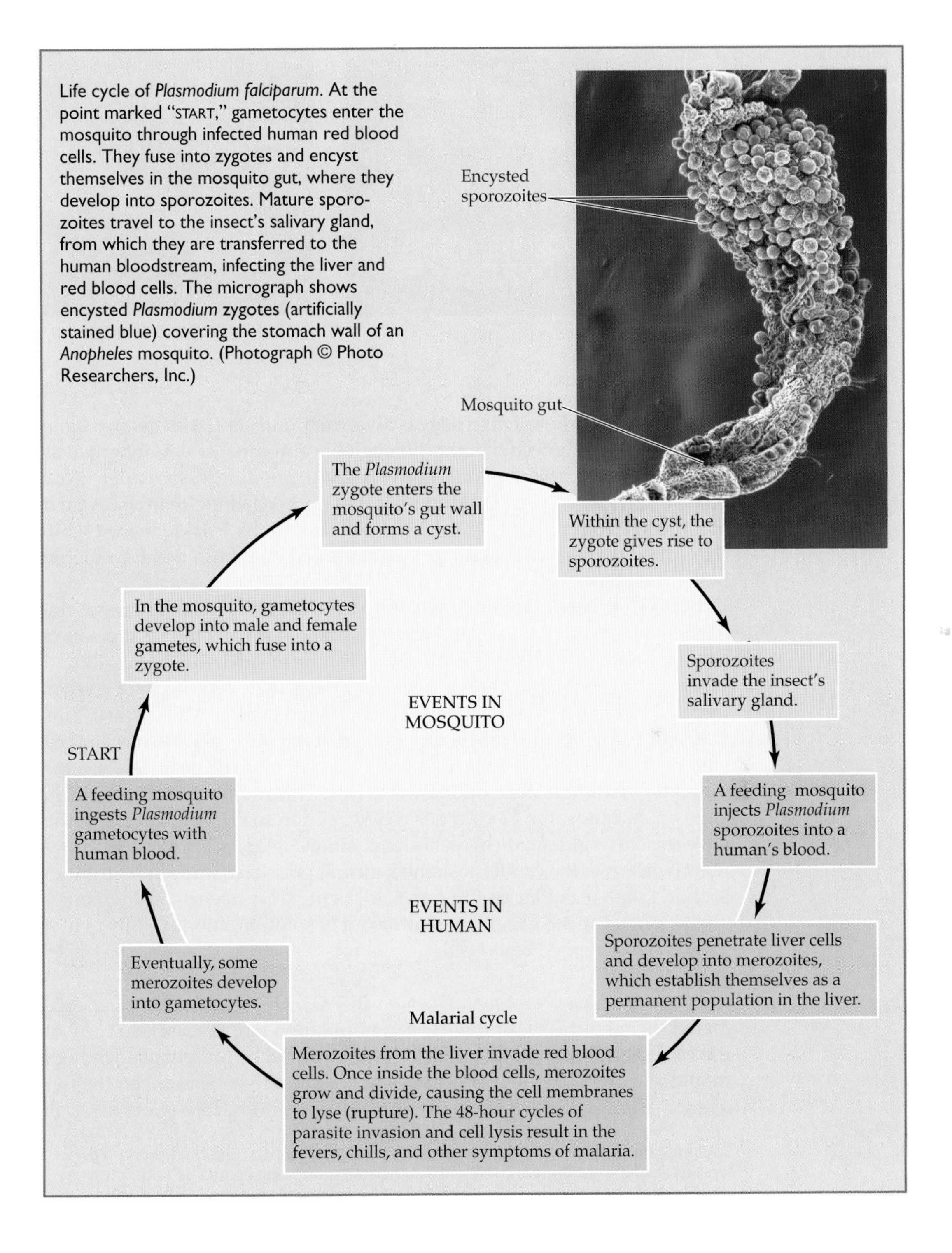

Life cycle of *Plasmodium falciparum*. At the point marked "START," gametocytes enter the mosquito through infected human red blood cells. They fuse into zygotes and encyst themselves in the mosquito gut, where they develop into sporozoites. Mature sporozoites travel to the insect's salivary gland, from which they are transferred to the human bloodstream, infecting the liver and red blood cells. The micrograph shows encysted *Plasmodium* zygotes (artificially stained blue) covering the stomach wall of an *Anopheles* mosquito. (Photograph © Photo Researchers, Inc.)

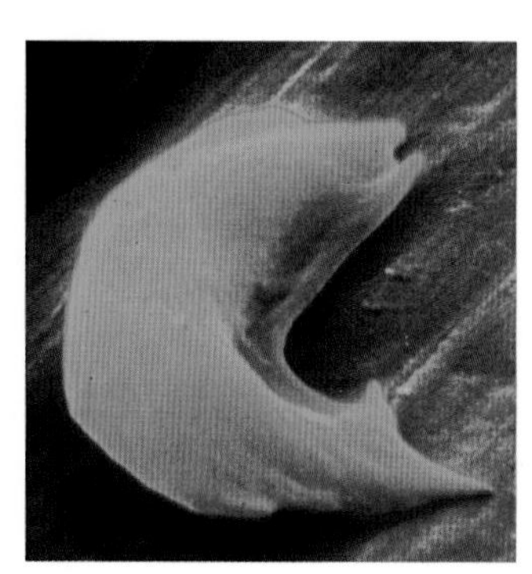
Sickle-cell phenotype

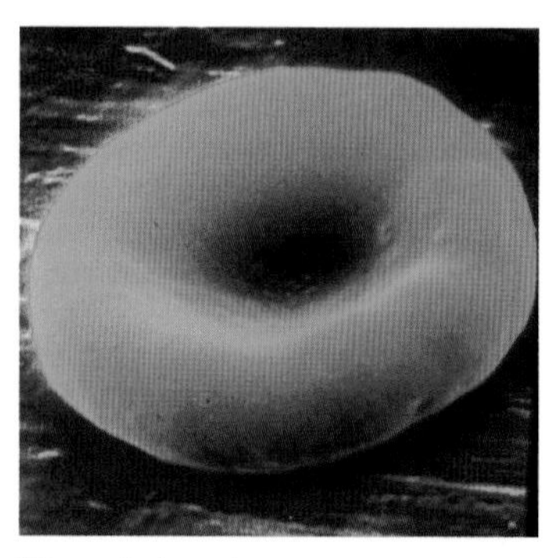
Normal phenotype

FIGURE 8.10 Sickled and normal red blood cells. The misshapen red blood cell on the left is the result of a mutation that changes the amino acid at a crucial position in the β subunit protein of hemoglobin. (Photographs © Stanley Flegler/Visuals Unlimited.)

patients with sickle-cell disease was abnormal, and shortly thereafter Itano and Neel (1950) showed that this hemoglobin abnormality was inherited as predicted by Mendelian genetics for an autosomal recessive allele. Red blood cells from wild-type individuals had normal hemoglobin (HbA); red blood cells from patients with sickle-cell disease had sickle hemoglobin (HbS); and the *parents* of sickle-cell patients had both HbA and HbS globin in their red blood cells.

In 1957, Vernon Ingram mapped the amino acids that compose hemoglobin and showed that glutamate, the amino acid normally found at position 6 in the β-globin chain, was replaced by valine in sickle-cell hemoglobin.* When information about amino acids and the genetic code became known, Ingram was able to predict the genetic mutation in sickle-cell disease. This was later confirmed by DNA sequencing, making sickle-cell disease the first genetic disorder whose molecular basis was known. The gene defect is the mutation of a single nucleotide from A to T. This mutation in the β-globin gene on chromosome 11 changes a GAG codon to GTG, resulting in the replacement of glutamate by valine at position 6. At low oxygen concentrations (such as in the capillaries), the valine at position 6 forms a hydrophobic bond with a similar region on an adjacent HbS protein, causing hemoglobin S molecules to aggregate, come out of solution, and form fibers that stretch across the red blood cell.

SELECTION FOR AN UNFAVORABLE GENE In 1933, Diggs and colleagues estimated that about 7.5% of the African population had at least one copy of the *HbS* allele. How could an allele so detrimental be present at such high frequencies? One would expect natural selection to have reduced the frequency of the sickle-cell allele to extremely low levels. Two scientists, J. B.

*Hemoglobin is made up of four globin protein subunits, two α-globins and two β-globins. Each of the subunits is complexed to a heme group that is capable of binding oxygen. Sickle-cell disease does not affect the fetus or newborn because while in utero we have γ-globin instead of β-globin.

S. Haldane and Anthony Allison, came up with a possible answer: to explain the high frequency of an allele that is deadly when homozygous, the *heterozygous* individuals must have a fitness advantage not only over those with the lethal homozygous condition, but also over "normal" homozygotes. In other words, *HbA/HbS* heterozygotes must be more fit than either *HbA/HbA* or *HbS/HbS* homozygotes.

Haldane, a pioneer of evolutionary biochemistry, found evidence that heterozygosity might indeed be favorable under certain conditions: "The corpuscles of the anaerobic heterozygote are smaller than normal, and more resistant to hypotonic solution. It is at least conceivable that they are also more resistant to attacks by the sporozoa that cause malaria" (Haldane 1949). This was remarkably good thinking. Once a method for culturing the malarial parasite had been perfected, it was found that *P. falciparum* infection increased the rate of sickling and dehydration of blood cells containing a mixture of HbS and HbA proteins. This led to the destruction of the blood cell—and of the parasite within it (Pasvol and Wilson 1980; Lew and Bookchin 2005). Thus it seems that the red blood cells of heterozygotes are not as good incubators for *Plasmodium* larvae as are normal erythrocytes. This would mean that, while *HbS/HbS* individuals are afflicted with sickle-cell disease and *HbA/HbA* "normals" are susceptible to malaria, *HbA/HbS* individuals would not have the disease and would be less susceptible to malarial infections.

While Haldane was looking at the biochemistry of the red blood cells, Allison, doing research in the months between university and medical school, was looking at African populations around Mt. Kenya. He found that different tribes had very different frequencies of the *HbS* allele. In some tribes, as many as 40% of the population was heterozygous for the *HbS* allele, while in others almost no one carried *HbS*. As he later wrote (Allison 2002):

> These observations raised questions; first, if there is strong selection against the sickle-cell homozygote, why is the frequency of the heterozygote high? Second, why is it high in some areas but not others? I formulated an exciting hypothesis; the heterozygotes have a selective advantage, because they are relatively resistant to malaria. This would operate only in areas of intense transmission of *Plasmodium falciparum*, and the selective advantage of the heterozygote would maintain a stable polymorphism. Testing the hypothesis had to wait until I had completed medical studies and received training in parasitology.

And in fact, once he had completed his studies, Allison (1954a) was able to show that heterozygous children were more resistant to induced malarial infections* than children with homozygous wild-type hemoglobin, and that these heterozygous individuals were common only in the regions where

*Infecting a person with malaria may seem unethical, but at that time inducing malarial infections was a normal treatment for syphilis (see Allison 2002).

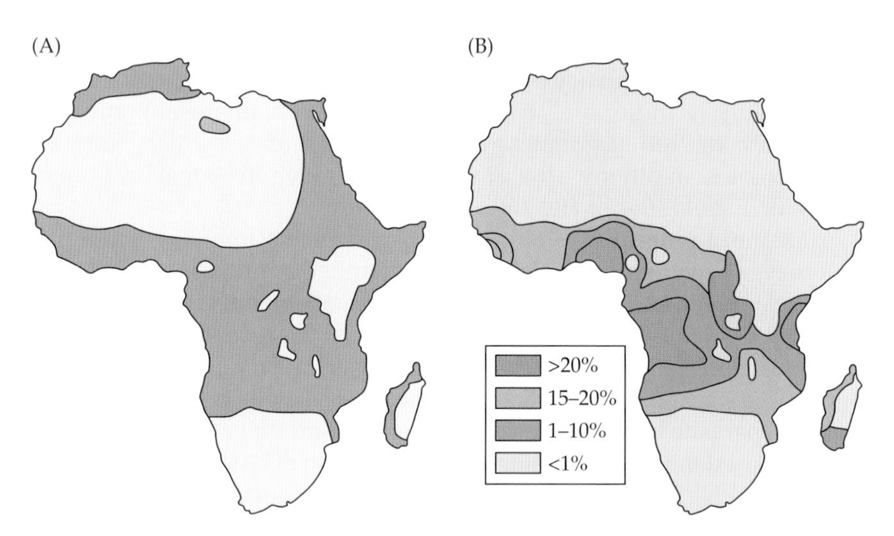

FIGURE 8.11 The sickle-cell allele (*HbS*) in regions of high malarial distribution. (A) The distribution of the malarial parasite *Plasmodium falciparum* in Africa before mosquito control was introduced. (B) Frequencies of *HbA/HbS* heterozygotes in different parts of Africa. (After Allison 2002.)

Plasmodium falciparum malaria was prevalent. Indeed, further studies showed that the only places where a high percentage of the population carried the *HbS* allele were those places where malaria was prevalent (Figure 8.11). Such areas included not only most of Africa, but also Greece and southern India.

Population genetics showed that, under normal selection, the *HbS* allele should be eliminated from the gene pool but that widespread presence of the malarial parasite favored selection for heterozygous individuals, who were more resistant to the parasite and thus more likely to survive to reproductive age (Allison 1954b; Vandepitte et al. 1955; Cavelli-Sforza and Bodmer 1971; Figure 8.12) Moreover, when the hemoglobin gene and the DNA adjacent to it were sequenced, it was determined that the *HbS* allele has evolved at least five separate times! There are separate GAG → GTG substitutions in the β-globin genes in the Bantu, Senegal, Benin, and Cameroon regions of Africa, and a fifth instance of the allele appearing in populations from India. Thus the same mutation appears to have arisen randomly and been selected in at least five instances (Labie et al. 1989; Lapoumeroulie et al. 1992).

Favism

There is another mutation, this time in the gene for an enzyme, that seems to render red blood cells inhospitable to *Plasmodium*. Interestingly, this

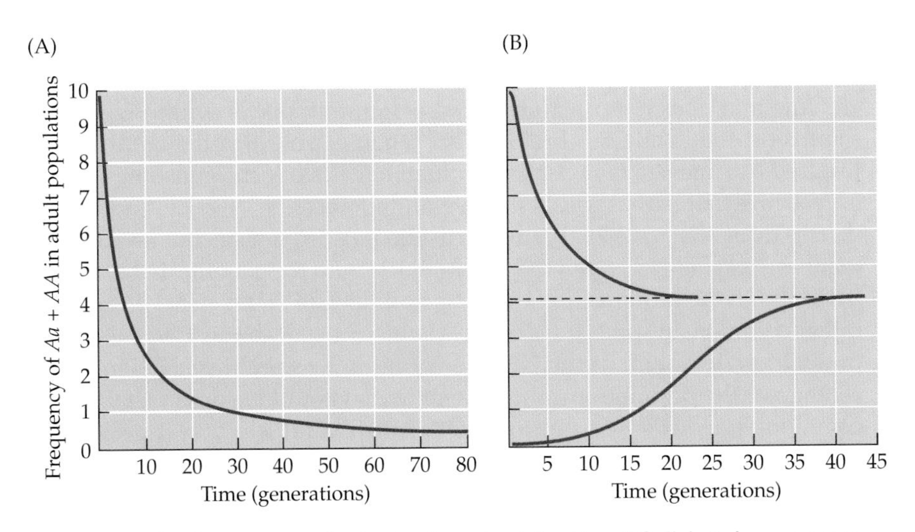

FIGURE 8.12 Importance of selection in maintaining the *HbS* allele in human populations. (A) The change in sickle-cell heterozygote + homozygote frequency that would occur in the absence of malaria, assuming that the fitness of the sickle-cell homozygote is 0.25 and that of sickle-cell heterozygotes and normal homozygotes is 1.0. (B) The change in sickle-cell heterozygote + homozygote frequency in populations that would occur from a high or a low level when the fitness of the normal homozygote, sickle-cell heterozygote, and sickle-cell homozygote are 0.95, 1.19, and 0.30, respectively. The interrupted horizontal line represents an equilibrium frequency of 40% heterozygotes, the highest observed. (From Allison 1954a.)

mutation can also be the cause of a disease, called favism, that manifests in certain environmental circumstances.

The ancient Greek mathematician Pythagoras, best known for his triangle theorem, was also a mystic and cult master who told his disciples that, although they must adhere to a vegetarian diet, they should avoid eating beans. By "beans," Pythagoras almost certainly meant fava beans (*Vicia faba*, sometimes called broad beans), as they were the only cultivated bean in the ancient Mediterranean. What could be the reason behind such a prohibition? There have been many speculations about this question (one of which involved the vegetable's supposed resemblance to the human testicle), but it is just possible that the ancient philosophers were aware that these beans could occasionally be lethal.

Fava beans contain high levels of oxidants such as divicine and isouramil that are capable of destroying the membranes of red blood cells. In most people, these compounds are harmless because the blood cells produce ample amounts of the enzyme G6PD (glucose-6-phosphate dehydrogenase). This enzyme is important in making glutathione, a reducing agent that combats environmental oxidants such as those found in fava beans

(Jollow and McMillan 2001; Ho et al. 2007). The gene for G6PD is carried on the X chromosome. Men with a mutation in this gene synthesize a mutant enzyme that does not function well. Less glutathione is synthesized, and their red blood cells can break apart when exposed to the compounds released by digested fava beans. (Women have two X chromosomes, and it is very rare that both copies will have mutations.)

This disease of red blood cell destruction is called favism, but favism only manifests as a disease when the diet interacts with the genome. Otherwise, a person with the mutant *G6PD* allele is normal. G6PD deficiency is actually rather common; in some areas of the Mediterranean, about 30% of the male population do not make the G6PD enzyme. Why should such a large percentage of the population contain a gene that could become lethal under commonplace dietary circumstances? Again, natural selection appears to be responsible and, again, malaria appears to be the agent of selection.

There is an advantage to being G6PD-deficient if one lives in a region where malaria is endemic. Figure 8.13 shows that the prevalence of the Mediterranean allele of *G6PD* (and hence favism) coincides with the areas where malaria has been endemic in a particular area of Italy. The relative lack of the G6PD enzyme and glutathione apparently make it more difficult for *Plasmodium* to reproduce in the red blood cells (see box). Thus, males with low *G6PD* gene activity appear to have an advantage in that they are less likely to die from a severe case of malaria. Even a lessening of severity of the disease could be an important factor in maintaining this mutant allele in the population (Mason et al. 2007).

Summary

The Modern Synthesis linking genetics and evolutionary biology demonstrated (1) that evolution within the species can be modeled mathematically, (2) that genes under selection were the units whose frequencies changed under Darwinian selection, (3) that those organisms possessing genes that allowed them to become fit would produce more fit offspring, and (4) that these offspring had a high likelihood of inheriting the genes that made their parents and grandparents fit. These genes were the material bases of Darwin's inherited traits. Moreover, the change in allelic frequencies predicted by the mathematics of population genetics was confirmed by DNA sequencing, making the Modern Synthesis one of the most successful and important explanatory theories in science.

However, the Modern Synthesis is mostly a set of theories about adults competing for reproductive success (i.e., who leaves the most offspring relative to others). There remains the notion that there is more to evolution than the frequency of alleles within a species. Natural selection can only

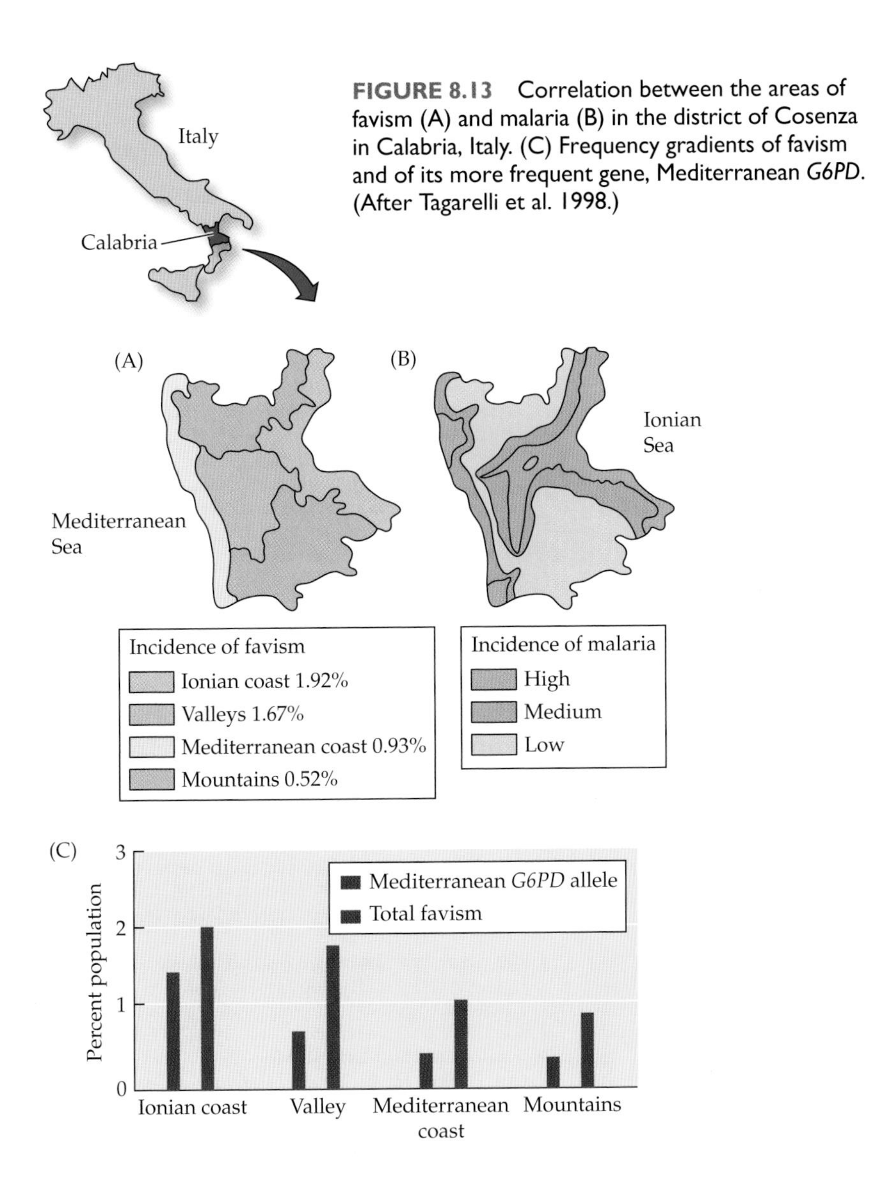

FIGURE 8.13 Correlation between the areas of favism (A) and malaria (B) in the district of Cosenza in Calabria, Italy. (C) Frequency gradients of favism and of its more frequent gene, Mediterranean *G6PD*. (After Tagarelli et al. 1998.)

work on existing variation, and the next two chapters will document that in addition to allelic variation in the structural genes, there are two other sources of variation that can be acted on by selection: allelic variation in the regulatory regions of genes and developmentally plastic variation. These may play critical roles in evolution within the species, in speciation events, and at higher taxonomic levels, producing the variations we associate with phyla and classes.

References

Allison, A. C. 1954a. Protection afforded by sickle-cell trait against subtertian malaria infection. *Br. Med. J.* 1: 290–294.

Allison, A. C. 1954b. Notes on sickle-cell polymorphism. *Ann. Hum. Genet.* 19: 39–57.

Allison, A. C. 2002. The discovery of the resistance to malaria in sickle-cell heterozygotes. *Biochem. Mol. Biol. Educ.* 30: 279–287.

Andersen, K. H., K. D. Farnsworth, U. H. Thyesen and J. E. Beyer. 2007. The evolutionary pressure from fishing on size at maturation of Baltic cod. *Ecol. Model.* 204: 246–252.

Axelrod R. and W. D. Hamilton 1981. The evolution of cooperation. *Science* 211: 1390 – 1396.

Balanyá, J., J. M. Oller, R. B. Huey, G. W. Gilchrist and L. Serra. 2006. Global genetic change tracks global climate warming in *Drosophila subobscura. Science* 313: 1773–1775.

Balfour, F. M. 1880–1881. *A Treatise on Comparative Embryology.* (2 vols.) Macmillan, London.

Bateson, W. 1894. *Materials for the Study of Variation: Treated with Especial Regard to Discontinuity in the Origin of Species.* Macmillan, London. (Repinted 1992, Johns Hopkins University Press, Baltimore.)

Bateson, W. 1900. Problems of heredity as a subject for horticultural investigation. *J. R. Hort. Soc.* 25: 1–8.

Bateson, W. 1922. Evolutionary faith and modern doubts. *Science* 55: 1412.

Berry, A. and J. Browne. 2008. The other beetle-hunter. *Nature* 453: 1188–1190.

Bowler, P. 1996. *Life's Splendid Drama.* University of Chicago Press, Chicago.

Boycott, A. E. and C. Diver. 1923. On the inheritance of sinistrality in *Limnaea peregra. Proc. R. Soc. Lond. B* 95: 207–213.

Boycott, A. E., C. Diver, S. L. Garstang and F. M. Turner. 1930. The inheritance of sinestrality in *Limnaea peregra* (Mollusca: Pulmonata). *Philos. Trans. R. Soc. Lond. B* 219: 51–131.

Bradshaw, W. E. and C. M. Holzapfel. 2001. Genetic shift in photoperiodic response correlated with global warming. *Proc. Natl. Acad. Sci. USA* 98: 14509–14514.

Bromham, L. and D. Penny. 2003. The modern molecular clock. *Nature Rev. Genet.* 4: 216–224.

Cavalli-Sforza, L. L. and W. Bodmer. 1971. *The Genetics of Human Populations.* W.H. Freeman, San Francisco.

Charmantier, A., R. H. McCleery, L. R. Cole, C. Perrins, L. E. B. Kruuk and B. C. Sheldon. 2008. Adaptive phenotypic plasticity in response to climate change in a wild bird population. *Science* 320: 800–803.

Coltman, D. W. P. O'Donoghue, J. T. Jorgensen, J. T. Hogg, C. Strobeck, and M. Festa-Blanchet. 2003. Undesirable evolutionary consequences of trophy hunting. *Nature* 426: 655 – 658.

Darwin, C. 1844. *Life and Letters* II: 28.

Darwin, C. 1859. *The Origin of Species.* John Murray, London.

Darwin, C. 1874. *The Descent of Man, and Selection in Relation to Sex*, 2nd Ed. (2 vols.) John Murray, London.

de Jong, P. W. S. W. S. Gussekloo, and P. M. Brakefield. 1996. Differences in thermal balance and activity between non-melanic and melanic two-spot ladybird beetles (*Adalia bipunctata*) under controlled conditions. *J. Exp. Biol.* 199: 2655 – 2666.

de Jong, P. W. and P. M. Brakefield. 1998. Climate and change in clines for melanism in the two-spot ladybird, *Adalia bipunctata* (Coleoptera: Coccinellidae). *Proc. R. Soc. Lond. B* 265: 39–43.

Diggs, L. W., G. F. Ahmann and S. Bibb. 1933. The incidence and significance of the sickle cell trait. *Ann. Intern. Med.* 7: 769–778.

Dobzhansky, Th. 1937. *Genetics and the Origin of Species.* Columbia University Press, New York.

Dobzhansky, Th. 1951. *Genetics and the Origin of Species*, 3rd Ed. Columbia University Press, New York.

Dobzhansky, Th. 1955. *Evolution, Genetics, and Man.* Wiley, New York.

Dobzhansky, Th. 1964. Biology, molecular and organismic. *Am. Zool.* 4: 443–452.

Edwards, A. W. F. 2001. Darwin and Mendel united: The contributions of Fisher, Haldane and Wright up to 1932. In E. C. R. Reeve (ed.), *Encyclopedia of Genetics.* Fitzroy Dearborn, London.

Fisher, R. A. 1930. *The Genetical Theory of Natural Selection.* Clarendon Press, Oxford.

Garstang, W. 1922. The theory of recapitulation: A critical restatement of the biogenetic law. *Proc. Linn. Soc. Lond.* 35: 81–101.

Gilbert, S. F. 1998. Bearing crosses: The historiography of genetics and embryology. *Am. J. Med. Genet.* 76: 168–182.

Gilbert S. F. 2003. *Developmental Biology*, 7th Ed. Sinauer Associates, Sunderland, MA.

Grant, P. 1986. *The Ecology and Evolution of Darwin's Finches.* Princeton University Press, Princeton.

Grant, B. S. 1999. Fine tuning the peppered moth paradigm. *Evolution* 53: 980–984.

Grant, B. S. and L. L. Wiseman. 2002. Recent history of melanism in American peppered moths. *J. Hered.* 93: 86–90.

Grant, B. S., D. F. Owen and C. A. Clarke 1996. Parallel rise and fall of melanic peppered moths in America and Britain. *J. Hered.* 87: 351–357.

Haeckel, E. 1866. *Generelle Morphologie der Organismen: Allegeneine Grundzüge der organischen Formen-Wissenschaft, mechanisch begründendet durch die von Charles Darwin reformite Descendenz-Theorie.* Vol. 1. G. Reimer, Berlin, pp. 43–60.

Halanych, K. M. 2004. The new animal phylogeny. *Annu. Rev. Ecol. Evol. Syst.* 35: 229–256.

Halanych, K. M., J. D. Bachelier, A. M. Aguinaldo, S. M. Liva, D. M. Hillis and J. A. Lake. 1995. Evidence from 18S ribosomal DNA that the lophophorates are protostome animals. *Science* 267: 1641–1643.
Haldane, J. B. S. 1949 Disease and evolution. *La Ricerca Scientifica*, Suppl. A 19: 68–76.
Haldane, J. B. S. 1953. Foreword. In R. Brown and J. F. Danielli (eds.), *Evolution: Society of Experimental Biology Symposium 7*. Cambridge University Press, Cambridge, pp. ix–xix.
Hamilton, W. D. 1964. The genetical evolution of social behaviour I and II. *J. Theor. Biol.* 7: 1–52.
Haugen, T. O. and L. A. Vøllestad. 2001. A century of life-history evolution in grayling. Genetica 112: 475–491.
Hamilton, W. D. 1967. Extraordinary sex ratios. *Science* 156: 477–488.
Hemingway, J. and S. H. Karunaratne. 1998. Mosquito carboxylesterases: A review of the molecular biology and biochemistry of a major insecticide resistance mechanism. *Med. Vet. Entomol.* 12: 1–12.
Ho, H. Y., M. L. Cheng, P. F. Cheng and D. T. Chu. 2007. Low oxygen tension alleviates oxidative damage and delays cellular senescence in G6PD-deficient cells. *Free Radical Res.* 41: 571–579.
Huxley, T. H. 1893. *Darwiniana: Collected Essays*, Vol. II. Macmillan, London.
Ingram, V. M. 1957. Gene mutations in human haemoglobins: The chemical difference between normal and sickle-cell haemoglobin. *Nature* 180: 326.
Itano, H. A. and J. V. Neel. 1950. A new inherited abnormality of human hemoglobin. *Proc. Nat. Acad. Sci. USA* 36: 613–617.
Jollow, D. J. and D. C. McMillan. 2001. Oxidative stress, glucose-6-phoosphate dehydrogenase, and the red cell. *Adv. Exp. Med. Biol.* 500: 595–605.
Judson, O. 2008. Optimism in evolution. New York Times 12 Aug. 2008. http://www.nytimes.com/2008/08/13/opinion/13judson.html
Kettlewell, H. B. D. 1959. Darwin's missing evidence. *Sci. Am.* 200(3): 48–53.
Kettlewell, H. B. D. 1973. *The Evolution of Melanism*. Clarendon Press, Oxford.
Kimura, M. 1983. *The Neutral Theory of Molecular Evolution*. Cambridge University Press, New York.
Kowalevsky, A. 1866. Entwickelungsgeschichte der einfachen Ascidien. *Mémoires de L'Académie Impériale des Sciences de St.-Pétersbourg*, VII Série. Tome X: 1–22.
Kreitman, M. 1983. Nucleotide polymorphism at the alcohol dehydrogenase locus of *Drosophila melanogaster*. *Nature* 304: 412–417.
Labie, D. and 11 others. 1989. Haplotypes in tribal Indians bearing the sickle gene: Evidence for the unicentric origin of the Hb^S mutation and the unicentric origin of the tribal populations in India. *Hum. Biol.* 61: 479–491.
Lapoumeroulie, C. and 9 others. 1992. A novel sickle-cell mutation of yet another origin in Africa: The Cameroon type. *Hum. Genet.* 89: 333–337.
Lancester, E. R. 1877. Notes on the embryology and classification of the animal kingdom: comprising a revision of speculations relative to the origin and significance of the germ layers. *Quart. Rev. Micr. Sci.* 2/17: 399–454.
Lankester, E. R. 1891. *Zoological Articles Contributed to the Encyclopedia Britannica*. A&C Black, London.
Lew, V. L. and R. M. Bookchin. 2005. Ion transport pathology in the mechanism of sickle cell dehydration. *Physiol. Rev.* 85: 179–200.
Lewontin, R. C. and J. L. Hubby. 1966. A molecular approach to the study of genic heterozygosity in natural populations. II. Amount of variation and degree of heterozygosity in natural populations of *Drosophila pseudoobscura*. *Genetics* 54: 595–609.
Lillie, F. R. 1898. Adaptation in cleavage. In *Biological Lectures from the Marine Biological Laboratory, Woods Hole, Massachusetts*. Ginn, Boston, pp. 43–67.
Lyons, S. 1999. *Thomas Henry Huxley: The Evolution of a Scientist*. Prometheus Books, Amherst, New York.
MacArthur, R. H. 1965. Ecological consequences of natural selection. In T. Waterman and H. Horowitz (eds.), *Theoretical and Mathematical Biology*. Blaisdell, New York, pp. 388–397.
Majerus, M. E. N. 1998. *Melanism: Evolution in Action*. Oxford University Press, Oxford.
Martinez-Torres, D. and 8 others. 1998. Molecular characterization of pyrethroid knockdown resistance (*kdr*) in the major malaria vector *Anopheles gambiae* s.s. *Insect Mol. Biol.* 7: 179–184.
Mason, P. J., J. M. Bautista and F. Gilsanz. 2007. G6PD deficiency: the genotype-phenotype association. *Blood Rev.* 21: 267–283.
Maynard Smith, J. and G. R. Price. 1973. The logic of animal conflict. *Nature* 246: 15–18.
Mayr, E. 1942. *Systematics and the Origin of Species*. Columbia University Press, New York.
Mayr, E. 1959. Where are we? *Cold Spring Harb. Symp. Quant. Biol.* 24: 1–14.
Mayr, E. 2004. *What Makes Biology Unique? Considerations of the Autonomy of a Scientific Discipline*. Cambridge University Press, Cambridge.
Mayr, E. and W. B. Provine (eds.). 1980. *The Evolutionary Synthesis: Perspectives on the Unification of Biology*. Harvard University Press, Cambridge, MA.
McElreath, R. and R. Boyd. 2007. *Mathematical Models of Social Evolution: A Guide for the Perplexed*. University of Chicago Press, Chicago, p. 82.
Miller-Rushing, A. J. and R. B. Primack. 2008. Global warming and flowering times in Thoreau's Concord: A community perspective. *Ecology* 89: 332–341.
Morgan, T. H. 1908. *Evolution and Adaptation*. Macmillan, New York.
Morgan, T. H. 1927. *Experimental Embryology*. Columbia University Press, New York.

Morgan, T. H. 1932a. The rise of genetics. *Science* 76: 261–267.
Morgan, T. H. 1932b. *The Scientific Basis of Evolution*. W.W. Norton, New York.
Müller F. 1869. *For Darwin*. (W. S. Dallas, transl.) John Murray, London.
Nei, M. 2005. Selectionism and neutralism in molecular evolution. *Mol. Biol. Evol.* 22: 2318–2342.
Nussey, D. H., E. Postma, P. Gienapp and M. E. Visser. 2005. Selection on hereditable phenotypic plasticity in a wild bird population. *Science* 310: 304–306.
Oppenheimer, J. M. 1940. The nonspecificity of the germ-layers. *Q. Rev. Biol.* 15: 1–27.
Oppenheimer, J. M. 1959. An embryological enigma in *The Origin of Species*. In B. Glass, O. Temkin and W. L. Straus, Jr. (eds.), *Forerunners of Darwin 1745–1859*. Johns Hopkins University Press, Baltimore, pp. 292–322.
Ospovat, D. 1981. *The Development of Darwin's Theory: Natural History, Natural Theology, and Natural Selection, 1838–1859*. Cambridge University Press, Cambridge.
Parmesan, C. and G. Yohe. 2003. A globally coherent fingerprint of climate change impacts across natural systems. *Nature* 421: 37–42.
Pasvol, G. and R. J. Wilson. 1980. The interaction between sickle-cell haemoglobin and the malarial parasite *Plasmodium falciparum*. *Trans. R. Soc. Trop. Med. Hyg.* 74: 701–705.
Pauling, L., H. Itano, S. J. Singer and I. C. Wells. 1949. Sickle-cell disease: A molecular disease. *Science* 110: 543.
Peñuelas, J. and I. Filella. 2001. Responses to a warming world. *Science* 294: 793–795.
Provine, W. B. 1986. *Sewall Wright and Evolutionary Biology*. University of Chicago Press, Chicago.
Raff, R. A. and T. C. Kaufman. 1983. *Embryos, Genes, and Evolution: The Developmental-Genetic Basis of Evolutionary Change*. Macmillan, New York.
Roux, W. 1894. The problems, methods, and scope of developmental mechanics. In *Biological Lectures of the Marine Biology Laboratory, Woods Hole, Massachusetts*. Ginn, Boston, pp. 149–190.
Sallares, R. 2002. *Malaria and Rome: A History of Malaria in Ancient Italy*. Oxford University Press, Oxford.
Shaw, R. F. and J. D. Mohler. 1953. The selective significance of the sex ratio. *Am. Natur.* 87: 337–342.
Smocovitis, V. B. 1996. *Unifying Biology: The Evolutionary Synthesis and Evolutionary Biology*. Princeton University Press, Princeton.
Stern, N. 2006. *Stern Review on the Economics of Climate Change*. Cambridge University Press, Cambridge.
Sturtevant, A. 1923. Inheritance of direction of coiling in *Limnaea*. *Science* 58: 269–270.
Tagarelli, A., L. Bastone, R. Cittadella, V. Calabro, M. Bria and C. Brancati. 1991. Glucose-6-Phosphate dehydrogenase (G6PD) deficiency in Southern Italy: A study on the population of the Cosenza province. *Gene Geog.* 5: 141–150.
Tran, C. V. and M. H. Saier, Jr. 2004. The principle chloroquine resistance protein of *Plasmodium falciparum* is a member of a drug/metabolite transporter superfamily. *Microbiology* 150: 1–3.
Trivers, R. L. 1972. Parental investment and sexual selection. In B. Campbell (ed.), *Sexual Selection and the Descent of Man, 1871–1971*. Aldine, Chicago, pp. 136–179.
Umina, P. A., A. R. Weeks, M. R. Keraney, S. W. McKechnie and A. A. Hoffmann. 2005. A rapid shift in a classic clinal pattern in *Drosophila* reflecting climate change. *Science* 308: 691–693.
Vandepitte, J. M., W. W. Zuelzer, J. V. Neel and J. Colaert. 1955. Evidence on the inadequacy of mutation as an explanation of the frequency of the sickle-cell gene in the Belgian Congo. *Blood* 10: 341.
Welsh, J. H. 1969. Mussels on the move. *Nat. Hist.* 78: 56–59.
Williams, N. 2008. Australia fears over climate change gloom. *Curr. Biol.* 18: R633–R634.
Wilson, E. B. 1898. Cell lineage and ancestral reminiscence. In *Biological Lectures from the Marine Biological Laboratories, Woods Hole, Massachusetts*. Ginn, Boston, pp. 21–42.
Wright, S. 1922. Coefficients of inbreeding and relationship. *Am. Natur.* 56: 330–338.
Yang, Z. and 8 others. 2007. Molecular analysis of chloroquine resistance in *Plasmodium falciparum* in Yunnan Province, China. *Trop. Med. Internatl. Health* 12: 1051–1060.

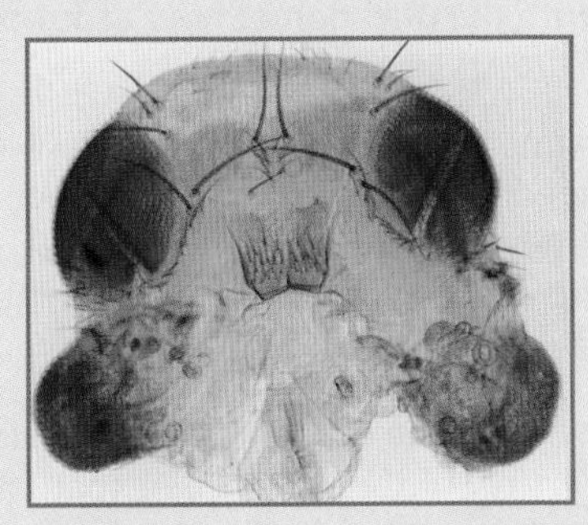

Chapter 9

Evolution through Developmental Regulatory Genes

Nature interests me because it's beautiful, complex, and robust. Evolutionary theory interests me because it explains why nature is beautiful, complex and robust.

David Quammen, 2007

[A] study of the effects of genes during development is as essential for an understanding of evolution as are the study of mutation and that of selection.

Julian Huxley, 1942

The Modern Synthesis explained the genetic mechanisms of natural selection, but it did not provide an adequate explanation for the anatomical variations for which natural selection selects. To the developmental biologist, the Modern Synthesis remains incomplete without a theory of how the construction of bodies can change (Amundson 2005). Indeed, without the integration of developmental genetics and developmental plasticity into it, evolutionary biology has no complete theory of variation. In 1964, Theodosius Dobzhansky proclaimed that "Nothing in biology makes sense except in the light of evolution." In 2006, evolutionary developmental biologist Jukka Jernvall paraphrased that famous statement, saying that "Nothing about variation makes sense except in the light of development."

For instance, when confronted with the question of how arthropod body plans arose, Hughes and Kaufman (2002) stated:

> To answer this question by invoking natural selection is correct—but insufficient. The fangs of a centipede ... and the claws of a lobster accord these organisms a fitness advantage. However, the crux of the mystery is this: From what developmental genetic changes did these novelties arise in the first place?

In other words, while the Modern Synthesis could explain the *survival* of the fittest, it could not explain the *arrival* of the fittest.* For that, one needed a theory of body construction and its possible changes, a theory of developmental change.

This chapter will discuss the developmental genetic sources of variation and how they function in evolution; the subsequent chapter will focus on the evolutionary importance of the variation derived from developmental plasticity. Thus, there are (at least) three types of variation important in evolution: allelic variation of structural genes, allelic variation of regulatory loci, and plasticity-derived epigenetic variation. Without development, there can be no complete theory of variation; and without an adequate theory of variation, there can be no adequate explanation of evolution that embraces changing genes and changing bodies.

The Origins of Evolutionary Developmental Biology

How does one bridge a theory of evolving bodies with a theory of changing genes? Perhaps the first stage is the realization that one needs to do so. This realization was proclaimed in a well-publicized 1975 research paper by Mary-Claire King and Alan Wilson. Entitled "Evolution at Two Levels in Humans and Chimpanzees," this study showed that despite the obvious anatomical differences between chimps and humans, their DNA was almost identical. Indeed, some scientists would opine that the paucity of differences places us as the third species of chimpanzee (Goodman 1999; Diamond 2002). In other words, the theory of evolving bodies and the the-

*The term "arrival of the fittest" is very apt, since it was one of the weakest points of the Modern Synthesis. Darwin alludes to this problem in the *Origin of Species*, noting the "characters may have originated from quite secondary sources, independently from natural selection." He discusses this problem at length in his book on variation, noting that natural selection must have something upon which to select and that the origin of variation appears to be internal to the organism and not a consequence of environmental factors. The term "arrival of the fittest" seems to have been coined several times, but one of its earliest uses was by Arthur Harris, who in 1904 wrote that "natural selection may explain the survival of the fittest, but it cannot explain the arrival of the fittest."

ory of genetic change were not lining up. The proposed solution was elegant and harkened back to those developmental biologists such as Conrad Waddington (see Appendix C), who felt that evolutionary change was predicated upon specific changes in gene regulation. Specifically, the King and Wilson paper states:

> The organismal differences between chimpanzees and humans would ... result chiefly from genetic changes in a few regulatory systems, while amino acid substitutions in general would rarely be a key factor in major adaptive shifts.

That is to say, the allelic substitutions of the genes that encode protein sequences—which seem to be pretty much the same for chimps and humans—are not what is important. The important differences are where, when, and how much the genes are activated.

In 1977, four publications further paved the way for evolutionary developmental biology. These publications were Stephen J. Gould's *Ontogeny and Phylogeny*, François Jacob's "Evolution and Tinkering," and two papers concerning the techniques for DNA sequencing. In *Ontogeny and Phylogeny*, Gould demonstrated how the German biologist Ernst Haeckel had misrepresented the field of evolutionary embryology and made it into an unscientific doctrine. Indeed, the first half of Gould's book exorcised Haeckel's ghost so that some other model of evolution and development could be substituted for Haeckel's famous (some would say infamous) biogenetic law wherein "ontogeny recapitulates phylogeny." Jacob's paper suggested a new model, generalizing the work of King and Wilson and focusing attention on how the same genes can create new types of body plans when their expression pattern is altered during development. This could become a testable theory if the genetic sequences could be read, and papers by Maxam and Gilbert (1977) and by the Sanger laboratory (Sanger et al. 1977) made this possible. Over the next 30 years, it became possible to study gene expression and gene sequence together. Recombinant DNA technology, polymerase chain reaction, in situ hybridization, and high-throughput RNA analysis enable us to look at gene sequences and to compare their expression between species. This ability has revolutionized the way we look at evolutionary processes. The new science of **evolutionary developmental biology** ("**evo-devo**") is looking at the mechanisms by which changes in gene regulation cause changes in anatomy. It looks at the mechanisms of tinkering.

Evolutionary developmental biology attempts to explain biodiversity in terms of developmental trajectories. It seeks to discern how changes in development create evolutionary novelties, how development might constrain certain phenotypes from arising, and how developmental mechanisms themselves evolve (Gilbert and Burian 2003; Müller 2007). As we will see in the next chapter, it also looks at the roles that environments can play in regulating developmental processes and how organisms may have evolved to incorporate environmental signaling.*

It used to be thought that changes in development would be detrimental to the organism because they would throw a finely honed harmonious system out of synchrony. Evolutionary developmental biologists have shown that there are three conditions that circumvent this problem and allow evolution to occur through changes in development: The first two, molecular parsimony and modularity, will be discussed here. The third condition, robustness and the adaptability of developmental processes, was discussed in Chapter 4 and will be revisited in the next chapter. Together, these processes allow the generation of diversity through changes in gene regulation.

Molecular Parsimony: "Toolkit Genes"

One of the surprising concepts that came out of evolutionary developmental biology is the notion of **molecular parsimony**, sometimes called the **small toolkit**. In other words, although development differs enormously from lineage to lineage, development within all lineages uses the same types of molecules. The transcription factors, paracrine factors, adhesion molecules, and signal transduction cascades are remarkably similar from one phylum to another. Whereas early evolutionary morphologists discussed homologies between bones and tissues, molecular biologists can now discuss an even deeper homology—the homologies of genes. The Hox genes and the genes for certain transcription factors such as Pax6 are found in all animal phyla, from sponges and cnidarians to insects and primates. In fact, there are "toolkit genes" that appear to play the same roles in all animal lineages.

Evolutionary developmental biologists such as Walter Gehring (1998, 2005) demonstrated that there is a developmental genetic pathway specifying photoreceptor development and that this is common to all animals with eyes. The basic genetic pattern for eye development is specified by homologues of the *Pax6* gene. This pathway probably evolved only once, in early metazoans, and every eye on the planet comes from a modification of this central scheme. Indeed, experimenters have taken the mouse *Pax6* gene and expressed it in flies, specifically in those cells that should give rise to the fly's jaw; the jaw then develops into a fly eye. If the fly *Pax6* gene is expressed in frog skin, the resultant fly Pax6 protein initiates eye develop-

*There are many variants of evolutionary developmental biology. The one that we are focusing on in this chapter is one where developmental genetics predominates. This is being emphasized because it links the genetics of the Modern Synthesis on the one hand with physiological ecology on the other. For other concerns and foci of evolutionary developmental biology, see Raff 1996; Hall 1999; Kirschner and Gerhart 2005; Carroll et al. 2001; Carroll 2005; Arthur 2004; and Pigliucci 2001.

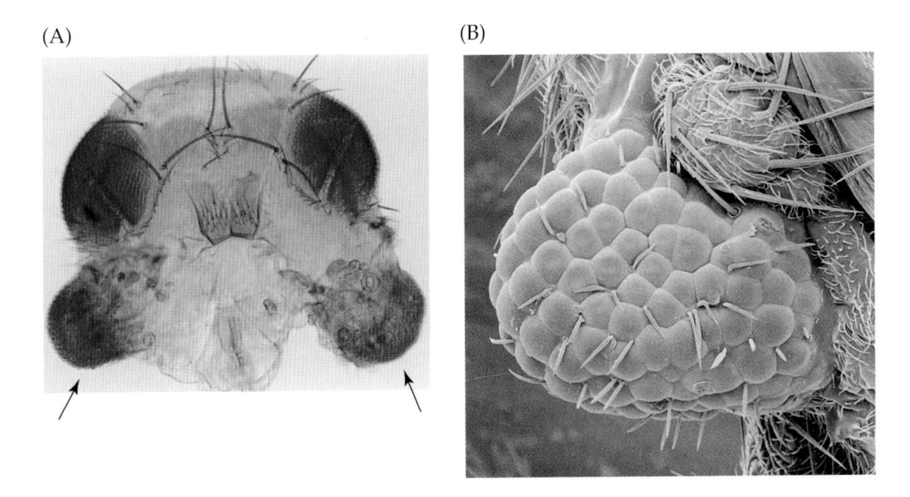

FIGURE 9.1 Evolutionary similarity of *Pax6* expression in vertebrates and insects. (A) Expression of a fly's *Pax6* gene in a non-eye (jaw) region of the fly produces ectopic fly eyes (arrows). The eye-specifying expression of the *Pax6* gene can interact with other components of the developing fly to induce the compound eye in new locations. (B) Expression of mouse *Pax6* in the leg region of a developing fly also causes an eye—an insect eye—to form there. Despite the enormous morphological differences in insect and vertebrate eyes, the mouse *Pax6* gene specifies insect eye development when placed in a fly embryo. (Photographs courtesy of W. Gehring and G. Halder.)

ment, causing the cells that should be the frog's epidermis to become an amphibian retina (Figure 9.1; Halder et al. 1995; Chow et al. 1999).

Even the known differences between major groups seem to be diminishing. Although vertebrates don't express juvenile hormone and insects don't express estrogens, the receptors for these hormones are found in both insects and vertebrates, and in some cases they will bind hormones from the other clade (Harmon et al. 1995; King-Jones and Thummel 2005) as well as hormones in their own clade. The fly eye and vertebrate eye are considered to be analogous, not homologous, structures; the vertebrate eye is not compounded of multitudinous ommatidia, nor do vertebrate eyes derive from imaginal discs activated by ecdysterone. Yet the *Pax6* gene *is* homologous in flies and vertebrates. In both groups, this same gene is essential for initiating eye formation, and in both groups it acts through the same pathway.

Duplication and divergence: The Hox genes

Studying the Hox gene complex is instructive, because Hox gene expression provides the basis for anterior-posterior axis specification throughout the animal kingdom. This means that the enormous variation of morphological forms among animals is underlain by a single common set of instructions.

The similarity of all the Hox genes is best explained by descent from a common ancestor (Figure 9.2). First, the different Hox genes in the homeotic gene cluster appear to have originated from duplications of an ancestral gene. This would mean that in fruit flies, the *Deformed*, *Ultrabithorax*, and *Antennapedia* genes all emerged as duplications of an original gene in some ancestor. The sequence patterns of these three genes (especially in the homeodomain region) are extremely well conserved. Such tandem gene duplications are thought to be the result of errors in DNA replication, and they are very common. Once duplicated, the gene copies can diverge by

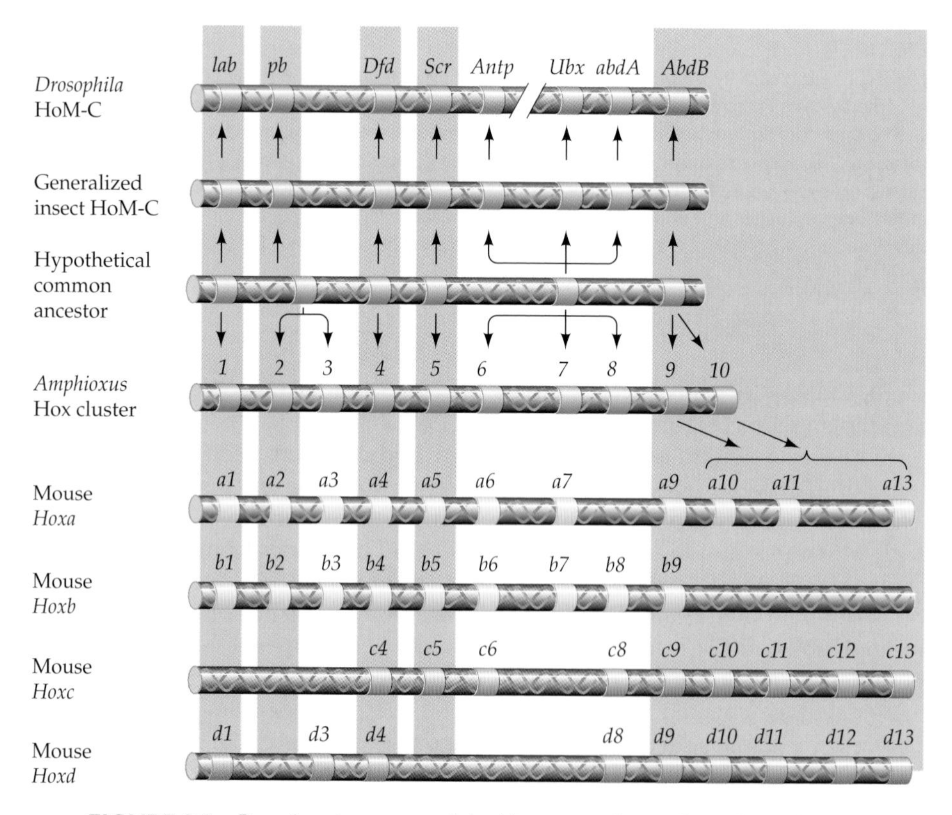

FIGURE 9.2 Postulated ancestry of the Hox genes from a hypothetical ancestor of both protostomes and deuterostomes. The deuterostome, *Amphioxus*, has only one set of Hox genes, just like the protostomes. Vertebrates, derived from an *Amphioxus*-like organism, have four Hox clusters, none of which is complete. (After Holland and Garcia-Fernández 1996.)

random mutations in their coding sequences and regulatory regions (such as enhancers), developing different expression patterns and new functions (Lynch and Conery 2000; Damen 2002; Locascio et al. 2002). Susumu Ohno (1970), one of the founders of the gene family concept, likened gene duplication to a method used by a sneaky criminal to circumvent surveillance. While the "police force" of natural selection makes certain that there is a "good" gene performing its function properly, that gene's duplicate—unencumbered by the constraints of selection—can mutate and potentially serve new functions.

This **duplication and divergence** scenario is seen in the Hox genes, the globin genes, the collagen genes, the *Distal-less* genes, and in many paracrine factor families (e.g., the Wnt genes). Each member of such a gene family is homologous to the others (that is, their sequence similarities are due to descent from a common ancestor and are not the result of convergence for a particular function), and they are called **paralogues**. Thus, the *Antennapedia* gene of *Drosophila* is a paralogue of the *Ultrabithorax* gene of *Drosophila*.

Moreover, each Hox gene in *Drosophila* has a homologue in vertebrates. In some cases, the homologies go very deep and can also be seen in the gene's functions. Not only is the vertebrate *Hoxb4* gene similar in sequence to its *Drosophila* homologue, *Deformed* (*Dfd*), but the human *HOXB4* gene can perform the functions of *Deformed* when introduced into *Dfd*-deficient *Drosophila* embryos (Malicki et al. 1992). Not only are the Hox genes of these different phyla homologous, but they are in the same order on their respective chromosomes. Their expression patterns are also remarkably similar: the genes at the 3′ end are expressed anteriorly, while those at the 5′ end are expressed posteriorly (Figure 9.3). Thus they are homologous genes between species (as opposed to members of a gene family being homologous within a species). Genes that are homologous between species are called **orthologues**.

All multicellular organisms—animals, plants, and fungi—have Hox-like genes,* so it is likely that an ancestral Hox gene existed that encoded a basic helix-loop-helix transcription factor in protozoans. In the earliest animal groups, this gene became duplicated. One of the two Hox genes present in some living cnidarians (such as jellyfish) corresponds to the anterior set of vertebrate Hox genes (and is expressed in the anterior portion of the cnidar-

*The Hox genes appear to have originated after the origin of multicellularity. Placozoans (very primitive metazoans) have Hox genes, and sponges have genes similar to Hox genes. However, Hox genes have not been found in the few protists whose genomes have been sequenced (Larroux et al. 2007; King et al. 2008.) Interestingly, the choanoflagellate genome contains dozens of genes whose protein products are similar to those used in vertebrate embryos for cell adhesion and intercellular signaling. And, although no Hox genes are present in these protists, homologues of the genes encoding the p53, Myc, and Sox transcription factors have been found.

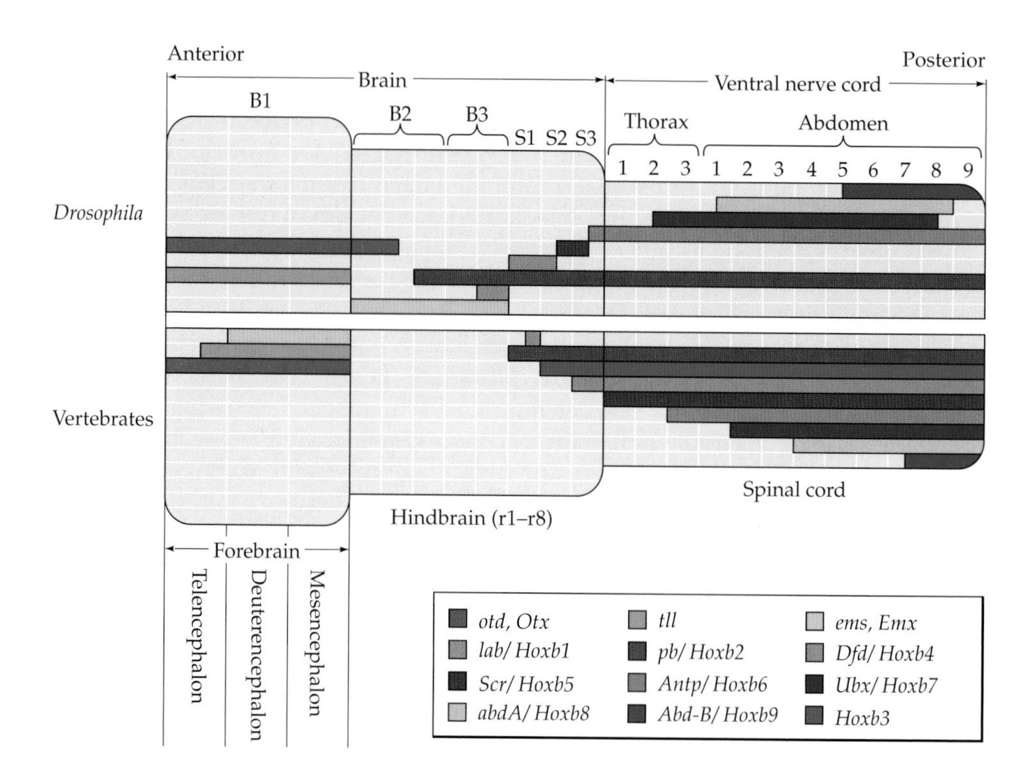

FIGURE 9.3 Expression of regulatory transcription factors in *Drosophila* and vertebrate embryos along the anterior-posterior axis. The Hox genes specify the anterior-posterior axis and are expressed in similar patterns from the hindbrain posteriorly through the spinal cord. The head regions are defined by homologues of the same three genes in both vertebrates and flies. (After Hirth and Reichert 1999.)

ian larva), while sequences in the other cnidarian gene correspond to a posterior-class Hox gene (Yanze et al. 2001; Hill et al. 2003; Finnerty et al. 2004). Perhaps even in the ancient cnidarians, Hox genes distinguished their anterior and posterior tissues. In bilateral phyla, the central Hox genes emerged as a duplication from one of the earlier genes (de Rosa et al. 1999).

In chordates, however, two large-scale duplications of the entire Hox cluster took place, such that vertebrates have four Hox clusters per haploid genome instead of just one. Thus, instead of having a single *Hox4* gene (orthologous to *Deformed* in *Drosophila*), vertebrates have *Hoxa4, Hoxb4, Hoxc4*, and *Hoxd4*. This constitutes the *Hox4* **paralogue group** in vertebrates. Such large-scale duplications have had several consequences. First, these duplications create much redundancy. It is difficult to obtain a loss-of-function mutant phenotype, since to do so means all copies of these paralogue group genes must be deleted or made nonfunctional (Wellik and

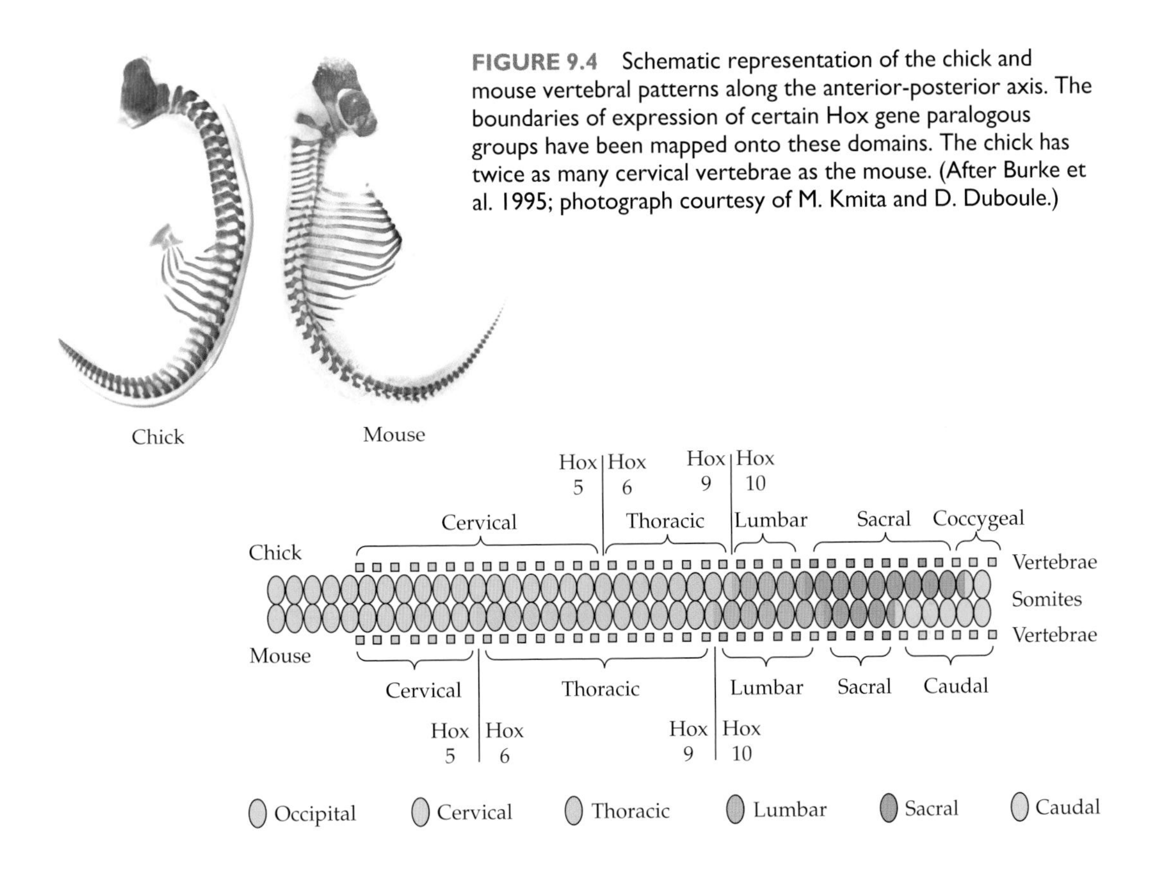

FIGURE 9.4 Schematic representation of the chick and mouse vertebral patterns along the anterior-posterior axis. The boundaries of expression of certain Hox gene paralogous groups have been mapped onto these domains. The chick has twice as many cervical vertebrae as the mouse. (After Burke et al. 1995; photograph courtesy of M. Kmita and D. Duboule.)

Capecchi 2003). However, in some instances, the genes *have* become specialized. *Hoxd11*, for instance, plays an important role in the mammalian limb bud, but not in the reproductive system. Mammalian *Hoxa-11*, on the other hand, plays roles in both the limb (where it is critical in specifying the zeugopod) and in the female reproductive tract (where it helps construct the uterus; Wong et al. 2004).

The evolutionary importance of Hox genes in the specification of the body axis can be seen in their evolutionary modification in vertebrates. Gaunt (1994) and Burke and colleagues (1995) have compared the vertebrae of the mouse and the chick. Although the mouse and the chick have a similar number of vertebrae, they apportion them differently. Mice (like all mammals, be they giraffes or whales) have only 7 cervical vertebrae. These are followed by 13 thoracic vertebrae, 6 lumbar vertebrae, 4 sacral vertebrae, and a variable (20+) number of caudal vertebrae (Figure 9.4). The chick, on the other hand, has 14 cervical vertebrae, 7 thoracic vertebrae, 12 or 13 (depending on the strain) lumbosacral vertebrae, and 5 coccygeal (fused tail) vertebrae. The researchers asked, "Does the constellation of Hox gene expression correlate with the type of vertebra formed (e.g., cervical or thoracic) or with the relative position of the vertebrae (e.g., number 8 or 9)?"

The answer is that the constellation of Hox gene expression predicts the type of vertebra formed. In the mouse, the transition between cervical and thoracic vertebrae is between vertebrae 7 and 8; in the chick, it is between vertebrae 14 and 15. In both cases, the *Hox5* paralogues are expressed in the final cervical vertebra, while the anterior boundary of the *Hox6* paralogues extends to the first thoracic vertebra. Similarly, in both animals, the thoracic-lumbar transition is seen at the boundary between the *Hox9* and *Hox10* paralogous groups. It appears there is a code of differing Hox gene expression along the anterior-posterior axis, and that code determines the type of vertebra formed rather than the number of segments.

One of the most radical modifications of the Hox gene patterning is seen in snakes. The forelimb usually forms just anterior to the most anterior expression domain of *Hoxc6*. Posterior to that point, *Hoxc6*, in collaboration with *Hoxc8*, helps specify the vertebrae to become thoracic (ribbed) vertebrae. During early snake development, *Hoxc6* is not expressed in the absence of *Hoxc8*, so the forelimbs do not form.* Rather, the combination of *Hoxc6* and *Hoxc8* expression is found for most of the length of the snake embryo, specifying nearly all of the vertebrae to be thoracic (Figure 9.5; Cohn and Tickle 1999). Thus, to understand the evolutionary mechanisms by which snakes lost their legs, one has to know the mechanisms of vertebrate axial development.

Homologous pathways of development

One of the most exciting findings of the past decade has been the discovery not only of homologous regulatory genes, but also of homologous developmental pathways. In different organisms, these pathways are composed of homologous proteins arranged in a homologous manner (Zuckerkandl 1994; Gilbert et al. 1996). In some instances, homologous pathways made of homologous parts are used for the same function in both protostomes and deuterostomes. Conserved similarities in both the pathway and its function over millions of years of phylogenetic divergence are considered to be evidence of "deep homology" between these structures. One example of deep homology is the chordin/BMP4 pathway for construction of the central nervous system and the patterning of the dorsoventral (back-belly) axis. In the mid 1990s, several investigators found that the same chordin/BMP4 interaction that specifies the formation of the neural tube in vertebrates also specified neural tissue in fruit flies (Holley et al. 1995; Schmidt et al. 1995; De Robertis and Sasai 1996; Hemmati-Brivanlou and

*The hindlimbs of snakes are inhibited in a different manner, probably by the loss of *sonic hedgehog* expression (Chiang et al. 1996.) Paleontological evidence also suggests that the hindlimbs and forelimbs were lost at different times, the loss of the forelimbs occuring earlier (Caldwell and Lee 1997).

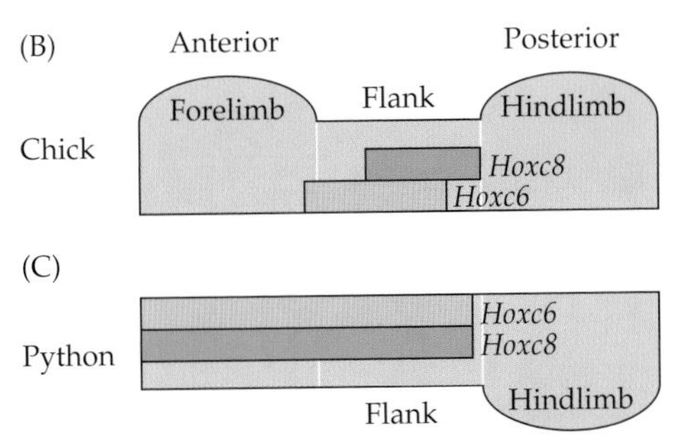

FIGURE 9.5 Loss of limbs in snakes. (A) Skeleton of the garter snake, *Thamnophis*, stained with alcian blue. Ribbed vertebrae are seen from the head to the tail. (B,C) Hox expression patterns in chick (B) and python (C). (Photograph courtesy of A. C. Burke; B, C after Cohn and Tickle 1999.)

Melton 1997). The same set of genes appears to instruct the formation of the vertebrate dorsal neural tube and the *Drosophila* ventral neural cord. In both vertebrates and invertebrates, chordin (called short gastrulation, or Sog, protein in insects) inhibits the lateralizing effects of Bmp4 (called Decapentaplegic, or Dpp, in insects). BMP4/Dpp will bind to ectodermal cells and instruct them to become epidermis unless chordin/Sog blocks the binding, thus permitting the ectoderm to become neural tissue. This protection occurs dorsally in vertebrates and ventrally in insects. These reactions are so similar that *Drosophila* Dpp protein can induce ventral fates in *Xenopus* and can substitute for Sog (Figure 9.6; Holley et al. 1995). Indeed, the Sog and chordin proteins are themselves regulated by homologous proteins (the Tolloid [fly] and Xolloid [frog] proteins) in arthropods and vertebrates. Thus the arthropod and vertebrate nervous systems, despite their obvious differences, seem to be formed by the same set of instructions. The plan for specifying the animal nervous system probably evolved only once (Mizutani and Bier 2008).*

*In 1822, the French anatomist Étienne Geoffroy Saint-Hilaire provoked one of the most heated and critical confrontations in biology when he proposed that the lobster was nothing more or less than a vertebrate upside down. He claimed that the ventral (belly) side of the lobster (with its nerve cord) was homologous to the dorsal (spinal) side of the vertebrate (Appel 1987). It seems that he was correct on the molecular level, if not on the anatomical. Indeed, despite the differences between arthropod and vertebrate neurons, the genes specifying neural cell type are also remarkably similar and play very similar roles in insects and mammals (see Bertrand et al. 2002).

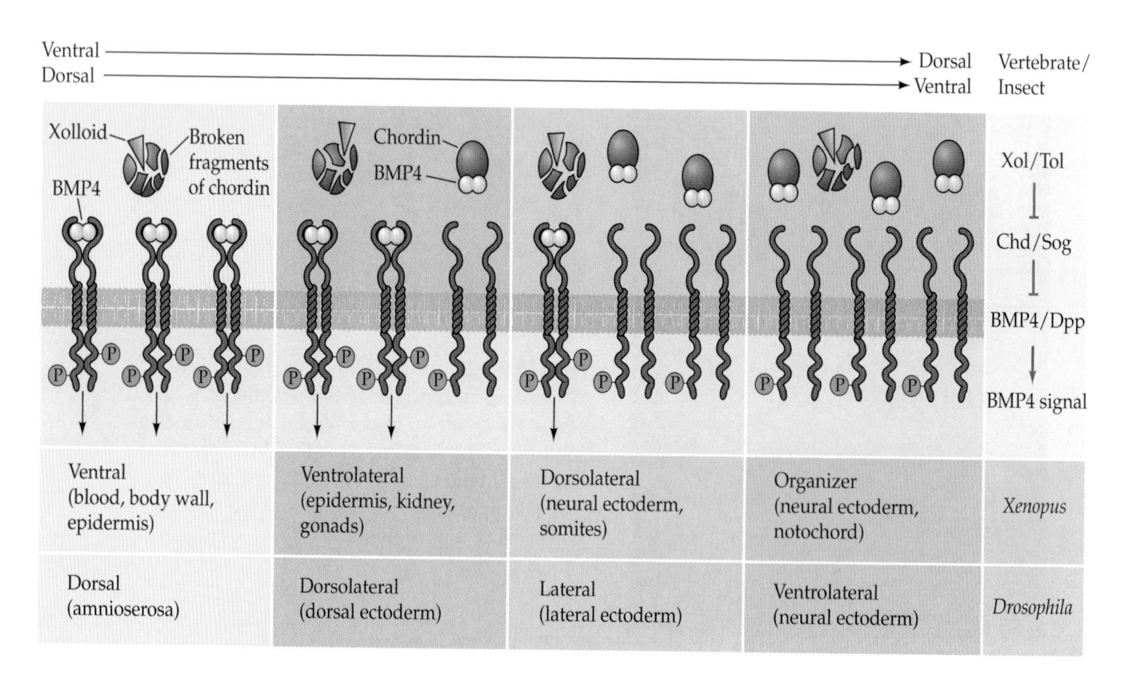

FIGURE 9.6 Homologous pathways specifying the central nervous system and the dorsal-ventral axes of flies and vertebrates. Both pathways involve a source of chordin/Sog protein (the most dorsal region of the vertebrate embryo; the most ventral region of the fly embryo) and a source of BMP4/Dpp (the ventral region of the vertebrate; the dorsal region of the fly). These two regions form antagonistic gradients. Those regions with the most chordin/Sog become central nervous tissue—on the dorsal side of the vertebrate body and on the ventral side of the fly body). The side with the BMP4/Dpp becomes epidermis. In both instances, the gradiant is shaped by a constant supply of Xolloid/Tolloid protein, which degrades chordin/Sog. (After Dale and Wardle 1999.)

Toolkit genes and evolution: A summary

To sum up, three important principles become manifest in the small toolkit for development. First, certain homologous (orthologous) regulatory genes are highly conserved, critical for the development of all animals (Table 9.1). Thus *Pax6* genes are important for eye development throughout the animal kingdom, *tinman* genes are responsible for heart development in flies, worms, and mammals, and Hox genes are critical for specifying the axes in all animals studied so far.

Second, gene duplication and divergence is an extremely important mechanism for evolution. Duplication allows the formation of homologous (paralogous) genes, and divergence allows these genes to assume new roles. Numerous transcription factors and paracrine factors are members of such paralogue families. Hox genes are used to pattern the body and limb

TABLE 9.1 Developmental regulatory genes conserved between protostomes and deuterostomes

Gene	Function	Distribution
achaete-scute group	Cell fate specification	Cnidarians, *Drosophila*, vertebrates
Bcl2/Drob-1/ced9	Programmed cell death	*Drosophila*, nematodes, vertebrates
Caudal	Posterior differentiation	*Drosophila*, vertebrates
delta/Xdelta-1	Primary neurogenesis	*Drosophila*, *Xenopus*
Distal-less/DLX	Appendage formation (proximal-distal axis)	Numerous phyla of protostomes and deuterostomes
Dorsal/NFκB	Immune response	*Drosophila*, vertebrates
forkhead/Fox	Terminal differentiation	*Drosophila*, vertebrates
Fringe/radical fringe	Formation of limb margin (apical ectodermal ridge in vertebrates)	*Drosophila*, chick
Hac-1/Apaf/ced 4	Programmed cell death	*Drosophila*, nematodes, vertebrates
Hox complex	Anterior-posterior patterning	Widespread among metazoans
lin-12/Notch	Cell fate specification	*C. elegans*, *Drosophila*, vertebrates
Otx-1, Otx-2/Otd, Emx-1, Emx-2/ems	Anterior patterning, cephalization	*Drosophila*, vertebrates
Pax6/eyeless; Eyes absent/eya	Anterior CNS/eye regulation	*Drosophila*, vertebrates
Polycomb group	Hox expression/cell differentiation control	*Drosophila*, vertebrates
Netrins, Split proteins, and their receptors	Axon guidance	*Drosophila*, vertebrates
RAS	Signal transduction	*Drosophila*, vertebrates
sine occulus/Six3	Anterior CNS/eye pattern formation	*Drosophila*, vertebrates
sog/chordin, dpp/BMP4	Dorsal-ventral patterning, neurogenesis	*Drosophila*, *Xenopus*
tinman/Nkx 2-5	Heart/blood vascular system	*Drosophila*, mouse
vnd, msh	Neural tube patterning	*Drosophila*, vertebrates

Source: After Erwin 1999.

axes, *Distal-less* genes are used to extend appendages and to pattern the vertebrate skull, and members of the *MyoD* family specify different stages of muscle development. These genes, each family derived from a single ancestral gene, are active in different tissues and provide instructions for the formation of different cell types. So perhaps "the most important difference between the genome of a fruit fly and that of a human is therefore not that the human has new genes but that where the fly only has one gene, our species has multigene families" (Morange 2001, p. 33).

Third, structural or regulatory modification of these genes can allow them to acquire different functions and to play important roles in evolution. Descent with modification is critical for developmental regulatory genes as well as for the organisms they help construct.

Modularity: Divergence through Dissociation

How can the development of an embryo change when development is so finely tuned and complex? How can such change occur without destroying the entire organism? Such changes can occur because development occurs through a series of discrete and interacting **modules** (Riedl 1978; Bolker 2000; Schlosser and Wagner 2004; Kirschner and Gerhart 2005). These modules can be tissues, fields (such as the limb field or the heart field, composed of cells committed to forming those structures), and even elements of gene enhancers. Such modules are a major factor enhancing the ability of organisms to evolve (see Pigliucci 2008).

Modularity has extremely important consequences in terms of how we look at natural selection. Darwin (following von Baer, Barry, and other embryologists) thought that the specific features distinguishing closely related species would emerge only at the end of development (see Ospovat 1981). However, modularity allows the selection of variants that involve even very early development. Michael Richardson (1999) noted that the "target of natural selection" can mean two things. On one level, it can mean the adult phenotype. Thus, the target of natural selection could be the limbs of adult bats, such that those bats who can fly and capture food most efficiently survive. On another level, the target of natural selection may be the genotype—for example, the genes involved in inducing the bat forelimb bud to form. We can appreciate this by comparing the limb development of the bat to that of a kiwi bird. Through natural selection, the forelimb of the bat has been enlarged in the adult. Similarly, the forelimb of the flightless kiwi has been reduced through selective pressures on the adult.

The means by which naturally selected phenotypes have evolved has been through changes in the embryo early in its development. The bat embryo doesn't make a long limb by sustaining limb development for a longer period of time after the bat has been born. Rather, these forelimb buds have started to develop earlier than normal, and by the time the bat is born, the forelimbs are much larger than the hindlimbs (Figure 9.7A). On the other hand (no pun intended), at the same early stage of development, the kiwi has a small forelimb bud and a large hindlimb bud (Figure 9.7B), even though most avian and mammalian embryos at this stage have a forelimb bud that is much larger than the hindlimb bud. In kiwis, the forelimb bud starts to develop *later* than the hindlimb. Thus, although the target of natural selection is an *adult* morphology, arriving at that adaptive phenotype involves changing the parameters of *early* development. What allows these early developmental changes is that forelimb development can change independently from that of the rest of the body. The entire body does not have to become small to produce a small limb bud; the limb bud behaves as a developmental module. This mosaic semiautonomy has been called **developmental modularity**.

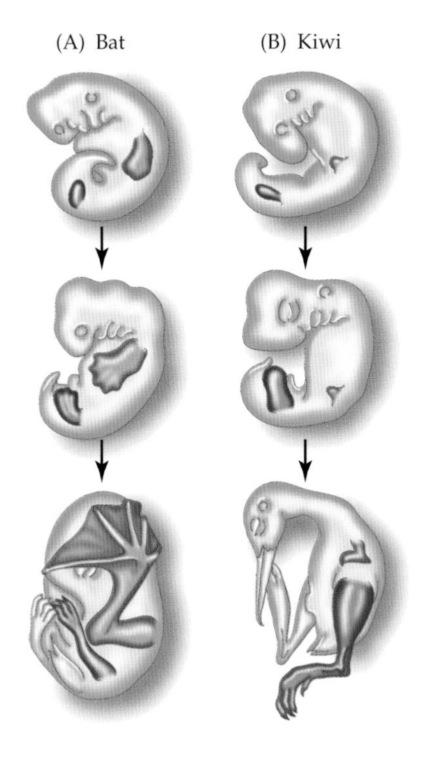

FIGURE 9.7 Limb development in a flying mammal and a flightless bird. (A) The bat *Rousettus amplexicaudatus* has a large forelimb bud that develops into a wing (orange), while the hindlimb (brown) is underdeveloped relative to other mammals. (B) The flightless kiwi (*Apteryx australis*) has much smaller forelimbs relative to the wings of most birds. (After Richardson 1999.)

Enhancer modularity

One of the most important insights of evolutionary developmental biology is that not only are the *anatomical* units modular (such that one part of the body can develop differently than the others), but the DNA regions that form the *enhancers* of genes are modular (see Davidson 2006; Gilbert 2006a). The modularity of development is determined by the modularity of the gene enhancers. Each enhancer element allows the gene to be expressed in a different tissue. As described in Chapter 2, the *Pax6* gene is expressed in the pancreas, the neural tube, the retina, and the cornea and lens. For each of these tissues, there is an enhancer element in the regulatory region of the *Pax6* gene (Figure 9.8A), so *Pax6* can be independently activated in each separate tissue typee (the Boolean "OR" condition).* Within each of these

*The Boolean logic terms used here can be understood conventionally, in that the "AND" cases demand that transcription factors A, B, and C all must be present for the gene to be activated. This is what happens within an enhancer unit. In the "OR" conditions, either A or B or C will function to activate the enhancer. Usually, within an enhancer element (say, for expressing the gene in the cornea), A, B, and C have to be present. Moreover, the same gene may be able to be expressed in the neural tube. Here, the enhancer element is activated by the "AND" conditions of R, S, and T. However, the "OR" condition applies between these enhancer elements; thus the gene can be expressed in those tissues having the A, B, AND C transcription factors (i.e., in the cornea) OR the R, S, AND T transcription factors (in the neural tube).

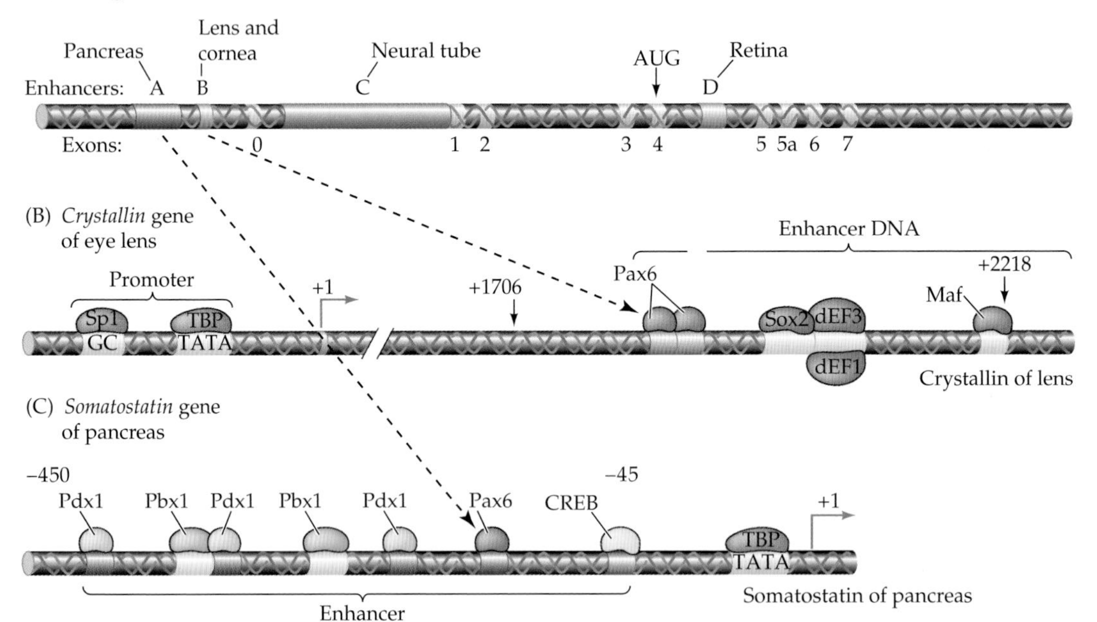

FIGURE 9.8 Modularity of enhancer regions. (A) The *Pax6* gene is expressed in the pancreas, lens/cornea, neural tube, and retina. There are specific regions of DNA that bind transcription factors in each of these primordia that act to allow the transcription of *Pax6* gene in these tissues. This is the "OR" condition. Pax6 is, itself, a transcription factor, and it binds to modular enhancers in genes that are expressed in the pancreas and eye. (B) In the eye it works with Maf and Sox2 to initiate the transcription of crystallin in the lens. (C) In the pancreas it works with Pbx1 and Pdx1 transcription factors to activate numerous pancreatic genes. (After Gilbert 2006a.)

enhancer elements are DNA binding sites that are recognized by transcription factor proteins. Often, several transcription factors are needed to bind in order to activate a gene (the Boolean "AND" function). For instance, the *Pax6* gene encodes a transcription factor. The Pax6 transcription factor that is produced in the lens can interact with other transcription factors (such as Maf and Sox2) to activate the transcription of the *crystallin* gene that encodes the transparent protein of the lens (Figure 9.8B). The Maf transcription factor is expressed in the presumptive lens cells only after these cells receive Fgf8 (a paracrine factor) from the neighboring optic cup. In this way, lens development is induced only in the head ectoderm (expressing the *Pax6* gene) that is also in contact with the optic cup (expressing *Fgf8*). Similarly, in the pancreas, Pax6 protein acts with other transcription factors (Pdx1, Pbx1) to activate the somatostatin gene of the islet cells (Figure 9.8C).

The modularity of enhancer elements allows particular sets of genes to be activated together and permits the same gene to become expressed in several discrete places. Thus, if a particular gene loses or gains a modular enhancer element, the organism containing that particular allele will express that gene in different places or at different times than those organisms retaining the original allele. This mutability can result in the development of different anatomical and physiological morphologies (Sucena and Stern 2000; Shapiro et al. 2004; McGregor et al. 2007).

The evolutionary importance of such enhancer modularity has been dramatically demonstrated by David Kingsley's analysis of evolution in three-spined stickleback fish (*Gasterosteus aculeatus*). Freshwater sticklebacks evolved from marine sticklebacks about 12,000 years ago, when marine populations colonized the newly formed freshwater lakes at the end of the last ice age. Marine sticklebacks have a pelvic spine that serves as protection against predation, lacerating the mouths of those predatory fish who try to eat the stickleback (Figure 9.9A). Freshwater sticklebacks, however, do not have pelvic spines (Figure 9.9B). This may be because freshwater species lack the piscine predators that the marine fish face but must deal instead with invertebrate predators that can easily capture them by grasping onto such spines. Thus, a pelvis without lacerating spines evolved in the freshwater populations of this species.

To determine which genes might be involved in this difference, researchers mated individuals from marine (with spines) and freshwater (without spines) populations. The resulting offspring were bred to each other and produced numerous progeny, some of which had pelvic spines and some of which didn't. Using molecular markers to identify specific regions of the parental chromosomes, Shapiro and coworkers (2004) found that the primary gene for pelvic spine development mapped to the distal end of chromosome 7. That is to say, nearly all the fish with pelvic spines inherited this "hindlimb-encoding" chromosomal region from the marine parent, while fish lacking pelvic spines inherited this region from the freshwater parent. Shapiro and coworkers then tested numerous candidate genes (e.g., genes known to be present in the hind limb structures of mice) and found that the gene encoding transcription factor Pitx1 was located on this region of chromosome 7.

When they compared the amino acid sequences of the Pitx1 protein between marine and freshwater sticklebacks, there were no differences. However, there was a critically important difference when they compared the *expression patterns* of *Pitx1*. In both species, *Pitx1* was expressed in the precursors of the thymus, nose, and sensory neurons. However, in the marine species, *Pitx1* was also expressed in the pelvic region, whereas pelvic expression of *Pitx1* was absent or severely reduced in the freshwater populations (Figure 9.9C). Since the coding region of *Pitx1* was not mutated (and since the gene involved in the pelvic spine differences maps to the site of the *Pitx1* gene, and the difference between the freshwater and marine

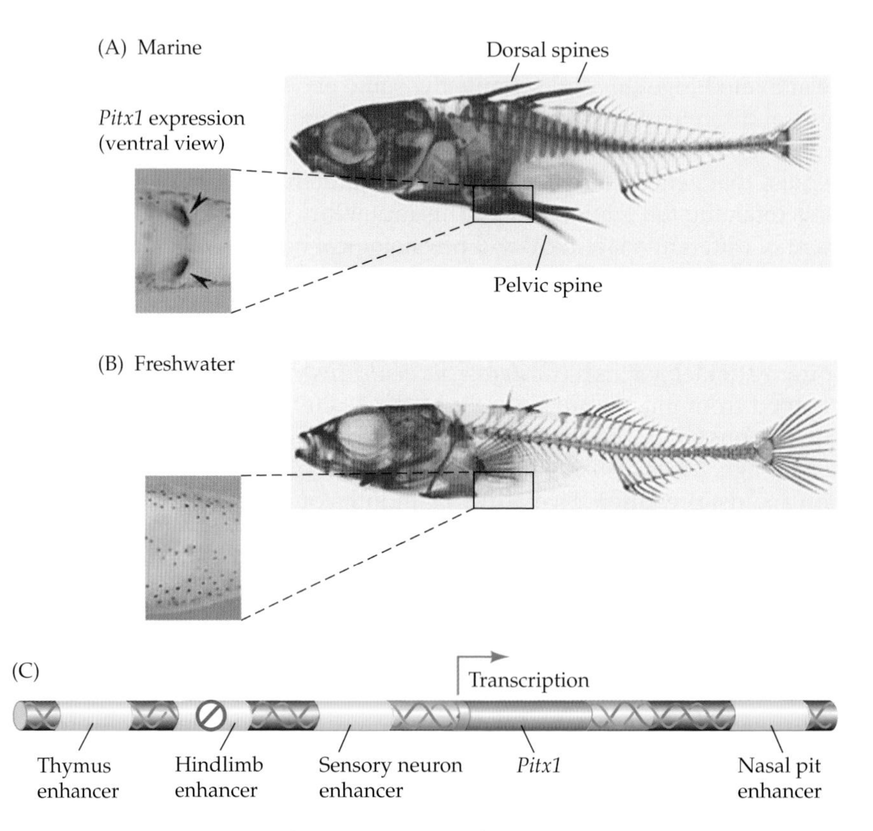

FIGURE 9.9 Modularity of enhancers. Loss of *Pitx1* expression in the pelvic region of freshwater three-spined sticklebacks. Bony plates cover more of the marine three-spined sticklebacks (A) than freshwater sticklebacks (B), and the marine forms have a prominent pelvic spine. The pelvic regions of the marine and freshwater populations show differences in *Pitxl* expression that can be readily observed at higher magnifications (from the area enclosed by the dashed lines). (C) Model for the evolution of pelvic spine loss in the freshwater three-spined stickleback. Four enhancers are postulated to reside near the coding region of the *Pitxl* gene. The enhancers direct the expression of this gene in the thymus, pelvic spine, sensory neurons, and nose, respectively. In the freshwater populations of this species, the pelvic spine enhancer module has been mutated so that it fails to function. (Photographs courtesy of D. M. Kingsley.)

species involves the expression of this gene at a particular site), one reasonable hypothesis is that the *enhancer region* allowing *Pitx1* to be expressed in the pelvic area no longer functions in the freshwater species. If this is the case, the modularity of the enhancer has enabled this particular expression domain to be lost, and with it the pelvic spine. No other function of the Pitx1 protein had to be disturbed.

Interestingly, the loss of the pelvic spines in other stickleback species appears to have been the result of independent losses of this *Pitx1* expression domain. The loss of hindlimbs in manatees may also be accounted for by the lack of pelvic *Pitx1* (Shapiro et al. 2006). This finding suggests that if the loss of *Pitx1* expression in the pelvis occurs, this trait can be readily selected (Colosimo et al. 2004). Thus we see that by combining population genetics approaches with developmental genetics approaches, one begins to understand the mechanisms by which evolution can occur.

Comparative developmental studies of the insect eye (Oakley and Cunningham 2002), stickleback fish armor plates and spines (Colosimo et al. 2004, 2005), and avian and *Drosophila* pigment patterns (Gompel et al. 2005; Mundy 2005) reinforce the fact that parallel evolution can result from the independent recruitment of similar developmental pathways by different organisms. Instead of the view that extrinsic selection pressures play the dominant role in such phenomena (see Schluter 2000; Meehan and Martin 2003), the current view is that intrinsic developmental factors are critical in producing these parallel variations. Such parallel evolution was once the justification for the "creativity" of natural selection. Now we can see that development is what is creative (Gilbert 2006b).

Malaria, again

Enhancer modularity can be seen in human evolution as well. In Chapter 8 we discussed at length the case of malaria and how a mutation in the gene that encodes the β-globin chain of hemoglobin provides resistance against the malarial agent *Plasmodium falciparum*. But not all malaria comes from this species of *Plasmodium*. A relative, *Plasmodium vivax*, causes about 75 million cases of malaria each year. This form of the disease is not as lethal as falciparum malaria, but it can be incapacitating, with severe pain, diarrhea, and fever. Some African populations are immune to *P. vivax* because their red blood cells lack a protein, the Duffy glycoprotein, that *P. vivax* needs in order to attach itself to the host's red blood cells. The Duffy glycoprotein is probably one of several receptors for interleukin-8 (IL8, a paracrine factor involved in the circulatory system), and it is found on the cells of the Purkinje neurons, veins, and red blood cells. People who lack Duffy glycoprotein on their red blood cells still have it on their veins and Purkinje neurons. So if one asks, "Why don't these people have Duffy glycoprotein on their red blood cells?" the ultimate answer is probably that the lack of the Duffy glycoprotein was selected for in these populations because it gives these people resistance to vivax malaria, given that *Plasmodium* uses the red blood cell protein to infect the cell. The proximate answer is that the lack of the Duffy glycoprotein is caused by a mutation in the enhancer, a C → G substitution at position –36, which prevents the binding of the GATA1 transcription factor present in red blood cell precursors (Tournamille et al. 1995). Thus, the mutation can be selected because it blocks only one of the enhancers (the one for its expression in red blood cells), while allowing the enhancers that allow the gene's expression in veins and Purkinje neurons to work.

Mechanisms of Macroevolutionary Change

In 1977, François Jacob, the Nobel laureate who helped establish the operon model of gene regulation, proposed that evolution rarely creates a new gene. Rather, evolution works with what it has: it combines existing parts in new ways rather than creating new parts. He predicted that such "tinkering" would be most likely to occur in those genes that construct the embryo, not in the genes that function in adults (Jacob 1977). Wallace Arthur (2004) has catalogued four ways in which Jacob's "tinkering" can take place at the level of gene expression:

- Heterotopy (change in location)
- Heterochrony (change in time)
- Heterometry (change in amount)
- Heterotypy (change in kind)

Heterotopy

Changing the location of a gene's expression (i.e., altering which cells it is expressed in) can result in enormous amounts of selectable variation. For instance, one of the biggest differences between the duck and the chicken is the webbed feet of the duck. This is an obviously important adaptation to swimming (as are the duck's broad beak and the oil glands of its skin.) Vertebrate embryonic limbs develop with their fingers and toes surrounded by a web of connective tissue and skin. In chicks (and humans), the cells that make up the webbing between the digits undergo apoptosis (programmed cell death) initiated by the paracrine factor BMP4. This cell death destroys the webbing and frees the digits. It turns out that ducks also express BMP4 in the webbing of their hindlimbs, but this protein is prevented from signaling cell death by the presence of another protein, the BMP inhibitor Gremlin (Laufer et al. 1997; Merino et al. 1999). Chick and human limbs also express Gremlin around the cartilaginous skeletal elements of the digits, but not in the webbing (Figure 9.10). Thus, the webbing in embryonic duck feet remains intact thanks to the heterotopy—the change in place of gene expression—of the *Gremlin* gene. Indeed, if one adds beads containing the Gremlin protein to the webbing of an embryonic chick limb, the limb will keep its webbing, just as a duck's does (Figure 9.11).

Bat forelimbs present a similar situation. Here the interdigital webbing that forms the wing has been maintained where it has disappeared in other mammals. Again, inhibition of BMP-induced apoptosis is critical. Instead of Grem-

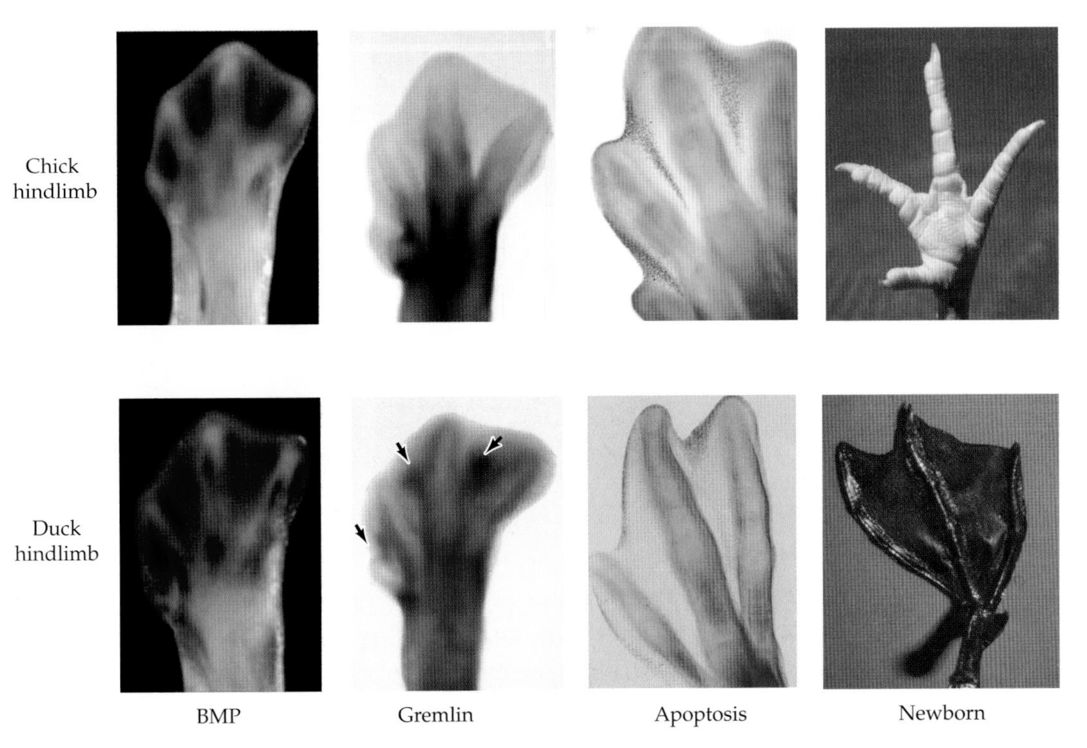

FIGURE 9.10 Heterotopy exemplified by the role of BMPs in the generation of webbed feet in ducks. BMPs cause apoptosis in the interdigital webbing. Autopods of chick feet (upper row) and duck feet (lower row) are shown at similar stages. The in situ hybridizations show that while BMPs are expressed in both the chick and duck hindlimb webbing, the duck limb shows expression of Gremlin protein (arrows) in the webbing as well. Gremlin is an inhibitor of BMPs. The pattern of cell death (shown by neutral red dye accumulation) becomes distinctly different in the two species. (Photographs courtesy of J. Hurle and E. Laufer.)

lin, however, the agent of BMP inhibition in bats appears to be both Gremlin and FGF signaling. Unlike other mammals, bats express *Fgf8* in their interdigital webbing, and this protein is critical for maintaining the cells there. If FGF signaling is inhibited in the bat by certain drugs, BMPs will induce the apoptosis of the forelimb webbing, just as in other mammals* (Weatherbee et al. 2006).

*Notice that BMP4 is being used in one part of the embryo to specify the central nervous system and in another part of the embryo to cause cell death. As we will see later, BMP4 is also used to construct the avian beak and to make bone tissue. Classical genetics usually studied proteins that have a single function; hemoglobin, insulin, and collagen, for instance, are each the end product of their cell lineage and each has a single function. Genes active in development, however, often have several different functions. Philosopher John Thorpe has called monotelism (the idea that each of nature's structures has but one purpose) "Aristotle's worst idea."

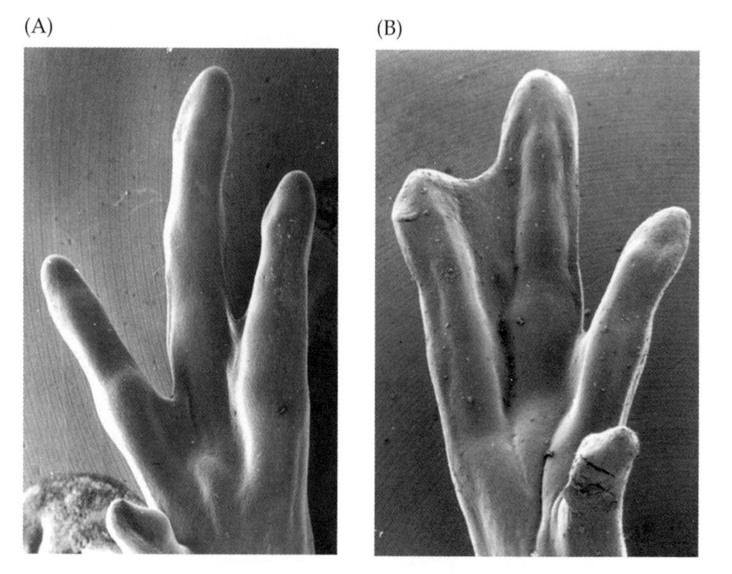

FIGURE 9.11 Inhibition of cell death by inhibiting BMPs. (A) Control chick limbs show extensive apoptosis in the space between the digits, leading to the absence of webbing. (B) However, when beads soaked with Gremlin protein are placed in the interdigital mesoderm, the webbing persists and generates a duck-like pattern. (After Merino et al. 1999; photographs courtesy of E. Hurle.)

Heterochrony

Heterochrony is a shift in the relative timing of two developmental processes from one generation to the next. Heterochrony can be seen at any level of development from gene regulation to adult animal behaviors (West-Eberhard 2003). In heterochrony, one module changes its time of expression or growth rate relative to the other modules of the embryo. We have already come across this concept in our discussion of the formation of the kiwi and bat forelimbs (see Figure 9.7), and heterochronic changes are also responsible for remodeling the jaws of cichlid fishes (see Figure 1.12). Limb heterochronies are in fact quite common. In marsupials, for instance, the forelimbs develop at a faster rate relative to the embryonic forelimbs of placental mammals, allowing the marsupial embryo to climb into the maternal pouch and suckle (Smith 2003; Sears 2004). The enormous number of ribs formed in embryonic snakes (more than 500 in some species; see Figure 9.5A) is likewise due to heterochrony: the segmentation reactions cycle nearly four times faster relative to tissue growth in snake embryos than they do in related vertebrate embryos (Gomez et al. 2008).

The elongated fingers in the dolphin flipper appear to be the result of heterochrony, wherein the region responsible for producing the growth factors for the developing limb (especially Fgf8) is present longer than in other mammals (Figure 9.12; Richardson and Oelschläger 2002). Another digit example of molecular heterochrony occurs in the lizard genus *Hemiergis*, which includes species with three, four, or five digits on each limb. The number of digits is regulated by the length of time the *sonic hedgehog* (*shh*) gene is active in the limb bud's zone of polarizing activity. The shorter the duration of *shh* expression, the fewer the number of digits (Shapiro et al. 2003).

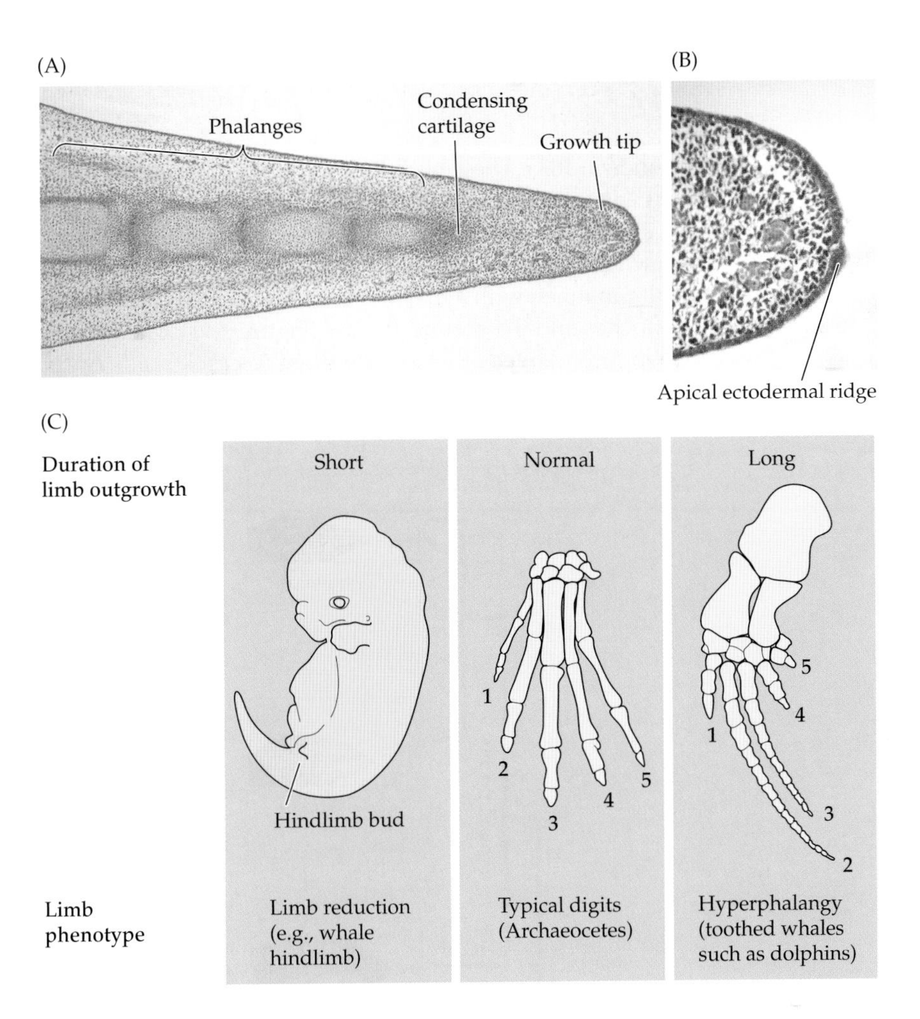

FIGURE 9.12 Heterochrony in the flipper development of the spotted dolphin (*Stenella attenuata*). (A) Dolphins show "hyperphalangy" of digits 2 and 3. That is to say, digits 2 and 3 continue growing long after the other digits, and they keep adding on new cartilaginous regions. (B) The mechanism for this appears to be the retention of the apical ectodermal ridge that secretes growth factors necessary for the digit growth. (C) A heterochronic hypothesis for cetacean limb development. When limb growth terminates early, the loss of distal structures is seen. This leads to rudimentary appendages in many whales. In Archaeocetes (the progenitors of whales), there appears to a be normal amount of limb growth. In toothed whales (such as dolphins), limb development terminates late, giving rise to extra long digits. (From Richardson and Oelschläger 2002.)

Heterometry

THE BEAKS OF THE FINCHES Heterometry is the change in the *amount* of a gene product. One of the best examples of heterometry involves Darwin's celebrated finches (mentioned in Chapter 8), a set of 14 closely related birds collected by Charles Darwin and his shipmates during his visit to the Galápagos and Cocos Islands in 1835. These birds helped him frame his evolutionary theory of descent with modification, and they still serve as one of the best examples of adaptive radiation and natural selection (see Grant 1999; Weiner 1994; Grant and Grant 2007). Systematists have shown that these finch species evolved in a particular manner, with a major speciation event being the split between the cactus finches and the ground finches (Figure 9.13). The ground finches evolved deep, broad beaks that enable them to crack seeds open,

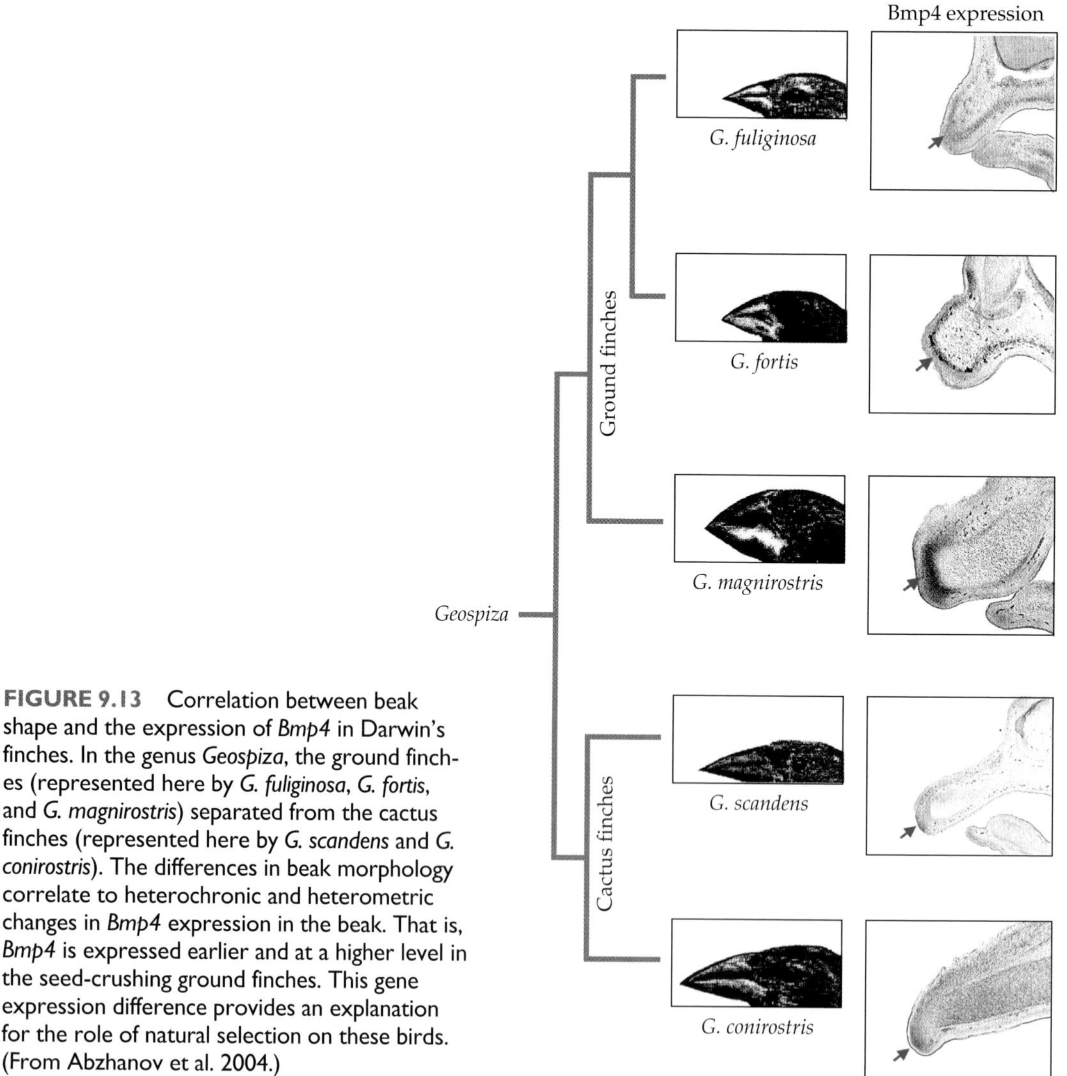

FIGURE 9.13 Correlation between beak shape and the expression of *Bmp4* in Darwin's finches. In the genus *Geospiza*, the ground finches (represented here by *G. fuliginosa*, *G. fortis*, and *G. magnirostris*) separated from the cactus finches (represented here by *G. scandens* and *G. conirostris*). The differences in beak morphology correlate to heterochronic and heterometric changes in *Bmp4* expression in the beak. That is, *Bmp4* is expressed earlier and at a higher level in the seed-crushing ground finches. This gene expression difference provides an explanation for the role of natural selection on these birds. (From Abzhanov et al. 2004.)

whereas the cactus finches evolved narrow, pointed beaks that allow them to probe cactus flowers and fruits for insects.

Research on other birds (Schneider and Helms 2003) showed that species differences in the beak pattern were caused by changes in the growth of the neural crest-derived mesenchyme of the frontonasal process (i.e., those cells that form the facial bones). Moreover, it had further been demonstrated that birds use BMP4 as a growth factor promoting cell division in their beaks, and that species-specific differences in beaks (such as those between the chick and the duck) were due to the placement of *Bmp4* expression (Wu et al. 2004, 2006).

Abzhanov and his colleagues (2004) found a remarkable correlation between the beak shape of the finches and the amount of *Bmp4* expression. No other paracrine factor showed such differences. The expression of *Bmp4* in the mesenchyme of embryonic *ground* finch beaks starts earlier and is much greater than *Bmp4* expression in *cactus* finch beaks. In all cases, the *Bmp4* expression pattern correlated with the breadth and depth of the beak.

The importance of these expression differences was confirmed experimentally by changing the *Bmp4* expression pattern in chick embryos (Abzhanov et al. 2004; Wu et al. 2004, 2006). When *Bmp4* expression was enhanced in the frontonasal process mesenchyme, the chick developed a broad beak reminiscent of the beaks of the ground finches. Conversely, when BMP signaling was inhibited in this region (by adding a BMP inhibitor to the developing beak primordium), the beak became narrow and pointed, like those of cactus finches. Thus, enhancers controlling the amount of beak-specific BMP4 synthesis may be critically important in the evolution of Darwin's finches.

But this was only the beginning of the story. While the ground finches were characterized by broad, blunt beaks, the cactus finches were known for their long, slender beaks, which they used to get seeds, nectar, and pollen out of cactus flowers. Gene chip technology showed that the cactus finch embryos had more calmodulin in their embryonic beaks than the blunt-beaked ground finches (Figure 9.14). Calmodulin is a protein that combines with many enzymes to make their activity dependent upon calcium ions. In situ hybridization and other techniques demonstrated that the calmodulin gene is expressed at higher levels in the cactus finch embryonic beaks than in the embryonic beaks of ground finches (see Figure 9.14). When calmodulin was upregulated in the embryonic chicken beak, the chicken, too, developed a long beak. Thus, the genes for BMP4 and calmodulin represent two targets for natural selection (Figure 9.15; Abzhanov et al. 2006). While natural selection will allow certain morphologies to survive, the generation of those morphologies depends upon variations of developmental regulatory genes such as those for BMP4 and calmodulin. It isn't the amount of adult beak components, but the amount of the embryonic construction units, that is being changed.

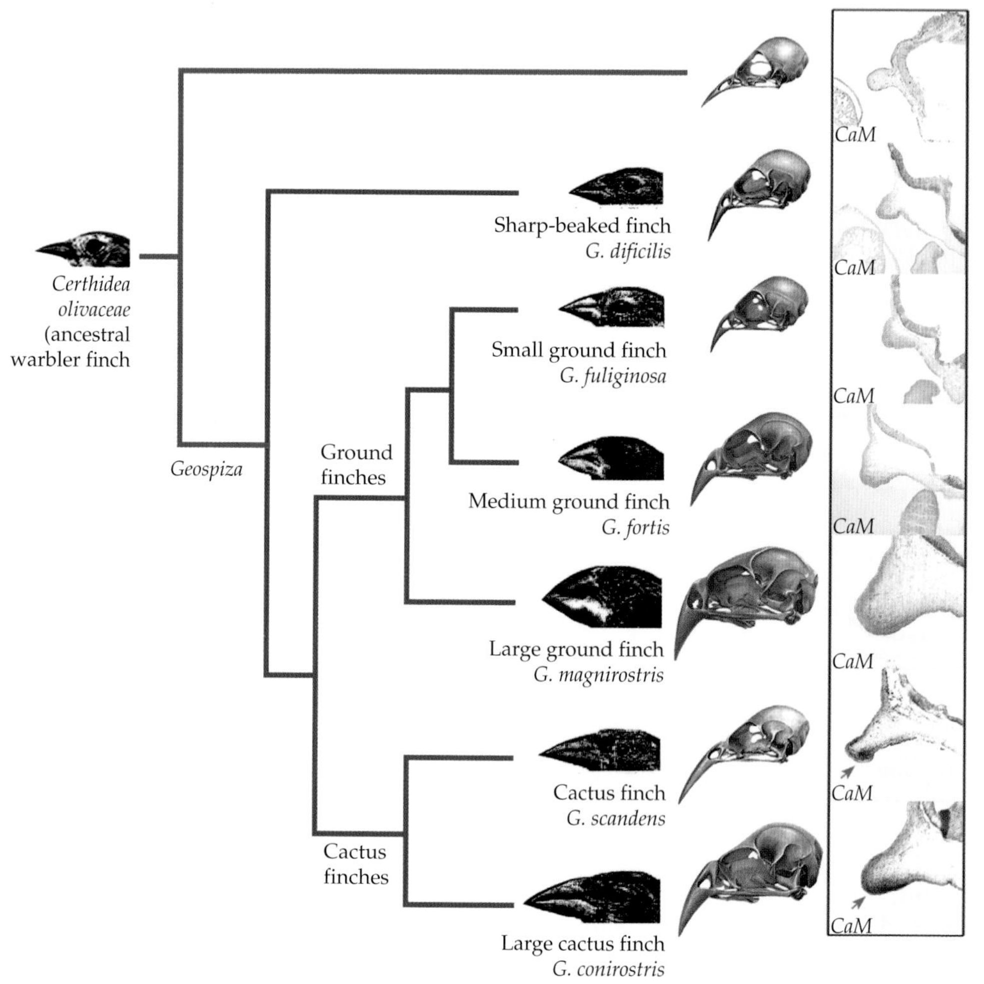

FIGURE 9.14 Correlation between the length of the beaks of Darwin's finches and the amount of calmodulin (*CaM*) gene expression. The various species of *Geospiza* are a monophyletic group (i.e., all are descended from a common ancestor); the differences in their beak morphologies can be seen skeletally. The *CaM* gene is expressed in a strong distal-ventral domain in the mesenchyme of the upper beak prominence of the large cactus finch *G. conirostris*, at somewhat lower levels in the cactus finch *G. scandens*, and at very low levels in the large ground finch and medium ground finch *G. magnirostris* and *G. fortis*, respectively. Very low levels of calmodulin gene expression were also detected in the mesenchyme of *G. difficilis*, *G. fuliginosa*, and the basal warbler finch *Certhidea olivacea*. (From Abzhanov et al. 2006.)

HUMAN IL4 Most of the genetic variation in humans, whether pathological or normal, does not come from changes in the structural genes. Rather, human variation arises from mutations in the regulatory regions of genes used during development (Rockman and Wray 2002). Very often, these

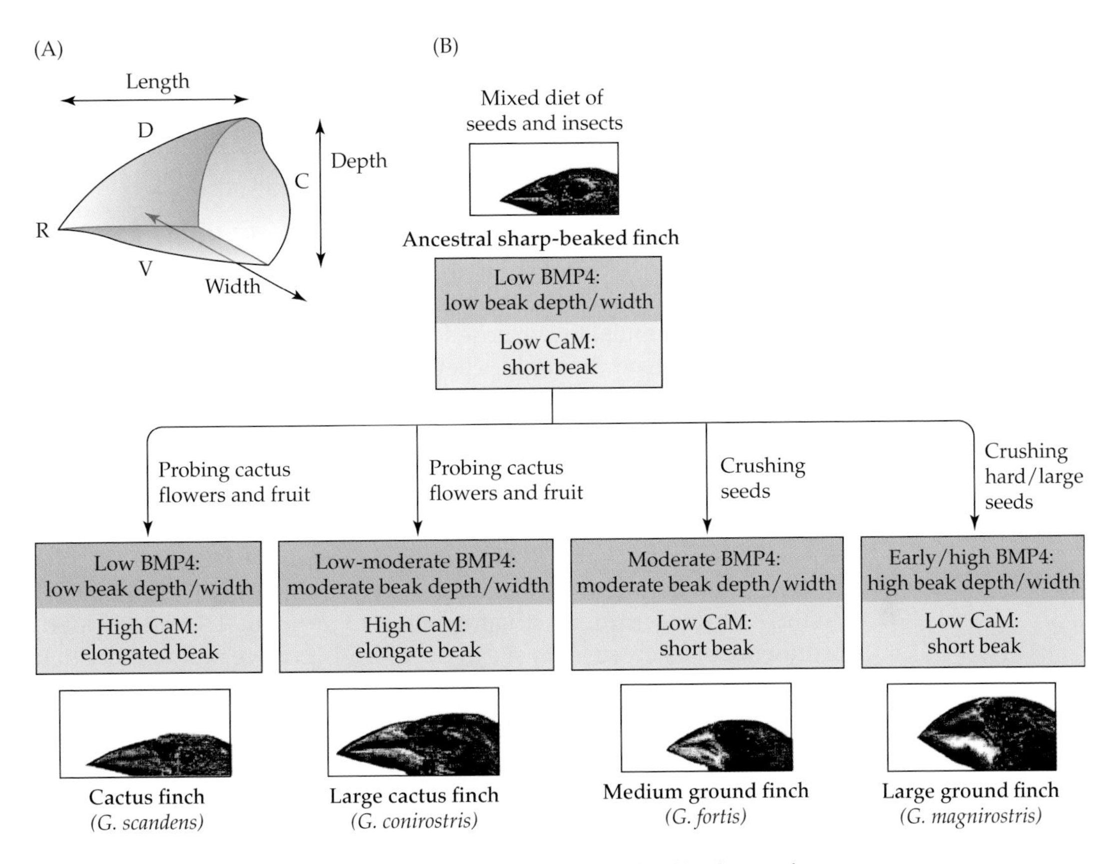

FIGURE 9.15 Beak evolution among Darwin's finches. (A) Bird beaks are three-dimensional structures that can change along any of the growth axes. C, caudal (tail-ward); D, dorsal (spineward); R, rostral (forward); V, ventral (belly). (B) A sharp-beaked finch represents the basal morphology for *Geospiza*, as the ancestral species is thought to have had such a morphology. The model for BMP4 and calmodulin involvement explains development of both the elongated and the deep/wide beaks of the derived species of *Geospiza*. (After Abzhanov et al. 2006.)

mutations involve changes in the rate of transcription (heterometry) of particular genes. One such example is the gene encoding interleukin-4 (IL4).

IL4 is a paracrine factor released by cells of the immune system to promote the differentiation of IgE-secreting B cells. These are the cells that initiate the cascade to remove pathogens, including parasitic worms, from our bodies. However, if these B cells recognize nonpathogenic substances, one gets allergies. Several alleles of the *IL4* gene are medically relevant. One of the most common alleles is *–524T* (that is, in this allele the nucleotide base thymine occupies the position 524 base pairs before the start of transcription). The *–524T* allele is recent and is found only among human beings (Rockman et al. 2003). The wild-type allele, *–524C* (which has cytosine in the –524 position) is found in all primate populations—and indeed in every other mammal tested. The *–524T* allele, therefore, emerged in the lineage separating *Homo* from the other great apes. Individuals carrying the *–524T* allele are subject to several disease states, including severe allergies, asthma, contact dermatitis, and subacute sclerosing panencephalitis. However, population genetic statistics show that while the *–524T* allele may cause lowered fitness in many habitats, this new allele appears to be advantageous in human populations exposed to certain parasitic worms. It allows these people to defend themselves better against helminth parasites. So in areas of filariasis (elephantiasis) and schistosomiasis, this allele appears to provides more fitness than the *–524C* allele and thus may be selected for in those populations.

What does the *–524T* allele do differently than the *–524C* allele? The C → T mutation creates a new binding site for the transcription factor NFAT (*n*uclear *f*actor for *a*ctivated *T* cells) in the IL4 enhancer (Figure 9.16). The wild-type promoters generally contain six NFAT binding sites that interact with the AP1 transcription activator factor to promote IL4 transcription. The presence of a new seventh site in *–524T* leads to more rapid production of IL4, yielding protein levels of at least threefold the norm. Thus, a single mutation in the regulatory region of a developmental regulatory gene can readily produce a selectable phenotype. This research also shows the beginnings of a new project in evolutionary biology—the population genetics of regulatory alleles.

Heterotypy

In heterochrony, heterotopy, and heterometry, the mutations affect the regulatory regions of the gene. The gene's product—the protein—remains the same, although it may be synthesized in a new place, at a different time, or in different amounts. In **heterotypy**, the changes affect the protein that binds to these regulatory regions. The changes of heterotypy affect the actual coding region of the gene, and thus they can change the functional properties of the protein being synthesized.

WHY INSECTS HAVE ONLY SIX LEGS Insects have six legs, whereas most other arthropod groups (spiders, millipedes, centipedes, and crustaceans) have many more. How is it that the insects came to form legs only in their three thoracic segments and have no appendages in their abdominal regions? The answer seems to reside in the relationship between Ultrabithorax (Ubx) protein and the *Distal-less* gene.

The *Ubx* gene is one of the Hox genes. It encodes a transcription factor that is used throughout the arthropods to activate those genes expressed in

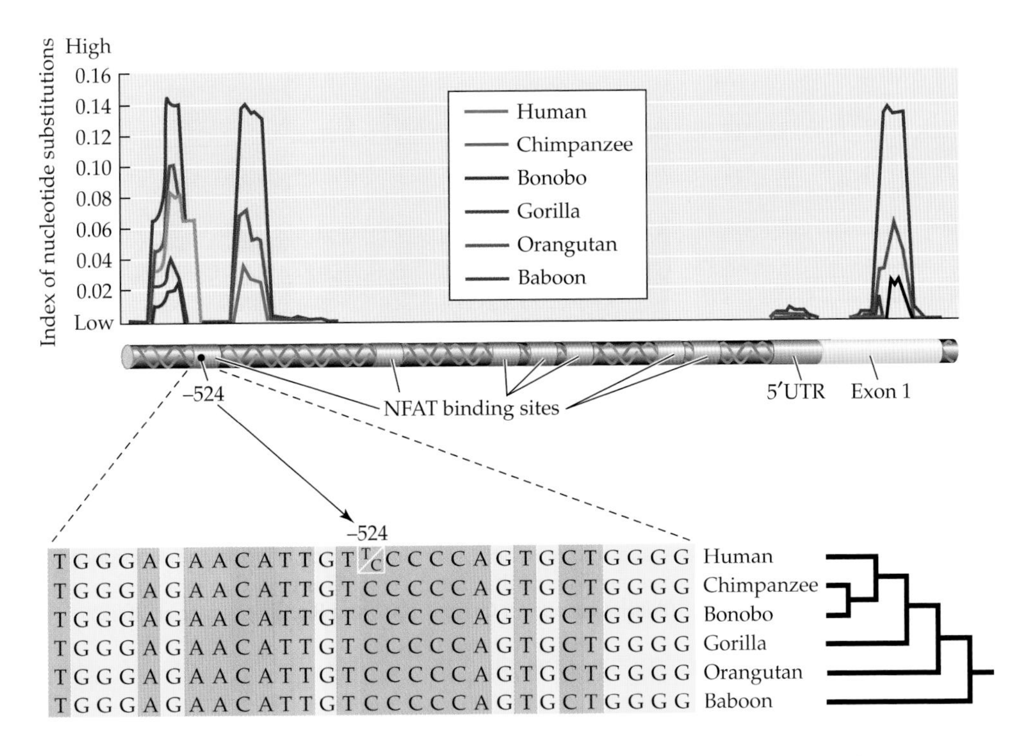

FIGURE 9.16 Heterometry (a change in the amount of a gene product expresses) can help drive evolution. (A) Regions of variability in the enhancer of the gene for interleukin-4 (IL4) among primates. Several regions are highly conserved, or invariable, indicating that any change (i.e., mutation) in these nucleotide sequences usually results in decreased fitness. These regions include the enhancers for the IL4 genes. (B) At position –524, in the midst of a highly conserved enhancer sequence, a mutation in the human population has created a new binding site for the NFAT transcription factor, enabling *IL4* to be transcribed in greater amounts. Although this has negative fitness consequences in many populations, it functions positively for the survival of individuals living in places where worm parasites are common. Heterometry is caused by mutation in the enhancer site of the *IL4* gene. (After Rockman et al. 2003.)

the lower thorax and abdomen. In most arthropod groups, Ubx protein does not inhibit the *Distal-less* gene. However, in the insect lineage, a mutation occurred in the *Ubx* gene wherein the original 3′ end of the protein-coding region was replaced by a group of nucleotides encoding a stretch of about 10 alanine residues (Figure 9.17; Galant and Carroll 2002; Ronshaugen et al. 2002). This polyalanine region represses *Distal-less* transcription. When a shrimp *Ubx* gene is experimentally modified to encode this polyalanine region, its Ubx, too, represses the *Distal-less* gene. The ability of insect Ubx to inhibit *Distal-less* thus appears to be the result of a gain-of-

Thus, form appears to evolve largely by altering the timing, location, and amount of gene expression. The heterotopy, heterochrony, and heterometry is accomplished by changing the enhancer elements regulating gene expression. Heterotypy can alter the transcription factors that bind to these *cis*-regulatory sequences, giving these proteins different properties. As Carroll (2008) has noted, "This constitutes a developmental genetics theory of morphological evolution which can supplement and extend the Modern Synthesis of evolutionary biology and population genetics."

Speciation

Natural selection explains how traits evolve within populations and among populations of the same species. But evolutionary theory also must explain how interbreeding populations diverge to the point that they can no longer interbreed and, at some point, become separate species. Despite the title of his best known work, Darwin was able to offer little insight into exactly how one species might diverge and become separate from another, and even the scientists of the Modern Synthesis were on firmer ground when discussing changes within populations rather than the divisions that might lead to new species. Armed with new insights from the molecular revolution, however, evolutionary biologists of the past quarter-century have approached "the species problem" with renewed vigor (see Coyne and Orr 2004).

Speciation in the Modern Synthesis

The Modern Synthesis explains speciation primarily in terms of physical, temporal, mechanical, behavioral, or genetic barriers to interbreeding. When geographic barriers divide a formerly interbeeding population, the isolated populations may eventually accumulate mutations such that they are no longer able to interbreed, even if they are secondarily reunited; this is called **allopatric speciation**. Disruptive natural selection or disruptive sexual selection can also bring about **sympatric speciation**—speciation within a territory, without physical isolation. Disruptive natural selection can occur when assortative mating occurs—that is, when mating is not random. Nonrandom mating can occur for a number of reasons. Sexual selection, for example, occurs when certain individuals mate preferentially with certain other individuals within a population. Mating can also become nonrandom when different niches within a territory select for different traits. All these mechanisms are supported by data, and no single scheme fits all organisms.

In addition, the Modern Synthesis concludes that **postzygotic isolation** can also result in speciation. In other words, given two populations A and B, A-A matings and B-B matings produce fit offspring, but the mating of A with B, although producing viable zygotes, does not produce fit offspring.

Developmental Mechanisms of Speciation: Gamete Recognition Proteins

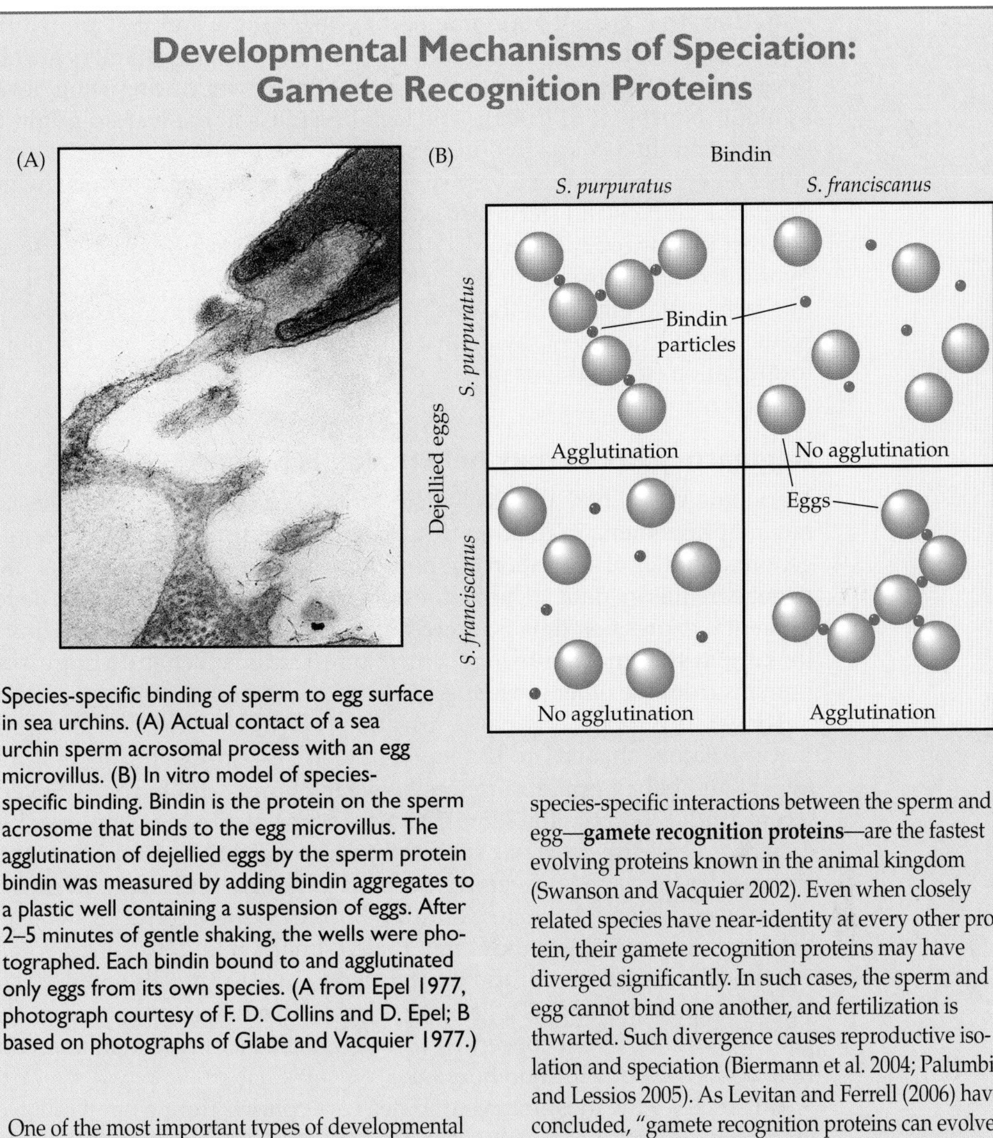

Species-specific binding of sperm to egg surface in sea urchins. (A) Actual contact of a sea urchin sperm acrosomal process with an egg microvillus. (B) In vitro model of species-specific binding. Bindin is the protein on the sperm acrosome that binds to the egg microvillus. The agglutination of dejellied eggs by the sperm protein bindin was measured by adding bindin aggregates to a plastic well containing a suspension of eggs. After 2–5 minutes of gentle shaking, the wells were photographed. Each bindin bound to and agglutinated only eggs from its own species. (A from Epel 1977, photograph courtesy of F. D. Collins and D. Epel; B based on photographs of Glabe and Vacquier 1977.)

One of the most important types of developmental barrier for speciation is provided by heterotypy of gamete recognition proteins. Proteins that mediate species-specific interactions between the sperm and egg—**gamete recognition proteins**—are the fastest evolving proteins known in the animal kingdom (Swanson and Vacquier 2002). Even when closely related species have near-identity at every other protein, their gamete recognition proteins may have diverged significantly. In such cases, the sperm and egg cannot bind one another, and fertilization is thwarted. Such divergence causes reproductive isolation and speciation (Biermann et al. 2004; Palumbi and Lessios 2005). As Levitan and Ferrell (2006) have concluded, "gamete recognition proteins can evolve at astonishing rates and lie at the heart of reproductive isolation and speciation in diverse taxa."

One sees this in *Heliconius* butterflies, where *Heliconius melpomene* mimics a black, red, and yellow model, while *H. cydno* mimics a black and white model. The two butterfly species are often seen together. While mating within the species produces mimics, mating between the species produces butterflies that are not good mimics of either model and that are readily eaten (Jiggins et al. 2001). The genes involved in *Heliconius* mimicry are also developmental genes involved in pigment patterning during wing development (Naisbit et al. 2003). This lack of fitness in the hybrid might be expected in the Galápagos finches, where there would be selection for either a very long beak or a very short beak. Other examples include hybrid death as embryos and sterility in adult hybrids (as in mules.)

Most postzygotic isolation is due to defective development. Reduced hybrid fertility is usually due to abnormal meiotic division or to cytoplasmic incompatability (as in the *Wolbachia* symbioses described in Chapter 3). So this is really a way of saying that developmental mechanisms contribute to speciation (see previous page).

Regulatory RNAs may help make us human

Mary Jane West-Eberhard (2005) noted that "Lack of attention to developmental phenomena in relation to speciation promises to change, because genomic studies of speciation can now contemplate gene expression as well as gene frequency data." The synthesis of comparative genomics and developmental expression data is already starting to yield remarkable fruits in looking at that most anthropocentric of questions: What distinguishes humans from the other great apes?

King and Wilson's original observations (1975) were the first to suggest that regulatory changes in development were responsible for the morphological and behavioral differences between chimps and humans. Several recent studies point in the same direction. Some researchers implicate heterometry. Using microarrays, several recent investigations found that while the quantity and types of genes expressed in human and chimp livers and blood were indeed extremely similar, human *brains* produced over five times as much mRNA as the chimp brains (Enard et al. 2002a; Preuss et al. 2004). Some researchers propose heterotypy: Enard and colleagues (2002b) have shown that the *FOXP2* gene is critical for language, and although this gene is extremely well conserved throughout most of mammalian evolution, it has a unique form in humans.

But some of the most interesting ideas are coming from a newly discovered type of heterotypy: changes in the sequence of regulatory RNAs. Pollard and her collaborators (2006) found this through a very logical and elegant path. First, they asked the DNA databases which areas of DNA (at least 100 base pairs long) are identical between the mouse, the rat, and the chimp. These **evolutionarily conserved regions** would be expected to encode genes whose functions are extremely important, and random mutations in these regions would be expected to have been strongly selected against. Such conserved regions often contain regulatory sequences that control the expression of transcription factors (Figure 9.19).

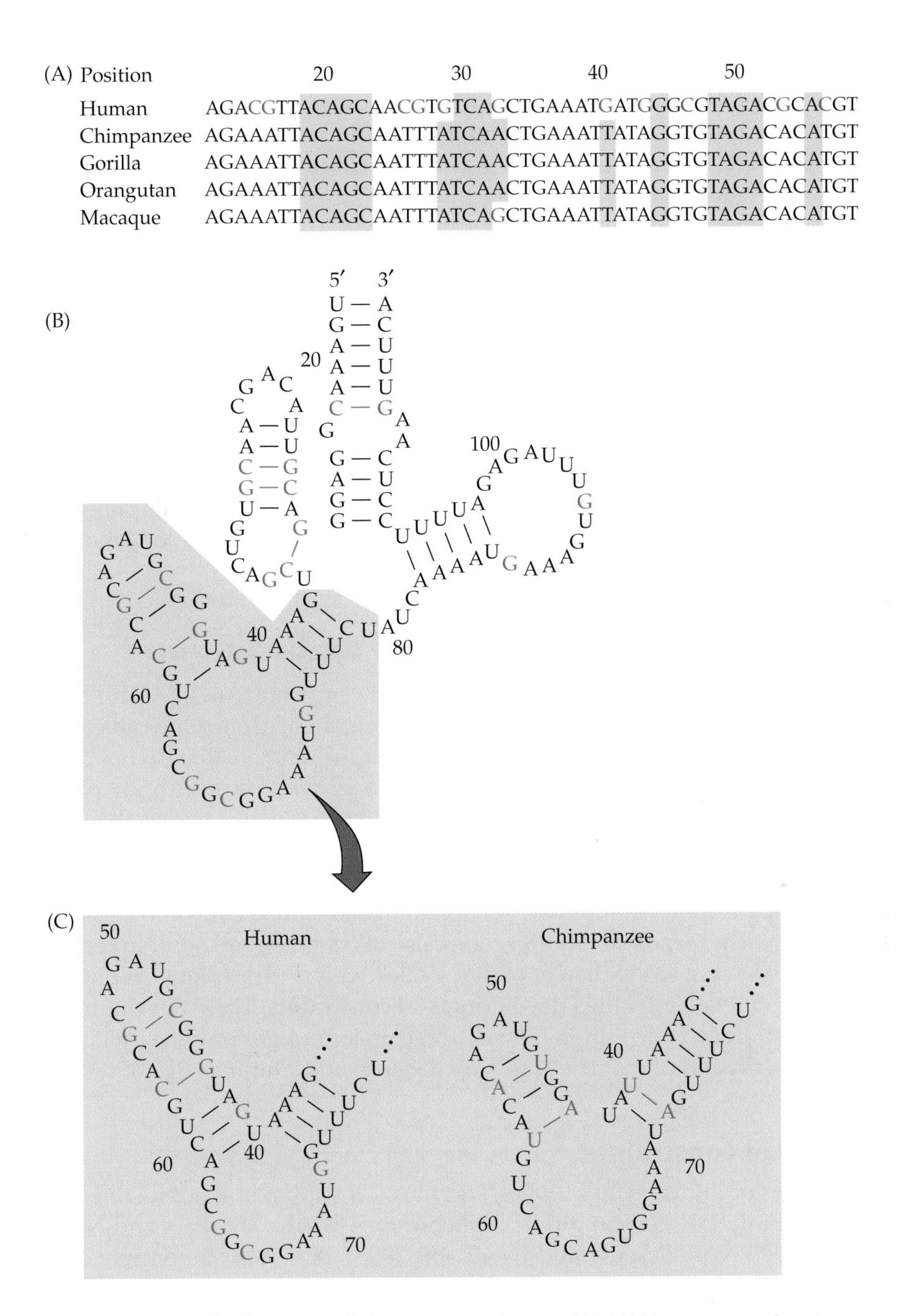

FIGURE 9.19 Molecular evolution among primates. (A) *HAR1* sequence showing sites where humans have accumulated mutations not found in other primates. (B) General secondary structure of the RNA in mammals. (C) How the folding of the chimp HAR1 and human HAR1 would differ. (After Pollard et al. 2006.)

Next, the researchers looked at about 35,000 of these conserved sequences and asked whether human homologues of any of these sequences diverge from the common mammalian norm. Their analyses revealed 202 **human accelerated regions**, or **HARs**. Two of these, *HAR1* on chromosome 20 and *HAR2* on chromosome 2, were especially interesting because these are dramatically distinct in humans, with DNA changes more than 10 times greater than changes found in chimp, orangutan, gorilla, spider monkey, or crab-eating macaque, in all of which the sequence is basically the same (see Figure 9.19). The *HAR1* region appears to be adjacent to a region of genes encoding transcription factors, and it encodes a regulatory RNA that is expressed in the cortex of the mammalian brain.

There are many components to being human, and most of them—the large and cognitive brain, the opposable thumb, the skeleton altered to allow walking, the musculature modified to allow language—have to do with changing development. This field of human evo-devo should continue to be a fascinating research area linking development, evolution, and the biological basis of becoming human.

Developmental Constraints on Evolution

There are only about three dozen animal lineages, and among them they represent all the major body plans of the animal kingdom. One can easily envision other types of body plans and imagine animals that do not exist (science fiction writers do it all the time). So why aren't more body plans found among the animals? To answer this, we have to consider the constraints that development imposes on evolution.

The number and forms of possible phenotypes that can be created are limited by the interactions that are possible among molecules and between modules. These molecular interactions also allow change to occur in certain directions more easily than in others. Collectively, the restraints on phenotype production are called **developmental constraints.** These developmental constraints on evolution fall into three major categories: physical, morphogenetic, and phylogenetic (see Richardson and Chipman 2003).

Physical constraints

The laws of diffusion, hydraulics, and physical support are immutable and will permit only certain physical phenotypes to arise (Forgacs and Newman 2005; von Dassow and Davidson 2008). A vertebrate on wheeled appendages such as Dorothy saw in Oz cannot exist because in reality blood cannot circulate through a rotating organ; this entire evolutionary avenue is closed off. Similarly, structural parameters and fluid dynamics would prohibit the existence of 6-foot-tall mosquitoes or 25-foot-long leeches.

Morphogenetic constraints

Rules for morphogenetic construction also limit the phenotypes that are possible (Oster et al. 1988). Bateson (1894) and Alberch (1989) noted that when organisms depart from their normal development, they do so in only a limited number of ways. Some of the best examples of these types of constraints come from the analysis of limb formation in vertebrates. Although there have been many modifications of the vertebrate limb over 300 million years, some modifications (such as a middle digit shorter than its surrounding digits) are not found (Holder 1983). Moreover, analyses of natural populations suggest that there is a relatively small number of ways in which limb changes can occur (Wake and Larson 1987). If a longer limb is favorable in a given environment, an elongated humerus variant may become favored by selection, but one never sees two smaller humeri joined together in tandem, although one could imagine the selective advantages that such an arrangement might have. This observation suggests a limb construction scheme that follows certain rules.

These rules are now becoming known. The chemical principles of the reaction-diffusion model (see box) appear to govern the architecture of the limb (Newman and Müller 2005). Oster and colleagues (1988) found that this model can explain the known morphologies of the limb, and it could explain why other morphologies are forbidden. The reaction-diffusion equations predicted the observed succession of bone development from stylopod (humerus/femur) to zeugopod (ulna-radius/tibia-fibula) to autopod (hand/foot). If limb morphology is indeed determined by the reaction-diffusion mechanism, then spatial features that cannot be easily generated by reaction-diffusion kinetics will not readily occur.

Phylogenetic constraints

Phylogenetic constraints on the evolution of new structures are historical restrictions based on the history of an organism's development (Gould and Lewontin 1979). Once a structure comes to be generated by inductive interactions, it is difficult to start over again. The notochord, for example, which is functional in adult protochordates such as *Amphioxus* (Berrill 1987), degenerates in adult vertebrates. Yet it is transiently necessary in vertebrate embryos, where it specifies the neural tube. Similarly, Waddington (1938) noted that although the pronephric kidney of the chick embryo is considered vestigial (since it has no ability to concentrate urine), it is the source of the ureteric bud that induces the formation of a functional kidney during chick development.

As genes acquire new functions during the course of evolution, they may become involved in more than one module, making change difficult. This ability of a gene to play multiple roles in different cells is called

Reaction-Diffusion Models

One of the most important mathematical models in developmental biology was formulated by Alan Turing (1952), one of the founders of computer science (and the mathematician who cracked the German "Enigma" code during World War II). He proposed a model wherein two homogeneously distributed substances would interact to produce stable patterns during morphogenesis. These patterns would represent regional differences in the concentrations of the two substances. Their interactions would produce an ordered structure out of random chaos.

Turing's reaction-diffusion model involves two substances. Substance P promotes the production of more substance P as well as substance S. Substance S, however, inhibits the production of substance P. Turing's mathematics show that if S diffuses more readily than P, sharp waves of concentration differences will be generated for substance P (see figure). These waves have been observed in certain chemical reactions (Prigogine and Nicolis 1967; Winfree 1974).

The reaction-diffusion model predicts alternating areas of high and low concentrations of some substance. When the concentration of such a substance is above a certain threshold level, a cell (or group of cells) may be instructed to differentiate in a certain way. An important feature of Turing's model is that particular chemical wavelengths will be amplified while all others will be suppressed. As local concentrations of P increase, the values of S form a peak centering on the P peak, but becoming broader and shallower because of S's more rapid diffusion. These S peaks inhibit other P peaks from forming. But which of the many P peaks will survive? That depends on the size and shape of the tissues in which the oscillating reaction is occurring. The mathematics describing which particular wavelengths are selected consist of complex polynomial equations. Such functions have been used to model the spiral patterning of slime molds, the polar organization of the limb, the radiation of mammalian teeth, and the pigment patterns of mammals, fishes, and snails.

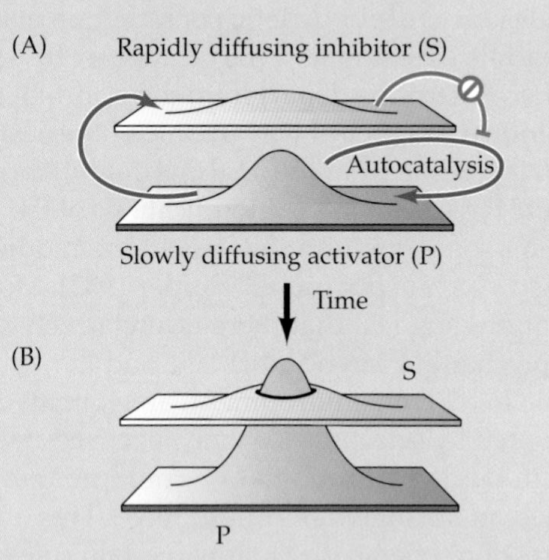

Reaction-diffusion (Turing model) system of pattern generation. Generation of periodic spatial heterogeneity can come about spontaneously when two reactants, S and P, are mixed together under the conditions that S inhibits P, P catalyzes production of both S and P, and S diffuses faster than P. (A) The conditions of the reaction-diffusion system yielding a peak of P and a lower peak of S at the same place. (B) The distribution of the reactants is initially random, and their concentrations fluctuate over a given average. As P increases locally, it produces more S, which diffuses to inhibit more peaks of P from forming in the vicinity of its production. The result is a series of P peaks ("standing waves") at regular intervals.

pleiotropy. In a sense, pleiotropy is the *opposite* of modularity, involving interdependent connections between parts rather than their independence. Galis and colleagues (2002) provide evidence that the reason the segment polarity gene network is conserved in all types of insects is that these genes play roles in several different pathways. Such pleiotropy constrains the possibilities for alternative mechanisms, since it makes change difficult.

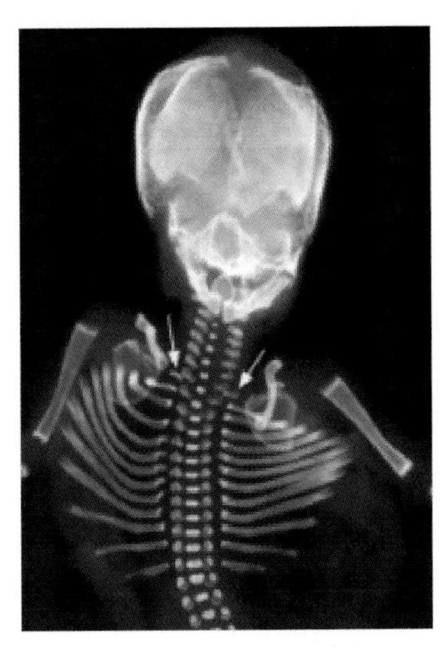

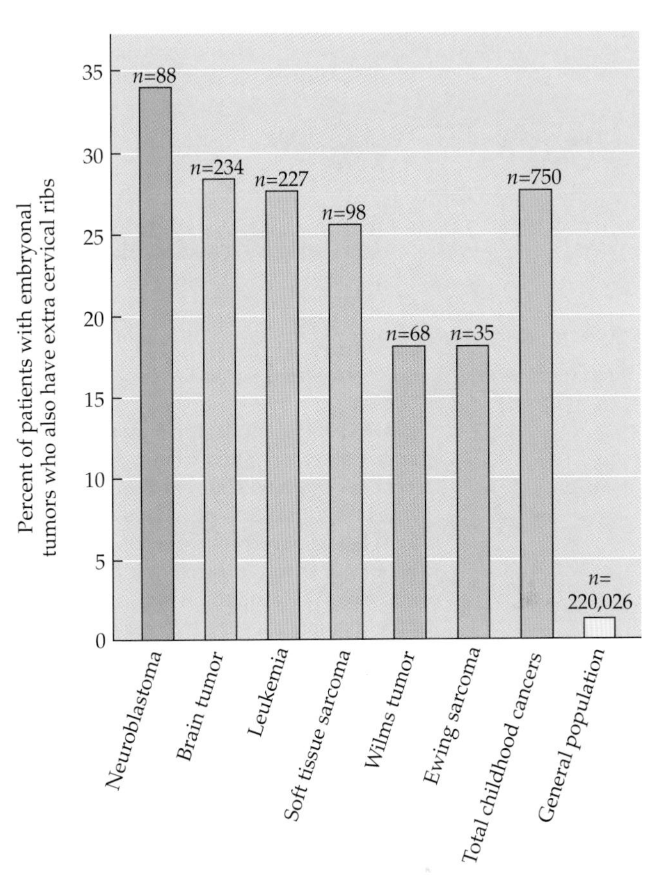

FIGURE 9.20 Nearly 80% of fetuses with extra cervical ribs die before birth. Those surviving have extremely high rates of cancers. The association of this axial abnormality with lower fitness suggests pleiotropy, namely that the specification of the cervical ribs is associated with the regulation of cell division. Thus, it is difficult to change axial specification (in this case, the number of cervical ribs) without causing abnormalities of mitosis that can lead to cancer. (From Galis et al. 2006.)

Pleiotropies may underlie some of the constraints seen in mammalian development. Galis speculates that mammals have only seven cervical vertebrae (while birds may have dozens) because the Hox genes that specify these vertebrae have become linked to cell proliferation in mammals (Galis 1999; Galis and Metz 2001). Thus, changes in Hox gene expression that might facilitate evolutionary changes in the skeleton might also misregulate cell proliferation and lead to cancers. She supports this speculation with epidemiological evidence showing that changes in skeletal morphology correlate with childhood cancer. Selection against embryos having more or fewer than seven cervical ribs appears to be remarkably strong. At least 78% of human embryos with an extra anterior rib (i.e., six cervical vertebrae) die before birth, and 83% die before they are a year old (Figure 9.20). These deaths appear to be the result of multiple congenital anomalies or cancers (Galis et al. 2006).

A third type of phylogenetic constraint involves the recruitment of existing genes for new functions. The pigmentation of *Drosophila* wings, for instance, is not random. Wing pigmentation is seen only in those places

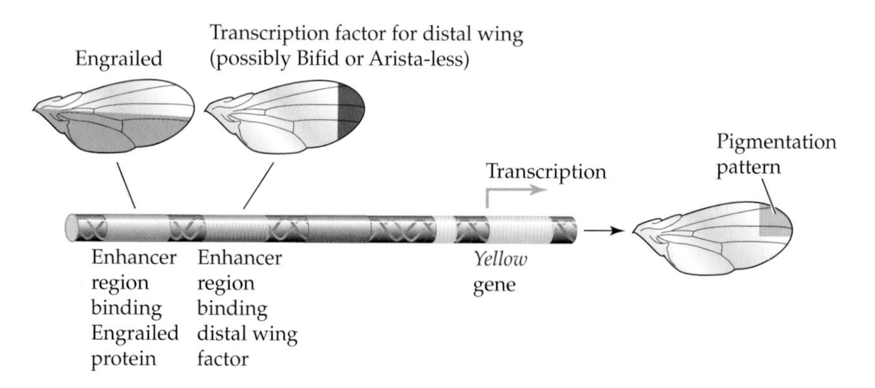

FIGURE 9.21 Model for pigmentation patterns in the fly wing depending on the pattern of transcription factors involved in wing development. The enhancers of different species bind different transcription factors. Here the wing pattern of a male *Drosophila biarmipes* is described. The *Yellow* gene (for the synthesis of melanin pigment) has acquired enhancer elements that cause it to be expressed in the wing tip. The enhancer for *Yellow* binds a transcriptional activator (possibly the Bifid protein) that allows the gene to be expressed in the wing tip; and the Engrailed transcription factor that inhibits gene expression in the ventral half of the wing. Together, they enable the activation of the *Yellow* gene in the anterior tip of the wing. (After Prud'homme et al. 2007.)

where enhancers have already established wing boundaries. Thus, the genes encoding the enzymes for pigment synthesis can acquire enhancers that respond to transcription factors involved in wing formation (Figure 9.21; Prud'homme et al. 2007; Jeong et al. 2008). Not only is this condition a remarkable case of enhancer modularity, but it is also an excellent example of why it is necessary to understand principles of development in order to explain adult phenotypes. As the example here shows, developmental constraints not only prevent evolution in some directions but also bias it in other directions (see Arthur 2004). If one does not know the means by which the wing is constructed, one cannot make sense of the various pigmentation patterns or understand why some seemingly obvious patterns are in fact not found.

Summary

Darwin said evolution is the result of natural selection. His friend and defender, Thomas Huxley, wrote in 1893 that "Evolution is not a speculation but a fact; and it takes place by epigenesis" (p. 202). Darwin was emphasizing a theory of population change; Huxley was emphasizing a theory of body construction. Both theories are correct.

Evolution depends on heritable changes in development. When we say that the contemporary one-toed horse has an ancestor with five toes, we are saying that during equine evolution, there have been changes in the placement and growth of its limb cartilage cells (Wolpert 1983). These changes may be caused by alterations in the timing, placement, and amount of gene expression. They can also be caused by changes in the protein being synthesized, and this is especially important when the protein is involved in organ formation or cell differentiation. Modularity and gene duplication have enabled these changes to occur without causing major disastrous changes in the developing organism. However, in some cases, the possible routes of evolutionary change are constrained by developmental and physical parameters.

Changes in development are necessary in order to explain the types of variation that can be challenged by natural selection. During the past few years, the population genetics of regulatory genes has emerged as a fertile field of study. This promises to be an exciting area that will further integrate evolutionary developmental biology into the evolutionary synthesis. Moreover, as gene expression patterns and comparative genomics make it possible to define precisely which sequences have been critical for phylogenetic change, population genetics may become integrated into a larger perspective of phylogenetic evolution.

So far, we have looked at two major sources of variation. The Modern Synthesis focused on the variation produced by differences in the protein-encoding sequences of DNA (such as those for globin and rhodopsin). Much of evolutionary developmental biology focuses on variation produced by differences in the regulatory regions of genes active in constructing the embryo. The next chapter continues this discussion of development in evolution by adding phenotypic plasticity into the merger of developmental biology and evolution.

References

Abzhanov, A., M. Protas, B. R. Grant, P. R. Grant and C. J. Tabin. 2004. *Bmp4* and morphological variation of beaks in Darwin's finches. *Science* 305: 1462–1465.

Abzhanov, A., W. P. Kuo, C. Hartmann, B. R. Grant, P. R. Grant and C. J. Tabin. 2006. The calmodulin pathway and evolution of elongated beak morphology in Darwin's finches. *Nature* 442: 563–567.

Alberch, P. 1989. The logic of monsters: Evidence for internal constraints in development and evolution. *Geobios* (Lyon), Mémoires Spécial 12: 21–57.

Amundson, R. 2005. *The Changing Role of the Embryo in Evolutionary Thought: Structure and Synthesis.* Cambridge University Press, Cambridge.

Appel, T. A. 1987. *The Cuvier-Geoffroy Debate: French Biology in the Decades before Darwin.* Oxford University Press, New York.

Arthur, W. 2004. *Biased Embryos and Evolution.* Cambridge University Press, Cambridge.

Bateson, W. 1894. *Materials for the Study of Variation.* Cambridge University Press, Cambridge.

Berrill, N. J. 1987. Early chordate evolution. I. *Amphioxus*, the riddle of the sands. *Int. J. Invert. Reprod. Dev.* 11: 1–27.

Bertrand, N., D. S. Castro and F. Guillemot. 2002. Proneural genes and the specification of neural cell types. *Nature Rev. Neurosci.* 3: 517–530.

Biermann, C. H., J. A. Marks, A. C. Vilela-Silva, M. O. Castro and P. A. Mourao. 2004. Carbohydrate-based species recognition in sea urchin fertilization: Another avenue for speciation? *Evol. Dev.* 6: 353–361.
Bolker, J. A. 2000. Modularity in development and why it matters to evo-devo. *Am. Zool.* 40: 770–776.
Burke, A. C., A. C. Nelson, B. A. Morgan and C. Tabin. 1995. Hox genes and the evolution of vertebrate axial morphology. *Development* 121: 333–346.
Caldwell, M. W. and M. S. Y. Lee. 1997. A snake with legs from the marine Cretaceous of the Middle East. *Nature* 386: 705–709.
Carroll, S. B. 2005. *Endless Forms Most Beautiful: The New Science of Evo Devo*. Norton, New York.
Carroll, S. B. 2008. Evo-devo and an expanding evolutionary synthesis: A genetic theory of morphological evolution. *Cell* 134: 25–36.
Carroll, S. B., J. K. Grenier and S. D. Weatherbee. 2001. *From DNA to Diversity: Molecular Genetics and the Evolution of Animal Design*. Blackwell, Malden, MA.
Chiang, C., Y. Litingtung, E. Lee, K. E. Young, J. L. Cordoen, H. Westphal and P. A. Beachy. 1996. Cyclopia and axial patterning in mice lacking *sonic hedgehog* gene function. *Nature* 383: 407–413.
Chow, R. L., C. R. Altmann, R. A. Lang and A. Hemmati-Brivanlou. 1999. *Pax6* induces ectopic eyes in a vertebrate. *Development* 126: 4213–4222.
Cohn, M. J. and C. Tickle. 1999. Developmental basis of limblessness and axial patterning in snakes. *Nature* 399: 474–479.
Colosimo, P. F., C. L. Peichel, K. Nereng, B. K. Blackman, M. D. Shapiro, D. Schluter and D. M. Kingsley. 2004. The genetic architecture of parallel armor plate reduction in threespine sticklebacks. *PLOS Biol.* 2: 635–641.
Colosimo, P. F. and 9 others. 2005. Widespread parallel evolution in sticklebacks by repeated fixation of ectodysplasin alleles. *Science* 307: 1928–1933.
Coyne, J. A. and H. A. Orr. 2004. *Speciation*. Sinauer Associates, Sunderland, MA.
Dale, L. and F. C. Wardle. 1999. A gradient of BMP activity specifies dorsal–ventral fates in early *Xenopus* embryos. *Sem. Cell Dev. Biol.* 10: 319–326.
Damen, W. G. 2002. *Fushi tarazu:* A Hox gene changes its role. *BioEssays* 24: 992–995.
Davidson, E. H. 2006. *The Regulatory Genome: Gene Regulatory Networks in Development and Evolution*. Academic Press, New York.
De Robertis, E. M. and Y. Sasai. 1996. A common plan for dorsoventral patterning in Bilateria. *Nature* 380: 37–40.
de Rosa, R. and 7 others. 1999. Hox genes in brachiopods and priapulids and protostome evolution. *Nature* 379: 772–776.
Diamond, J. 2002. *The Third Chimpanzee: The Evolution and Future of the Human Animal.* HarperCollins, New York.
Dobzhansky, Th. 1955. *Evolution, Genetics, and Man.* Wiley, New York.
Dobzhansky, Th. 1964. Biology, molecular and organismic. *Amer. Zool.* 4: 443–452 (p. 443).
Enard, W. and 12 others. 2002a. Intra- and interspecific variation in primate gene expression patterns. *Science* 296: 340–343.
Enard, W. and 7 others. 2002b. Molecular evolution of *FOXP2*, a gene involved in speech and language. *Nature* 418: 869–872.
Finnerty, J. R., K. Pang, P. Burton, D. Paulson and M. Q. Martindale. 2004. Origins of bilateral symmetry: *Hox* and *Dpp* expression in a sea anemone. *Science* 304: 1335–1337.
Forgacs, G. and S. A. Newman. 2005. *Biological Physics of the Developing Embryo.* Cambridge University Press, Cambridge.
Galant, R. and S. B. Carroll. 2002. Evolution of a transcriptional repression domain in an insect Hox protein. *Nature* 415: 910–913.
Galis, F. 1999. Why do almost all mammals have seven cervical vertebrae? Developmental constraints, Hox genes, and cancer. *J. Exp. Zool./Mol. Dev. Evol.* 285: 19–26.
Galis, F. and J. A. Metz. 2001. Testing the vulnerability of the phylotypic stage: On modularity and evolutionary conservation. *J. Exp. Zool./Mol. Dev. Evol.* 29: 195–204.
Galis, F., T. J. M. van Dooren and J. A. Metz. 2002. Conservation of the segmented germband stage: Robustness or pleiotropy? *Trends Genet.* 18: 504–509.
Galis, F. and 7 others. 2006. Extreme selection against homeotic transformations of cervical vertebrae in humans. *Evolution Int. J. Org. Evolution* 60: 2643–2654.
Gaunt, S. J. 1994. Conservation in the Hox code during morphological evolution. *Int. J. Dev. Biol.* 38: 549–552.
Gehring, W. J. 1998. *Master Control Genes in Development and Evolution: The Homeobox Story*. Yale University Press, New Haven.
Gehring, W. 2005. New perspectives on eye development and the evolution of eyes and photoreceptors. *J. Heredity* 96: 171–184.
Gilbert, S. F. 1996. Cellular dialogues in organogenesis. *In* M. E. Martini-Neri, G. Neri and J. M. Opitz (eds.), *Gene Regulation and Fetal Development: Proceedings of the Third International Workshop on Fetal Genetic Pathology, June 3-6, 1993.* Wiley-Liss, New York, pp. 1–12.
Gilbert, S. F. 2006. The generation of novelty: The province of developmental biology. *Biological Theory* 1: 209–212.
Gilbert, S. F. and R. Burian. 2003. Development, evolution, and evolutionary developmental biology. In B. K. Hall and W. M. Olson (eds.), *Key Concepts and Approaches in Evolutionary Developmental Biology.* Harvard University Press, Cambridge, MA, pp. 61–68.

Gilbert, S. F., J. M. Opitz and R. A. Raff. 1996. Resynthesizing evolutionary and developmental biology. *Dev. Biol.* 173: 357–372.

Gomez, C., E. M. Özbudak, J. Wunderlich, D. Baumann and O. Pourquié. 2008. Control of segment number in vertebrate embryos. *Nature* 454: 335–338.

Gompel, N., B. Prud'homme, P. J. Wittkop, V. A. Kassner and S. B. Carroll. 2005. Chance caught on the wing: *Cis*-regulatory evolution and the origin of pigment patterns in *Drosophila*. *Nature* 433: 481–487.

Goodman, M. 1999. The genomic record of humankind's evolutionary roots. *Am. J. Hum. Genet.* 64: 31–39.

Gould, S. J. 1977. *Ontogeny and Phylogeny*. Harvard University Press, Cambridge, MA.

Gould, S. J. and R. C. Lewontin. 1979. The spandrels of San Marcos and the Panglossian paradigm: A critique of the adaptationist programme. *Proc. R. Soc. Lond. B* 205: 581–598.

Grant, P. 1999. *The Ecology and Evolution of Darwin's Finches*, Rev. Ed. Princeton University Press, Princeton, NJ.

Grant, P. R. and B. R. Grant. 2007. *How and Why Species Multiple: The radiation of Darwin's Finches*. Princeton University Press, Princeton, NJ.

Halder, G., P. Callaerts and W. J. Gehring. 1995. Induction of ectopic eyes by targeted expression of the *eyeless* gene in *Drosophila*. *Science* 267: 1788–1792.

Hall, B. K. 1999. *Evolutionary Developmental Biology*. Kluwer Academic Publishers, Dordrecht.

Harmon, M. A., M. F. Boehm, R. A. Heyman and D. J. Mangelsdorf. 1995. Activation of mammalian retinoid X receptors by the insect growth regulator methoprene. *Proc. Natl. Acad. Sci. USA* 92: 6157–6160.

Harris, A. 1904. Quoted in H. DeVries. 1904. *Species and Varieties: Their Origin by Mutation*. Open Court Publishing, Chicago, p. 401.

Hemmati-Brivanlou, A. and D. A. Melton. 1997. Vertebrate embryonic cells will become nerve cells unless told otherwise. *Cell* 88: 13–17.

Hill, A., A. Wagner and M. Hill. 2003. Hox and paraHox genes from the anthozoan *Parazoanthus parasiticus*. *Mol. Phylog. Evol.* 28: 529–535.

Hirth, F. and H. Reichert. 1999. Conserved genetic programs in insect and mammalian brain development. *BioEssays* 21: 677–684.

Holder, N. 1983. Developmental constraints and the evolution of vertebrate limb patterns. *J. Theor. Biol.* 104: 451–471.

Holland, P. W. H. and J. Garcia-Fernández. 1996. Hox genes and chordate evolution. *Dev. Bio.* 173: 382–395.

Holley, S., P. D. Jackson, Y. Sasai, B. Lu, E. M. De Robertis, F. M. Hoffmann and E. L. Ferguson. 1995. A conserved system for dorsal-ventral patterning in insects and vertebrates involving sog and chordin. *Nature* 376: 249–253.

Hughes, C. L. and T. C. Kaufman. 2002. Hox genes and the evolution of the arthropod body plan. *Evol. Dev.* 4: 459–499.

Huxley, T. 1893. *Darwiniana: Collected Essays*, Vol. II. Macmillan, London.

Jacob, F. 1977. Evolution and tinkering. *Science* 196: 1161–1166.

Jeong, S., M. Rebeiz, P. Andolfatto, T. Werner, J. True and S. B. Carroll. 2008. The evolution of gene regulation underlies a morphological difference between two *Drosophila* sister species. *Cell* 132: 783–793.

Jernvall, J. 2006. *The curious incident of the seal in the pond*. Plenary session. European Society for Evolutionary Developmental Biology meeting, Prague, August 2006.

Jiggins, C. D., R. E. Naiit, R. L. Coe and J. Mallet. 2001. Reproductive isolation caused by colour pattern mimicry. *Nature* 411: 302–305.

King, M.-C. and A. C. Wilson. 1975. Evolution at two levels in humans and chimpanzees. *Science* 188: 107–116.

King, N. and 35 others. 2008. The genome of the choanoflagellate *Monosiga brevicollis* and the origins of metazoans. *Nature* 451: 783–788.

King-Jones, K. and C. S. Thummel. 2005. Nuclear receptors: A perspective from *Drosophila*. *Nature Rev. Genet.* 6: 311–323.

Kirschner, M. W. and J. C. Gerhart. 2005. *The Plausibility of Life: Resolving Darwin's Dilemma*. Norton, New York.

Larroux, C., B. Fahey, S. M. Degnan, M. Adamski, D. S. Rokhsar and B. M. Degnan. 2007. The NK homoeobox gene cluster predates the origin of Hox genes. *Curr. Biol.* 17: 706–710.

Laufer, E., S. Pizette, H. Zou, O. E. Orozco and L. Niswander. 1997. BMP expression in duck interdigital webbing: A reanalysis. *Science* 278: 305.

Levitan, D. R. and D. L. Ferrell. 2006. Selection on gamete recognition proteins depends on sex, density, and genotype frequency. *Science* 312: 267–269.

Locascio, A., M. Manjanares, M. J. Blanco and M. A. Nieto. 2002. Modularity and reshuffling of *Snail* and *Slug* expression during vertebrate evolution. *Proc. Natl. Acad. Sci. USA* 99: 16841–16846.

Lynch, M. and J. S. Conery. 2000. The evolutionary fate and consequences of duplicate genes. *Science* 290: 1151–1155.

Malicki, J., L. C. Cianetti, C. Peschle and W. McGinnis. 1992. Human *HOX4B* regulatory element provides head-specific expression in *Drosophila* embryos. *Nature* 358: 345–347.

Maxam, A. and W. Gilbert. 1977. A new method for sequencing DNA. *Proc. Natl. Acad. Sci. USA* 74: 560–564.

McGregor, A. P., V. Orgogozo, I. Delon, J. Zanet, D. G. Srinivasan, F. Payre and D. L. Stern. 2007. Morphological evolution through multiple *cis*-regulatory mutations at a single gene. *Nature* 448: 587–590.

Meehan, T. J. and L. D. Martin. 2003. Extinction and re-evolution of similar adaptive types (ecomorphs) in Cenozoic North American ungulates and carnivores reflect van der Hammen's cycles. *Naturwiss.* 90: 131–135.

Merino, R., J. Rodríguez-Leon, D. Macias, Y. Ganan, A. N. Economides and J. M. Hurle. 1999. The BMP antagonist Gremlin regulates outgrowth, chondrogenesis and programmed cell death in the developing limb. *Development* 126: 5515–5522.

Mizutani, C. M. and E. Bier. 2008. EvoD/Vo: The origins of BMP signaling in the neuroectoderm. *Nature Rev. Genet.* 9: 663–675.

Morange, M. 2001. *The Misunderstood Gene.* Harvard University Press, Cambridge, MA.

Müller, G. B. 2007. Evo-devo: Extending the evolutionary synthesis. *Nature Rev. Genet.* 8: 943–949.

Mundy, N. I. 2005. A window on the genetics of evolution: MC1R and plumage colouration. *Proc. Biol. Soc.* 272: 1633–1640.

Naisbit, R. E., C. D. Jiggins and J. Mallet. 2003. Mimicry: Developmental genes that contribute to speciation. *Evol. Dev.* 5: 269–280.

Newman, S. A. and G. B. Müller. 2005. Origination and innovation in the vertebrate limb skeleton: An epigenetic perspective. *J. Exp. Zool.* 304B: 593–609.

Oakley, T. H. and C. W. Cunningham. 2002. Molecular phylogenetic evidence for the independent evolutionary origin of an arthropod compound eye. *Proc. Natl. Acad. Sci. USA* 99: 1426–1430.

Ohno, S. 1970. *Evolution by Gene Duplicaiton.* Springer, Berlin.

Ospovat, D. 1981. *The Development of Darwin's Theory: Natural History, Natural Theology, and Natural Selection.* Cambridge University Press, New York.

Oster, G. F., N. Shubin, J. D. Murray and P. Alberch. 1988. Evolution and morphogenetic rules: The shape of the vertebrate limb in ontogeny and phylogeny. *Evolution* 42: 862–884.

Palumbi, S. R. and H. A. Lessios. 2005. Evolutionary animation: How do molecular phylogenies compare to Mayr's reconstruction of speciation patterns in the sea? *Proc. Natl. Acad. Sci. USA* 102 (suppl. 1): 6566–6572.

Pigliucci, M. 2001. *Phenotypic Plasticity: Beyond Nature and Nurture.* Johns Hopkins University Press, Baltimore.

Pigliucci, M. 2008. Is evolvability evolvable? *Nature Rev. Genet.* 9: 75–82.

Pollard, K. S. and 15 others. 2006. An RNA gene expressed during cortical development evolved rapidly in humans. *Nature* 443: 167–172.

Preuss, T. M., M. Caceres, M. C. Oldham and D. H.Geschwind. 2004. Human brain evolution: Insights from microarrays. *Nature Rev. Genet.* 5: 850–860.

Prigogine, I. and G. Nicolis. 1967. On symmetry-breaking instabilities in dissipative systems. *J. Chem. Phys.* 46: 3542–3550.

Prud'homme, B., N. Gompel and S. B. Carroll. 2007. Emerging principles of regulatory evolution. *Proc. Natl. Acad. Sci. USA* 104: 8605–8612.

Raff, R. A. 1996. *The Shape of Life: Genes, Development, and the Evolution of Animal Form.* University of Chicago Press, Chicago.

Richardson, M. 1999. Vertebrate evolution: The developmental origins of adult variation. *BioEssays* 21: 604–613.

Richardson, M. D. and H. H. Oelschläger. 2002. Time, pattern and heterochrony: A study of hyperphalangy in the dolphin embryo flipper. *Evol. Dev.* 4: 435–444.

Richardson, M. K. and A. D. Chipman. 2003. Developmental constraints in a comparative framework: A test case using variations in phalanx number during amniote evolution. *J. Exp. Zool. (MDE) B* 296: 8–22.

Riedl, R. 1978. *Order in Living Systems: A Systems Analysis of Evolution.* John Wiley and Sons, New York.

Rockman, M. V. and G. A. Wray. 2002. Abundant raw material for *cis*-regulatory evolution in humans. *Mol. Biol. Evol.* 19: 1991–2004.

Rockman, M. V., M. W. Hahn, N. Soranzo, D. B. Goldstein and G. A. Wray. 2003. Positive selection on a human-specific transcription factor binding site regulating IL4 expression. *Curr. Biol.* 13: 2118–2123.

Ronshaugen, M., N. McGinnis and W. McGinnis. 2002. Hox protein mutation and macroevolution of the insect body plan. *Nature* 415: 914–917.

Sanger, F., S. Nicklen and A. R. Coulson. 1977. DNA sequencing with chain-terminating inhibitors. *Proc. Natl. Acad. Sci. USA* 74: 5463–5467.

Schlosser, G. and G. P. Wagner. 2004. *Modularity in Development and Evolution.* University of Chicago Press, Chicago.

Schluter, D. 2000. *The Ecology of Adaptive Radiation.* Oxford University Press, Oxford.

Schmidt, J., V. Francoise, E. Bier and D. Kimelman. 1995. *Drosophila* short gastrulation induces an ectopic axis in *Xenopus:* Evidence for conserved mechanisms of dorsoventral patterning. *Development* 121: 4319–4328.

Schneider, R. A. and J. A. Helms. 2003. The cellular and molecular origins of beak morphology. *Science* 299: 565–568.

Sears, K. E. 2004. Constraints on the evolution of morphological evolution of marsupial shoulder girdles. *Evolution* 58: 2353–2370.

Shapiro, M. D., J. Hanken and N. Rosenthal. 2003. Developmental basis of evolutionary digit loss in the Australian lizard *Hemiergis. J. Exp. Zool.* 279B: 48–56.

Shapiro, M. D., and 7 others. 2004. Genetic and developmental basis of evolutionary pelvic reduction in threespine sticklebacks. *Nature* 428: 717–723.

Shapiro, M. D., M. A. Bell and D. M. Kingsley. 2006. Parallel genetic origins of pelvic reduction in vertebrates. *Proc. Natl. Acad. Sci. USA* 103: 13753–13758.

Smith, K. 2003. Time's arrow: Heterochrony and the evolution of development. *Int. J. Dev. Biol.* 47: 613–621.

Sucena, E. and D. Stern. 2000. Divergence of larval morphology between *Drosophila sechellia* and its sibling species caused by *cis*-regulatory evolution of *ovo/shaven-baby*. *Proc. Natl. Acad. Sci. USA* 97: 4530–4534.

Swanson, W. J. and V. D. Vacquier. 2002. The rapid evolution of reproductive proteins. *Nature Rev. Genet.* 3: 137–144.

Tournamille, C., Y. Colin, J. P. Cartron and C. Le Van Kim. 1995. Disruption of a GATA motif in the *Duffy* gene promoter abolishes erythroid gene expression in Duffy-negative individuals. *Nature Genet.* 10: 224–228.

Turing, A. M. 1952. The chemical basis of morphogenesis. *Philos. Trans. R. Soc. Lond. B* 237: 37–72.

von Dassow, M V. and L. A. Davidson. 2008. Variation and robustness of the mechanics of gastrulation: The role of tissue mechanical properties during morphogenesis. *Birth Def. Res. C: Embryo Today* 81: 253–269.

Waddington, C. H. 1938. The morphogenetic function of a vestigial organ in the chick. *J. Exp. Biol.* 15: 371–376.

Wake, D. B. and A. Larson. 1987. A multidimensional analysis of an evolving lineage. *Science* 238: 42–48.

Wang, H. and 8 others. 2005. The origin of the naked grains of maize. *Nature* 436: 714–719.

Weatherbee, S. D., R. R. Behringer, J. J. Rasweiler, IV, and L. A. Niswander. 2006. Interdigital webbing retention in bat wings illustrates genetic changes underlying amniote limb diversification. *Proc. Natl. Acad. Sci. USA* 103: 15103–15107.

Weiner, J. 1994. *The Beak of the Finch: A Story of Evolution in Our Time.* Random House, New York.

Wellik, D. M. and M. R. Capecchi. 2003. *Hox10* and *Hox11* genes are required to globally pattern the mammalian skeleton. *Science* 301: 363–367.

West-Eberhard, M. J. 2003. *Developmental Plasticity and Evolution*. Oxford University Press, New York.

West-Eberhard, M. J. 2005. Developmental plasticity and the origin of species differences. *Proc. Natl. Acad. Sci. USA* 102: 6543–6549.

Winfree, A. T. 1974. Rotating chemical reactions. *Sci. Am.* 230(6): 82–95.

Wolpert, L. 1983. Constancy and change in the development and evolution of pattern. *In* B. C. Goodwin, N. Holder and C. C. Wylie (eds.), *Development and Evolution*. Cambridge University Press, Cambridge, pp. 47–57.

Wong, K. H., H. D. Wintch and M. R. Capecchi. 2004. Hoxa11 regulates stromal cell death and proliferation during neonatal uterine development. *Mol. Endocrinol.* 18: 184–193.

Wu, P., T. X. Jiang, S. Suksaweang, R. B. Widelitz and C. M. Chuong. 2004. Molecular shaping of the beak. *Science* 305: 1465–1466.

Wu, P., T. X. Jiang, J. Y. Shen, R. B. Widelitz and C. M. Chuong. 2006. Morphoregulation of avian beaks: Comparative mapping of growth zone activities and morphological evolution. *Dev. Dyn.* 235: 1400–1412.

Yanze, N., J. Spring, C. Schmidli and V. Schmid. 2001. Conservation of Hox/ParaHox-related genes in the early development of a cnidarian. *Dev. Biol.* 236: 89–98.

Zuckerkandl, E. 1994. Molecular pathways to parallel evolution. I. Gene nexuses and their morphological correlates. *J. Mol. Evol.* 39: 661–678.

Chapter 10

Environment, Development, and Evolution

Toward a New Synthesis

In my opinion, the greatest error which I have committed, has not been allowing sufficient weight to the direct action of the environment, i.e. food, climate, etc., independently of natural selection.

Charles Darwin to Moritz Wagner, 1876

A plausible argument could be made that evolution is the control of development by ecology. Oddly, neither area has figured importantly in evolutionary theory since Darwin, who contributed much to each.

Leigh Van Valen, 1973

The importance of development for evolutionary biology is not limited to the role of developmental regulatory genes discussed in Chapter 9. The study of developmental plasticity points to something quite unexpected in evolutionary theory: that the environment not only *selects* variation, it helps *construct* variation. This integration of ecological developmental biology into evolutionary biology has sometimes been referred to as **ecological evolutionary developmental biology**, or "eco-evo-devo."

Eco-evo-devo doesn't question that evolution occurs, any more than one would question that gravity exists. However, it does recognize that ecological developmental biology has called into question three of the assumptions of the Modern Synthesis upon which the genetic theory of evolution has been predicated:

- *"All evolutionarily significant variation is heritable, and therefore caused by genetic variation in alleles."* Ecological developmental biology demonstrates that *epigenetic variation* can also be transmitted from generation to generation and is therefore selectable.
- *"Organisms are genetically single individuals."* Ecological developmental biology shows that organisms may be more like ecosystems, composed of numerous genotypes that interact with each other. This may allow natural selection to favor "teams" rather than particular individuals and may also privilege "relationships" as a unit of selection.
- *"The environment is a selective agent but is of no consequence in producing the phenotype."* Ecological developmental biology has shown that the environment can instruct which phenotype can be produced from the genetic repertoire of the nucleus. It helps produce phenotypes that might be adaptive in a specific environment.

Several historians and biologists have pointed out that the Modern Synthesis has been a constriction as well as an integration, in that it has constrained the direction of evolutionary research (see Provine 1989; Gould 2002). By applying knowledge gained from the studies now being done in ecological developmental biology, a new and more inclusive evolutionary theory may be forged. So far, eco-devo has contributed at least three components to this nascent evolutionary synthesis. The first of these is the concept of **epigenetic inheritance systems**, the idea that inherited traits can be passed on through changes in chromatin structure as well as changes in DNA sequence. The second, **heterocyberny**, is the concept that environmentally induced changes to a phenotype, when adaptive over long periods of time, can become the genetic norm for a species. The third concept is that of **niche construction**, wherein the developing organism can modify its environment due to plastic features in its habitat.

EPIGENETIC INHERITANCE SYSTEMS

We have seen throughout this book that there are, in addition to the genetic inheritance system, certain epigenetic inheritance systems, defined as systems that enable the "phenotypic expression of the information in a cell or an individual to be transmitted to the next generation" (Jablonka and Lamb 1995; Jablonka and Raz 2009). We have discussed several examples of transgenerational, epigenetically inherited variation, including:

- In *Daphnia*, the predator-induced morph is propagated in subsequent generations even when the predator is not present (Agrawal et al. 1999).

- The gregarious and solitary morphs of the migratory locust are transmitted from one generation to another through the foam placed on the eggs by the mother (McCaffery and Simpson 1998).
- The inheritance of symbionts that carry instructions for development can be through the germline (as in the vertical inheritance of *Wolbachia* through the insect egg cytoplasm) or by infection from the mother or the environment (as in the acquisition of gut symbionts in mammals).
- In the *viable Agouti* phenotype in mice, methylation differences affect coat color and obesity. When a pregnant female is fed a diet that contains substantial levels of methyl donors, the specific methylation pattern at the *Agouti* locus is transmitted not only to the progeny developing in utero, but also to the progeny of those mice and to their progeny (Jirtle and Skinner 2007).
- Enzymatic and metabolic phenotypes are established in utero by protein-restricted diets in mice. Protein restriction during a female mouse's pregnancy leads to a specific methylation pattern in her pups and grandpups (Burdge et al. 2007).
- The endocrine disruptors vinclozolin, methoxychlor, and bisphenol A have the ability to alter DNA methylation patterns in the germline, thereby causing illnesses and predispositions to diseases in the grandpups of mice exposed to these chemicals in utero (Anway et al. 2005, 2006a,b; Newbold et al. 2006; Chang et al. 2006; Crews et al. 2007).
- Stress-resistant behavior of rats was shown to be due to methylation patterns, induced by maternal care, in the glucocorticoid receptor genes. Meaney (2001) found that rats who received extensive maternal care had less stress-induced anxiety and, if female, developed into mothers who gave their offspring similar levels of maternal care.

The latter four examples demonstrate that altered chromatin, once formed, can be inherited from one generation to the next. These altered chromatin configurations act just as genetic alleles might act and are therefore referred to as **epialleles**. It does not matter to the developing system whether a gene has been inactivated by a mutation or by an altered chromatin configuration; the effect is the same (Figure 10.1). Epialleles have been found in many instances of inheritance. The epiallele of toadflax (see Figure 2.19) that so confounded Linnaeus is only one of numerous plant epialleles known, and epialleles have been described in many animals, including humans (see Appendix D; Cabej 2008; Jablonka and Raz 2009).

In addition, epialleles appear to play roles in sexual selection. The fungicide vinclozolin is an anti-androgenic endocrine disruptor, inducing DNA methylation changes that can last for several generations (see Chapter 6). Crews and colleagues (2007) have shown that female mice three genera-

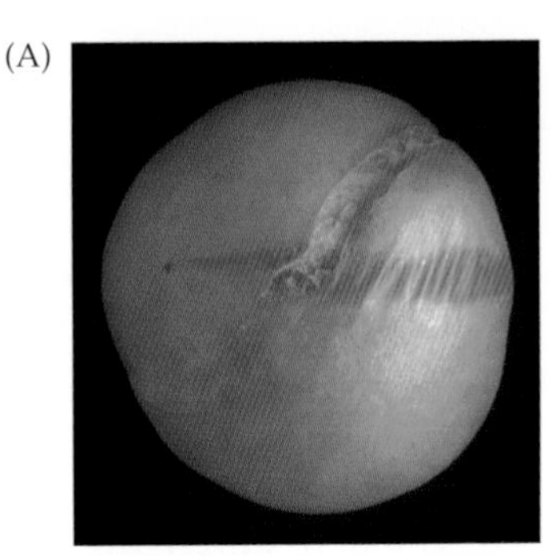

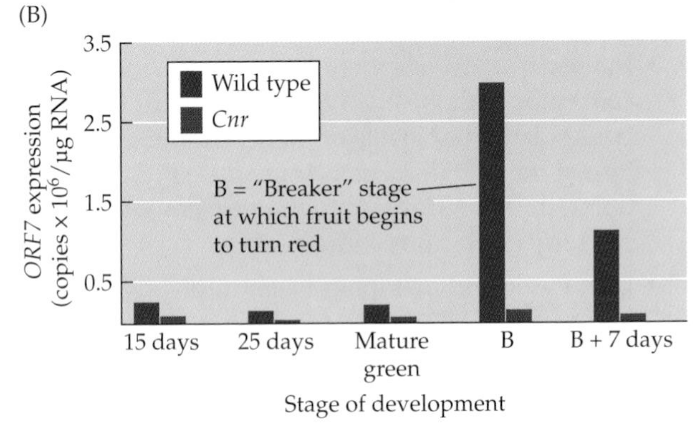

FIGURE 10.1 The *Cnr* (colorless, non-ripening) mutation. An allele of *Cnr*, a gene involved in fruit ripening, causes lack of pigmentation and loss of cell adhesion in the fruits. The sequences of the wild-type and the "mutant" are identical; the "mutant" allele turns out to be an epiallele caused by extensive DNA methylation, which turns the gene off. (A) A *Cnr* mutant tomato. The red "stripe" represents a reversion to the wild-type. (B) In wild-type plants, *ORF7* mRNA from the *Cnr* gene is expressed at the time the fruit ripens. *ORF7* is not expressed in plants whose fruit is colorless and does not ripen. (After Manning et al. 2006; photograph courtesy of K. Manning.)

tions removed from in utero exposure to vinclozolin prefer males who have no history of vinclozolin exposure. In other words, males whose parents or grandparents had been exposed to vinclozolin were less attractive to normal females. (Males had no preference for females with or without a history of vinclozolin exposure.) Thus, conclude the researchers, "an environmental factor can promote a transgenerational alteration in the epigenome that influences sexual selection and could impact the viability of population and evolution of the species."

The concept of epiallelic inheritance brings back into evolutionary biology the notion of the inheritance of acquired characters. While this model does not represent the Lamarckian "use and disuse" principle (see Appendix D), the above cases indicate that if the germline DNA is susceptible to methylation or other epigenetic changes, then mutation is not needed to inactivate these genes, and the inheritance of these environmentally induced chromatin alterations is possible. Thus epiallelic inheritance (as well as the inheritance of symbionts) must be included in a new evolutionary synthesis.

HETEROCYBERNY: PLASTICITY-DRIVEN ADAPTATION

One set of proposals taking seriously the ideas of developmental plasticity involves the production of adaptive phenotypes through developmental plasticity followed by the "fixing" or "stabilization" of those phenotypes through natural selection. This idea has a long history. Even before the rediscovery of Mendel's laws in 1900, biologists such as Gulick (1872), Spalding (1873), Baldwin (1896), Lloyd Morgan (1896), and Osborn (1897) had been impressed by developmental and behavioral plasticities. These scientists envisioned ways by which physiological responses to environmental cues might become genetically fixed and thus would not need the environmental inducer in order to be expressed. In other words, if a physiological response to the environment became adaptive in all situations an organism was likely to meet, this trait would become expressed through genes rather than by environmental agents. The response then would be initiated within the organism by cell-cell interaction rather than by interaction between the organism's cells and the environment. However, without a theory of gene transmission to undergird such a concept, it remained merely an interesting speculation.

The concept of one of the morphs of a phenotypically plastic trait becoming the genetically transmitted standard ("wild-type") for that species has gone under many names. In accordance with the four mechanisms of evolutionary "tinkering" discussed in Chapter 9—heterochrony, heterotopy, heterometry, and heterotypy—we wish to group under the heading of *heterocyberny** (change in governance) those mechanisms whereby, through selection, environmentally induced phenotypes become stabilized in the genome such that the phenotypes arise even in the absence of the environmental inducer.

Within the Modern Synthesis, developmental plasticity was marginalized to the point that the notion that the environment could lead to the substitution of one allele for another was seen as heretical. There are still evolutionary biologists who maintain this view (see Appendices A and D). However, during the last decade, developmental plasticity has come to be seen as an important part of normative development, and a large number of studies have indicated that other plasticity-driven evolutionary schemes might also be normative for evolution (see West-Eberhard 2003; Pigliucci et al. 2006; Pigliucci 2007). We begin with a topic that startled and excited some of the early Darwinian naturalists: the phenomenon of phenocopies.

Phenocopies and Ecotypes

Phenocopies are environmentally produced phenotypes that mimic the phenotypes produced by mutant alleles. One of the simplest ways of inducing phenocopies in insects is by cold or heat shock at a particular time during development. Each of the environmentally induced traits has a particular time when the environmental factor works. This could be at the embryonic, larval, or pupal stage, depending on the trait. During that criti-

*It is likely that heterocyberny is underlain by the other four "hetero's" and as such it represents a different level of change. However, since the processes of heterocyberny are at the level of the gene regulation networks used by the other four mechanisms, it should be discussed as such, but separately (as we do here). (We thank W. Arthur for interesting discussions of these ideas.)

cal period, however, several types of trauma may be equally effective at eliciting the same phenotype.

As early as the 1890s, scientists used heat shock to disrupt the pattern of butterfly wing pigmentation (Merrifield 1890, 1893). This thermal technique has provided some surprising results. Color patterns that develop after temperature shock sometimes mimic the normal genetically controlled patterns of subspecies (races) or of related species living at different temperatures. A subspecies whose phenotype is characteristic of the species in one particular geographic range is called an **ecotype** (Turesson 1922). One of the first experiments linking environment to heritable variation was that of Standfuss (1896), who demonstrated that a heat-shocked phenocopy of the Swiss subspecies of *Iphiclides podalirius* resembled the normal form of the Sicilian subspecies of that butterfly. Similarly, heat shocking the central European form of the swallowtail butterfly *Papilio machaon* produced some individuals that resembled the Syrian subtype and some that resembled the Turkish variety. Richard Goldschmidt (1938) produced one of the most striking phenocopies. He observed that heat-shocked specimens of the central European subspecies of *Aglais urticae* produced wing patterns that resembled the Sardinian subspecies, while cold-shocked individuals of the central European variety developed the wing patterns of the subspecies from northern Scandinavia (Figure 10.2).

FIGURE 10.2 Temperature shocking *Aglais urticae* produces phenocopies of geographic variants shown in Goldschmidt's original illustrations. (A) Usual central European variant. (B) Heat-shocked phenocopy resembling the Sardinian form. (C) The Sardinian form of the species. (From Goldschmidt 1938.)

Further observations by Shapiro (1976) on the mourning cloak butterfly (*Nymphalis antiopa*) and by Nijhout (1984) on the buckeye butterfly (*Precis coenia*) have confirmed the view that temperature shock can produce phenocopies that mimic genetically controlled patterns of related races or species existing in colder or warmer conditions. Chilling the pupa of *Pieris occidentalis* will cause it to have the short-day phenotype (Shapiro 1982), and this phenotype is similar to that of the northern subspecies of pierids. Even "instinctive" behavioral phenotypes associated with these color changes (such as mating and flying) are phenocopied (see Burnet et al. 1973; Chow and Chan 1999). Thus, an *environmentally* induced phenotype might become the standard *genetically* induced phenotype in one part of the range of that organism.

In Chapter 2, we mentioned that August Weismann found that the seasonal color pattern of the wings of the European map butterfly was induced by temperature. Indeed, although Weismann is usually credited with disproving the notion that acquired characteristics could be inherited (see Appendix D), he noted that this was proven only in the case of accidents and functional abilities (e.g., an amputee's child would have normal limbs, and a carpenter's child would have only normal arm muscles). However, from his studies on the geographical and seasonal variation in butterfly wing coloration, he noted that germline modification by the environment was possible and that "in many other animals and plants, influences of temperature and environment may very possibly produce hereditary variations" (Weismann 1892, p. 405).

In order for an environmentally induced trait to become genetically fixed, the population must be exposed to environmental conditions that repeatedly induce the phenotype; there must be selective pressure such that the induced phenotype results in higher fitness in that environment; and there must be sufficient genetic variation within the population to stabilize this particular phenotype. This process, whereby an environmentally induced change can be genetically stabilized, is often called **genetic assimilation**.

Genetic Assimilation

Ivan Schmalhausen, C. H. Waddington, and Mary Jane West-Eberhard, among others, have proposed various models by which novel phenotypes could first be generated by developmental plasticity. If this plastic response is adaptive, and if it continues to be induced by the environment, it will spread under continued selection. Moreover, if this environmentally induced phenotype confers greater fitness, and if having the phenotype produced from the genome provides more fitness than acquiring it from the environment, then it should become stabilized by genetic means, if possible. This often means that "modifier" genes will be selected, such that *any* environment can consistently induce the phenotype. If the genes allowing this phenotype become fixed (i.e., genetically assimilated), then the phenotype becomes independent of the environment and is always produced. This scheme, originally introduced by Ivan Schmalhausen (1949) and Conrad Waddington (1952, 1953), concluded that genes stabilized a preexisting phenotype. These genes did not originate these phenotypes.

It is important to remember that phenotypes come not from the environment (or from the genes), but from the organisms. The organisms, due to their genotype and developmental flexibility, can produce a novel response to the environment. West-Eberhard (2005) stated her belief that "genes are probably more often followers than leaders in evolutionary change."* The allelic frequency change follows repeated phenotype induction and the selection of genotypes that are capable of generating the favored response (phenotype). In other words, the phenotype is seen first (as it is induced physiologically by the environment); if the phenotype is beneficial, the genes subsequently stabilize it.

There are several variations on this theme of heterocyberny, variously called the "Baldwin effect," "genetic assimilation," "stabilizing selection," and "genetic accommodation" (see King and Stanfield 1985; Gottlieb 1992; Hall 2003; Bateson 2005). The differences are generally in the details of how the genetic structure can replace the environment-induced plasticity. The variations share many ideas in common, including (1) that environmentally induced phenotypes are seen first, and (2) there is selection for those phenotypes that are most adaptive. The advantages of such scenarios are:

- **The phenotype is not random.** The environment elicited the novel phenotype, and the phenotype has already been tested by natural selection. This would eliminate a long period of testing phenotypes derived by random mutations. As Garson and colleagues (2003) note, although mutation is random, developmental parameters may account for some of the directionality in morphological evolution.
- **The phenotype already exists in a large portion of the population.** In the Modern Synthesis, one of the problems of explaining new phenotypes is that the bearers of such phenotypes are "monsters" compared with the wild type. How would such mutations, perhaps present only in one individual or one family, become established and eventually take over a population? The developmental model solves this problem: this phenotype has been around for a long while, and the capacity to express it is widespread in the population; it merely needs to be genetically stabilized by modifier genes that already exist in the population.

*According to Konstantin Anokhin (pers. comm.), Schmalhausen had written precisely the same thing in 1942.

In the Baldwin effect, the emphasis is on the fixation of the environmentally induced phenotype by a new mutation. In genetic accommodation, the plasticity itself is seen as beneficial, so genes that produce plasticity are selected (see the next section). In **genetic assimilation**, there is an emphasis on the fixation of the environmentally induced phenotype by preexisting genotypic variation in the population.

Genetic assimilation in the laboratory

Genetic assimilation is readily demonstrated in the laboratory. In fact, the idea of genetic assimilation was introduced independently by Waddington (1942, 1953, 1961) and Schmalhausen (1949) to explain the remarkable outcomes of artificial selection experiments in which an environmentally induced phenotype became expressed even in the *absence* of the external stimulus that was initially necessary to induce it.

Waddington (1952, 1953) found that when pupae from a laboratory population of wild-type *Drosophila melanogaster* were exposed to a heat shock of 40°C, the wings of some of the emerging adults had a gap in their posterior crossveins. This gap is not normally present in untreated flies. Two selection regimens were followed, one in which only the aberrant flies were bred to one another, and another in which only non-aberrant flies were mated to one another. After some generations of selection in which only the individuals showing the gap were allowed to breed, the proportion of adults with broken crossveins induced by heat shock at the pupal stage was raised to above 90%. Moreover (and significantly), by generation 14 of such inbreeding, a small proportion of individuals were crossveinless *even among flies of this line that had not been exposed to temperature shock*. When Waddington extended this artificial selection by breeding together only those adults that had developed the abnormality without heat shock, the frequency of crossveinless individuals among untreated flies became very high, reaching 100% in some lines. The phenotypically induced trait had become genetically assimilated into the population.*

Schmalhausen explained such results by saying that the environmental perturbation unmasked genetic heterogeneity for modifier genes which had already existed in the population. In other words, selection on existing genetic variability can stabilize the environmentally induced phenotype. Waddington emphasized that these modifier genes could act by changing the activation threshold for the phenotype, allowing it to become expressed over a wider set of environmental conditions (see Pigliucci et al. 2006). He explained his results by saying that the development of crossveins can be

*Note that in these artificial selection experiments, the original phenotype induced by the environment was not intrinsically adaptive. Only the hand of the experimenter choosing which flies mated made it so.

influenced by environmental disturbances above a certain threshold of intensity. But individuals from wild-type populations have a threshold so high that only an unusually strong stimulus, such as a heat shock, can effectively induce a modified expression. Thus, phenotypic variation does not arise if all the fly embryos have a threshold too high to be affected by the disturbances prevailing in the usual environment. However, when an exceptionally severe disturbance occurs, it is those individuals in which a phenotypic change is induced (those with the most sensitive genotypes for responding to the environmental stimulus) that are favored under artificial selection. As we will see, Waddington's and Schmalhausen's explanations appear to fit different cases of genetic assimilation.

In addition to finding the crossveinless phenotype upon exposure to heat shock, Waddington showed that his laboratory strains of *Drosophila* had a particular reaction norm in the fly's response to ether. Embryos exposed to ether at a particular stage developed a phenotype similar to the *bithorax* mutation and had four wings instead of two. (The flies' halteres—balancing structures on the third thoracic segment—were transformed into wings.) Generation after generation was exposed to ether, and individuals showing the four-winged state were selectively bred each time. After 20 generations, the selected *Drosophila* strain produced the mutant phenotype even when no ether was applied (Figure 10.3; Waddington 1953, 1956).

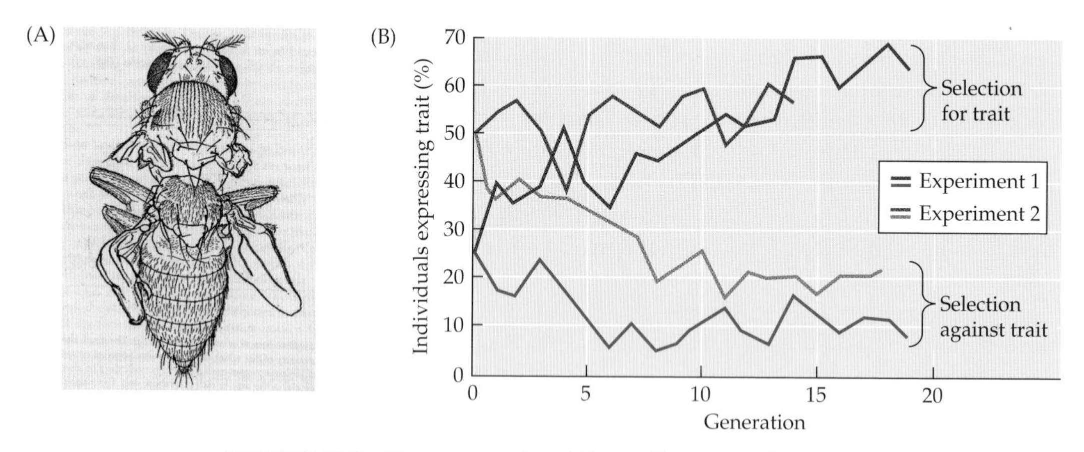

FIGURE 10.3 Phenocopy of the *bithorax* (four-winged) mutation. (A) Waddington's drawing of a *bithorax* mutant phenotype produced after treatment of the fly embryo with ether. The forewings have been removed to show the aberrant metathorax. This particular individual was in fact taken from "assimilated" stock that produced this phenotype without being exposed to ether. (B) Selection experiments for or against *bithorax*-like response to ether treatment. Two experiments are shown (red and blue lines). In each experiment, one group of flies was selected for the trait while the other was selected against the trait. (From Waddington 1956.)

Subsequent experiments have borne out Waddington's findings (see Bateman 1959a, b; Matsuda 1982; Ho et al. 1983). In 1996, Gibson and Hogness repeated Waddington's experiments and "saw a steady increase in the frequency of thoracic abnormalities called phenocopies in each generation from 13 percent in the starting population to a plateau of 45 percent" (Gibson 1996). Conversely, when Gibson and Hogness selectively bred non-transformed flies resistant to ether treatment, the frequency of thoracic abnormalities dropped steadily. Gibson and Hogness (1996) found that four alleles of the *Ultrabithorax* (*Ubx*) gene had existed in the population and were critical for the genetic assimilation of the ether-induced *bithorax* phenotype.* "Waddington's experiment showed some fruit flies were more sensitive to ether-induced phenocopies than others, but he had no idea why," Gibson said. "In our experiment, we show that differences in the *Ubx* gene are the cause of these morphological changes." In other words, the phenocopies were stabilized by alleles of regulatory genes such as those described in Chapter 9. In this case, then, Schmalhausen's model of preexisting regulatory alleles within the population appears to be the correct explanation.

Genetic assimilation in nature: Mechanisms, models, and inferences

While genetic assimilation is readily shown to occur in the laboratory, the idea that it could occur in nature remained controversial until Rutherford and Lindquist (1998) demonstrated a possible molecular mechanism unmasking cryptic genetic variation through stress. They were looking at the effects of mutations of the *Drosophila* heat shock protein gene *Hsp83* and the inactivation of its protein product, Hsp90. When this gene or its protein product were inactivated, a whole range of phenotypes appeared. Several mutations that were preexisting in the population were allowed to become expressed. Moreover, the resulting mutants could be bred such that, within several generations, almost all the flies expressed the mutant phenotype, and some of the flies had that phenotype even if they contained a functional *Hsp83* gene. This phenomenon looked a great deal like genetic assimilation.

Hsp90 is a molecular "chaperone" that is required to keep many proteins in their proper three-dimensional conformation. It is especially important for the structure of several signaling molecules. There are numerous genetic mutations that are usually not expressed as a mutant phenotype because Hsp90 can correct the small changes in mutant protein structure. However, when heat shock (or ether) is applied, there are numerous other proteins

*This finding confirmed genetic mapping studies, which had already shown that mutant alleles of the *Ubx* gene were responsible for producing the *bithorax* phenotype.

Thresholds of Genetic Assimilation

One mechanism for genetic assimilation is based on changing the threshold at which a given environmentally induced phenotype is seen. Bateman (1959a) proposed that genetic assimilation can occur by three major routes. Molecular biology has since found evidence that each of these mechanisms may work to produce the genetic fixation of a phenotype that was originally induced by environmental factors.

Bateman's models posit an original population in which the norm and most deviation fall short of a threshold value needed to see a particular phenotype (Figure A). In the first model, stress shifts the norm toward the threshold value (Figure B). For example, if a certain Hsp90 expression phenotype occurs at a particular threshold temperature, no fly in the original population will express that phenotype at a lower temperature. But stress, leading to Hsp90 deficiency, causes numerous mutant individuals to express the induced phenotype, altering the distribution of flies in the populations. Now there are some individuals who are over the threshold and who will express the phenotype under certain conditions.

In the second model, the threshold is shifted toward the mean of the variation (Figure 10.4C). In other words, the variation is still centered around the same point, but the threshold temperature for the production of the new phenotype is genetically changed. In the third model, the threshold and the norm remain the same, but the *amount* of variation increases such that some fraction of the population is over the threshold value (Figure 10.4D).

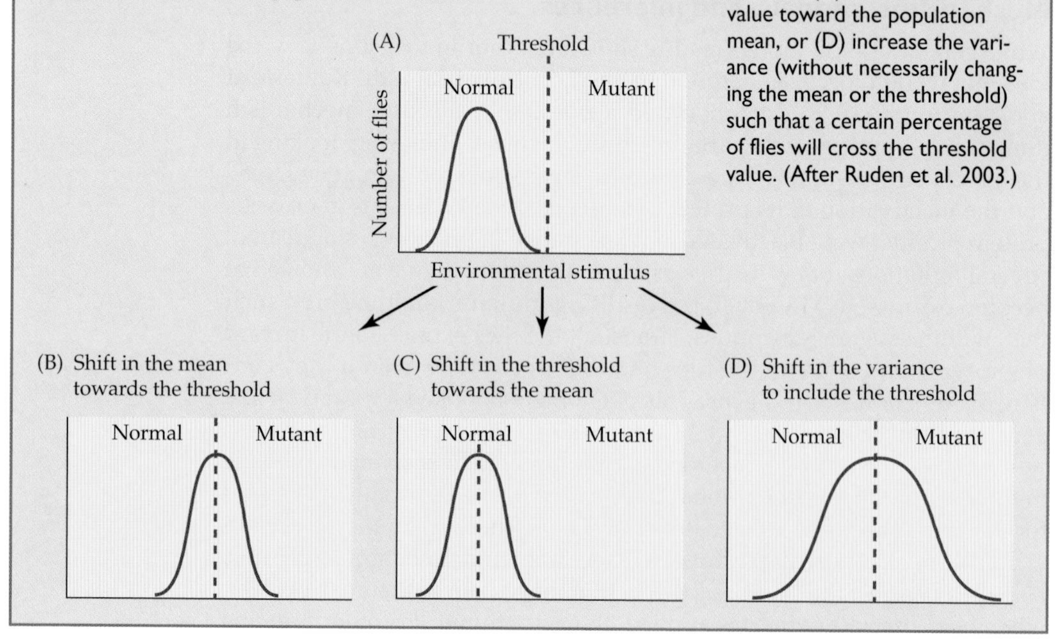

Three models of genetic assimilation. (A) Initial, unstressed situations in which the phenotype is not expressed in the population. (B–D) The stress-induced phenotype can be fixed genetically by genes that (B) shift the mean of the population toward the threshold value, (C) shift the threshold value toward the population mean, or (D) increase the variance (without necessarily changing the mean or the threshold) such that a certain percentage of flies will cross the threshold value. (After Ruden et al. 2003.)

that need the attention of Hsp90, and there is not enough Hsp90 to go around. Thus, the variation that has existed but has not been expressed now becomes expressed. In this way, Hsp90 appears to be a "capacitor" for evolutionary change, allowing genetic changes to accumulate until environmental stress releases them to reveal their effects on phenotype. This allows a wide range of genetic mixtures to become possible and allows these combinations to accumulate. When the environment changes and these combinations are expressed, most of them will probably be lethal. But some might be beneficial and selected in the new environment. Continued selection would enable the genetic fixation of the adaptive physiological response (Rutherford and Lindquist 1998; Queitsch et al. 2002).

In two recent papers, Sangster and colleagues (2008a, b) demonstrated that Hsp90-buffered variation is so widespread through the plant species *Arabidopsis thaliana* that every quantitative trait can be predicted to have at least one major component buffered by Hsp90. Moreover, they found that relatively slight environmental changes result in stress conditions that lower the levels of Hsp90 available and thereby cause the expression of hitherto hidden variations. It appears, then, that Hsp90 is a major cause of the canalization of phenotypes, enabling the same "wild-type" phenotype to be displayed across a range of genetic and environmental conditions (see Chapter 4). Environmental stress can cause the appearance of variants that can then be challenged by natural selection and selected such that they occur even when the environmental stressor is not present. Sangster and colleagues conclude that "Hsp90 appears to fulfill Waddington's concept of a developmental buffer or molecular canalization mechanism; one which can lead to the assimilation of novel traits on a large scale."

Genetic assimilation and natural selection

Both Waddington and Schmalhausen independently proposed genetic assimilation to explain how some species have evolved rapidly in particular directions (see Gilbert 1994). For instance, both scientists were impressed by the calluses of the ostrich. Most mammalian and avian skin has the ability to form calluses on areas that are abraded by the ground or some other surface.* The skin cells respond to friction by proliferating. While such examples of environmentally induced callus formation are widespread, the ostrich is born with calluses already present where it will touch the ground

*Until the widespread use of typewriters (and later computers) in the twentieth century, writers could be easily recognized by the calluses on their fingers. Thus, in "The Red Headed League," Sherlock Holmes correctly surmised that the red-headed man had worked as a scrivener.

FIGURE 10.4 Ventral side of an ostrich; arrows mark calluses that are present on the animal from the time it is hatched. (From Waddington 1942.)

(Figure 10.4; Duerden 1920). Waddington and Schmalhausen hypothesized that since the skin cells are already competent to be induced environmentally by friction, they could be induced by other things as well. As ostriches evolved, a mutation or a particular combination of alleles appeared that enabled the skin cells to respond to some substance *within* the embryo. Waddington (1942) wrote:

> Presumably its skin, like that of other animals, would react directly to external pressure and rubbing by becoming thicker. ... This capacity to react must itself be dependent upon genes. ... It may then not be too difficult for a gene mutation to occur which will modify some other area in the embryo in such a way that it takes over the function of external pressure, interacting with the skin so as to "pull the trigger" and set off the development of callosities.

As Waddington (1961) pointed out, a combination of orthodox Darwinism and orthodox embryology can enable the inheritance of what had been an environmentally induced phenotype.

There are numerous examples of evolution by natural selection that may be explained by genetic assimilation (see Matsuda 1987; Hall 1999). Jean Piaget (who began his career as a biologist and only later became an educator who applied much of his biological knowledge to educational theory) noted that the pond snail *Lymnaea stagnalis* had two phenotypes. There were snails with elongated shells and snails with flattened shells. A flatter shell developed in those snails that were raised in turbulent water, as this shape was better adapted for the rough environment. Moreover, after several generations, this environmentally induced phenotype became genetically fixed and would continue to be formed even if the snails were placed into still water (Piaget 1929a, b).

Brakefield and colleagues (1996) showed that they could genetically fix the different morphs of the adaptive polyphenism of *Bicyclus* (see Chapter 2), and (as mentioned above) Shapiro (1976) showed that the short-day (cold-weather) adaptive phenotype of several butterflies is the same as the single genetically produced phenotype of related species or subspecies living at higher altitudes or latitudes. The genetic assimilation of morphs originally produced through developmental plasticity may accelerate the divergence of the fixed morphs, thereby contributing to the origin of new species (West-Eberhard 1989, 2003).

West-Eberhard (2003) has noted that environmentally induced novelties may be extremely important in evolution. Mutationally induced novelties would occur in only a family of individuals, whereas environmentally induced novelties would occur throughout a population. Moreover, the inducing environment would most often also be a selecting environment. Thus, there would be selection immediately for this trait, and the trait would be continuously induced. Computer modeling has shown that such an environmentally induced genetic assimilation can produce rapid evolutionary change as well as speciation (Kaneko 2002; Behara and Nanjundiah 2004). Indeed, West-Eberhard (2005) has proposed that studying gene expression and plasticity will help us understand the genetic divergence that gives rise to speciation. "Contrary to common belief, environmentally initiated novelties may have greater evolutionary potential than mutationally induced ones. Therefore, the genetics of speciation can profit from studies of changes in gene expression as well as changes in gene frequency and genetic isolation."

Epigenetic Assimilation: Another Hsp90 Story

As has been stressed throughout this book, variation need not be allelic. It can also be epigenetic. One variant on the theme of genetic modification is the notion of **epigenetic assimilation**. Hsp90 protein provides a model here, too, but in a different capacity. In addition to serving as a "capacitor" for the accumulation of genetic variation, Hsp90 can also work directly on altering gene expression through an epigenetic pathway. Sollars and colleagues (2003) have shown that reduced activity of Hsp90 induces a heritable alteration in the epigenetic state of chromatin. Moreover, this altered pattern of gene expression was able to be selected such that it occurred in the absence of reduced Hsp90. In this way, different heritable phenotypes can be passed on in organisms having the same genotypes (see the figure).

This model enables an adaptive response to be fixed epigenetically, even without the presence of genetic variation in the population. Indeed, epigenetic fixation may be a rapid evolutionary process and may be an intermediate step toward genetic assimilation.

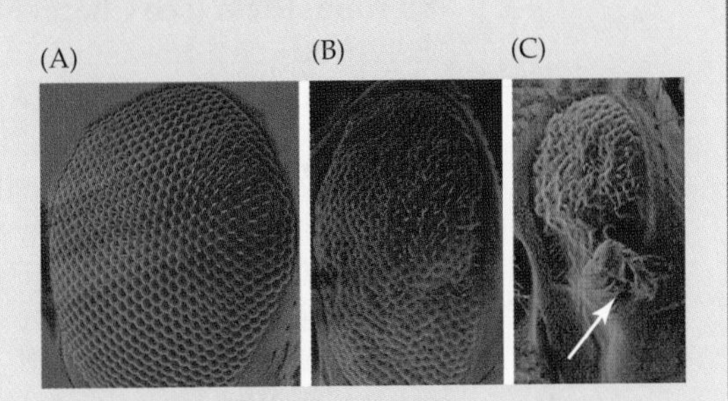

Examples of epigenetically selected flies with an inducible phenotype. (A) Wild-type eye showing normal facets. (B) *Kruppel-Ifl* mutant phenotype showing reduced and rough eye facets. (C) Ventral outgrowths from the eyes of flies heterozygous for *Krlfl* and treated with geldamycin, a drug that inhibits *Hsp90*. When these aberrant flies were mated together and selected for ventral outgrowths in the absence of the drug for 13 generations, over 50% of the flies had ectopic ventral outgrowths. Thus, after only one generation in which *Hsp90* gene expression was reduced, the mutant phenotypic effect persisted for many generations. (After Sollars et al. 2003.)

Thus, genetic assimilation is a mechanism whereby environmentally induced traits can become internally induced during embryonic development through genomic influences. In this process, previously hidden (cryptic) genetic variation becomes important for the stabilization of that phenotype or for the regulation of that expression after an environmental stimulus overcomes the threshold for the expression of these phenotypes. Selection in the presence of this environmental factor enriches the gene pool for the cryptic alleles that would determine this trait, and eventually these alleles become so frequent that the trait appears even in the absence of the environmental stimulus. In this way, a phenotypically plastic trait can be converted into a genetically fixed trait that is constantly produced under a wide range of environmental conditions.

Genetic Accommodation

Plasticity can be enhanced or it can be repressed. Genetic assimilation represses plasticity, narrowing the range of variation from a plastic to a fixed state. That is, the original organisms have a plastic phenotype that varies with the environment, but the descendant population has a phenotype that has been selected for one of the possible variants. Waddington called this process of fixation **canalization**; it is also known as **developmental robustness** (see Chapter 4 and Gilbert 2002). The processes of canalization produce a robust phenotype that is buffered against environmental perturbations. In genetic accommodation, however, it is not canalization but plasticity that is selected for.

Numerous studies of wild populations have shown that an adaptive potential for plasticity can evolve (Nussey et al. 2005; Danielson-François et al. 2006; Gutteling et al. 2007). This plasticity can be especially important for those species inhabiting several environments, where one variant is not optimal over all habitats. If a trait is inducible by the environment, one of several results can follow (West-Eberhard 2003). The first is genetic assimilation of an advantageous trait, as described in the previous section. Alternatively, the trait may not be advantageous in the particular environment and its frequency diminishes. Another path is the stabilization of plasticity, wherein cryptic variants are selected that can direct the phenotype in either direction; this is **genetic accommodation**.

Genetic assimilation and genetic accommodation can be viewed as opposite sides of the same coin. Indeed, Suzuki and Nijhout (2006) have shown genetic accommodation and assimilation in the larvae of the tobacco hornworm moth *Manduca sexta*. By judicious selection protocols, they were able to breed one line in which plasticity decreased (genetic assimilation) and another line in which plasticity increased (genetic accommodation).

The hornworm moth *Manduca quinquemaculata* has a temperature-dependent color polyphenism that appears to be adaptive. When they hatch at 20°C, these caterpillars have a black phenotype, which enables them to absorb sunlight more efficiently. During the warmer season, with temperatures around 28°C, the hatching caterpillars are green, which affords them camouflage in their environment. A trade-off between cryptic coloration and the need to absorb heat appears to drive this phenotypic plasticity. Caterpillars of the related species *M. sexta* (the tobacco hornworm moth), however, are not phenotypically plastic in coloration. The *M. sexta* caterpillar is green no matter what the temperature, although mutant black forms exist (Figure 10.5A). This black mutation is due to reduced levels of juvenile hormone, which leads to increased levels of melanin production in the caterpillar's skin.

Suzuki and Nijhout tested whether they could induce phenotypic plasticity in *M. sexta* by artificial selection. When green larvae were heat

FIGURE 10.5 Effect of selection on temperature-mediated larval color change in the *black* mutant of the moth *Manduca sexta*. (A) The two color morphs of *Manduca sexta*. (B) Changes in the coloration of heat-shocked larvae in response to selection for increased (black) and decreased (green) color response to heat shock treatments, compared with no selection (red). The color score indicates the relative amount of colored regions in the larvae. (C) Reaction norm for generation 13 flies reared at constant temperatures between 20°C and 33°C, and heat shocked at 42°C. Note the steep polyphenism at around 28°C. (After Suzuki and Nijhout 2006; photograph courtesy of Fred Nijhout.)

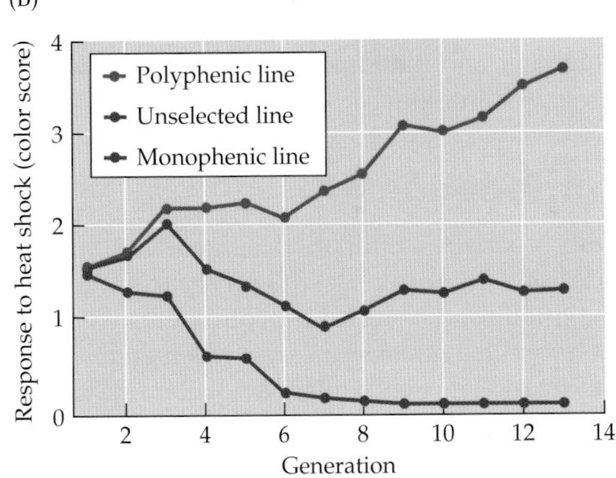

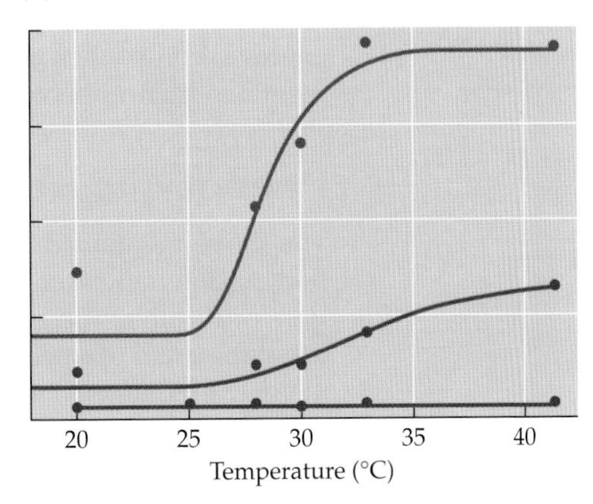

shocked, they remained green. However, when the mutant black larvae were heat shocked at 42°C for 6 hours, they developed phenotypes ranging from black to green. These mutant caterpillars were then subjected to three different regimens. In one case, the adults developing from those black larvae that became green with heat shock were bred to one another. This scheme selected for plasticity. In a second case, a "monophenic" line was selected by mating together those moths resulting from black larvae that did not turned green after heat shock. A third group had no selection applied to it.

Within 13 generations, all individuals selected for the polyphenism (i.e., turning green) always developed a green phenotype after heat shock. Within 7 generations, all the individuals in the monophenic line remained black after heat shock. In the 13th generation, larvae from both lines were subjected not to heat shock but to the mild heat difference (28°C) that would initiate the polyphenism in *M. quinquemaculata*. The monophenic strain stayed black at all temperatures. In other words, this strain had gone from a slightly polyphenic population to a monophenic population by genetic assimilation. Selection for green coloration by heat shock, however, produced a polyphenic population. In this population, the larvae were usually black in temperatures below 28.5°C. Above this temperature, individuals were mostly green. In other words, the selection of this polyphenic line resulted in a temperature-dependent polyphenic switch (Figure 10.5B,C). Although the polyphenic response was not directly selected for (by selecting individuals in different environments), the color response evolved through the artificial selection. Moreover, Suzuki and Nijhout showed that the target of this selection was most likely the titer of juvenile hormone, just as in *M. quinquemaculata*.

Thus, at least in the laboratory, genetic accommodation can occur through the interaction of a sensitizing mutation (which brings the phenotype closer to a threshold), environmental change (which can allow it to cross that threshold), and cryptic genetic changes, which can stabilize this plastic phenotype. Genetic accommodation allows plasticity to persist such that it is expressed only when advantageous, and it demonstrates how the environmental influences become integrated into developmental trajectories.

Phenotypic Accommodation

The above-mentioned processes—genetic assimilation, epigenetic assimilation, and genetic accommodation—might work well on a particular cell type or tissue to increase the likelihood of novel morphologies through developmental plasticity. But how can a novel *organ* arise when there are so many components in a single organ? The evolution of a novel morphology may be facilitated by a phenomenon called **phenotypic accommodation**, which is the adaptive mutual adjustment, *without genetic change*, among variable aspects of the phenotype following a novel input during development (West-Eberhard 2003, 2005).

Evolutionary considerations

Phenotypic accommodation has a long history in both embryology and evolutionary biology, but it is only recently that such a notion has been revived in modern evolutionary biology. Darwin (1868, p. 312) wrote extensively about "correlated variation," noting that "when one part is modified through continued selection, either by man or under nature, other parts of the organization will be unavoidably modified." Darwin studied the correlations between the parts of the skull, documenting the morphological changes that occurred when rabbits were under selection for long ears. He even recorded (p. 122) the asymmetric anatomy of the skull of "half-lop" rabbits, in which one ear is longer than the other.

The dramatic changes in bone arrangement from agnathans to jawed fishes, from jawed fishes to amphibians, and from reptiles to mammals were coordinated with changes in jaw structure, jaw musculature, tooth deposition and shape, and the structure of the cranial vault and ear (Kemp 1982; Thomson 1988; Fischman 1995). On a less sweeping scale, the results of artificial selection in domestic dogs also demonstrate correlated development. Although the skulls of the many dog breeds range from the pointed snouts of collies to the blunt snouts of bulldogs, in each case the jaw muscles are coordinated with jaw cartilage and bone to allow the dog to properly grasp and chew its food. Goldschmidt (1940) and Schmalhausen (1938, 1949) recognized that the regulative ability of the embryo allowed it to accept changes in its structure, and Schmalhausen (1938) used this notion of integration to criticize evolutionary biologists who thought of organisms as mere mosaics of independent characters (see Levit et al. 2006).

One fascinating case of how powerful phenotypic accommodation can be is illustrated in a handicapped goat originally studied by the Dutch veterinary anatomist E. J. Slijper (1942a,b; West-Eberhard 2003). Slijper's goat was born with paralyzed forelimbs and could not walk on four legs. Rather, it hopped on its two functional hind legs.* When it died, Slijper dissected the animal and published an account of the phenotypic changes in its hindlimbs and torso. Its pelvis had changed shape to accommodate an upright posture, especially in the ischium (Figure 10.6). This accommodation went along with changes in the skeletal musculature around the pelvis, including an elongated and thickened gluteal tongue whose attachment to

*Although rare, bipedal quadrupeds are occasionally observed. One of the most famous is Faith, a dog born with congenitally deformed forelimbs. Born in 2002, Faith has been walking bipedally for several years, and a video of her can be seen on YouTube (www.youtube.com/watch?v=MexyuH-m97A).

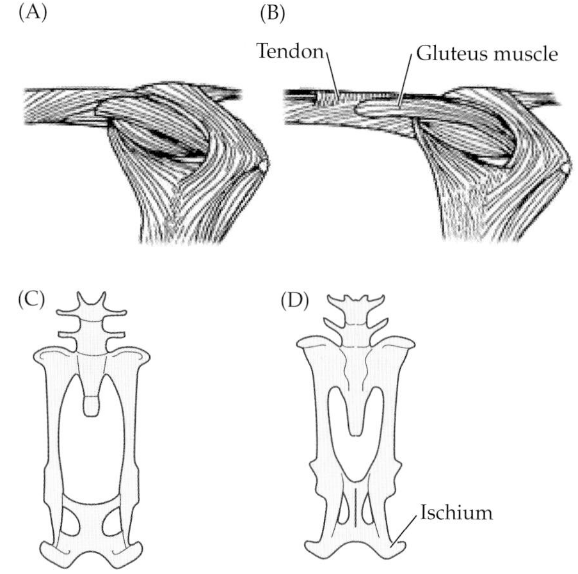

FIGURE 10.6 Pelvic musculature and skeleton of two-legged goat co-adapt to produce novel phenotypes not seen in normal goats. These changes are achieved without any genetic alterations. (A) Pelvic musculature of a normal goat. (B) Pelvic musculature of a two-legged goat. The gluteus muscle has a long anterior "tongue" that has been reinforced by novel tendons. (C, D) Skeletal changes in pelvic bones between normal goats (C) and a two-legged goat (D) are likely due to unusual stresses on the bones of the two-legged goat. In the two-legged goat, the dorsal-ventral flattening and the elongation of the ischium resemble those of kangaroos. (After West-Eberhard 2003.)

the pelvis was mediated by tendons that do not exist in normal goats. The bones of hindlimb, thorax, and sternum were also changed. Thus, a complete set of muscles, skeleton, and tendons co-adapted to form an anatomical novelty. Whether the trigger was environmental or genetic doesn't matter; what does matter is that the initiator acted to control an entire suite of anatomical changes. These novel morphologies actually resembled changes seen in the evolution of bipedal species. The compressed thorax and elongated ischium are similar to those of kangaroos, and the wide sternum resembles that bone of the orangutan (which, like the goat, lacks a supportive tail). Indeed, Marks (1989) has suggested the possibility that genetic assimilation and phenotypic accommodation contributed to the origin of human bipedalism.

In the model for the evolution of new phenotypes proposed by West-Eberhard (2003), phenotypic accommodation plays a central role. In this view, there are four steps by which an environmentally induced trait can come to characterize a species:

1. **Developmental variation.** An environmental change produces changes in development leading to the appearance of a new trait.
2. **Phenotypic accommodation.** The regulative plasticity of the developing organism adapts to this new trait.
3. **Spread of the new variant.** If the initial change is environmentally induced, the variant trait has already occurred in a large part of the population.
4. **Genetic fixation.** Allelic variation in the population and natural selection allows for the genetic fixation (assimilation) of the trait so that it is produced irrespective of the environment.

Developmental mechanisms of phenotypic accommodation

The underlying cause for such correlated changes involves the inductive processes in embryonic development (Waddington 1957). Experimental embryologist Hans Spemann (1901, 1907) reviewed the field of "developmental correlations."* He noted that when he put a small piece of tadpole tissue into a salamander embryo, the salamander formed a tadpole jaw, and that the musculature and skeletal elements were all coordinated. Similarly, Twitty (1932) found that when he transplanted the eye-forming region from the embryos of large salamanders into embryos of smaller species, the midbrain region of the side innervated by the larger eye grew bigger to accommodate the larger number of neurons. This coordination of head morphogenesis, where changes in one element cause reciprocal changes on other elements, has been more recently verified in chimeras made between chick and quail embryos (Köntges and Lumsden 1996).

Phenotypic accommodation has also been shown in the limb, and this accommodation has important evolutionary implications. Repeating the earlier experiments of Hampé (1959), Gerd Müller (1989) inserted barriers of gold foil into the prechondrogenic hindlimb buds of a 3.5-day chick embryo. This barrier separated the regions of tibia formation and fibula formation. The results of these experiments were twofold. First, the tibia was shortened, and the fibula bowed and retained its connection to the fibulare (the distal portion of the tibia). Such relationships between the tibia and fibula are not usually seen in birds, but they are characteristic of reptiles (Figure 10.7). Second, the musculature of the hindlimb underwent changes in parallel with the bones. Three of the muscles that attach to these bones showed characteristic reptilian patterns of insertion. It seems, therefore, that experimental manipulations that alter the development of one part of the mesodermal limb-forming field also alter the development of other mesodermal components. This was crucial in the evolution of the bird hindlimb from the reptile hindlimb. As with the correlated progression seen in facial development, these changes all appear to be due to interactions within a module, in this case, the chick hindlimb field. These changes are not global effects and can occur independently of the other portions of the body.

*This concept of correlated development and phenotypic accommodation has gone under many aliases, each emphasizing either the developmental or evolutionary aspects of this process. Frazetta (1975) called this "phenotypic compensation," and Müller (1990) has named it "ontogenetic buffering." Its basis is the normal reciprocal embryonic induction that is responsible for forming complex organs such as the eye or kidney, making certain that the proportion of cells allocated for each structure will result in an appropriately formed organ. This is another excellent example of the concept that it is not possible to explain evolutionary change without knowing the underlying developmental principles.

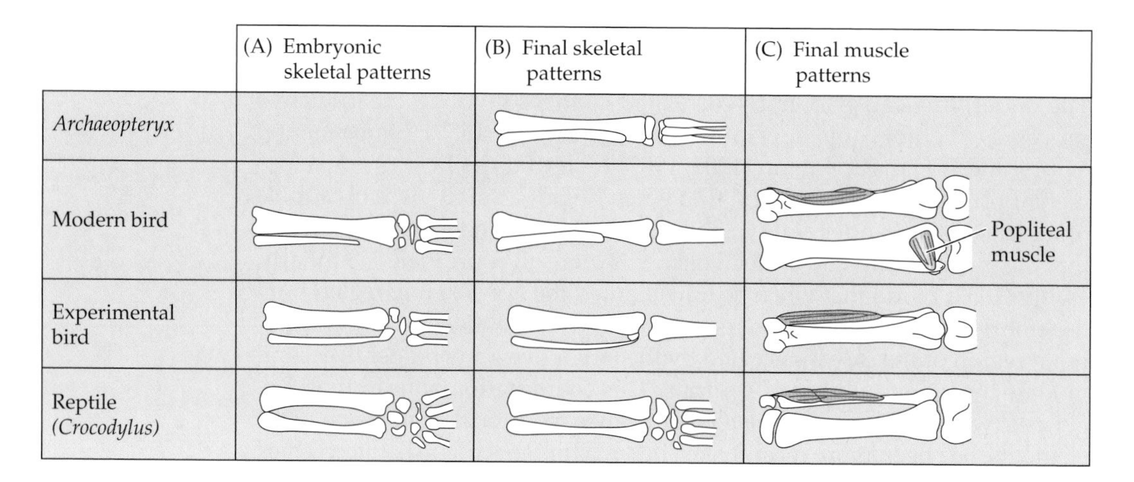

FIGURE 10.7 Experimental "atavisms" produced by altering embryonic fields in the limb. Results of Müller's experiments using gold foil to split the chick hindlimb field. (A, B) The embryonic and final bone patterns, indicating that the fibulare structure was retained by the experimental chick limb, as it is in extant reptiles and as it is thought to have been in *Archaeopteryx*, the earliest known bird. (C) Some of the correlated muscle patterns. The popliteal muscle is present in the normal chick limb but is absent from reptile limbs and from the experimental limb. The fibularis brevis muscle, which normally originates from both the tibia and fibula in chicks, takes on the reptilian pattern of originating solely from the fibula in the experimental limb. (After Müller 1989.)

Reciprocal accommodation

As we have seen in earlier chapters, there can be reciprocal accommodation between the environment and the developing organism. Diet, for example, can cause changes in jaw structure in mammals and fishes. Cichlids that feed on molluscs usually have large jaws; but if these fish are reared on soft diets, they develop the jaw morphology characteristic of closely related species that eat soft insects (Greenwood 1965; Meyer 1990; Galis et al. 1994).

Liem and Kaufman (1984) showed that there was a positive feedback loop established between jaw morphology and diet. When two specialized morphs, one with a large (mollusc-eating) jaw and one with a small (insect-eating) jaw were given an abundant supply of both types of food, both types of cichlid preferred insects over molluscs. But when food was scarce, they specialized according to jaw type. So in conditions of scarce food supply, the morphology affects the diet, and feeding behavior reinforces selection in divergent directions. Not only does diet cause morphological changes, but these morphological changes can lead to specialization and to natural selection. Phenotypic accommodation, then, is a direct consequence of reciprocal embryonic induction.

West-Eberhard (2005) has pointed out that phenotypic accommodation, like genetic assimilation and genetic accommodation, is in no way in conflict with the standard view of adaptive evolution as it concerns variation, selection, and changes in allele frequency. What phenotypic accommodation *does* show is that a novel morphological structure, whether its initiation was environmental or genetic, can be complex and adaptive from the beginning rather than needing to be built up under selection for small changes over long periods of time. Moreover, environmentally induced traits can spread rapidly even without positive selection, as long as the inducing factor persists. Since environmental factors can affect entire populations composed of individuals with many genetic backgrounds (unlike a rare mutation that appears on a single genetic background), there is a high likelihood that some backgrounds will lead to the stabilization of the trait. Such stabilization becomes more probable when those organisms that can produce the phenotype are likely to mate with one another, further increasing the prevalence of the right genetic background. Thus, environmentally induced changes may be more efficient in producing selectable evolutionary novelties than mutations, which originate in only one individual or one family.

NICHE CONSTRUCTION

Phenotypic accommodation is predicated on the embryonic interactions called **reciprocal embryonic induction**. Reciprocal embryonic induction between two adjacent tissues is the fundamental principle for organ formation. For instance, the presumptive retina of the mammalian eye is a bulge of cells from the forebrain, while the presumptive lens is a subset of epithelial cells in the surface ectoderm of the head. When the two tissues meet, a complex dialogue is established whereby the presumptive lens cells tell the brain bulge to become the retina, and the presumptive retinal cells tell the placodal epithelium of the head to become the lens (Figure 10.8). Similarly, when the ureteric bud grows into and contacts the nephrogenic mesenchyme, the nephrogenic mesenchyme is told to become the nephron, and at the same time it tells the ureteric bud to become the renal pelvis and collecting ducts of the kidney.

But, as we have seen, the external environment can also be a source of signals that affect development, and one developing organism can produce factors that alter the development of other organisms. As environmental signals have been found to be critical in induction, it is not surprising that such environmental signals can also be accommodated. But what if these developmental interactions seen in reciprocal embryonic induction extended to interactions between organisms? This question is being addressed by a relatively new branch of evolutionary biology called **niche construction** (see Laland et al. 2008).

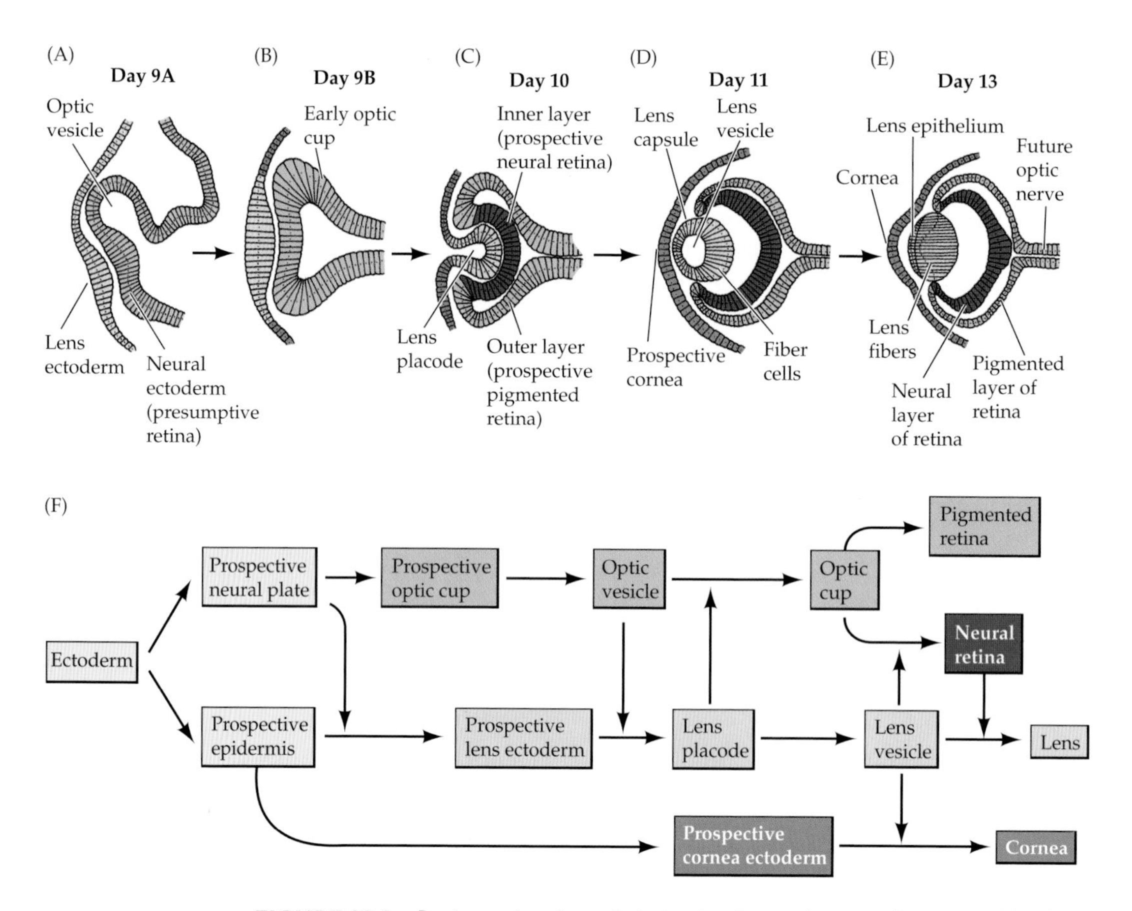

FIGURE 10.8 Reciprocal embryonic induction forms the eye of a mouse. (A, B) At embryonic day 9, a region surface ectoderm (destined to become the lens and cornea of the eye) comes in contact with a bulge of neuroectoderm from the forebrain, which is induced to form an optic cup. (C) By day 10, the lens-forming cells have been induced to invaginate, and two layers of retinal cells have been distinguished. (D) By day 11 of gestation, contact with the optic vesicle has induced differentiation of the lens from the presumptive corneal cells. (E) By day 13, two types of lens cells—cuboidal cells and elongated fiber cells—have been established and the cornea has developed in front of them. (F) Summary of some of the inductive interactions during mammalian eye development. (After Cvekl and Piatigorsky 1996.)

Niche construction is an evolutionary idea that emphasizes the capacity of organisms to modify selection pressures and thereby act as codirectors of their own evolution and that of other species. The importance of this perspective to evolutionary biology was recognized by Richard Lewontin (1982, 1983), who noted that "organisms fit the world so well because they have constructed it." He argued that the organism was not just a passive entity being acted upon by selective forces in the environment, but also an active agent capable of constructing an environment suited to its own ends. This idea has recently been extended by Lewontin (2000) and by John Odling-Smee (1988) and has become strengthened by theoretical population genetic and experimental findings that show niche construction to be an important factor in an organism's fitness (Laland et al. 1996; Odling-Smee et al. 1996; Laland 1999; Odling-Smee 2003; Donohue 2005). Niche construction can even counteract natural selection in instances where adult organisms can modify the environment to constrain or overcome selective forces. For instance, earthworms (*Lumbricus terrestris*) make burrows that provide an aqueous niche within the terrestrial environment. As a result, this species has evolved very little since migrating onto land more than 50 million years ago (Turner et al. 2000).

Niche construction takes on an even more important role when it is linked to development. Developmental environments become coupled to developing organisms by the niche-constructing activities of their organisms. Here, the niche construction seen during co-development (see Chapter 3) becomes the macroscopic analogue to reciprocal embryonic induction (Waddington 1953, 1957; Gilbert 2001; Laland et al. 2008). Niche construction similarly extends the dialogues of organ formation from within the embryo to outside it. The "circuitry diagrams" of organ formation within the embryo become united with the "circuitry diagrams" of organisms within the ecosystem.

We have already mentioned some interesting cases of niche construction. Some of the symbiotic microbes in the mouse intestine, for instance, induce gene expression in the gut epithelia not only to help the host, but to help themselves. The normal gut microbes, such as *Bacteroides*, induce gene expression in the Paneth cells of the intestine, instructing these cells to produce two compounds—angiogenin-4 and RegIII—that prevent the colonization of the intestine by *other* species of microbes. *Bacteroides*, *Escherichia coli*, and other symbiotic species are impervious to this compound, while several pathogenic Gram-positive bacteria (*Enterococcus faecalis* and *Listeria monocytogenes*) are wiped out by it (Hooper et al. 2003; Cash et al. 2006). Thus, the microbial species is modifying its niche, causing its environment to change in such a way that they can better survive.

Such niche-constructing behavior is also seen in relationships between symbiotic bacteria and their arthropod hosts. Indeed, sometimes the interaction is more one-sided than in the mammalian example. We saw in Chapter 3 how the bacterium *Wolbachia* can skew the sex ratio of the progeny of its host insects by feminizing genetically male embryos, resulting in all or mostly female progeny (see Figures 3.4 and 3.5). This feminization of insect populations has a clear advantage for the infecting *Wolbachia* because it directly increases the number of females, the only sex capable of passing the bacteria on to the insect progeny.

FIGURE 10.9 Niche-constructing behavior of *Eurosta solidaginis*, the goldenrod gall fly. (A) A female fly deposits her eggs inside the goldenrod's stem, within which the eggs hatch. (B) As the larva feeds, proteins in its saliva induce the formation of a gall on the plant's stem. (C) An opened gall reveals the fly larva at the center. (A, C photographs by K. and J. M. Storey, www.carleton.ca/~kbstorey; B photograph by David McIntyre.)

Another example of niche-constructing behavior is found in the goldenrod gall fly (*Eurosta solidaginis*). The female fly lays her eggs inside the goldenrod's stem, within which the eggs hatch into larvae (caterpillars). As a caterpillar eats the stem, proteins in the larval saliva induce cell proliferation in the goldenrod, resulting in the formation of a gall (Figure 10.9). The caterpillar enters the gall and continues eating from within it. As winter approaches, the larva, in danger of freezing to death, begins to produce sorbitol and trehalose sugars that act to prevent ice formation inside the cells. The trigger for synthesis of this "antifreeze" is not temperature but aromatic substances produced by the desiccating plant tissues of the gall (Williams and Lee 2005).

Here we see reciprocal induction on the ecological level. The fly larva creates a niche (the gall) by causing the plant to change its development. The niche provides not only nutrition but also the signal that allows the larva to change its development as winter approaches. Reviewing the literature on gall development, West-Eberhard (2003) concludes, "In gall production, the plant is a crucial and specific influential element of the insect's development, and the insect is a specific form-inducing element of the plant's developmental environment… ." There is a co-developmental relationship much like that of the organs in the embryo.

These examples are not isolated or marginal cases. Mammalian development is a case par excellence of an organism creating a niche and having the niche modify and permit the development of the organism. Mammalian embryos construct their niche by instructing the uterus to alter its cell cycles and its adhesion proteins and by inducing angiogenesis and a barrier to the immune system. The placenta induces the decidua reaction in the uterus, causing the uterus to become a habitat for the developing embryo. In so doing, the placenta instructs the uterus to bring blood vessels into the region and also inhibits the maternal immune system from acting against the fetus. Hormones from the embryo itself help construct an embryonic niche from the developmentally plastic anatomy of its mother's reproductive tract. The uterus reciprocally helps induce the formation of the placental tissues of the embryo (see Gluckman and Hanson 2005).

In addition, pollinators and flowers must co-evolve developmental patterns in which the eclosion of pollinators is in synchrony with the emergence of floral organs; and the timing of seed production with the foraging habits of seed dispersers coordinates the growth of forests. Thus reciprocal developmental interactions can coordinate ecological rhythms.

In short, when the organism alters its development in response to environmental cues such that the organism is more fit in the particular environment, it is called *adaptive developmental plasticity*; when a developing organism induces changes in its physical environment in ways that make the environment more fit for the organism, it is called *niche construction*. Moreover, there can be reciprocity between these two sets of inductive phenomena.

Summary: Eco-Evo-Devo

The renaissance of research in developmental plasticity is causing a need to reflect on and critique our gene-based models of biology. Phenotype is not merely the unrolling of genotype. The variations that provide the raw material for evolution need not be solely genetic but can be environmentally induced, as well. Even "the inheritance of acquired characteristics"—the whipping boy for generations of biologists—may have a mechanism if the inherited morphological alteration can be mediated by epigenetic changes in the DNA methylation of germ cells.

All of this has profound implications for evolution. In the material presented here, we have shown that environmentally induced variation can be fixed into the genome (genetic assimilation; epigenetic assimilation; genetic accommodation) and that the variation of one part of the body can have important changes on other parts of the body, so selection need not be on several separate components (bone size, muscle volume, muscle insertion) but only on one co-developing suite (including all components, as in phenotypic accommodation.) Even epigenetic phenomena can be selected, as in the plant (*Sagittaria sagittifolia*) that develops one set of leaves underwater and a different set on land (West-Eberhard 2003).

In 1986, population geneticist Bruce Wallace wrote an article entitled "Can Embryologists Contribute to an Understanding of Evolutionary Mechanisms?" His answer was a definite "No." Development, he asserted, was irrelevant to evolution. Echoing the opinions of many evolutionary biologists, he claimed that genetic assimilation had nothing to do with changing the genetics of populations. Moreover, he felt that development could never have anything to do with evolution because the germline could not be modified either within the embryo or by the environment. "How does a developing organism alter the genetic program carried in its originating germ cells?" he asked. And because it cannot alter that program, he thought, development would always be irrelevant to evolution. But as we have seen here, there is no single "genetic program." There are reaction norms such that the genotype can produce a range of phenotypes. Second, we now know that the germline is not so tightly sequestered as had been believed. While mutations may be rare, induced methylations and other chromatin modifications might be relatively common. The embryo does not need to modify any trajectory; the environment can signal the alteration. The presence of environmentally induced methylation in a promoter can silence a gene as well as a mutation, and methylation differences in germ cells can be transmitted between generations—for hundreds of years in the *Peloria* variant of toadflax, and for at least four generations in mice exposed to endocrine disruptors.

But nothing here negates the importance of natural selection. Evolution by natural selection is the way of the world, and this is Charles Darwin's key contribution to the intellectual patrimony of humankind. Even plasticity is a selectable trait (as we saw in the experiment by Suzuki and Nijhout; see Figure 10.5). But the explanation of biodiversity demands more than natural selection. It also demands a mechanism for the production of variation. But variation, as Darwin knew well, is *not* produced by natural selection. In his book dealing specifically with variation, Darwin acknowledges that evolutionary variations are produced by changes in development independently of natural selection. "The external conditions of life," he wrote, "are quite insignificant, in relationship to any particular variation, in comparison with the organization and constitution of the being which varies. We are thus driven to conclude that in most cases the conditions of life play a subordinate part in causing any particular modification…" (Darwin 1883, p. 282).

Genetics provided the first concrete evidence for such variation, and the Modern Synthesis focused almost exclusively on variations produced by gene mutations. There were reasons for this. First, genetics was much better understood than developmental biology. Second, genetics provided a mechanism for the transmission of phenotypic traits from one generation to the other as well as mathematical rules for understanding the production of phenotypes from one generation to the next. Third, genetics provided a fundable way to study evolution, especially in the United States (Beatty 2006). Fourth, neo-Darwinism linked evolution with population genetics, and to get a good genetic story, one does not want to deal with organisms whose phenotypes are significantly controlled by the environment. As Sonia Sultan (2003) points out, "neo-Darwinian botanists were often quite frustrated in their attempts to discern genetically based local adaptations through this 'environmental noise,'" and this led them to overlook the adaptive nature of these plastic responses. And fifth, the fear of Lysenkoism caused a backlash against any other type of hereditary variation that was not linked directly to allelic mutation (Lindegren 1966; Lewontin and Levins 1976; Sapp 1987). We now know that genetic variation is only a subset of heritable variation. Epigenetic variation can be inherited through the germline, and phenotypic plasticity can be inherited from generation to generation. Neither genetic nor epigenetic inheritance can become manifest except through development. Developmental variation consists of genetic variation, epigenetic variation, phenotypic plasticity, and the interactions among these three components.

The Modern Synthesis is a theory of genes, but the phenomena to be explained by evolution include the anatomical change of organisms (see Popper 1987; Amundson 2005). Development is the cascade of events linking genotype with phenotype, connecting genes with morphology, physiology, and behaviors. To get from the genes within the nucleus to the adult animal that eats, competes, mates, and defecates, one needs to link genotype to phenotype. One cannot have an evolutionary theory without an understanding of the mechanisms through which tissues, organs, and organ systems are constructed and can change. And, as Johannsen said in 1909, these mechanisms involve both genome and environments.

The history of evolutionary biology can be seen as nested subsets of syntheses, each one explaining more phenomena. Darwin's synthesis of biogeography, breeding, paleontology, and taxonomy was superseded by the Fisher-Wright version of the Modern Synthesis, which integrated Darwinian evolution with population genetics and showed the mechanisms through which Darwin's natural selection can take place through changes in alleles within a population. As population genetics matured, the Mayr-Dobzhansky version of the Modern Synthesis explained more about how speciation events could arise in natural populations. This version remains the core of evolutionary biology onto which other evolutionary theories have been appended.

But the Mayr-Dobzhansky version of the Modern Synthesis is more than 50 years old. It was generated before the advent of modern molecular biology, before developmental genetics, and before techniques such as the polymerase chain reaction, in situ hybridization, and high-throughput mRNA analysis revolutionized biology. It was formulated at a time when developmental biology was in its infancy and the idea of environmental regulation of phenotype production was suspect. In short, the Mayr-Dobzhansky synthesis is due for a major revision, after which it, too, will be a nested subset of another synthesis. Several writers (see Gilbert et al. 1996) have called for a "developmental synthesis." Massimo Pigliucci (2007) and Gerd Müller (2007) have proclaimed an "extended evolutionary synthesis," and Sean Carroll (2008) has called for an "expanding evolutionary synthesis." Eva Jablonka and Marion Lamb (2007) point toward a new theory of evolution "informed by developmental studies and epigenetic inheritance." Whatever its name, developmental biology will be the core of such a new synthesis. And whatever form it takes, ecological developmental biology will be a major bridge linking development and evolution.

Indeed, writing in 1894, Wilhelm Roux predicted that developmental biology would leave evolutionary biology and become allied with physiology. But sometime in the future, Roux predicted, developmental biology would return to evolutionary biology and bring with it an understanding of developmental mechanisms that would help complete the theory of evolution. That time has come.

References

Agrawal , A. A., C. Laforsch and R. Tollrian. 1999. Transgenerational induction of defences in animals and plants. *Nature* 401: 60–63.

Amundson, R. 2005. *The Changing Role of the Embryo in Evolutionary Thought: Structure and Synthesis.* Cambridge University Press, Cambridge.

Anway, M. D., A. S. Cupp, M. Uzumcu and M. K. Skinner. 2005. Epigenetic transgenerational actions of endocrine disruptors and mate fertility. *Science* 308: 1466–1469.

Anway, M. D., C. Leathers and M. K. Skinner. 2006a. Endocrine disruptor vinclozolin induced epigenetic transgenerational adult-onset disease. *Endocrinology* 147: 515–5523.

Anway, M. D., M. A. Memon, M. Uzumcu and M. K. Skinner. 2006b. Transgenerational effect of the endocrine disruptor vinclozolin on male spermatogenesis. *J. Andrology* 27: 868–879.

Baldwin, J. M. 1896. A new factor in evolution. *Am. Nat.* 30: 441–451; 536–553.

Baldwin, J. M. 1902. *Development and Evolution.* Macmillan, New York.

Bateman, K. G. 1959a. Genetic assimilation of the *dumpy* phenocopy. *J. Genet.* 56: 341–352.

Bateman, K. G. 1959b. Genetic assimilation of four venation phenocopies. *J. Genet.* 56: 443–474.

Bateson, P. 2005. The return of the whole organism. *J. Biosci.* 30: 31–39.

Beatty, J. 2006. Masking disagreement among experts. *Episteme: J. Soc. Epist.* 3: 52–67.

Brakefield, P. M., F. Kesbeke and P. B. Koch 1996. The regulation of phenotypic plasticity of eyespots in the butterfly *Bicyclus*. *Am. Nat.* 152: 853–860.

Burdge, G. C., J. Slater-Jefferies, C. Torrens, E. S. Phillips, M. A. Hanson and K. A. Lillycrop. 2007. Dietary protein restriction of pregnant rats in the F_0 generation

induces altered methylation of hepatic gene promoters in the adult male offspring in the F_1 and F_2 generations. *Br. J. Nutr.* 97: 435–439.
Burnet, B., K. Connolly and B. Harrison. 1973. Phenocopies of pigmentary and behavioural effects of the *Yellow* mutant in *Drosophila* induced by ?-dimethyltyrosine. *Science* 181: 1059–1060.
Cabej, N. 2008. *Epigenetic Principles of Evolution.* Albenet, Dumont, NJ.
Carroll, S. B. 2008. Evo-devo and an expanding evolutionary synthesis: A genetic theory of morphological evolution. *Cell* 134: 25–36.
Cash, H. L., C. V. Whitman, C. L. Benedict and L. V. Hooper. 2006. Symbiotic bacteria direct expression of an intestinal bactericidal lectin. *Science* 313: 1126–1130.
Chang, H. S., M. D. Anway, S. S. Rekow and M. K. Skinner M K. 2006. Transgenerational epigenetic imprinting of the male germline by endocrine disruptor exposure during gonadal sex determination. *Endocrinology* 147: 5524–5541.
Chow, K. L. and K. W. Chan. 1999. Stress-induced phenocopy of *C. elegans* defines functional steps of sensory organ differentiation. *Dev. Growth Diff.* 41: 629–637.
Crews, D. and 7 others. 2007. Transgenerational epigenetic imprints on mate preference. *Proc. Nat. Acad. Sci. USA* 104: 5942–5946.
Cvekl, A. and J. Piatigorsky. 1996. Lens development and crystallin gene expression: Many roles for Pax6. *BioEssays* 18: 621–630.
Danielson-François, A. M., J. K. Kelly and M. D. Greenfield. 2006. Genotype x environment interaction for mate attractiveness in an acoustic moth: evidence for plasticity and canalization. *J. Evol. Biol.* 9: 532–542.
Darwin, C. 1868. *The Variation of Animals and Plants under Domestication*. Appleton, New York.
Darwin, C. 1883. *The Variation of Animals and Plants under Domestication*, 2nd Ed., Appleton, New York.
Donohue, K. 2005. Niche construction through phonological plasticity: Life history dynamics and ecological consequences. *New Phytologist* 166: 83–92.
Duerden, J. E. 1920. Inheritance of callosities in the ostrich. *Am. Nat.* 54: 289–312.
Fischman, J. 1995. Why mammalian ears went on the move. *Science* 270: 1436.
Frazetta, T. H. 1975. *Complex Adaptations in Evolving Populations.* Sinauer Associates, Sunderland, MA.
Galis, F., A. Terlouw and J. W. M. Osse. 1994. The relation between morphology and behaviour during ontogenetic and evolutionary changes. *J. Fish. Biol.* 45(suppl. A): 13–26.
Garson, J., L. Wang and S. Sarkar. 2003. How development may direct evolution. *Biol. Philos.* 18: 353–370.
Gibson, G. 1996. Quoted in S. Pobojewsky, *The University Record*, January 16.
Gibson, G., and D. S. Hogness. 1996. Effect of polymorphism in the *Drosophila* regulatory gene *Ultrabithorax* on homeotic stability. *Science* 271: 200–203.
Gilbert, S. F. 1994. Dobzhansky, Waddington, and Schmalhausen: Embryology and the modern synthesis. In M. B. Adams (ed.), *The Evolution of Theodosius Dobzhansky*. Princeton University Press, Princeton, NJ, pp. 143–154.
Gilbert, S. F. 2001. Ecological developmental biology: Developmental biology meets the real world. *Dev. Biol.* 233: 1–12.
Gilbert, S. F. 2002. Canalization. In M. Pagel (ed.), *Encyclopedia of Evolution*, Vol. 1. Oxford University Press, New York, pp. 133–135.
Gilbert, S. F., J. Opitz and R. A. Raff. 1996. Resynthesizing evolutionary and developmental biology. *Dev. Biol.*173: 357–372.
Gluckman, P. D. and M. Hanson. 2005. *The Fetal Matrix: Evolution, Development, and Disease*. Cambridge University Press, Cambridge.
Goldschmidt, R. B. 1938. *Physiological Genetics*. McGraw-Hill, New York.
Goldschmidt, R. B. 1940. *The Material Basis of Evolution*. Yale University Press, New Haven.
Gottlieb, G. 1992. *Individual Development and Evolution*. Oxford University Press, New York.
Gould, S. J. 2002. *The Structure of Evolutionary Theory*. Harvard University Press, Cambridge, MA.
Greenwood, P. H. 1965. Environmental effects on the pharyngeal mill of a cichlid fish, *Astatoreochromis alluaudi*, and their taxonomic implications. *Proc. Linn. Soc. Lond.* 176: 1–10.
Gulick, J. T. 1872. *Nature* 6: 222–224. Quoted in B. K. Hall, 2006, Evolutionist and missionary, the Reverend John Thomas Gulick (1832–1923). II. Coincident or Ontogenetic Selection: The Baldwin Effect. *J. Exp. Zool. MDB* 306B: 489–495.
Gutteling, E. W., J. A. G. Riksen, J. Bakker and J. E. Kammenga. 2007. Mapping phenotypic plasticity and genotype-environment interactions affecting life history traits in *Caenorhabidis elegans*. *Heredity* 98: 28–37.
Haldane, J. B. S. 1932. *The Causes of Evolution*. Longmans, London.
Hall, B. K. 1999. *Evolutionary Developmental Biology*, 2nd Ed. Kluwer, Dordrecht.
Hall, B. K. 2003. Baldwin and beyond: Organic selection and genetic assimilation. In B. H. Weber and D. J. Depew (eds.), *Evolution and Learning: The Baldwin Effect Reconsidered*. MIT Press, Cambridge, MA, pp. 141–168.
Hampé, A. 1959. Contribution à l'étude du développement et la régulation des déficiences et excédents dans la patte de l'embryon de poulet. *Arch. Anat. Microsc. Morphol. Exp.* 48: 347–479.
Hansell, M. H. 2004. *Animal Architecture*. Oxford Animal Biology Series, Oxford.
Ho, M.-W., E. Bolton and P. T. Saundres. 1983. The bithorax phenocopy and pattern formation. 1. Spatiotem-

poral characteristics of the phenocopy response. *Exp. Cell Biol.* 51: 282–290.

Hooper, L. V., T. S. Stappenbeck, C. V. Hong and J. I. Gordon. 2003. Angiogenins: A new class of microbicidal proteins involved in innate immunity. *Nature Immunol.* 4: 269–273.

Jablonka, E. and M. J. Lamb. 2005. *Evolution in Four Dimensions: Genetic, Epigenetic, Behavioral, and Symbolic Variation in the History of Life.* MIT Press, Cambridge, MA.

Jablonka, E. and M. J. Lamb. 2007. The expanded evolutionary synthesis: A response to Godfrey-Smith, Haig, and West-Eberhard. *Biol. Philos.* 22: 453–472.

Jablonka, E. and G. Raz. 2009. Transgenerational epigenetic inheritance: Prevalence, mechanisms, and implications for the study of heredity. *Q. Rev. Biol.* In press.

Jirtle, R. L. and M. K. Skinner. 2007. Environmental epigenomics and disease susceptibility. *Nature Rev. Genet.* 8: 253–262.

Johannsen, W. 1909. *Elemente der Exakten Erblichkeitslehre.* Gustav Fischer Verlag, Jena.

Kaneko, K. 2002. Symbiotic sympatric speciation: consequence of interaction–driven phnotype differentiation through developmental plasticity. *Pop. Ecol.* 44: 71–85.

Kemp, T. S. 1982. *Mammal-Like Reptiles and the Origin of Mammals.* Academic Press, New York.

King, R. C. and W. D. Stanfield. 1985. *A Dictionary of Genetics*, 3rd Ed. Oxford University Press, Oxford.

Köntges, G. and A. Lumsden. 1996. Rhombencephalic neural crest segmentation is preserved throughout craniofacial ontogeny. *Development* 122: 3229–3242.

Laland, K. N. 1999. Evolutionary consequences of niche construction and their implications for ecology. *Proc. Natl. Acad. Sci. USA* 96: 10242–10247.

Laland, K. N., F. J. Odling-Smee and M. W. Feldman. 1996. On the evolutionary consequences of niche construction. *J. Evol. Biol.* 9: 293–316.

Laland, K. N., J. Odling-Smee and S. F. Gilbert. 2008. Evo-devo and niche construction: Building bridges. *J. Exp. Zool. B.* in press.

Levit, G. S., U. Hossfeld and L. Olsson. 2006. From the "modern synthesis" to cybernetics: Ivan Ivanovich Schmalhausen (1884–1963) and his research program for a synthesis of evolutionary and developmental biology. *J. Exp. Zool.* 306B: 89–106.

Lewontin, R. C. 1982. Organism and environment. In H. C. Plotkin (ed.), *Learning, Development and Culture.* Wiley, New York.

Lewontin, R. 1983. Gene, organism, and environment. In D. S. Bendall (ed.), *Evolution from Molecules to Men.* Cambridge University Press, Cambridge.

Lewontin, R. 2000. *The Triple Helix: Gene, Organism, and Environment.* Harvard University Press, Cambridge MA.

Lewontin, R. and R. Levins. 1976. The problem of Lysenkoism. In H. Rose and S. Rose (eds.), *The Radicalisation of Science.* Macmillan, London, pp. 32–65.

Liem, K. F. and L. S. Kaufman. 1984 Intraspecific macroevolution: Functional biology of the polymorphic cichlid species *Cichlasoma minckleyi.* In A. A. Echelle and I. Kornfield (eds.), *Evolution of Fish Species Flocks.* University of Maine Press, Orono, pp. 203–215.

Lindegren, C. C. 1966. *The Cold War in Biology.* Planarian Press, Ann Arbor, MI.

Lloyd Morgan, C. 1896. *Habit and Instinct.* Arnold, London.

Manning, K. and 7 others. 2006. A naturally occurring epigenetic mutation in a gene encoding an SBP-box transcription factor inhibits tomato fruit ripening. *Nature Genet.* 38: 948-952.

Marks, J. 1989. Genetic assimilation in the evolution of bipedalism. *Hum. Evol.* 4: 493–499.

Matsuda, R. 1982. The evolutionary process in talitrid amphipods and salamanders in changing environments, with a discussion of "genetic assimilation" and some other evolutionary concepts. *Can. J. Zool.* 60: 733–749.

Matsuda, R. 1987. *Animal Evolution in Changing Environments with Special Reference to Abnormal Metamorphosis.* Wiley, London.

McCaffery, A. and S. J. Simpson. 1998. A gregarizing factor present in the egg pod foam of the desert locust *Schistocerca gregaria. J. Exp. Biol.* 201: 347–363.

Meaney, M. J. 2001. Maternal care, gene expression, and the transmission of individual differences in stress reactivity across generations. *Annu. Rev. Neurosci.* 24: 1161–1192.

Merrifield, F. 1890. Systematic temperature experiments on some Lepidoptera in all their stages. *Trans. Entomol. Soc. Lond.* 131–159.

Merrifield, F. 1893. The effects of temperature in the pupal stage on the colouring of *Pieris napi, Vanessa atalanta, Chrysophanus phloeas*, and *Ephyra punctaria. Trans. Entomol. Soc. Lond.* 425–438.

Meyer, A. 1990. Morphometrics and allometry in the trophically polymorphic cichlid fish, *Cichlasoma citrinellum* (Pisces: Cichlidae). *Biol. J. Linn. Soc.* 39: 279–299.

Müller, G. B. 1989. Ancestral patterns in bird limb development: A new look at Hampé's experiment. *J. Evol. Biol.* 1: 31–47.

Müller, G. B. 1990. Developmental mechanisms at the origin of morphological novelty: A side-effect hypothesis. In M. H. Nitecki (ed.), *Evolutionary Innovations.* University of Chicago Press, Chicago.

Müller, G. B. 2007. Evo-devo: Extending the evolutionary synthesis. *Nature Rev. Genet.* 8: 943–949.

Newbold, R. R., E. Padilla-Banks and W. N. Jefferson. 2006. Adverse effects of the model environmental estrogen diethylstilbestrol are transmitted to subsequent generations. *Endocrinology* 147: S11–S17.

Nijhout, H. F. 1984. Color pattern modification by cold shock in Lepidoptera. *J. Embryol. Exp. Morphol.* 81: 287–305.

Nussey, D. H., E. Postma, P. Gienapp and M. E. Visser. 2005. Selection on hereditable phenotypic plasticity in a wild bird population. *Science* 310: 304–306.

Odling-Smee, F. J. 1988. Niche constructing phenotypes. In H. C. Plotkin (ed.), *The Role of Behavior in Evolution*. MIT Press, Cambridge, MA, pp. 73–132.

Odling-Smee F. J. 2003. *Niche Construction. The Neglected Process in Evolution*. Monographs in Population Biology 37, Princeton University Press, Princeton, NJ.

Odling-Smee, F. J., K. N. Laland and M. W. Feldman. 1996. Niche construction. *Am. Nat.* 147: 641–648.

Osborn, H. F. 1897. Organic selection. *Science* 15: 583–587.

Piaget, J. 1929a. Les races lacustres de la *Limnaea stagnalis* L.: Recherches sur les rapports de l'adaptation hereditaires avec le milieu. *Bull. Biol. Fr. Belg.* 63: 424–455.

Piaget, J. 1929b. L'adaptation de la *Limnaea stagnalis* au milieu lacustres de la Suisse Romande: Etude biométrique et génétique. *Rev. Suisse Zool.* 36: 263–531.

Pigliucci, M. 2007. Do we need an extended evolutionary synthesis? *Evolution* 61: 2743–2749.

Pigliucci, M., C. J. Murren and C. D. Schlichting. 2006. Phenotypic plasticity and evolution by genetic assimilation. *J. Exp. Biol.* 209: 2362–2367.

Popper, K. 1987. Quoted in N. I. Platnick and D. E. Rosen, Popper and evolutionary novelties. *Hist. Philos. Life Sci.* 9: 5–16

Provine, W. B. 1989. Progress in evolution and meaning in life. In M. Nitecki (ed.), *Evolutionary Progress*. University of Chicago Press, Chicago, pp. 49–74.

Queitsch, C., T. A. Sangster and S. Lindquist. 2002. Hsp90 as a capacitor of phenotypic variation. *Nature* 417: 618–624.

Roux, W. 1894. Einleitung. *Roux's Arch. Entwicklungsmech. Org*. Translated (W. M. Wheeler, tr.) as "The problems, methods, and scope of developmental mechanics," W. M. Wheeler (trans.) in J. Maienschein (ed),1986. *Defining Biology*. Harvard University Press, Cambridge, MA, pp. 107–148.

Ruden, D. M., M. D. Garfinkel, V. E. Sollars and X. Lu. 2003. Waddington's widget: Hsp90 and the inheritance of acquired characteristics. *Sem. Cell Dev. Biol.* 14: 301–310.

Rutherford, S. L. and S. Lindquist. 1998. Hsp90 as a capacitor for morphological evolution. *Nature* 396: 336–342.

Sangster, T. A., N. Salathia, S. Uneurraga, K. Schellenberg, S. Lindquist and C. Queitsch. 2008a. Hsp90 affects the expression of genetic variation and developmental stability in quantitative traits. *Proc. Natl. Acad Sci USA* 105: 2963 –2968.

Sangster, T. A. and 9 others. 2008b. Hsp90-buffered genetic variation is common in *Arabidopsis thaliana*. *Proc. Natl. Acad. Sci. USA* 105: 2969–2974.

Sapp, J. 11987. *Beyond the Gene: Cytoplasmic Inheritance and the Struggle for Authority in Genetics*. Oxford University Press, New York.

Schmalhausen, I. I. 1938. Organizm kak tseloje v individual'nom I istoricheskom razvitii. ("The Organism as a Whole in its Individual and Historical Development"). AN SSR, Leningrad.

Schmalhausen, I. I. 1949. *Factors of Evolution: The Theory of Stabilizing Selection*. Blakiston, Philadelphia.

Shapiro, A. M. 1976. Seasonal polyphenism. *Evol. Biol.* 9: 259–333.

Shapiro, A. M. 1982. Redundancy in pierid polyphenisms: Pupal chilling induces vernal phenotype in *Pieris occidentalis* (Pieridae). *J. Lepidopt. Soc.* 36: 174–177.

Slijper, E. J. 1942a. Biologic-anatomical investigations on the bipedal gait and upright posture in mammals, with special reference to a little goat, born without forelegs. I. *Proc. Konink Ned. Akad. Wet.* 45: 288–295.

Slijper, E. J. 1942b. Biologic-anatomical investigations on the bipedal gait and upright posture in mammals, with special reference to a little goat, born without forelegs. II *Proc. Konink Ned. Akad. Wet.* 45: 407–415.

Sollars, V., X. Liu, L. Xiao, X. Wang, M. D. Garfinkel, and D. M. Ruden. 2003. Evidence for an epigenetic mechanism by which Hsp90 acts as a capacitor for morphological evolution. *Nature Genet.* 33: 70–74.

Spalding, D. 1873. Instinct with original observations on young animals. *MacMillan's Magazine* 27: 282–293.

Spemann, H. 1901. Über Correlationen in der Entwicklung des Auges. *Verhand. Anat. Ges.* 15: 61–79.

Spemann, H. 1907. Zum Problem der Correlation in der tierischen Entwicklung. *Verhadl. Deutsche Zool. Gesell.* 17: 22–49.

Standfuss, M. 1896. *Handbuch der palearctischen Gross-Schmetterlinge für Forscher und Sammler*. Gustav Fischer, Jena.

Sultan, S. E. 2003. Phenotypic plasticity in plants: A case study in ecological development. *Evol. Dev.* 5: 25–33.

Suzuki, Y. and H. F. Nijhout. 2006. Evolution of a polyphenism by genetic assimilation. *Science* 311: 650–652.

Thomson, K. S. 1988. *Morphogenesis and Evolution*. Oxford University Press, New York.

Turesson, G. 1922. The geotypical response of a plant species to the habitat. *Hereditas* 3: 211–350.

Turner, J. S. 2000. The extended organism: The physiology of animal-built structures. Harvard University Press, Cambridge, MA.

Twitty, V. C. 1932. Influence of the eye on the growth of its associated structures, studied by means of heteroplastic transplantation. *J. Exp. Zool.* 61: 333–374.

Waddington, C. H. 1942. The canalization of development and the inheritance of acquired characters. *Nature* 150: 563.

Waddington, C. H. 1953. Epigenetics and evolution. In *Symposia for the Society for Experimental Biology VII: Evolution*. Cambridge University Press, pp. 186–189.

Waddington, C. H. 1956. Genetic assimilation of the *bithorax* phenotype. *Evolution* 10: 1–13.

Waddington, C. H. 1957. *The Strategy of the Genes*. Allen and Unwin, London.

Waddington, C. H. 1952 Selection of the genetic basis for an acquired character. *Nature* 169: 278.
Waddington, C. H. 1961. Genetic assimilation. *Adv. Genet.* 10: 257–290.
Wallace, B. 1986. Can embryologists contribute to an understanding of evolutionary mechanisms? In *Integrating Scientific Disciplines*. (W. Bechtel). Martinus Nijhoff, Dordrecht. Pp. 149–163.
Weismann A 1892. *The Germ Plasm: A Theory of Heredity.* Scribners. New York.
West-Eberhard, M. J. 1989. Phenotypic plasticity and the origins of diversity. *Annu. Rev. Ecol. Syst.* 20: 249–278.
West-Eberhard, M. J. 2003. *Developmental Plasticity and Evolution*. Oxford University Press, Oxford.
West-Eberhard, M. J. 2005. Phenotypic accommodation: adaptive innovation due to developmental plasticity. *J. Exp. Zool. (MDE)* 304B: 610–618.
Williams, J. B. and R. E. Lee Jr. 2005. Plant senescence cues entry into diapause in the gall fly *Eurosta solidaginis*: Resulting metabolic depression is critical for water conservation. *J. Exp. Biol.* 208: 4437–4444.

Coda

Philosophical Concerns Raised by Ecological Developmental Biology

Hearing about the great Nunobiki waterfall
Did not prepare me
For hearing the great Nunobiki waterfall.

Fujiwara Yoshikiyo, 1187

This book has attempted to present an integrated developmental perspective on the origins of animal diversity, the origins of certain diseases, and the integration of living communities through co-development. In a sense, it has proposed a revised view of the organismal world. So how (if at all) does this developmental perspective change our views concerning the living world and how it can be appropriately studied?

The common denominator of all the components of this perspective on life is development: the emergence of form, the processes of becoming. In this view of living beings (note the gerund case of both these words), the processes of development percolate through evolution, genetics, and ecology. In effect, evolution is seen as the *outcome* of development, while genetics is seen as those processes through which similar developmental trajectories are transmitted between generations, a notion proposed by biologists as early as E. B. Wilson (1896) and revived by philosophers of science such as J. Griesemer (2002). Ecology becomes a science focusing on developmental and adult interactions in time and space, which in turn produce and sustain the patterns of interspecies interactions at particular locations.

Philosophy is often subdivided into *ontology* (a theory of what is real), *epistemology* (a theory of how we perceive and organize data), and *ethics* (a theory of how to live a righteous life.) The applications of these branches of philosophy include, respectively, pedagogy (what and how we teach), methodology (how we acquire data), and public policy (how we make social decisions consistent with our values). The revised view of nature presented in this volume impacts each of these areas.

Indeed, we expect that this developmental view of nature will elicit new questions from philosophers, or at least cause them to ask questions that haven't been asked for a while.

Ontology

What is an "individual" in terms of its developmental and ecological history?

Ecological developmental biology has some extremely significant propositions concerning ontology. At the very least, the prevalence of polyphenisms and reaction norms instructed by environmental agents abolishes any notion of a genetic determinism. We can give a definite answer to the question posed by Lewis Wolpert (1994):

> Will the egg be computable? That is given a total description of the fertilized egg—the total DNA sequence and location of all proteins and RNA—could we predict how the embryo will develop?

The answer is a resounding "No—and thank goodness." The trajectory of development can depend to a significantly large degree on environmental agents.

Moreover, developmental symbioses are no longer relegated to marginal exceptional cases. Rather, they are the norm. The developmental interconnections between organisms are so intimate that, as Lewis Thomas pointed out, our biological individuality is illusionary. Our self becomes a semi-permeable self. This metaphor is, of course, that of a cell, an "individual" entity that is a functioning unit limited by a membrane, but which is also a member of a larger community, receiving signals from that larger group.

In their groundbreaking theory of symbiosis, Sagan and Margulis (1991) saw symbiosis as occurring between autopoietic ("self-developing", *qua* Maturana and Varela 1980) entities. They wrote, "What is remarkable is the tendency of autopoietic entities to interact with other autopoietic entities." However, if developmental symbioses are the rule, then the entire notion of an "autopoietic individual" has to be questioned. We are not adults entering into symbioses with other consenting adults or microbes. Rather, the processes that construct a particular adult organism involve symbioses between *developing* organisms (Gilbert 2002). We are talking "co-construction" on an enormous scale. Developmental symbiosis and phenotypic plasticity demonstrate that development is co-development. Epigenesis is interspecies as well as within the embryo. This does not necessarily mean that the entire biosphere is the unit of development, but just that the borders of development do not end at the ectoderm; once outside the ectoderm, there can be a steep gradient of influence.

These ideas concerning the importance of the environment in phenotype production have been fundamental platforms of *developmental systems theory* (DST), a school of philosophy that takes developmental biology very seriously and that incorporated the epigenetic ideas of Waddington, Whitehead, and J. M. Baldwin at about the same time that evolutionary developmental biology was being formed (see Gottlieb 1992; Robert 2004). Susan Oyama (1985, p. 123), one of the founders of DST, has written that the unit of development is not the organism but, rather, a developmental system encompassing "not just genomes with cellular structures and processes, but intra- and interorganismic relations, including relations with members of other species and interactions with the inanimate surround as well." J. S. Haldane (1933, p. 74) had the same intuition, declaring that "there is no spatial limit to the life of an organism."

What is an "individual" in terms of its developmental and evolutionary history?

The ability of species to direct the development of other species into more fit trajectories has implications for evolutionary biology. Evolutionary biologists, at least in Britain and the United States, have tended to find in nature the ontology of Locke, Mills, and Hobbes—philosophers of individual rights and of a nature characterized by Hobbe's "war of all against all." Moreover, Darwin's and A. R. Wallace's pioneering insights into natural selection were historically and philosophically grounded in Thomas Malthus' political economy of individual competition, which they transferred from European economics into the natural world. As historian Sylvan Schweber (1985, p. 38) noted, in *The Origin of Species*, "biology joined hands with Scottish political economy, sociology, and historiography, and with English philosophy of science. The political economy was that of Adam Smith and his disciples." This philosophy assumes that each individual is a unit of self-interest that interacts with other units of self-interest to yield cooperation and wealth. Thus Darwin concluded the *Origin* with praise to a nature that can form a "tangled bank" of harmonious interactions from competition, death, and destruction.

Modern evolutionary theorists have taken evolutionary biology from metaphors of mercantile capitalism to metaphors of investment capitalism, positing genes as the unit of currency (see Haraway 1979; Young 1985). Indeed, most evolutionary biology has accepted the individualistic premises by which each unit is competing for limited resources against all others.* Thus, for classical Darwinism, the harmonious interactions one sees in the biosphere are the result of underlying competitive interactions. Just as the wooden pencil was created from self-interested individuals, so was the tree from which the wood came.

Embryologists, however, have had different philosophical inputs. Centered on the European continent, classical embryology had as its underpinnings the philosophies of emergent form found in Immanuel Kant and Wolfgang Goethe. These philosophies allowed a scientist to hold both mechanistic and teleological views (Lenoir 1982; Gilbert and Faber 1996). Lenoir (1982) and Huneman (2006) have argued that the founders of modern embryology—Dollinger, Pander, von Baer and Rathke—subscribed to the organicism set forth in Kant's *Critique of Judgement*:

*Several of Darwin's contemporaries, especially those in Russia, felt that Darwin had made an error in borrowing so strongly from British economic theorists and philosophers, making evolution a particularly British rather than universal theory. Chernyshevskii (1888), for instance, wrote that Darwin's view of nature was that to be expected "if Adam Smith had taken it upon himself to write a course in zoology." Darwin was also partial to the philosophy of Gottfried Leibniz, who also saw cooperation coming from competition (it is from Leibniz that Darwin got the idea that nature doesn't make jumps). Leibniz went on to say that not all things that can be present simultaneously. While Darwin thought of this in terms of adult competition, this idea, which Leibniz called "compossibility," can be even more important when thinking in terms of ecological developmental biology.

> The first principle required for the notion of an object conceived as a natural purpose is that the parts, with respect to both form and being, are only possible through their relationship to the whole ... Secondly, it is required that the parts bind themselves mutually into the unity of a whole in such a way that they are mutually cause and effect of one another.

According to this view, the parts not only determine the properties of the whole, but the whole reciprocally determines the properties of the parts. These philosophical views of causation were not competitive but cooperative and integrative, where the parts helped form one another and were dependent upon each other. Embryologists took them to be the natural way of life. Writing about experimental embryology in Weimar Germany, Hans Holtfreter (1968, p. xi) claimed:

> We managed more or less successfully to keep our work undisturbed by humanity's strife and struggle around us and proceeded to study the plants and animals, and particularly, the secrets of amphibian development. Here, at least, in the realm of undespoiled Nature, everything seemed peaceful and in perfect order. It was from our growing intimacy with the inner harmony, the meaningfulness, the integration, and the interdependence of the structures and functions as we observed them in dumb creatures that we derived our own philosophy of life. It has served us well in this continuously troublesome world.

Holtfreter's American contemporary Ernest E. Just, who quoted Goethe and who came from a tradition of observing rather than experimenting, also claimed, "Whether we study atoms or stars or that form of matter, known as living, always must we reckon with inter-relations" (Just 1939, p. 368). Indeed, Just routinely saw developing organisms as being integrated into their ecological contexts (see Byrnes and Eckberg 2006).

Interdependence, harmony, and integration: a different perspective on nature than the autonomous, competitive nature of classical evolutionary biology. In addition to Locke and Hobbes, we have Kant and Goethe.* These developmental views do not invalidate Charles Darwin's or Thomas Huxley's competitive perceptions of nature. (It was Huxley who compared nature to gladiatorial combat and who claimed Hobbes to be one of the three greatest philosophers.) Nature abounds in cruel and "heartless" behaviors. Anyone who has studied in the Galápagos Islands can discourse on birds that throw their younger sibs out of the nest to starve (with their mothers' complicity) and on female lizards that fight viciously for limited nesting sites. Darwin (see Gould 1982; Gillespie 1979) wrote that a "devil's chaplain" could write a great book concerning the "horribly cruel works of nature," and he felt that the life cycles of parasitoid wasps (see Chapter 3) were incompatible with a deity who was both benevolent and omnipotent.

However, what we have seen here is that there is also a cooperative account of evolution. Such an account has been popularized by Ashley Montague (1955) and more recently by Lynn Margulis and Dorian Sagan (2002), who emphasize symbiosis as demonstrating the cooperative interconnectedness of life. "Life," they claim,

*Perhaps it is through evolutionary developmental biology that we can best appreciate Haeckel's (1882) intuition that Darwin had solved the great conundrum posed by Kant, namely, "how a purposefully directed form of organization can arise without the aid of a purposefully effective cause."

"did not take over the globe by combat, but by networking" (Sagan and Margulis 1986). They argue that symbiosis is responsible for the origin of life as we know it as well as for speciation events. We have argued here that the symbiosis between developing organisms shows that each organism develops as an interconnected community and that the environment provides normative cues for embryogenesis. Thus, developmental symbioses and environmentally induced polyphenisms support Margulis' (1998) notion of a "symbiotic planet," and the argument that cooperation is critical in the history and perseverance of life on Earth.

Evolutionary biologist and embryologist Conrad Waddington noted (1953) that evolution takes place in two areas—in the embryo and in the environment—and that the parts of an embryo have to function cooperatively to form the organism that can then compete.* This notion of cooperation now extends beyond the embryo to the realms of "interspecies epigenesis," where development is coordinated between two or more species. Therefore, instead of demanding that the external (competitive) mode of evolution serve as a model for the embryo (see Buss 1987 and Gilbert 1992), ecological developmental biology suggests that the internal, cooperative embryonic mode of evolution be seen between species. With signals coming from outside the embryo, the game of embryonic induction is played between organisms. Symbiotic bacteria help form our guts and immune systems; predators, competitors, and food change developmental trajectories by the evolved mechanisms of developmental plasticity; signals from the warmth of the sun or from the number of hours the sun is above the horizon determine the sex of some vertebrates and the pigmentation of others. In some cases, the induction is reciprocal, allowing each organism to find its place within the ecosystem as it develops. This has important implications for evolutionary theory. If the environment is giving instructive information as well as selective pressures, and if relationships, rather than individuals, are being selected (that is, if the metaphor of evolution is more like embryogenesis than it is a group of wedges), then group selection, until recently an "outlier," is brought to the center of evolutionary theory. Interestingly, group selection is being seen as increasingly important by recent evolutionary theorists (Wilson and Sober 1994; Wilson 2005).

The data from ecological developmental biology also have profound philosophical implications for the question of "who we are." One of the insights of twentieth-century philosophy was that we are "defined by the 'other.'" However, this definition has been depicted as being a confrontation between the "self" and the "other"; in this way, it is similar to traditional evolutionary view of competition between separate autonomous entities. Both traditional evolutionary biology and existentialism grow from the soil of Hobbesian competition and individuality. While ecological developmental biology similarly postulates that we are defined, in part, by the "other," it depicts our identities as *becoming with* the "other." The relationship between "self" and "other" is not that between autonomous and antagonistic indi-

*In this view, Waddington reflects T. H. Huxley's ideas about inner unity and external struggle. Postulating an early version of group selection, Huxley (1894) noted that a society must cooperate internally to be strong enough to successfully compete against other societies. Huxley's Russian contemporary, Petr Kropotkin (1902), noted that other animals form societies and that those who cooperate best would survive. Notions of "inner" and outer," however, become problematic when discussing certain ecological developmental notions such as bacterial symbiosis. Here, the bacteria and the host organism are both environment for each other.

viduals, but that of non-autonomous agents-in-the-making that inter- and intra-act to form both themselves and other novel patterns. (See Barad 2007 for similar conclusions emanating from quantum physics.)

The eco-evo-devo model of evolution does not claim that competition is unimportant in the animal kingdom any more than it could claim the absence of predation. It does, however, claim that alongside this competitive model of evolution there belongs a cooperative model of evolution. "Becoming with" does not exclude self-defense strategies; a mosquito seeking a person's blood to complete her oogenesis puts herself and her lineage in mortal peril. However, the older idea of evolution as intraspecies competition may be a small, fine-tuning, part of a larger view of evolution, wherein natural selection favors certain relationships both between embryonic cells (to generate novelty) and between species (to stabilize ecosystems). The biological models of the twenty-first century should be richer than that of the twentieth.

Integrative philosophical traditions

Several philosophical systems of ontology have also come to this conclusion. Historically, two important philosophies have been dialectical materialism and process philosophy. Dialectical materialism was prominent in Russian biology even before the advent of Communism, and it is not an accident that so many of the early researchers mentioned in this book were Russian. The Russian school of evolutionary biology thought that competition played only a relatively minor role in the origin of species, and one of the leading books on the history of Russian evolutionary thought (Todes 1989) is provocatively titled *Darwin without Malthus*. Several British embryologists, notably Joseph and Dorothy Needham, thought that the interactivism of dialectical materialism provided a framework for animal development as well (Werskey 1978; Haraway 1976; Winchester 2008).

In the 1930s, another philosophy of emergence, the "process philosophy" of Alfred North Whitehead, became popular among embryologists (see Haraway 1976; Gilbert 1991). Although the totality of Whitehead's philosophy and theology was not taken up by the scientists, several embryologists were impressed by Whitehead's (1929) concepts of developing systems, processes, and the creative advance into novelty. Speaking for a group of such embryologists, Waddington (1975, pp. 3,5) reflected, "As far as scientific practice is concerned, the lessons to be learned from Whitehead were ... his replacement of 'things' by processes which have an individual character which depends upon the 'concrescence' into a unity of very many relations with other processes." Indeed, Waddington reflected, "I tried to put the Whiteheadian outlook to actual use in particular experimental situations."

Affinities between Whitehead's ontology and those of classical Asian philosophers have long been noted (see, for instance, Streng 1975; Odin 1982). Chief among these ancient Asian thinkers was the Buddhist sage Nargarjuna. Nagarjuna is regarded as one of the most important interpreters of what the Buddha said, and he emphasized that "Things derive their being and nature by mutual dependence and are nothing in themselves" (quoted in Murti 1955, p. 138). This doctrine, called *Pratītyasamutpāda*, is variously translated as "dependent origination," "conditioned genesis," "interdependent co-origination," and "interdependent arising." Thus, any phenomenon comes into being ("be-comes") only because of the coming into existence of other phenomena in a network of mutual cause and effect. Indeed, according to Nagarjuna, what the Buddha awakened *to* (*bodhi* means "to awaken") was the truth of interdependent co-origination.

In the twentieth century, another group of philosophers, who became known as phenomenologists, believed that the physical sciences had become too abstract—that they had sacrificed the physical nature of the perceived universe for particles too small to be observed. Atoms and protons had become more real than chairs and trees, the parts more real than the whole. The phenomenologists proposed that there was a reciprocal interaction between the knower and the known. The sensing body was seen as not being pre-programmed by its genes or by its brain but being constantly in the act of improvising its relationship to other factors in the world. According to Merleau-Ponty (1962, p. 52), the body and its environment are in constant and reciprocal exchange, both creating and changing itself and the other. Indeed, in his *Phenomenology of Perception,* Merleau-Ponty (1962, p. 317) uses symbiosis as a metaphor of the reciprocity he is trying to express. Ecological philosophers such as David Abram (1996) and Bruce Foltz (1995) have applied phenomenological analysis to biology, claiming that it had made genes more real than organisms. As developmental biology begins to forms its relationships with ecology, phenomenology might prove to be a good starting point for a philosophy that can organize a theory of reality around co-development. Here the organism can be seen as the concrete, fleshy nexus integrating (in time and space) the internal networks of developmental genetic interactions with the external networks of ecological interactions.

But the phenomenological approach to development should neither supersede nor eclipse the molecular. Interdependent origination and the context-dependent co-construction of identity is to be found at each biological level, from the molecular through the ecological. Interactions within and between each of these levels should reinforce rather than exclude one another. Whitehead (1920, p. 152) chided philosophers for dividing reality into phenomenal and causal realms, saying that they were all concretions of processes. And indeed, developmental biologists (including the most molecularly oriented) have found that even the properties of proteins depend upon their contexts: A repressor of gene transcription in one cell may activate the same gene in another cell; the same gene that produces an enzyme in the liver may encode a structural protein in the lens. BMP4, a protein that changes cartilage into bone in one part of the limb, tells cells to die in another part of the limb. In the head, BMP4 tells a different set of cells to divide to form the face, and earlier in development, it instructed the heart to form in a particular area of the embryo. Its identity is dependent on its context (see Gilbert 2006).

One philosopher who has been able to fuse phenomenological analyses with dialectic materialism, process philosophy, and developmental systems theory is Donna Haraway. Trained in both developmental biology and ecology, her philosophy reflects epigenetic, interactionist, and process-oriented perspectives. Indeed, Haraway has kept up with epigenetic biology and has used it extensively in her philosophical critiques of society. More than 15 years ago, she used the termite *Myxotricha paradoxa* as "an entity that interrogates individuality and collectivity at the same time" (Haraway 1991), and she has extended this interactive epigenetic account to all creatures. "Organisms are ecosystems of genomes, consortia, partly digested dinners, mortal boundary formations," Haraway says (2008, p. 31), and she quotes Margulis and Sagan's notion that the organism "is a co-option of strangers."

Haraway is probably the only philosopher who can write, "Reciprocal induction is the name of the game" (Haraway 2008, p. 228). She cites this process, along with developmental symbioses, as examples of "becoming with." Indeed, "becoming with" is a central feature of Haraway's philosophy and one of its major sources is developmental (and especially ecological developmental) biology. Haraway (2003, pp. 20, 24) has written that "The relation is the smallest unit of analysis," but in 2008, she preferred not to use the word "unit" because she felt that this word misled one into thinking that there is an "atom" that instigates identity. She now writes (Haraway 2008, p. 26) that "relationships are the smallest possible pattern for analysis." This is a concept that seems fundamental to ecological developmental biology at all levels: the combinatorial relationship of transcription factors to DNA; the relationship of different cells forming an organ; the relationship of bacteria (both beneficial and harmful) to the host organism; the relationship of organisms to other developing organisms, etc. She has developed this into her own brand of interspecies symbiosis. Haraway doesn't so much ground her beliefs in nature as use natural examples as a way to visualize principles and ground them in reality, especially the notion of "becoming with." These principles are not mere metaphors; they have substance, and these substances interact.

Emergence

Emergence is normative in developmental biology and has been an important concept since the late 1800s (see Gilbert and Sarkar 2000). The fertilized egg has no hemoglobin, tooth enamel, liver enzymes, or even the rudiments of a heart. Rather, these entities emerge. And they emerge through reciprocal embryonic induction. Classical embryology has insisted that an entity is defined both by its parts and by its context. The genes → organ approach must be complemented by the organ → gene approach. Thus, the structure and function of a liver cell depends not only upon the properties of its genetic products, but also on the properties of the liver in which it resides and on the history of that organ. A teratocarcinoma stem cell can become a tumor in one tissue and a set of organs in another. Experimental embryologist Oskar Hertwig (1894), who catalogued the ways in which the environment controlled development (see Chapter 1), claimed explicitly that "The parts of the organism develop in relation to each other, that is, the development of the part is dependent on the development of the whole." And embryologist Hans Spemann (1943, p. 219) remarked, "We are standing and walking with parts of our body which could have been used for thinking had they developed in another part of the embryo."

When reciprocal induction is applied between and among organisms, this concept expands into evolution. Entities need to be thought of in terms of several geometries at the same time. They are defined by the braiding of down-top and top-down (as well as lateral and temporal) networks built from patterns of reciprocal causation. Evolutionary biologist Ernst Mayr (1988) recognized that the characteristics of living wholes "cannot be deduced (even in theory) from the most complete knowledge of the components, new characteristics of the whole emerge that could not have been predicted from a knowledge of the constituents." How they emerge can now be seen in terms of reciprocal inductions.

Pedagogy

A discipline can be defined both by the set of questions it hopes to answer and by the content of its textbooks and journals. Pedagogical issues thus reflect views of scientific reality. It should be obvious that this book contains a sustained argument that the present disciplinary boundaries need to be expanded in order to apprehend the realities that science is discovering. Developmental biology needs to be expanded to overlap and to include both evolutionary and ecological dimensions. Student field trips should not be only the province of ecologists; developmental biology students need to see frog eggs in ponds as well as in the laboratory. Moreover, developmental biology courses should include discussions of philosophical issues attendant upon its discoveries.*

Evolutionary biology needs to be similarly expanded to intersect developmental biology. Indeed, the last three chapters of this book attempt to show the importance and relevance of such an expansion. The *explananda* (that which needs to be explained) of evolutionary biology must include both a theory of change *and* a theory of body construction. And ecology must see itself not as an isolated discipline but as providing wider bridges connecting it to both developmental and evolutionary biology. Harkening back to Van Valen's (1973) notion that evolution is development controlled by ecology, ecologists must embrace cell cycles as well as nutrient cycles and not allow their discipline to become a refuge for biologists proud to neglect cells and the contents of their nuclei. None of these disciplines should subsume the other, for their questions are, for the most part, different ones. Yet, together, they constitute Waddington's "diachronic biology" as well as Severtsov's (1935; see Chapter 10) recipe for a complete evolutionary science.

In addition to ecology, evolutionary biology, and developmental biology, numerous medical disciplines should be affected by ecological developmental biology. Certainly oncology should include development as a major component. Not only are epigenetic mechanisms capable of causing cancers, but numerous tumors have now been shown to arise from the misregulation of adult stem cells. Medical embryology and human genetics courses might consider uniting to include not only the developmental genetic mechanisms of human disease, but also the mechanisms of teratogenesis and endocrine disruption.

Even the metaphors we use to explain and describe evolution should change. Evolution is real, to be sure; but the metaphors we use to describe it need revision. Darwin used the metaphor of metallic wedges. Only so many wedges fit into a piece of wood, and when one fits better, it forces another out. "Fit" is the root metaphor here, and it is similar to the lock-and-key model of enzyme activity that was being proposed at a similar time in biochemistry (see Gilbert and Greenberg 1984). However, the lock-and-key model in biochemistry doesn't work. It doesn't have good "fit" to nature. Rather, the current model is that of "induced fit," where-

*This is not a new idea. The first meeting of the Growth Society (later the Society for Developmental Biology), organized by N. J. Berrill in 1939, had representatives from numerous areas of biology., as well as the philosopher J. H. Woodger, who spoke on what it all meant. This notion has been revised and revitalized (see Gilbert and Fausto-Sterling 2003) for the inclusion of social issues concerning stem cell research, abortion, cloning, environmental justice (such as the effects of putting incinerators and toxic waste dumps in low-income areas), sustainability, and consumer safety.

in there is an interaction between the substrate and the catalytic site. Developmental plasticity indicates that there are interactions between the animal and its environment such that the phenotype can become fit for the environment. It is a kind of induced fitness.

Another metaphor concerns the environment as a selector. Sometimes thought of as a policeman, a filter, or a sieve, we now see that the environment is both selective *and* instructive. Although "instructive" is in fact a technical term that describes certain developmental processes, one can use it in its pedagogic sense as well. The environment can be likened to an instructor. The instructor has two roles: one is to impart information; the other is to test the student. So it is with the environment. It both imparts information to the developing organisms and simultaneously challenges them to see if they have integrated that information in a useful way. Our ways of visualizing evolution will have to change to fit these new perceptions.

Epistemology and Methodology

How we study development

The demands of eco-devo require that developmental biologists get beyond the easily accessible model systems in order to study the effects of the environment on development. The six model organisms of animal developmental biology (the fly *Drosophila*, the frog *Xenopus*, the nematode *Caenorhabditis*, the mammal *Mus*, the bird *Gallus*, and the zebrafish *Danio*) were each selected in part because of the absence of major environmental factors in their early development; the absence of significant plasticity makes these model organisms easier to study because environmental variables are not important. Model organisms play a large role in defining the reality of a discipline, and these model organisms make it appear that the fertilized egg might contain everything needed for phenotype production. This is because the model organisms derived in the 1970s were constructed for looking at the roles of genes during development* (Bolker 1995). Getting away from "model systems" will take a great deal of effort and a change in funding priorities (see NSF 2005; Collins et al. 2007).

Although the development of these model organisms provides an excellent approximation of development and a needed first step in figuring out the proteins and genes involved in forming complex structures, the model system approach does not give complete answers. As McFall-Ngai (2002) has noted, our epistemology is flawed. "The implicit assumption that has accompanied the study of animal development is that only 'self' cells (i.e., those containing the host genome) communicate to induce developmental pathways." The presence of other organisms and abiotioc signals can cause the phenotypes that develop in nature to differ significantly from those obtained in the laboratory. Relyea and Mills (2001) found that

*The model organisms of the earlier part of the twentieth century (the chick, the flatworm, sea urchins, and salamanders) had been constructed for the ease of tissue transplantation and the explication of inducing centers and morphogenetic fields. The model organisms of evolutionary developmental biology and ecological developmental biology are not single organisms but collections of organisms which allow comparisons along phylogenetic or ecological axes (Collins et al. 2007).

when tadpoles were raised in isolation, they were relatively insensitive to the herbicide carbaryl. However, when they were simultaneously exposed to that herbicide and made to undergo predator-induced polyphenism by placing a dragonfly nymph in the water, the same concentration of the herbicide could wipe out the population.. What had seemed a harmless chemical in laboratory tests could wipe out the frog population under more natural conditions. We now have to include the environment in our concepts of development. This has not been the standard way of studying development, nor have we usually looked at developmental differences within populations. Developmental biologists have tended to look at species as a whole rather than looking at the variations within a species. This is a more evolutionary approach, and it is one that developmental biologists are just beginning to adopt (see Kopp et al. 2000).

Another epistemological change may concern the autonomy of developmental biology. The past decade has seen the emergence and rise in popularity of a "new" field called "systems biology"—but this approach already had a longstanding history in developmental biology. Systems biology views an organism as an integrated and interacting network of genes, gene products, cellular components, physiological coordinating systems, and ecological agents. The interactions themselves are seen as temporally patterned entities. Starting in the 1920s, several developmental biologists, notably Paul Weiss (1971) and Ludwig von Bertalanffy (1968), put forth the notion of systems biology. That Weiss and von Bertalanffy both trained in Vienna was probably not incidental to their coming up with such ideas (see Drack et al. 2008), since the research program of Vienna's Prater Vivarium has been seen as a precursor to eco-evo-devo (Wagner and Laubichler 2004).

The systems biology approach gained momentum only after the various genome projects failed to deliver the expected goal of reading the "Book of Life" from the accumulated DNA sequences. It was then realized that epigenetic approaches were also required. Ecological developmental biology provides one kind of systems biology approach, through which the neuroendocrine and paracrine signaling cascades integrate the gene wiring diagrams of developmental biology with the biotic and abiotic flow diagrams of ecology. Historically dynamic processes such as cell cycles, life cycles, nitrogen cycles, and solar cycles interact to create a new systems biology wherein developmental physiology is linked to ecology. And when one has a theory of development that involves signal transduction both within and between embryos, one can ask new questions. Marc Kirschner (2005) highlights some new, big questions raised by the systems approach:

> The big question to understand in biology is not regulatory linkage but the nature of biological systems that allows them to be linked together in many nonlethal and even useful combinations…These circuits may have certain robustness, but more important they have adaptability and versatility. The ease of putting conserved processes under regulatory control is an inherent design feature of the processes themselves. Among other things, it loads the deck in evolutionary variation and makes it more feasible to generate useful phenotypes upon which selection can act.

However, the nascent field of systems biology is still basing its modeling on concepts of autopoesis and homeostasis. But autopoiesis is probably very rare, and homeostasis is underlain by metabolism, a temporally dynamic set of processes. Because metabolism enables the stabilization of an "individual" by permitting the organism to retain constancy while constantly changing its component parts, "individuals" are always relational processes in time (Jonas 1966; Gilbert 1982). Individuals are always in the making, and are always in relation with other becoming entities over many simultaneous wavelengths of time—metabolic, developmental, ecological, and evolutionary. The questions of developmental biology will be placed into new contexts, and the *explananda* of the field are expected to change accordingly.

How we study evolution and ecology

In such a systems view of nature, development is critical for discussions of adaptability and variation. "Evolution is not a speculation but a fact, and it takes place by epigenesis," wrote Thomas Huxley (1893, p. 202)—but our notions of both evolution and epigenesis have become much different from those known in Victorian England. What does evolution look like when the proper unit of analysis is not the individual but the relationship (at each different level)? What does evolution look like when selection may be on "teams" of organisms and on the relationships between these teams? What does natural selection mean when the environment is not only an agent that selects adaptive phenotypes but also contains agents that help instruct the formation of adaptive phenotypes (and may undergo changes itself because of it)? Moreover, how do we revise our views about the environment and evolution when germline DNA methylation can effect the transmission of environmentally induced characters from one generation to the next? Chapter 10 concerned some of the possible frameworks on which to build a more comprehensive theory of evolution. This should be a highly productive field during the next decades.

One area highlighted by ecological developmental biology concerns the origins of variation. In addition to the origins of variation within an organism, there can be selectable variation in the relationships between organisms. There can be differences in how organisms perceive such signals. Auletta and colleagues (2008) define the situation as follows:

> According to the traditional information (communication) theory, the main problem is reliability, understood as the matching between input and output. However, in biological processes we are much more interested in situations in which the receiver does not have full control over the input and is therefore *forced to guess the nature of the input by taking the received partial information as a sign of it* (revealing its nature).

Thus, prey species undergoing predator-induced morphological changes have developed a means of deciphering a compound produced by the predator as an honest sign of the predator; and the mammalian fetus recognizes a food cue as a sign of likely future environmental conditions. There can be selectable variation within a species as to how well individuals of that species send, receive, and interpret environmental signals (see Dodd et al. 2006). Moreover, one would expect there to be mimicry and deception in the signaling as well. The ability of the environment to instruct the organisms phenotype thus adds a thick new layer of contingency to the study of evolution.

Ethics and Policy

Various policy questions, especially those involving chemical testing, have been mooted in the course of this book. These, of course, do not limit the types of questions that ecological developmental biology may address. Certainly ecological developmental biology highlights areas of environmental justice and other topics concerning ecosystems in general. Ecological developmental biology, in particular, focuses on the notions that genes do not control destiny. Thus, the notion of improving the world by selecting people with "good genes" is not so much good science as bad fantasy. Geneticist Theodosius Dobzhansky pointed out that the world's problems are not caused by people with inferior genomes, but by people with perfectly good genomes using their talents for improper ends. Cor van der Weele (1999) has noted that polyphenisms and epigenetics mean that genes do not determine our lives, and calls for an "ethics of attention." Science, she points out, helps determine what we see and what we hope to understand. For example, since 88% of breast cancers are thought to be caused by "the environment" (American Cancer Society 2006) whereas the "cancer genes" *BRCA1* and *BRCA2* account for only 5–10% of breast tumors, our emphasis on the genetic component of this disease appears to be dangerously misplaced.

The genetic account of our personhood has ethical consequences. Abraham J. Heschel (1965, pp. 7–8) noted that whereas "a theory about the stars never becomes part of the being of the stars," a theory about human beings enters our consciousness, determines our self-image, and modifies our very existence. "The image of man affects the nature of man ... We become what we think of ourselves." If we think of ourselves as killer apes, certain behavioral phenotypes are acceptable that would not be socially allowed if we view humans as the current apex of an evolutionary trend towards cooperation.

Storytelling matters. If there is anything that is appreciated by all cultures, it is a good story, and storytelling is a critical part of the scientific enterprise. Scientists have a moral imperative to tell accurate stories. Scientific stories must always fall within the limits of the existing data.* Evolutionary narratives are the most critical stories in biology, in science, and perhaps in Western civilization, so we had better get them in line with the biological data. The competitive story of evolution has given rise to stories such as social Darwinism in the past and to the popularization of sociobiology, evolutionary psychology, and "sperm warfare" in the present. Our narratives tell us what science thinks is normal, and we have been told that selfishness is normal and adaptive. Michael Ghiselin (1974) informs us of the perception that

> The evolution of society fits the Darwinian paradigm in its most individualistic form. Nothing in it cries out to be otherwise explained. The economy of nature is competitive from beginning to end ... No hint of genuine charity ameliorates our vision of society, once sentimentalism has been laid aside.

One need only read David Brooks's column in the February 18, 2007, *New York Times*, to see how ingrained these ideas have become: "From the content of our

*The stories promulgated by intelligent design proponents, for example, are not bounded by scientific data, are frequently based on nonscientific assumptions, and freely ignore known science when it conflicts with previously held views (see Gilbert et al. 2007).

genes, the nature of neurons, and the lessons of evolutionary biology, it has become apparent that nature is filled with competition and conflicts of interest." According to Brooks, the idea that human nature was inherently good was finally destroyed by "the hands of science." This account leaves out almost everything that this present book says about cooperativity and evolution. It gives an incomplete story and partial account of nature that makes humans seem completely self-interested, atomic individuals. So one underappreciated moral imperative is highlighted here: we must tell better stories.

Indeed, the imperative to educate has never been greater. The Jewish ecological action group Shomrei Adamah ("Guardians of the Earth") published a poster saying. "Noah was a righteous man in his generation. Will you be in yours?" However, the rabbis had a difficult time figuring out exactly why the Bible called Noah a righteous man. One of the rabbinic commentaries (Tanhuma 2,5; see Zornberg 1995) suggests that Noah was righteous because he went out of his way to painstakingly acquire detailed knowledge of the habits and feeding schedules of the animals so that he could house them properly on the ark. Here we have the beginnings of the notion that one has to know what is true in order to do what is good. The agnostic evolutionist Thomas Huxley (1870) made this an explicit principle: "Learn what is true in order to do what is right." The imperative is both to discover and to teach.

This imperative links intimately with another, the ethic of sustainability. Here, the insights of ecological developmental biology can play a role in conservation efforts. As mentioned earlier in this book, knowledge of developmental ecology has been important in attempts to conserve turtle populations that have temperature-dependent sex determination and certain butterflies whose larval stages have symbiotic relationships with other organisms. Michael Pollan (2006) has publicized the ecologically devastating effects of industrialized farming practices and has raised important questions concerning the morality of many agribusiness procedures. One of the ecological solutions mentioned in his book is "grass farming," an ecologically efficient approach to livestock production. This management-intensive program is an ecological developmental approach, based on synchronizing the development of grass, insects, poultry, and mammals in a manner that takes advantage of their mutual interactions (Ekarius 1999; Nation 2007; Wilkins 2008). Thus, when the grass reaches a certain height, cows are allowed graze. The cows then defecate in the field and insect larvae grow in the cow feces. At a certain time, the larvae will be at the appropriate stage for chickens to feed on them. By rotating the placement of the cows and the chickens, livestock grows with minimal investment in chemical fertilizers, antibiotics, or pesticides.

But as Elliot Sober (1995) has pointed out, the sustainability ethic has problems being accepted. Why should species and ecosystems be preserved if they are of no use to humans? And, while individual animals can suffer, neither species nor mountain forests do. The arguments made are often utilitarian (maybe the species contains a chemical that would cure cancer), aesthetic (these are beautiful places that refresh the spirit), or religious (we have no right to destroy God's creation or cause the extinction of any of God's creatures).

Ecological developmental biology gives some new answers to that question. One answer borrows from the utilitarian argument, but makes it immediate and intimate: We should preserve ecosystems because we are part of them. As much as we have separated ourselves from nature, we are intimately connected to the rest of the biological world through symbioses. We are built by symbiotic relationships, and our air and food are produced by such relationships. We are intimately connected to nature, and even the algae in the ocean are essential to our living.

The second answer relates to who we are as people. Heschel (1955) noted that we are human because we are able to experience wonder. But wonder has a short half-life, and it decays rapidly into awe and curiosity. Awe, he points out, is the font of religion; while curiosity gives rise, as Plato and Aristotle said it did, to philosophy and science (see Gilbert and Faber 1996). So humans have a vested interest in preserving sources of wonder. Wonder is where religion, philosophy, and science—the "grandchildren of wonder"—have a natural alliance. So in this sense, retaining the mountain forests, the meadows, and rivers is indeed valuable to humans, for they maintain our humanity.

We must also recognize that nature, as we know it, is rarely untouched. While we should try to preserve the few expanses of "pristine" nature left, it is critical to remember that just as we are constantly being touched by "external" nature, nature is constantly touched by everything we place into the water and air (see McKibben 1989). Keeping sources of wonder alive does not necessitate their being pristine; indeed, by the very nature of the interactions between organisms, we could hardly expect them to be. Ecological developmental biology fosters an ethic that can integrate both selfishness and otherness, as one might expect of a discipline where self and other mutually construct each another.

Ethics for the Anthropocene

In February 2008, a group of 21 scientists of the Geological Society of London concluded that we and the other organisms with whom we share this planet, and indeed, the planet itself, are no longer in the Holocene epoch. The interglacial age is over, and we are now living in the Anthropocene, an epoch characterized by a human-dominated environment (Zalasiewicz et al. 2008). For better or worse, humans have become, as Julian Huxley (1963, p. 139) predicted we would, "business managers for the cosmic process of evolution." However, there are no graduate schools or training manuals to prepare us for this job. We must use our combined experience wisely in this most important on-the-job training of all time.

The first two things a manager must learn are the rules of the process and what the goals of the process should be. As we hope to have demonstrated throughout this book, the rules of the evolutionary process include the crucial notion of "becoming with." Co-development is normative, and evolution is not merely the battle of each against all. As to the goals, Chris Cuomo (1998, p. 62) and Donna Haraway (2008, p. 134) conclude that sustainability is not enough, and stewardship is only the beginning. They propose that the ethical starting point of ecological theory is "a commitment to the *flourishing* or well-being, of individuals, species, and communities." This ethic of flourishing and well-becoming are perhaps visualized best in that very union we had mentioned earlier in the book—the orchid seed. The beauty of the orchid is made possible by the invasion of its seed coat by a symbiotic fungus. Alliances are strange and wonderful things in the living kingdom. We "become with" the world, and that is our enormous responsibility.

References

Abram, D. 1996. *The Spell of the Sensuous: Perception and Language in a More-than-Human World*. Vintage Books, New York.

American Cancer Society. 2006. Frequently requested statistics in cancer. www.cancer.org.

Auletta, G., G. F. Ellis and L. Jaeger. 2008. Top-down causation by informational control: From a philosophical problem to a scientific research programme. *J. Roy. Soc. Interface*, doi 10.1098/rsif.2008.0018.

Barad, K. 2007. *Meeting the Universe Half-Way: Quantum Physics and the Entanglement of Matter and Meaning*. Duke University Press, Durham, NC.

Bolker, J. A. 1995. Model systems in developmental biology. *BioEssays* 17: 451–455.

Brooks, D. 2007. Human nature redux. *New York Times* Feb. 18, 2007. http://select.nytimes.com/2007/02/18/opinion/18brooks.html?_r=1&oref=slogin.

Buss, L. W. 1987. *The Evolution of Individuality*. Princeton University Press, Princeton, NJ.

Byrnes W. M. and W. R. Eckberg. 2006. Ernest Everett Just (1883–1941): An early ecological developmental biologist. *Dev. Biol*. 296: 1–11.

Chernyshevskii, N. G. 1888. Quoted in Todes, op. cit. 1989.

Collins, J. P., S. F. Gilbert, M. D. Laubichler and G. B. Müller. 2007. How to integrate development, evolution, and ecology: Modeling in evo-devo. In M. D. Laubichler and G. B. Müller (eds.), *Modeling Biology: Structures, Behaviors, Evolution*. MIT Press, Cambridge, MA, pp. 355–378.

Cuomo, C. 1998. *Feminism and Ecological Communities: An Ethic of Flourishing*. Routledge, New York.

Dodd, K. A., C. Murdock and T. Wibbels. 2006. Interclutch variation in sex ratios produced at pivotal temperature in the red-eared slider, a turtle with temperature-dependent sex determination. *J. Herp*. 40: 544–549.

Drack M., W. Apfalter, and D. Pouvreau. 2007. On the making of a system theory of life: Paul A. Weiss and Ludwig von Bertalanffy's conceptual connection. *Q. Rev. Biol*. 82: 349–373.

Ekarius, C. 1999. *Small-Scale Livestock Farming: A Grass-Based Approach for Health, Sustainability, and Profit*. Storey Books, North Adams, MA.

Foltz, B. V. 1995. *Inhabiting the Earth: Heidegger, Environmental Ethics and the Metaphysics of Nature*. Prometheus Books, Amherst, NY.

Ghiselin, M. T. 1974. *The Economy of Nature and the Evolution of Sex*. University of California Press, Berkeley.

Gilbert, S. F. 1982. Intellectual traditions in the life sciences: Molecular biology and biochemistry. *Persp. Biol. Med*. 26: 151–162.

Gilbert, S. F. 1991. Induction and the origins of developmental genetics. In S. F. Gilbert (ed.), *A Conceptual History of Modern Embryology*. Plenum Press, NY, pp. 181–206.

Gilbert, S. F. 1992. Cells in search of community: Critiques of Weismannism and selectable units of ontogeny. *Biol. Phil*. 7: 473–487.

Gilbert, S. F. 2002. The genome in its ecological context: Philosophical perspectives on interspecies epigenesis. *Ann. NY Acad. Sci*. 981: 202–218.

Gilbert, S. F. 2006. *Developmental Biology*, 8th Ed. Sinauer Associates, Sunderland, MA.

Gilbert, S. F. 2008. When "personhood" begins in the embryo: avoiding a syllabus of errors. *Birth Defects Res. C: Embryo Today*. 84: 164–173.

Gilbert, S. F. and M. Faber. 1996. Looking at embryos: The visual and conceptual aesthetics of emerging form. In A. I. Tauber (ed.), *The Elusive Synthesis: Aesthetics and Science*. Kluwer, Dordecht, pp. 125–151.

Gilbert, S. F. and A. Fausto-Sterling. 2003. Educating for social responsibility: Changing the syllabus of developmental biology. *Int. J. Dev. Biol*. 47: 327–244.

Gilbert, S. F. and J. Greenberg. 1984. Intellectual traditions in the life sciences. II. Stereospecificity. *Persp. Biol. Med*. 28: 18–34.

Gilbert, S. F. and S. Sarkar. 2000. Embracing complexity: Organicism for the twenty-first century. *Dev. Dynamics* 219: 1–9.

Gilbert, S. F. and the Swarthmore College Evolution and Development Seminar. 2007. The aerodynamics of flying carpets: Why biologists are loath to "teach the controversy." In N. C. Comfort (ed.) *The Panda's Black Box: Opening Up the Intelligent Design Controversy*. Johns Hopkins University Press, Baltimore, pp. 40–62.

Gillespie, N. C. 1979. *Charles Darwin and the Problem of Creation*. University of Chicago Press, Chicago.

Gottlieb, G. 1992. *Individual Development and Evolution: The Genesis of Novel Behavior*. Oxford University Press, New York.

Gould, S. J. 1982. Nonmoral Nature. *Natural History* 91(Feb.): 19–26.

Griesemer, J. 2002. What is "Epi" about epigenetics? *Ann. NY Acad. Sci*. 981: 97–110.

Haldane, J. S. 1933. *The Philosophical Basis of Biology*. Hodder and Stoughton, London.

Haraway, D. J. 1976. *Crystals, Fabrics, and Fields: Metaphors of Organicism in Twentieth-Century Biology*. Yale University Press, New Haven.

Haraway, D. J. 1979. The biological enterprise: Sex, mind and profit from human engineering to sociobiology. *Rad. Hist. Rev*. 20: 206–37.

Haraway, D. J. 1991. Otherworldly conversations, Terran topics, local terms. *Science as Culture* 3: 64–98.

Haraway, D. J. 2003. *The Companion Species Manifesto: Dogs, People, and Significant Otherness*. Prickly Paradigm Press, Chicago.

Haraway, D. J. 2008. *When Species Meet*. University of Minnesota Press, Minneapolis.

Hertwig, O. 1894. *Zeit- und Streitfragen der Biologie I. Präformation oder Epigenese? Grundzüge einer Entwicklungstheorie der Organismen*. Gustav Fischer Jena. Translated by P. C. Mitchell as *The Biological Problem of tToday: Preformation or Epigenesis?* Macmillan, New York.

Heschel, A. J. 1955. *God in Search of Man*. Harper, New York.

Heschel, A. J. 1965. *Who Is Man?* University of California Press, Berkeley.

Holtfreter, J. 1968. Address in honor of Viktor Hamburger. In M. Locke (ed.), *The Emergence of Order in Developing Systems*. Academic Press, New York.

Huneman, P. 2006. Naturalising purpose: From comparative anatomy to the "adventure of reason." *Stud. Hist. Phil. Biol. Biomed. Sci.* 37: 649–674.

Huxley, J. 1963. *Evolution in Action*. Chatto and Windus, London.

Huxley, T. H. 1870. On Descartes' "Discourse touching the method of using one's reason rightly and of seeking scientific truth." *Macmillan's Magazine* 24 March 1870; reprinted in T. H. Huxley 1877. *Lay Sermons, Addresses, and Reviews*. Appleton, New York.

Huxley, T. H. 1893. Evolution in biology. In *Darwiniana: Collected Essays* II. Macmillan, London.

Huxley, T. H. 1894. Prolegomena. In T. H. Huxley, 1896. *Evolution and Ethics and Other Essays*. Appleton, New York, pp. 1–45.

Jonas, H. 1966. *The Phenomenon of Life: Toward a Philosophical Biology* University of Chicago Press, Chicago.

Just, E. E. 1939. *The Biology of the Cell Surface*. Blakiston Press, Philadelphia.

Kirschner, M. W. 2005. The meaning of systems biology. *Cell*. 121: 503–504.

Kopp, A., I. Duncan, D. Godt and S. B. Carroll. 2000. Genetic control and evolution of sexually dimorphic characters in *Drosophila*. *Nature* 408: 553–559.

Kropotkin, P. 1902/1955. *Mutual Aid: A Factor of Evolution*. Porter-Sargent, Boston.

Lenoir, T. 1982. *The Strategy of Life: Teleology and Mechanics in Nineteenth Century German Biology*. D. Reidel, Boston.

Margulis, L. 1998. *Symbiotic Planet: A New Look at Evolution*. Basic Books, New York.

Margulis, L. and D. Sagan. 2002. *Acquiring Genomes: A Theory of the Origins of Species*. Basic Books, New York.

Maturana, H. and F. Varela.1980. Autopoiesis and cognition: The realization of the living. *Boston Studies in the Philosophy of Science*. D. 42. Reidel, Dordrecht.

Mayr, E.1988. *Toward a New Philosophy of Biology: Observations of an Evolutionist*. Harvard University Press, Cambridge.

McFall-Ngai, M. J. 2002. Unseen forces: the influence of bacteria on animal development. *Dev. Biol*. 242: 1–14.

McKibben, B. 1989. *The End of Nature*. Anchor Books, New York.

Merleau-Ponty, M. 1962. *Phenomenology of Perception*. Translated by C. Smith. Humanities Press, New York.

Montague, A. 1955. Foreword. In P. Kropotkin's *Mutual Aid: A Factor of Evolution*. Porter-Sargent, Boston.

Murti, T. R. V. 1955. The *Central Philosophy of Buddhism*. Allen and Unwin, London.

Nation, A. 2007. Getting started in grass farming. The Stockman Grassfarmer. http://stockmangrassfarmer.net.

National Science Foundation (NSF). 2005. *An Integrated Developmental Biology*: Workshop Report. Booklet 08053.

Odin, S. 1982. *Process Metaphysics and Hua-Yen Buddhism*. SUNY Press, Albany, NY.

Oyama, S. 1985. *The Ontogeny of Information*. Cambridge University Press, Cambridge.

Pander, C. 1817. *Beiträge zur Entwickelungsgeschichte des Hühnchens um Eye*. H. L. Brönner, Würzburg, p. 12; quoted in F. B. Churchill, 1991. The rise of classical descriptive embryology. In S. F. Gilbert (ed.), *A Conceptual History of Modern Embryology*. Plenum Press, NY, pp. 1–19.

Pollan, M. 2006. *The Omnivore's Dilemma: A Natural History of Four Meals*. Penguin Press, New York.

Relyea, R. A. and N. Mills 2001. Predator-induced stress makes the pesticide carbaryl more deadly to gray treefrog tadpoles (*Hyla versicolor*). *Proc. Natl. Acad. Sci. USA*. 98: 2491–2496.

Robert, J. S. 2004. *Embryology, Epigenesis, and Evolution: Taking Development Seriously*. Cambridge University Press, Cambridge.

Sagan, D. and L. Margulis. 1986. *Origins of Sex: Three Billion Years of Genetic Recombination*. Yale University Press, New Haven.

Sagan, D. and L. Margulis. 1991. Epilogue: The uncut self. In A. I. Tauber (ed.), *Organism and the Origins of Life*. Kluwer, Dordrecht, pp. 361–364.

Schweber, S. S. 1985. The wider British context in Darwin's theorizing. In D. Kohn (ed.), *The Darwinian Heritage*. Princeton University Press, Princeton, NJ.

Severtsov, A. N. 1935. *Modes of Phyloembryogenesis*. Quoted in M. B. Adams, 1980. Severtsov and Schmalhausen: Russian morphology and the evolutionary synthesis, In E. Mayr and W. B. Provine (eds.). *The Evolutionary Synthesis: Perspectives on the Unification of Biology*. Harvard University Press, Cambridge, MA, p. 193–225 (p. 217).

Sober, E. 1995. Philosophical problems for environmentalism. In R. Elliot (ed.), *Environmental Ethics*. Oxford University Press, New York.

Spemann, H. 1943. *Forschung und Leben*. Quoted in Spemann and the organizer, 1986. In T. J. Horder, J. A. Witkowski, and C. C. Wylie (eds.). *A History of Embryology*. Cambridge University Press, New York.

Streng, F. J. 1975. Metaphysics, negative dialectic, and the expression of the inexpressible. *Philos. East West* 25: 429–447.

Todes, D. P. 1989. *Darwin without Malthus: The Struggle for Existence in Russian Evolutionary Thought*. Oxford University Press, New York.
van der Weele, C. 1999. *Images of Development: Environmental Causes in Ontogeny.* SUNY Press, Albany, NY.
Van Valen, L. 1973. Festschrift. *Science* 180: 488.
von Bertalanffy, L. 1968. *General System Theory: Foundations, Development, Applications.* George Braziller, New York.
Waddington, C. H. 1953. Epigenetics and evolution. In R. Brown and J. F. Danielli (eds.). *Evolution* (SEB Symposium VII). Cambridge University Press, Cambridge, pp. 186–199.
Waddington, C. H. 1975. The practical consequences of metaphysical beliefs on a biologist's work: An autobiographical note. In *The Evolution of an Evolutionist,* Cornell University Press, Ithaca, NY.
Wagner, G. P. and M. D. Laubichler 2004. Rupert Reiedl and the re-synthesis of evolutionary and developmental biology: Body plans and evolvability. *J. Exp. Zool. (MDB)*, 302B: 92–102.
Weiss, P. A. (ed.). 1971. *Hierarchically Organized Systems in Theory and Practice.* Hafner, New York.
Werskey, G. 1978. *The Visible College: A Collective Biography of British Scientists and Socialists of the 1930s.* Viking Press, New York.
Whitehead, A. N. 1920. *The Concept of Nature*. Cambridge University Press, Cambridge.
Whitehead, A. N. 1929. *Process and Reality.* Macmillan, New York.
Wilkins, R. J. 2008. Eco-efficient approaches to land management: a case for increased integration of crop and animal production systems. *Philos. Trans. Roy. Soc. B* 363: 517–525.
Wilson, D. S. and E. Sober. 1994. Reintroducing group selection to the human behavioral sciences. *Behavioral and Brain Sciences* 17: 585–654.
Wilson, E. B. 1896. *The Cell in Development and Inheritance*. Macmillan, New York.
Wilson, E. O. 2005. Kin selection as the key to altruism: Its rise and fall. *Social Research* 72: 159–166.
Winchester, S. 2008. *The Man Who Loved China*. Harper Collins, New York.
Wolpert, L. 1994. Do we understand development? *Science* 266: 571–572.
Young, R. M. 1985. *Darwin's Metaphor*. Cambridge University Press, New York.
Zalasiewicz, J. and 20 others. 2008. Are we now living in the Anthropocene? *GSA Today* 18(2): 4–8.
Zornberg, A. G. 1995. *The Beginning of Desire: Reflections on Genesis*. Doubleday, New York.

Appendix A

Lysenko, Kammerer, and the Truncated Tradition of Ecological Developmental Biology

The notion that embryos receive cues for normal development from the environment; that a particular embryo could develop different phenotypes depending upon the particular environmental circumstances it was in; and that such specific environmentally induced phenotypes could sometimes be inherited from one generation to the next was an important part of late nineteenth century and early twentieth century biology. It formed a major part of the biology of central Europe (see Herbst 1894), although it was of lesser importance in Great Britain and the United States.

The Continental synthesis of ecology, development, and evolution was largely demolished in the 1940s. The Prater Vivarium of Vienna, which had been one of its leading research centers, was bombed during World War II, and many of its scientists either died in concentration camps or committed suicide. The German biological enterprise was almost completely wiped out. Meanwhile, in the Soviet Union, the concept of phenotypic plasticity became so malignantly hegemonic that its application took it far beyond the realm of scientific practice. The Soviet biological establishment, headed by Trofim Lysenko, considered the phenotype to be totally at the mercy of the environment and asserted that the genes did nothing. The firing, exile, and imprisonment of Soviet geneticists caused a huge reaction in other countries, ultimately leading to the dismissal of environmental causes of phenotype production in most of Western biology.

In the eyes of Western geneticists, two related scandals severely weakened the hypothesis that phenotypic plasticity was an important source of hereditable variation: Lysenko's failed attempts to vernalize wheat, and the disastrous outcome of Kammerer's experiments on developmental plasticity in the midwife toad.

Lysenko's Vernalized Wheat

A major reason for the marginalization of phenotypic plasticity in the study of evolution and development is almost certainly the spectre of Trofim Denisovich Lysenko. Lysenko was an opportunistic plant breeder whose anti-genetics rhetoric resonated with Josef Stalin's views on the malleability of human nature. Lysenko rose through the ranks of Soviet biology, largely due to the patronage of particular Soviet officials and his ability to obtain the support of influential scientists who thought Lysenko had the potential for greatness. (He would turn on those scientists later.) Lysenko capitalized on his peasant background, castigating the academic scientists as neither knowing nor caring about the practical value of science. How can one study eye color in fruit flies when people are suffering from food shortages? He cast his agricultural argument against genetics as a struggle of true, useful, Soviet science against false, racist, capitalist metaphysics (see Medvedev 1969; Joravsky 1970; Soyfer 1994).

From the 1930s to the early 1960s, Lysenko's program declared that genetics was a capitalist product and that the gene was a metaphysical entity with little, if any, role in the production of an organism's traits. Rather, the organism was plastic and the environment determined its specific traits. In the early 1930s, the Soviet Union had one of the most advanced genetics programs in the world, with luminaries such as Chetverikov, Severtsov, Philipchenko, Karpechenko, Timofeev-Ressovsky, and Vavilov. Lysenko's campaign against these geneticists was waged both scientifically and politically, and his techniques were vicious in both contexts. In 1940, the leading plant geneticist in the Soviet Union, Nikolai Vavilov (one of the scientists who at first promoted Lysenko) was seized on a field trip and charged with sabotaging Soviet agriculture, spying for England, and belonging to a right-wing conspiracy. A military tribunal sentenced him to death, and although this was commuted to a sentence of life imprisonment, Vavilov died of starvation in the Gulag in 1943 (see Pringle 2008). Other scientists were dismissed from their posts. Initially, some biologists felt secure enough to combat Lysenko, but in 1948, Lysenko announced at a scientific congress that he had Stalin's full support. He and his followers removed the remaining geneticists from their positions, imprisoning some and causing others to flee (see Joravsky 1970; Medvedev 1969; Pringle 2008).

Lewontin and Levins (1976) have shown that Lysenkoism was nothing less than an aborted cultural revolution and that no analysis of Lysenkoism can be made without understanding Russia's ecology. Nearly all of Russia lies above 40°N (the latitude of Minneapolis), so its climate is more similar to that of Canada than that of the United States. However, Russia has even more extreme weather conditions than Canada, and the temperature gap between summer and winter varies enormously from year to year. The growing season in its chief agricultural belt is short and prone to unpredictable fluctuations in rainfall and frosts.

The big problem for Soviet wheat production was that the crop had to be planted late enough in the spring to avoid chilling frosts, but early enough to get the full benefit of the comparatively short growing season. Rather than breed genetically hybrid wheat strains that might be fit for particular environments, Lysenko claimed that he could adapt Russian wheat to the harsh climate through a process called vernalization.

Vernalization is a natural attribute of certain cereal plants that can be artificially enhanced by chilling and wetting the seeds to extend their growing season. Lysenko was far from the first to describe the phenomenon of vernalization, although he did coin the name. Plant physiologists of the 1920s and 1930s were especially interested in how light and temperature could direct plant development in different directions. In particular, there was widespread interest in the notion of "aftereffects," cases in which specific temperature or light regimes focused on seeds or seedlings would cause changes in development weeks or months later (see Roll-Hansen 2005). For instance, temperature had long been used to cause plants to flower "out of season." Winter cereals need a period of cold in order to flower. They are usually sown in late autumn and flower the next summer. If sown in the spring, they will stay in vegetative growth and not flower until after the winter. But Gassner (1918) had shown that if given a pulse of low temperature immediately after germination in the spring, winter cereals can flower that season. Gassner initiated a research program to find the physiological mechanism by which the norms of reaction (the ability to flower early or late) became directed into one path or the other by temperature (see Chapter 2).

Lysenko claimed to have done this on a grand scale and for the sake of Soviet agriculture. He said that his results showed that if seeds are chilled and wetted, these "vernalized" seeds can be planted in the spring, rapidly develop, and complete their growth cycle before the winter frosts. In other words, Lysenko claimed that he was able to turn a winter grain into a spring grain through chilling the seed. In a letter of recommendation for Lysenko, Vavilov (1933) praised Lysenko:

> For the first time, with such penetrating profoundness and scope, Lysenko managed to find ways to cope with vegetation control, shift vegetation phases and transform winter crops into spring ones, or late ripening into early. His work is a discovery of primary importance, as it is opening a new sphere for research, and quite an attainable sphere. Undoubtedly Lysenko's work will entail development of the whole branch of plant physiology; this discovery would provide for an opportunity of wide-scale utilization of the world's plant diversity in hybridization for shifting their areas to more remote northern territories. Even the current phase of Lysenko's dicoveries is of paramount interest.

So even though Lysenko later made exaggerated claims for his work, it started off within the realm of science. Similarly, Lysenko's antagonism to genetics was not outside of scientific debate, especially in the Soviet Union. In 1920s and 1930s, many biologists did not think that classical genetics gave natural selection enough variation to work on. Among taxonomists, agronomists, systematists, ecologists, and paleontologists, there was a widespread dissatisfaction with the genetic approach to evolution (Roll-Hansen 2005).

Moreover, in the 1920s, genetics had become increasingly tied to eugenics, the "science" of breeding better people. As a social movement that had a great deal of popular support in the West, eugenics had become increasingly tied to conservative and even fascist politics (Ludmerer 1972; Kevles 1998; Selden 1999; Carlson 2001). As such, the value of genetics in the Soviet Union diminished, especially in the botanical community, which did not see the gene as a particularly valid explanation of variation. Indeed, when mass selection experiments were performed on populations of plants (where all plants surviving selection were able to breed rather than mating only the most outstanding individuals), they usually failed to select anything. Leading plant ecologist Vladimir Komarov (1944, quoted in Roll-Hansen

2005) noted that botanists saw that a large amount of variability was due to the plants' responses to environmental factors. Such adaptive variability, he noted, could be heritable if it was maintained through the plants' germ cells. The germ cells are not as rapidly segregated in plants as they were thought to be in animals, suggesting that traits acquired by the young plant could be transmitted to its progeny.

However, Lysenko took this idea of vernalization and phenotypic plasticity far beyond the bounds of science. His well-publicized experiments were criticized as not having been well controlled. For instance, Konstantinov and colleagues (1936) complained that Lysenko threw out data that did not confirm his ideas, and that he gave vernalized seeds better growing conditions than nonvernalized seeds. Lysenko responded that such criticism betrayed the capitalist notions of its authors. In 1935, Lysenko and I. I. Prezent published a textbook that contradicted genetics in favor of an environmental view wherein the genotype was described as a set of possibilities for development in different directions, dependent upon the conditions of the environment. In defiance of the plant breeding experiments of Mendel or the work being done with *Drosophila* in Thomas Hunt Morgan's laboratory, this book claimed that the genome did not contain instructive information. When well-known plant breeders pointed out the deficiency of Lysenko's views and the critical nature of genes as shown by careful experiments, Lysenko responded with theoretical comments that often had no basis in fact (such as the notion that early ripening was always dominant over late ripening).

Outside the botanical community, there were many evolutionary geneticists in the Soviet Union who felt that the environment could provide guidance for the phenotype through its interaction with genes. These geneticists tried to find a common dialogue between Lysenko and the rest of the genetics community. However, Lysenko would have no compromise. In the late 1930s, Lysenko and Prezent initiated a campaign in Soviet newspapers, linking geneticists to traitors, and calling their scientific opponents "Trotskyites-Bukharinites." Here, historian Roll-Hansen notes (2005), "The links to eugenics and race theories became fatal for genetics. It got caught up in the campaign against Fascism." In 1939, Lysenko and Prezent succeeded in blocking the election of two eminent biologists (N. K. Koltsov and M. M. Zavadovskii) to the Academy of Science while assuring Lysenko's election. In this campaign, they mentioned the mild support for eugenics that these geneticists had shown decades earlier. In 1940, with the arrest of Vavilov and other geneticists (and with Koltsov dying of a heart attack), Lysenko had consolidated his position as the head of biology in the Soviet Union. Eventually, when Lysenko was unable to show significant gains in Soviet wheat production and his followers (who were not the brightest of scientific lights) fumbled at meetings, Lysenko's position grew more unstable. Some scientists, including the brilliant geneticist Ivan I. Schmalhausen (see Chapter 10 and Appendix D), worked to push Lysenko and his followers out and fill their positions with competent scientists; but it wasn't until 1964, when Nikita Khrushchev was ousted from power, that the Lysenkoists began to be removed from their posts.

Kammerer's Midwife Toads

The last bastion of phenotypic plasticity research in Western biology was probably the Prader Vivarium in Vienna. In the 1920s, one of its biologists, Paul Kammerer, attempted to synthesize Darwinism and chromosome theory while highlighting (1) phenotypic plasticity as a major source of variation, and (2) the inheritance of acquired characteristics as an experimentally demonstrable effect. He claimed that animal form could be modified by environmental influences, and that these variations were passed on from generation to generation. Thus, striped salamanders became spotted and blind cave salamanders developed functional eyes.

Kammerer's most famous (or infamous) experiment involved changing the morphology and behaviors of midwife toads (genus *Alytes*). These toads normally mate on land, and the eggs become attached to the hindlimbs of the male. Kammerer kept the toads in water and induced them to mate there (where the eggs floated away). From the few eggs that survived, Kammerer claimed to have bred a line of midwife toads that habitually bred in the water. Moreover, these toads developed "nuptial pads" on the male forelimbs. These were rough dark patches of skin that enabled the water-bound male to grab his slippery female partner. A scandal broke in 1926 when one of the last remaining water-habituated midwife toads was demonstrated to have had India ink injected where the nuptial pads were supposed to be. Kammerer denied faking the nuptial pad, but a few weeks later committed suicide. Numerous scientists had seen the toads and attested to the presence of nuptial pads, but G. K. Noble and William Bateson doubted their existence.

There have been numerous attempts to explain this work and the scandal. Arthur Koestler (1971) suggests that an enemy of Kammerer (and he had several) framed him or that a well-meaning laboratory assistant touched up a deteriorating specimen. He claims that Kammerer's suicide was not due to his despondency over a supposed fraud, but to his failure to win the beautiful Alma Mahler as his mistress. More recently, Sander Gliboff (2006) provides evidence that Kammerer had a difficult time photographing his slimy subjects, and that he may well have inked the nuptial pad to make it more visible on film:

> I agree with Iltis that the pad was not fabricated. Kammerer worked with highly plastic organisms that probably did exhibit the traits he reported, and there was more evidence for the existence of the nuptial pad than just the doctored specimen. However, I suspect Kammerer of injecting the specimen anyway, only to enhance, not create, its appearance.

Whatever the reason, the discovery that Kammerer's celebrated toads had inked forelimbs, along with Kammerer's suicide coming so soon afterward, created a sense of scientific fraud that severely damaged the notion of developmental plasticity.* The toad's case was one of the few publicly known examples of non-Mendelian inheritance, and with its demise another objection to the rule of genetics was demolished.

The Lysenkoist revolution and Kammerer's fraudulent toad had a major effect on Western biology. First, they were seen as object lessons: This is what happens when you let science be subservient to ideology and politics. (Kammerer was an outspoken Socialist who was thinking about moving his laboratory to Moscow; see Goldschmidt 1949, for a view of these episodes as a warning to the West.) This was a somewhat ironic twist, since the generation of geneticists in America and Great

*Most of Kammerer's experiments, including the nuptial pad investigations, have never been re-attempted. However, embryologist J. R. Whittaker (1985) was unable to reproduce Kammerer's results when he re-did the experiments on lengthening in tunicate siphons.

Britain in the 1940s and 1950s were trying to free Western genetics from its own political involvement with the eugenics and anti-immigration movements. The reaction against Lysenkoism also led to a hardening of a genetic position that did not include phenotypic plasticity. The American Society for Genetics hired a public relations firm and began a Golden Jubilee Program for 1950, to celebrate the 50 years of genetics since the rediscovery of Mendel's paper (and enshrine Mendel as the founder of the new science; see Sapp 1987; Wolfe 2002; Gormley 2007). But in placing itself in opposition to Lysenkoism, Western genetics had also placed itself in opposition to phenotypic plasticity. As historian Jan Sapp (1987) has documented, the West also became ideologically committed. The biologists' anti-Communism became manifest in anti-plasticity.

Biology in the Context of Ideological Struggle

Biology, as constructed after World War II, has seen the dominance of American traditions. Much of what we now consider to be the basic structure of the biological sciences (and indeed the basic structure of the natural world) was formulated in the context of the Cold War. Nowhere is this more clearly seen than in the formulation of contemporary genetics in the context of Lysenkoism. Genetics, in the 1930s, was America's and Britain's contribution to biology. Moreover, the geneticists provided a mechanism for inheritance that was seen as being as independent of development and also independent of the environment. The genotype directed the phenotype, and each character was seen as individually inherited (see Appendix C). As Lewontin and Levins (1976) have pointed out, the reduction of character to gene and the atomicity (separateness) of each trait was consistent with the Anglo-American notions of individuality and analytic philosophy.

Another reason for the marginalization of developmental plasticity was that it did not fit into the paradigms of evolutionary biology or developmental biology. Evolutionary biology was defined by variation that was inherited through genes. Environmentally induced variation was not considered heritable, and those schemes that proposed models for the inheritance of environmentally induced variation (such as Waddington's genetic assimilation or Schmalhausen's stabilizing selection) were looked upon as either bad science, politically motivated science, or a combination of both (see Appendix D).

One of the few Western evolutionary biologists who did refer to norms of reaction was Theodosius Dobzhansky. He came from the Russian school of evolution, but he stayed in the United States while Lysenkoism eclipsed Soviet evolutionary genetics. In 1926 (while still in Russia), he equated the reaction norm with the genotype (Sarkar 1999), and his 1937 book, *Genetics and the Origin of Species,* reintroduced the idea of the reaction norm to Western science. However, rather than seeing it as a developmental phenomenon, Dobzhansky (1955; Dobzhansky and Spassky 1963) redefined the norm of reaction to refer to the adaptive plasticity of the entire population's collection of genomes. As Sarkar notes, "Dobzhansky, unlike Schmalhausen, and like a true geneticist from that period, generally ignored embryology." Thus the notion of developmental plasticity in individuals became lost from genetics and evolutionary biology.

It also became lost from developmental biology. As developmental biology became molecularized, this branch of science became more dependent on model systems—those organisms from which a large community of scientists could obtain detailed information. In the field of animal developmental biology spanning the 1960s to the present, research on *Xenopus leavis, Drosophila melanogaster, Caenorhabditis elegans*, and *Mus musculus* has been ascendant. As Bolker (1995) pointed out, these model organisms converged on a particular suite of phenotypes that allowed them to be used for the genetic analysis of development: small size, fast growth to sexual maturity, large litter size, and a "rapid, highly canalized development" that reduces the effects of environment on phenotype production. "These biases," she notes, "influence both data collection and interpretation, and our views of how development works and which aspects of it are important." Indeed, they led to an erroneous perception that the organism, in the words of Jacques Monod (1971, p. 87), "came into being spontaneously and autonomously, without outside help and without the injection of additional information."

The Return of Plasticity

The return of developmental plasticity into discussions of evolution, development, and health has several sources. First, of course, it was never totally absent from discussions. Scientists such as C. H. Waddington and Richard Lewontin (a student of Dobzhansky's) kept alive discussions of the possible roles of plasticity in evolution. The controversy over IQ testing made plasticity a fundamental concern for Lewontin (1972, 1976). When some social and physical scientists (Arthur Jensen and William Shockley being the most prominent) wrote popular books claiming that intelligence was a genetically fixed trait that was distributed unequally between races, Lewontin used the concept of developmental plasticity to show that genetics and environment interact to make phenotypes. (Indeed, this argument had been used previously by Lancelot Hogben to counter eugenic proposals claiming that intelligence was hereditary.) Lewontin continued to bring this line of reasoning into evolutionary theory and to popularize it as well (Lewontin 1993, 2001, 2002).

Lewontin was also important in keeping alive a reaction against the view that genetics (and the human genome project in particular) could explain all the important parts of one's phenotype. In the latter part of the twentieth-century, genetic reductionism, the view that the complete phenotype of an organism can be traced back solely to the genome, became part of the rhetoric of evolutionary biology and later, the Human Genome Project (see Burian 1982; Keller 1992; Nelkin and Lindee 1995). Dawkins (1979) extended this rhetoric to the conclusion that the genome was the book of life and that our bodies are merely the transient vehicles for the survival and propagation of our immortal DNA. This genocentric view of life came under fire from several biologists and philosophers (see Malacinski 1990). Some critics were scientists who were actively pursuing research into phenotypic plasticity or epigenesis. David Nanney (1957, 1985) viewed the cytoplasm as critical in heredity and saw in the rhetoric of DNA the image of a totalitarian state. H. Frederik Nijhout (1990a,b) and Richard Strathmann (1985) criticized the genocentric view of inheritance from their studies of the hormonal control of larval development; Brian Goodwin (1982) and Stuart Newman (1979, 1990), developmental biologists concerned with organizational patterning, also formulated critiques of genetic reductionism. In 1983, Newman co-founded the Council for Responsible Genetics, a major source of information against genetic reductionism.

Other critiques of genetic reductionism and determinism came from feminist scientists who saw that the genes-are-destiny rhetoric was being used against women. Donna Haraway (1979), Evelyn Fox Keller (1985), Anne Fausto-Sterling (1985), and Bonnier Spanier (1995), used their scientific training to criticize the genetic determinism in the field. The Biology and Gender Study Group (1988) was likewise constituted of scientists and science students. Philosophers of biology (e.g., Tauber and Sarkar 1992) and developmental systems theorists starting with Gilbert Gottlieb (1971, 1992) and Susan Oyama (1985) made the critique of genetic determinism a central part of their analyses. This provided another cause for the rebirth of interest in phenotypic plasticity. By 1999, developmentally plastic phenomena (seasonal polyphenisms, temperature-sensitive and context-dependent sex determination, nutritional polyphenisms, predator-induced polyphenisms, and neural plasticity) were linked to criticisms of genetic determinism to provide a positive alternative to the genocentric approach (see Gilbert 2002).

Another positive alternative came from newly discovered epigenetic phenomena. One major contribution to the re-emergence of plasticity was a set of studies in mammalian molecular biology that demonstrated the contextual nature of genetics. In particular, the discovery of chromosome imprinting demonstrated that developmental as well as genetic parameters determined whether a gene functioned (Barton et al. 1984; McGrath and Solter 1984; DeChiara et al. 1991). Indeed, as early as 1991, a "Human Epigenome Project" was mooted to balance the genetic deterministic rhetoric of the Human Genome Project (Gilbert 1991a).

Conservation biology provided another goad for studying plasticity. Conservation programs require information not just about adult survival in limited or compromised environments, but also about the development and vulnerabilities of embryos and larvae. Such data can have profound practical implications. For example, Morreale and coworkers (1982) revealed that sea turtle conservation programs had failed to take into account the temperature-dependent mechanism of turtle sex determination and as a result were releasing thousands of hatchling turtles—all of the same sex. In addition, Mead and Epel (1995) showed that fertilization phenomena studied in the laboratory may not be adequate descriptions of what occurs in nature. Indeed, few criticisms of developmental biology are as strong as "It doesn't tell us about how animals develop in nature." To correct this, we must forge an ecological developmental biology. A related reason for the increase in interest in the ecological aspects of development has been the increasing recognition of (and alarm about) environmental contamination by anthropogenic teratogens. Environmental chemicals that we once assumed were harmless (at least to adults) may be dangerous to developing organisms and may also threaten adult fertility. Some recent studies have been popularized by the media (Colburn et al. 1996; Hayes et al. 2002), and several others (Relyea and Mills 2001) demonstrate the need to study these compounds in the "real world" as opposed to just the laboratory.

For these reasons, and perhaps other reasons as well, phenotypic plasticity is coming back to play important roles in developmental biology, ecology, genetics, and evolutionary developmental biology (evo-devo). Waddington had a chapter on developmental plasticity in *The Strategy of the Genes* (1957), but thereafter it had been ignored in Anglophone embryology and developmental biology textbooks. Pritchard's *Foundations of Developmental Genetics* (1986) had sections on plasticity and on genetic assimilation, but it wasn't until 2000 that an English-language developmental biology textbook (see Gilbert 2000) had a chapter devoted exclusively to ecological developmental biology. In genetics, the field of quantitative genetic analysis began to analyze reaction norms, demonstrating how allelic variation might allow for different responses to environmental signals (Scheiner 1993; Roff 1997). In ecology, the field of life history strategies found plasticity to be central to mating and adaptation strategies throughout the animal and plant kingdoms (see Stearns 1992; Mead and Epel 1995; Pigliucci 2001). Moreover, evolutionary development biology saw itself as being able to explain the proximate causes of these life history strategies in cellular and molecular terms. Viewing the organism as being constructed by interacting modules that could receive environmental and well as internal signals, the ideas of Schmalhausen and Waddington were being re-read and reinterpreted in light of newly discovered molecular and cellular mechanisms (Hall 1992a,b; Gilbert 1991b, 2000; West-Eberhard 2003). Brian Hall's *Evolutionary Developmental Biology* textbook had material on polyphenisms and an entire chapter on genetic assimilation as early as 1992. Environmental regulation of phenotype production was to become a normal part of developmental biology and a critical part of evolutionary developmental biology. Within the past decade, numerous books concerning plasticity have been published, and Mary Jane West-Eberhard's *Developmental Plasticity and Evolution* (2003) brought developmental plasticity into the heart of evolutionary theory. The first symposium in ecological developmental biology (Gilbert and Bolker 2003) was conducted at the Society for Integrative and Comparative Biology in January 2002.

References

Barton, S. C., M. A. Surani and M. L. Norris. 1984. Role of paternal and maternal genomes in mouse development. *Nature* 311: 374–376.

Biology and Gender Study Group. 1988. The importance of feminist critique for contemporary cell biology. *Hypatia* 3: 172–186.

Bolker, J. 1995. Model systems in developmental biology. *Bioessays* 17: 451–455.

Bradshaw, A. D. 1965. Evolutionary significance of phenotypic plasticity in plants. *Adv. Genet.* 13: 115–155.

Burian, R. 1982. Human sociobiology and genetic determinism. *Philos. Forum* 13: 40–66.

Carlson, E. A. 2001. *The Unfit: A History of a Bad Idea.* Cold Spring Harbor Press, Cold Spring Harbor, NY.

Collins, J. P., S. F. Gilbert, M. D. Laubichler and G. B. Müller. 2007. How to integrate development, evolution, and ecology: modeling in evo-devo. In M. D. Laubichler (ed.), *Roots of Theoretical Biology: The Prater Vivarium Centenary.* MIT Press, Cambridge, MA, pp. 355–378.

Colburn, T., D. Dumanoski and J. P. Myers. 1996. *Our Stolen Future.* Dutton, New York.

Dawkins, R. 1976. *The Selfish Gene.* Oxford University Press, New York.

DeChiara, T. M., E. J. Robertson and A. Efstratiadis. 1991. Parental imprinting of the mouse insulin-like growth factor II gene. *Cell* 64: 849–859.

Dobzhansky, Th. 1955. *Evolution, Genetics, and Man.* John Wiley, New York.

Dobzhansky, Th. and B. Spassky. 1963. Genetics of natural populations. XXXIV. Adaptive norm, genetic load, and genetic elite in *Drosophila pseudoobscura.* *Genetics* 48: 1467–1485.

Fausto-Sterling, A. 1985. *Myths of Gender: Biological Theories about Men and Women.* Basic Books, New York.

Gassner, G. 1918. Beiträge zur physiologischen Charakterisik sommer-und winterannueller Gewächse, insbesondere der Getreidepflanzen. *Zeit. Botan.* 10: 417–480.

Gilbert, S. F. 1991a. Cytoplasmic action in development. *Q. Rev. Biol.* 66: 309–316.

Gilbert, S. F. 1991b. Induction and the origins of developmental genetics. In S. F. Gilbert (ed.), *A Conceptual History of Modern Embryology.* Plenum, New York, pp. 181–206.

Gilbert, S. F. 2000. Diachronic biology meets evo-devo: C. H. Waddington's approach to evolutionary developmental biology. *Am. Zool.* 40: 729–737.

Gilbert, S. F. 2002. Genetic determinism: The battle between scientific data and social image in contemporary developmental biology. In A. Grunwald, M. Gutmann and E. M. Neumann-Held (eds.), *On Human Nature: Anthropological, Biological, and Philo-*

sophical Foundations. Springer-Verlag, New York, pp. 121–140.
Gilbert, S. F. and J. Bolker. 2003. Ecological developmental biology: Preface to the symposium. *Evol. Dev.* 5: 3–8.
Gliboff, S. 2006. The case of Paul Kammerer : Evolution and experimentation in the early twentieth Century. *J. Hist. Biol.* 39: 525–563.
Goldschmidt, R. B. 1949. Research and politics. *Science* 109 : 219–227.
Goodwin, B.C. 1982. Genetic epistemology and constructionist biology. *Rev. Int. Philo.* 36: 527–548.
Gormley, M. 2007. L. C. Dunn and the reception of Lysenkoism in the United States. Ph.D. thesis, Oregon State University.
Gottlieb, G. 1971. *Development of Species Identification in Birds: An Inquiry into the Prenatal Determinants of Perception*. University of Chicago Press, Chicago.
Hall, B. K. 1992a. Waddington's legacy in development and evolution. *Amer. Zool.* 32: 113–122.
Hall, B. K. 1992b. *Evolutionary Developmental Biology*. Chapman and Hall, New York.
Haraway, D. J. 1979. The biological enterprise: Sex, mind, and profit from human engineering to sociobiology. *Rad. Hist. Rev.* 20: 206–237.
Hayes, T. B. and 6 others. 2002. Hermaphroditic, demasculinized frogs after exposure to the herbicide atrazine at low ecologically relevant doses. *Proc. Natl. Acad. Sci. USA* 99: 5476–5480.
Hertwig, O. 1894. *The Biological Problem of Today: Preformationism or Epigenesis ?* (P.C. Mitchell, transl.) Macmillan, New York.
Joravsky, D. 1970. *The Lysenko Affair*. Harvard University Press, Cambridge, MA.
Keller, E. F. 1985. *Reflections on Gender and Science*. Yale University Press, New Haven.
Kevles, D. 1998. *In the Name of Eugenics*. Harvard University Press, Cambridge, MA.
Koestler, A 1971. *The Case of the Midwife Toad*. Random House, New York.
Konstantinov, P. N., P. I. Lisitsyn and D. Kostov. 1936. Neskol'koslov o rabotakh odesskogo instituta seleksii I genetikii. *SRSKh* 11: 121–130.
Lewontin, R. C. 1972. The apportionment of human diversity, *Evolutionary Biology* 6: 381–398.
Lewontin, R. C. 1976. The fallacy of biological determinism. *The Sciences* 16: 6–10.
Lewontin, R. C. 1993. *Biology as Ideology: The Doctrine of DNA*. Harper, New York.
Lewontin, R. C. 2001. *It Ain't Necessarily So*. Granta Press, Cambridge University Press, Cambridge.
Lewontin, R. C. 2002. *The Triple Helix: Gene, Organism, and Environment*. Harvard University Press, Cambridge, MA.
Lewontin, R. C. and R. Levins 1976. The problem of Lysenkoism. In H. Rose and S. Rose (ed.), *The Radicalisation of Science. Ideology of/in the Natural Sciences*. Macmillan, London, pp. 32–64.
Ludmerer, K. M. 1972. *Genetics and American Society*. Johns Hopkins University Press, Baltimore.
Lysenko, T. D. and I. I. Prezent. 1935. *Seleksiia i Teoriia Stadiinogo Razvitiia Rasteniia*. Sel'khozgiz, Moscow.
Malacinski, G. 1990. *Cytoplasmic Organization Systems*. McGraw Hill, New York.
McGrath, J. and D. Solter. 1984. Completion of mouse embryogenesis requires both the maternal and paternal genomes. *Cell* 37: 179–183.
Mead, K. S. and D. Epel. 1995. Beakers versus breakers: How fertilization in the laboratory differs from fertilization in nature. *Zygote* 3: 95–99.
Medvedev, Z. 1969. *The Rise and Fall of T. D. Lysenko*. Columbia University Press, New York.
Monod, J. 1971. *Chance and Necessity*, Knopf, New York.
Morreale, S. J., G. J. Ruiz, J. R. Spotila and E. A. Standora. 1982. Temperature-dependent sex determination: Current practices threaten conservation of sea turtles. *Science* 216: 1245–1247.
Nanney, D. L. 1957. The role of the cytoplasm in heredity. In W. D. McElroy and B. Glass (eds.), *The Chemical Basis of Heredity: A Symposium*. Johns Hopkins University Press, Baltimore, pp. 134–166.
Nanney, D. L. 1985. Heredity without genes: Ciliate explorations of clonal heredity. *Trends Genet.* 1: 295–298.
Nelkin, D. and M. S. Lindee. 1995. *The DNA Mystique: The Gene as a Cultural Icon*. W.H. Freeman, New York.
Newman, S. A. and H. L. Frisch, H. L. 1979. Dynamics of skeletal pattern formation in developing chick limb. *Science* 205: 662–668.
Newman, S. A., and W. D. Comper. 1990. "Generic" physical mechanisms of morphogenesis and pattern formation. *Development* 110: 1–18.
Nijhout, H. F. 1990a. Metaphors and the role of genes in development. *BioEssays* 12: 441–446.
Nijhout, H. F., 1990b. A comprehensive model for color pattern formation in butterflies. *Proc. Roy. Soc.* B 239: 81–113.
Nyhart, L. 1995. *Biology Takes Form*. University of Chicago Press, Chicago.
Oyama S. 1985. *The Ontogeny of Information: Developmental Systems and Evolution*. Duke University Press, Durham, NC.
Pigliucci, M. 2001. *Phenotypic Plasticity: Beyond Nature and Nurture*. Johns Hopkins University Press, Baltimore.
Pringle, P. 2008. *The Murder of Nikolai Vavilov*. Simon and Schuster, New York.
Pritchard D. J. 1986, *Foundations of Developmental Genetics*. Taylor and Francis, London
Relyea, R. A. and N. Mills. 2001. Predator-induced stress makes the pesticide carbaryl more deadly to grey treefrog tadpoles (*Hyla versicolor*). *Proc. Natl. Acad. Sci. USA* 98: 2491–2496.
Roff, D. A. 1997. *Evolutionary Quantitative Genetics*. Springer-Verlag, New York.
Roll-Hansen, N. 2005. *The Lysenko Effect: The Politics of Science*. Humanity Books, Amherst , NY.

Sapp, J. 1987. *Beyond the Gene: Cytoplasmic Inheritance and the Struggle for Authority in Genetics.* Oxford University Press, Oxford.

Sarkar, S. 1999. From the *Reaktionsnorm* to the Adaptive Norm: The norm of reaction, 1906–1960. *Biol. Philos.* 14: 235–252.

Scheiner, S. M. 1993. Genetics and evolution of phenotypic plasticity. *Annu. Rev. Ecol. System.* 24: 35–68.

Selden, S. 1999. *Inheriting Shame: The Story of Eugenics and Racism in America.* Teachers College Press, New York.

Soyfer, V. 1994. *Lysenko and the Tragedy of Soviet Science.* Transl. L. Gruliow and R. Gruliow. Rutgers University Press, New Brunswick, NJ.

Spanier, B. 1995. *Im/Partial Science: Gender Ideology in Molecular Biology.* Indiana University Press.

Stearns, S. C. 1992. *The Evolution of Life Histories.* Oxford University Press, New York.

Strathmann, R. R. 1985. Feeding and non-feeding larval development and life history evolution in marine invertebrates. *Annu. Rev. Ecol. Syst.* 16: 339–361.

Tauber, A. I. and S. Sarkar. 1992.The human genome project: Has blind reductionism gone too far? *Perspect Biol Med.* 35: 220–235.

Vavilov, N. K .1933. Quoted in Ni. I. Vavilov Institute of Plant Industry. Lysenko's role in the development of agricultural science in the USSR. http://www.vir.nw.ru/history/lysenko.htm.

Waddington, C. H. 1957. *The Strategy of the Genes.* George Allen and Unwin, London.

Waddington, C. H. 1968. The basic ideas of biology. In C. H. Waddington (ed.), *Towards a Theoretical Biology.* Edinburgh: Edinburgh University Press, pp. 1–32.

Weismann A. 1875. Über den Saison-Dimorphismus der Schmetterlinge. In *Studien zur Descendenz-Theorie.* Engelmann, Leipzig.

West-Eberhard, M. J. 2003. *Developmental Plasticity and Evolution.* Oxford University Press, Oxford.

Whittaker, J R. 1985. Paul Kammerer and the suspect siphons. *MBL Science;* http://www.mbl.edu/publications/pub_archive/Ciona/Kammerer/index.html

Wolfe, A, J. 2002. Speaking for Nature and Nation: Biologists as public intellectuals in cold war culture. Ph.D. thesis, University of Pennsylvania.

Woltereck, R. 1909. Weitere experimentelle Untersuchungen über Artveränderung, speziel über das quantitativer Artunterschiede bei Daphnien. *Verhandlungen der Deutschen zoologischen Gessellschaft* 19: 110 – 173.

Appendix B

The Molecular Mechanisms of Epigenetic Change

Epigenetics has been redefined many times over the past half century. Originally, C. H. Waddington coined the term to refer to what today we would call developmental genetics—how the genotype gives rise to the phenotype through developmental interactions. Waddington modified the term *epigenesis* (development) by melding it with the word *genetics*. However, the *epi* in epigenesis could also mean "that which is above" (as in *epi*dermis or *epi*glottis). So in recent years, epigenetics has sometimes been used to mean that which is "above" the genetic sequences—in other words, the "formatting" of the genome.

As Eccleston and colleagues (2007) wrote, "Epigenetics is typically defined as the study of heritable changes in gene expression that are not due to changes in DNA sequence." However, this definition depends on what one means by "heritable." If one is thinking in terms of cells, it means that the gene expression pattern of one cell is transmitted accurately to that cell's descendants. This is important in the stability of differentiated states. We don't want our skin cells turning into gut cells while we sleep. But, as this book demonstrates (see Chapters 2, 6, and 10), we are coming to realize how agents causing epigenetic changes in germ cells can create stable patterns of gene expression that become transmitted from an organism to its progeny.

Understanding epigenetics requires that we understand how the chromatin becomes altered to activate or suppress transcription (gene expression). We also have to understand how these alterations in chromatin structure become stabilized. That is to say, how is each gene region given a "transcriptional memory" that it can rely on after replication so that alterations in the chromatin are repeated and gene expression patterns are stable? Although we lack anything near a complete understanding of these events, we do know that both of these involve modifications of the nucleosomes.

Three integrated systems may be involved in stabilizing chromatin conformations between cell generations. It is possible that each progresses to the same end from different starting points. One system starts with methylation of cytosine nucleotides in the organism's DNA. The other two start with nucleosome alterations.

DNA Methylation

In many animals (including mammals, but not usually in flies), the nucleotide sequence CG has the potential to become methylated (that is, to have methyl groups attached) at its C residue. (In plants, methylation is often on a CNG-sequence cytosine.) But how is the methyl group restored each time a new DNA strand is replicated during mitosis? The answer lies in the fact that the other C of the opposing CG pair would be methylated, and a particular enzyme (DNA methyltransferase-1 DNMT1) recognizes hemimethylated (half-methylated) sites and "completes" them so that each methyl-CG has a corresponding GC-methyl group. DNMT1 does not recognize unmethylated Cs. In that way, the pattern of methylated cytosines is repeated from cell generation to generation.

This copying of methylated sites is far from exact. While DNA replication is so accurate (due to the editing subunit of DNA polymerase) that only 1 base in 10^8 is copied incorrectly, the error rate for DNA methylation is about 1 in every 25 cytosines. This means that methylated sites are often bunched close together so that the effect of errors is reduced. Even so, random errors in methylation can create changes in gene expression within clonal populations of cells (Laird et al. 2004; Bird 2007).

The importance of methylation in stably inhibiting transcription was originally shown in two mammalian conditions in which genes are silenced during development. First, in mammalian dosage compensation, one X chromosome is "turned off" in every female (XX) cell. Once this is accomplished, that same X chromosome—either the one from the father or the one from the mother—remains inactive in all the progeny of that cell (see Brockdorff and Turner 2007; Migeon 2007). This is why calico cats (where a major pigmentation gene is located on the X chromosome) have separate areas of black and orange pigmentation; which color is expressed depends on which X chromosome was inactivated in the progenitor pigment cell. It also explains why calico cats are always females (or sterile XXY males.) The maintenance (but not the origin) of this pattern depends on DNA methylation of clustered CG pairs (Venolia et al. 1982; Wolf et al. 1984).

The second phenomenon explained by DNA methylation is genomic imprinting, where the chromosomes from the male and the female are not equivalent. At about 80 known loci, the genes contributed by the sperm differ in activity from the genes contributed by the egg. In some of these cases, the nonfunctioning gene has been rendered inactive by DNA methylation.* As mentioned before, methylated DNA is associated with stable DNA silencing, either (1) by interfering with the binding of gene-activating transcription factors, or (2) by recruiting repressor proteins that stabilize nucleosomes in a restrictive manner along the gene. The presence of a methyl group can prevent transcription factors from binding to the DNA, thereby preventing the gene from being activated. For instance, during early embryonic development in mice, the *Igf2* (insulin-like growth factor-2) gene is active only from the father's chromosome 7; the egg-derived *Igf2* gene does not function. This is because in the female-derived chromosome 7, the CTCF protein—an inhibitor that can block a gene's promoter region from receiving activation signals from enhancers—binds to a region near the maternal *Igf2* gene. Once bound, it prevents the maternally

*This means that mammals must have both a male and a female parent. Unlike sea urchins, flies, or frogs, parthenogenesis—virgin birth—cannot happen naturally in mammals.

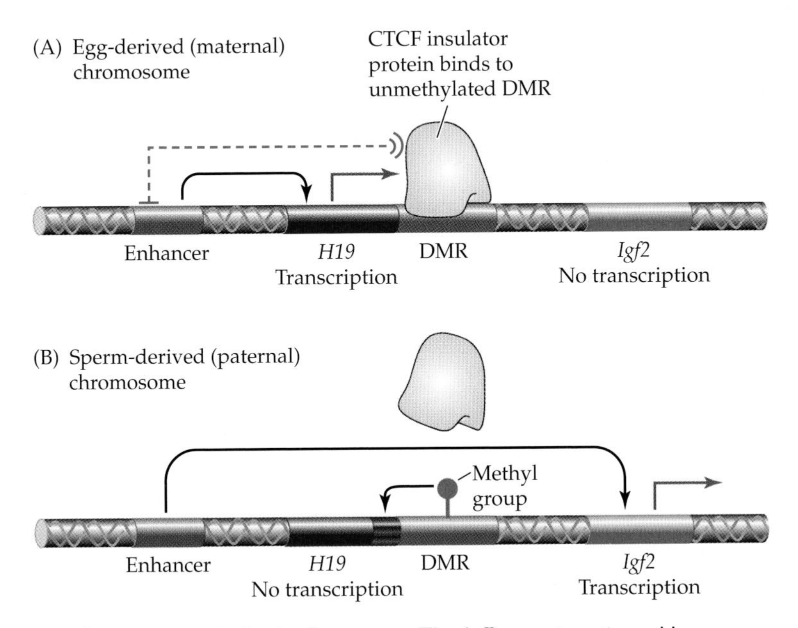

FIGURE 1 Genomic imprinting in the mouse. The *Igf2* gene is activated by an enhancer element it shares with the *H19* gene. A differentially methylated region (DMR) lies between the enhancer and the *Igf2* gene, adjacent to the promoter region of the *H19* gene. (A) In the egg-derived chromosome, the DMR is unmethylated and binds the CTCF insulator protein, thus blocking the enhancer signal and preventing *Igf2* expression. (B) In the sperm-derived chromosome, the DMR is methylated, with two consequences. First, the CTCG insulator protein cannot bind to the DMR and the enhancer can activate *Igf2*. Second, the methylated DMR alters the chromatin of the *H19* promoter, preventing its activation by the enhancer. Therefore, the maternal chromosome expresses *H19* but not *Igf2*, whereas the paternally derived chromosome expresses *Igf2* but not *H19*.

derived *Igf2* gene from functioning. But it is only in the female-derived homologue that this binding region is unmethylated and available. In the sperm-derived chromosome 7, the region where CTCF would bind is methylated. CTCF cannot bind and thus the gene is not inhibited and expresses a functional protein (Figure 1; Bartolomei et al. 1993; Ferguson-Smith et al. 1993; Bell and Felsenfeld 2000).

The second mechanism is the converse of the first. Instead of impeding proteins from binding, the methyl-CG actually is bound by specific proteins. Proteins such as MeCP2, MBD1, and Kaiso have a methyl-CpG-binding domain (MBD) that recognizes methyl-CpG sequences on the DNA (Bird and Wolffe 1999). When these proteins bind to DNA, they can associate with histone deacetylases and histone methyltransferases that stabilize nucleosomes and prevent transcription (Jones et al. 1998; Nan et al. 1998; Sarraf and Stancheva 2004). Methylated DNA may also bind (probably indirectly) histone 1, the "linker" histone subunit found between the nucleosomes, thus providing another mechanism through which the nucleosomes become tightly packed on methylated DNA (Rupp and Becker 2005).

Histone Modification and Histone Substitution

Histone modification is a way of getting nucleosomes into a conformation that will either permit or block RNA polymerase from either initiating or elongating messenger RNA at a particular gene. Some modifications appear to be critical for initiating transcription—that is, for getting RNA polymerase poised on the promoter (see Figure 2.1). These modifications often include a trimethylated lysine on the fourth position of histone H3 (HeK4me3) and the acetylation of histone H3 at lysines on positions 9 and 14. Other histone modifications, such as H3K36me3 (translated as "trimethylated lysine at position 36 of histone H3") appear to be necessary within the gene to allow the RNA polymerase to pass through. Both modifications might be required to achieve transcription of the specific gene (Guenther et al. 2007). Figure 2 shows a nucleosome and its H3 tail, which contains certain amino acids whose modification can regulate transcription.

Many transcription factors work by binding one site to the DNA and using another site on the protein to recruit histone-modifying enzymes. For instance, the Pax7 transcription factor that activates muscle-specific genes binds to the enhancer

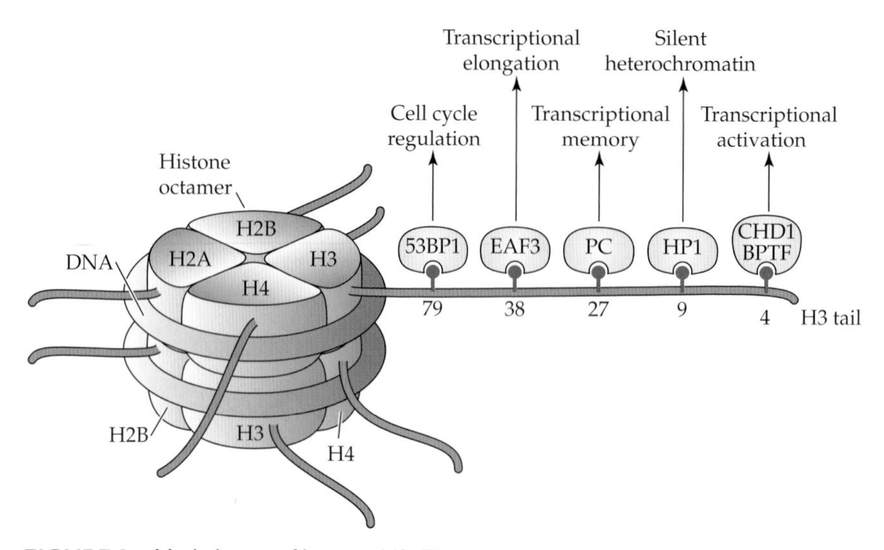

FIGURE 2 Methylation of histone H3. The tail of histone H3 sticks out from the nucleosome and is capable of being methylated In this diagram, methylated lysine residues are recognized by particular proteins. The methylated residues at positions 4, 38, and 79 are associated with gene activation, whereas methylated lysines at positions 9 and 27 are associated with repression. The specific proteins binding these sites, represented above the methyl groups, are not shown to scale. (After Kouzarides and Berger 2007.)

region of these genes within the muscle precursor cells and recruits a histone methyltransferase to that region. The histone methyltransferase methylates the lysine in the fourth position of histone H3 (H3K4), resulting in trimethylation of this lysine and the activation of gene transcription (McKinnell et al. 2008). The repair and displacement of nucleosomes along the DNA make it possible for the transcription factors to find their binding sites (Adkins et al. 2004; Li et al. 2007).

PROTEIN RECRUITMENT Histone modifications can also signal the recruitment of the proteins that can retain the memory of transcriptional state from generation to generation through mitosis. These are the proteins of the Trithorax and Polycomb families. Trithorax proteins, when bound to nucleosomes of active genes, keep these genes active; Polycomb proteins, which bind to condensed nucleosomes, keep the genes in an inactive state.

The Polycomb proteins fall into two categories that act sequentially in repression. The first set has histone methyltransferase activities that methylate lysines H3K27 and H3K9 to repress gene activity. In many organisms, this repressive state is stabilized by the activity of a second set of Polycomb factors that bind to the methylated tails of histone 3 and keep the methylation active and which also methylate adjacent nucleosomes, thereby forming tightly packed repressive complexes* (Grossniklaus and Paro 2007).

The Trithorax proteins act to counter the effect of the Polycomb proteins and are necessary to retain the memory of activation. They can either modify nucleosomes or alter the positions of nucleosomes on the chromatin. Thus, some of the Trithorax proteins become parts of complexes that use the energy from ATP hydrolysis to move nucleosomes along the chromatin or to remodel them, allowing transcription factors to bind. Other Trithorax group proteins keep the H3.K4 lysine trimethylated (Kingston and Tamkun 2007).

REINFORCED REPRESSION Reinforcement between repressive chromatin and repressive DNA has also been observed. Just as methylated DNA is able to attract proteins that deacetylate histones and attract H1 linker histones (both of which stabilize nucleosomes), so repressive states of chromatin are able to recruit enzymes that methylate DNA. DNA methylation patterns during gametogenesis depend in part on the DNA methyltransferase DNMT3L. Although this protein has lost its enzymatic activity, it can still bind to the amino end of histone H3. Once bound, it will recruit and/or activate the DNA methyltransferase DNMT3A2 to methylate the cytosines on nearby CG pairs (Fan et al. 2007; Ooi et al. 2007). However, if the lysine at H3.K4 is methylated, DNMT3L will not bind.

*Indeed, the maintenance of active and repressed states of the FLC gene involved in vernalization (see Chapter 2, p. 47) involves by chromatin modification by polycomb proteins. Cold temperature (or artificial vernalization) results in the inactivation of the FLC gene by the cold-induced activation of the genes producing VIN3 and Polycomb group proteins. These proteins deacetylate histone H3 and methylate it on specific regions. This methylation recruits the HP1 proteins that aggregate nucleosomes together through histone H3. This nucleosome aggregation is maintained in the nuclei of the apical meristem cells through mitotic divisions. In this way, when warm weather returns, the wheat is ready to flower (see Sung and Amasino 2005).

HISTONE SUBSTITUTION The epigenetic memory of an active gene state can also be maintained by histone substitution. When nuclei from differentiated tadpole cells are placed into enucleated oocytes, they can form complete, cloned tadpoles; however, some of the cloned tissues can retain transcription characteristic of the original cell from which the nucleus was taken. For example, Ng and Gurdon (2008) found that when the donor nuclei came from muscle cells, the muscle-specific gene *MyoD* was expressed in nonmuscle cell lineages—even after the cell nuclei had gone through 28 replication cycles in which *MyoD* transcription had been suppressed. The marker of this transcriptional memory was neither DNA methylation nor a covalent modification of the histones. Rather, it was the substitution of a variant form (H3.3) of histone 3. In several species studied, the H3.3 variant appears to be enriched in active gene loci (Ahmad and Henikoff 2002; McKittrick et al. 2004).

Epigenetic Repression by Small RNAs

In addition to proteins, small, non-coding RNAs can also inhibit transcription. In yeasts, heterochromatin (large regions of repressed chromatin) is induced by the modification of histones through RNA-induced transcriptional silencing (RITS). The small interfering RNA (siRNA) forms a duplex with complementary DNA where it recruits proteins that stabilize nucleosomes and that degrade any new transcripts. In female mammals, a similar mechanism may initiate X-chromosome inactivation (see above), where the non-coding *Xist* RNA appears to bind proteins that maintain the methylated state of chromatin throughout most of the inactivated chromosome (Chang et al. 2006; Migeon 2007).

Recent studies have demonstrated that siRNAs can regulate transcription from mammalian germline cells by initiating methylation in these regions (Kuramochi-Migawa et al. 2008; Watanabe et al. 2008). The possibility that siRNAs can regulate gene transcription in mammals supports reports that RNAs can be inherited through the germline and that these RNAs maintain an inhibited pattern of gene expression from one generation to the next (see Rassoulzadegan et al. 2006).

Cell-Cell Interactions and Chromatin Remodeling

One of the significant recent breakthroughs of developmental biology has been the linking of cell-cell interactions with chromatin remodeling. These two phenomena had always been considered as separate areas. The first was concerned with how changes at the cell surface cause cells to differentiate along particular pathways. The second concerned the mechanisms by which transcription factors in the nucleus cause new patterns of gene expression. We now have dozens of examples of the following paradigm that links the two (see Gilbert 2006).

Paracrine factors from one cell bind to receptors on another cell. Once these receptors have bound the factor, they gain enzymatic activity (often becoming kinases) which then activate certain enzymes within the cytoplasm. These cytoplasmic enzymes trigger a signal transduction cascade that ends in the activation of a dormant transcription factor. This transcription factor, now active, can bind to DNA and recruit chromatin-modification enzymes that will regulate gene transcription. Thus, signal transduction cascades bring together events occurring at the cell membrane with events occurring in the cell nucleus.

Summary

The mechanisms of epigenetic control restrict which genes are able to become active in each cell and also determine which genes remain active through rounds of mitosis. In this way, cell types are specified and stabilized. These mechanisms largely involve DNA methylation and the modification of nucleosomes by enzymes that can methylate or acetylate particular histone residues. The epigenetic changes in DNA are crucial to our understanding of the mechanisms by which the environment can alter organismal phenotypes, the origin of cancerous cellular phenotypes, the origin of the aging syndrome, and the production of selectable traits in evolution. Indeed, since the Human Genome Project has not clarified how the various genetic components are regulated, a Human Epigenome Project has been proposed (Beck et al. 1999) that would attempt to identify and catalogue the sites of variable human DNA methylation in normal and cancerous cells.

References

Adkins, C. C., S. R. Howar and J. K. Tyler. 2004. Chromatin disassembly mediated by histone chaperone Asf1 is essential for transcriptional activation of the yeast *PHO5* and *PHO8* genes. *Mol. Cell* 14: 657–666.

Ahmad, K. and S. Henikoff. 2002. The histone variant H3.3 marks active chromatin by replication-independent nucleosome assembly. *Mol. Cell* 9: 1191–1200.

Bartomomei, M. S., A. L. Webber, M. E. Brinow and S. M. Tilghman. 1993. Epigenetic mechanisms underlying the imprinting of the mouse *H19* gene. *Genes Dev*. 7: 1663–1673.

Beck, S., A. Olek and J. Walter. 1999. From genomics to epigenomics: A loftier view of life. *Nature Biotechnol.* 17: 1144.

Bell, A. C. and G. Felsenfeld. 2000. Methylation of a CTCF-dependent boundary controls imprinted expression of the *Igf2* gene. *Nature* 405: 482–485.

Bird, A. 2007. Perceptions of epigenetics. *Nature* 447: 396–398.

Bird, A. and A. P. Wolffe. 1999. Methylation-induced repression: Belts, braces, and chromatin. *Cell* 99: 451–454.

Brockdorff, N. and B. M. Turner. 2007. Dosage compensation in mammals. In C. D. Allis, T. Jenuwein and D. Reinberg (eds.), *Epigenetics*. Cold Spring Harbor Press, NY, pp. 321–340.

Chang, S. C., T. Tucker, N. P. Thoprogood and C. J. Brown. 2006. Mechanisms of X-chromosome inactivation. *Frontiers Biosci.* 11: 852–866.

Eccleston, A., N. DeWitt, C. Gunter, B. Marte and D. Nath. 2007. Epigenetics. *Nature* 447: 395.

Fan, L. and 7 others. 2007. Recognition of unmethylated histone H3 lysine 4 links BHC80 to LSD1-mediated gene repression. *Nature* 448: 718–722.

Ferguson-Smith, A. C., H. Sasaki, B. M. Cattanach and M. A. Surani. 1993. Paternal-origin-specific epigenetic modification of the mouse *H19* gene. *Nature* 362: 751–755.

Gilbert, S. 2006. *Developmental Biology*, 8th Ed. Sinauer Associates, Sunderland, MA.

Grossniklaus, U. and R. Paro. 2007. Transcriptional silencing by Polycomb group proteins. In C. D. Allis, T. Jenuwein and D. Reinberg (eds.), *Epigenetics*. Cold Spring Harbor Press, NY, pp. 211–230.

Guenther, M. G., S. S. Levine, L. A. Boyer, R. Jaenisch and R. A. Young. 2007. A chromatin landmark and transcription initiation at most promoters in human cells. *Cell* 130: 77–88.

Jones, P. L. and 7 others. 1998. Methylated DNA and MeCP2 recruit histone deacetylase to repress transcription. *Nature Genet*. 19: 187–191.

Kingston, R. E. and J. W. Tamkun. 2007. Transcriptional regulation by Trithorax group proteins. In C. D. Allis, T. Jenuwein and D. Reinberg (eds.), *Epigenetics*. Cold Spring Harbor Press, NY, pp. 231–248.

Kouzarides, T. and S. L. Berger. 2007. Chromatin modifications and their mechanisms of action. In C. D. Allis, T. Jenuwein and D. Reinberg (eds.), *Epigenetics*. Cold Spring Harbor Press, New York, pp. 191–209.

Kuramochi-Miyagawa, S. and 17 others. 2008. DNA methylation of retrotransposon genes is regulated by Piwi family members MILI and MIWI2 in murine fetal testes. *Genes Dev*. 22: 908 – 917.

Laird, C. D. and 9 others. 2004. Hairpin-bisulfite PCR: Assessing epigenetic methylation patterns on complementary strands of individual DNA molecules. *Proc. Natl. Acad. Sci. USA* 101: 204–209.

Li, B., M. Carey and J. L. Workman. 2007. The role of chromatin during transcription. *Cell* 128: 707–719.

McKinnell, I. W. and 7 others. 2008. Pax7 activates myogenic genes by recruitment of a histone methyltransferase complex. *Nature Cell Biol*. 10: 77–84.

McKittrick, E., P. R. Gafken, K. Ahmad and S. Henikoff. 2004. Histone H3.3 is enriched in covalent modifications associated with active chromatin. *Proc. Natl. Acad. Sci. USA* 101: 1525–1530.

Migeon, B. R. 2007. *Females Are Mosaics: X-inactivation and Sex Differences in Disease*. Oxford University Press, New York.

Nan, X., H.-H. Ng, C. A. Johnson, C. D. Laherty, B. M. Turner, R. N. Eisenman and A. Bird. 1998. Transcriptional repression by the methyl-CpG-binding protein MeCP2 involves a histone deacetylase complex. *Nature* 393: 386–389.

Ng, R. K. and J. B. Gurdon. 2008. Epigenetic memory of an active gene state depends on histone H3.3 incorporation into chromatin in the absence of transcription. *Nature Cell Biol.* 10: 102–109.

Ooi, S. K. T. and 11 others. 2007. DNMT3L connects unmethylated lysine 4 histone H3 to de novo methylation of DNA. *Nature* 448: 714–717.

Rassoulzadegan, M., V. Grandjean, P. Gounon, S. Vincetn, I. Gillot and F. Cuzin. 2006. RNA-mediated non-Mendelian inheritance of an epigenetic change in the mouse. *Nature* 441: 469–474.

Rupp, R. A. W. and P. B. Becker. 2005. Gene regulation by histone H1: New links to DNA methylation. *Cell* 123: 1178–1179.

Sarraf, S. A. and I. Stancheva. 2004. Methyl-CpG-binding protein MBD1 couples histone H3 methylation at lysine 9 by SETDB1 to DNA replication and chromatin assembly. *Mol. Cell* 15: 595–605.

Sung, S. and R. M. Amasino. 2005. Remembering winter: Toward a molecular understanding of vernalization. *Annu. Rev. Plant Biol.* 56: 491–508.

Venolia, L., S. M. Gartler, E. R. Wasserman, P. Yen, T. Mohandas and L. J. Shapiro. 1982. Transformation with DNA from 5-azacytidine reactivated X chromosomes. *Proc. Natl. Acad. Sci. USA* 79: 2352–2354.

Watanabe, T. and 12 others. 2008. Endogenous siRNAs from naturally formed dsRNAs regulate transcripts in mouse oocytes. *Nature* 453: 539–543.

Wolf, S. F., D. J. Jolly, K. D. Lunnen, T. Friedman and B. R. Migeon. 1984. Methylation of the hypoxanthine guanosine phosphoribosyltransferase locus on the human X chromosome: Implications for X-chromosome inactivation. *Proc. Natl. Acad. Sci. USA* 81: 2806–2810.

Appendix C

Writing Development Out of the Modern Synthesis

As described in Chapter 9, what was left out of the Modern Synthesis was embryology (Harrison 1937; Hamburger 1980; Amundson 2000, 2001). Embryology is the study of how the zygote gives rise to the interconnected organs of the body, and for many embryologists, no theory of evolution made sense without a theory of body construction. For instance, Lillie (1927) could say that the methods and theories used by geneticists "have no place among their categories for the ontological process and *a fortiori* for the phylogenetic." Thus, for Lillie (and for many other embryologists, especially John Berrill and Gavin de Beer), if genetics could not explain development, then it could not explain evolution. De Beer (1954) would write that "the processes whereby the structures are formed are as important as the structures themselves from the point of view of evolutionary morphology and homology."

But until recently, embryologists themselves did not have an adequate theory of body construction. There was no coherent theory of differentiation or morphogenesis until the 1980s, and to geneticists, this was reason enough for excluding embryology from the Modern Synthesis. In this sense, it was important for geneticists that evolution be a theory of change, independent of a theory of body construction. Most geneticists (with a few exceptions such as Sewall Wright) did not think that development was needed to explain evolution. Development happened, and that was enough. The genotype was expressed as a phenotype, and development was merely the way it got there. In contradistinction to de Beer, Ernst Mayr (1980) would write, "... the clarification of the biochemical mechanism by which the genetic program is translated into the phenotype tells us absolutely nothing about the steps by which natural selection had built up the particular genetic program."*

*Mayr's "genetic program" metaphor had a lot to do with taking development out of the modern synthesis. Bruce Wallace's argument that embryology could play no role in evolution (Wallace 1986; see Chapter 10) is also based on the notion that there is a genetic program for development and that nothing comes between the genotype and the phenotype.

Related to this issue was that genetics and developmental biology explained different things. The *explanandum* of the Modern Synthesis was natural selection; the population geneticists did not need knowledge of development to get their answers. The *explanandum* for developmental biologists interested in evolution was anatomical change; natural selection was secondary in that it allowed some of these changes to persist. Rudy Raff, one of the earliest advocates of a developmental genetic approach to evolution, said that "they're interested in species; we're interested in bodies" (Amundson 2005, p. 253). Indeed, most early embryologists were not interested in variations within species, and this disinterest allowed the evolutionary biologists to ignore developmental considerations in their work.

Goldschmidt: Macromutations and "Hopeful Monsters"

Some biologists thought development and evolution should be reconnected and so began constructing their own theories. In the 1940s, Richard Goldschmidt criticized the Modern Synthesis, writing that their paradigm of the accumulation of small genetic changes was not sufficient to explain the generation of evolutionarily novel anatomical structures such as teeth, feathers, cnidocysts (the stinging cells of cnidarians), or the shells of snails and other molluscs. He claimed that evolution could only occur through heritable changes in those genes that regulated development (Goldschmidt 1940).

In *The Material Basis of Evolution*, Goldschmidt (1940) presented two models relating gene activity, development, and evolutionary dynamics. The first model proposed that new species might originate as "hopeful monsters" that were the result of mutations in developmentally important loci ("developmental macromutations"). In the second model, Goldschmidt argued that chromosomal rearrangements ("systemic mutations") would have the effect of many cumulative developmental macromutations and cause even larger phenotypic changes. Goldschmidt's view of systemic mutations did not win much favor at the time, nor did his idea that "a single mutational step affecting the right process at the right moment can accomplish everything, providing that it is able to set in motion the ever present potentialities of embryonic regulation" (p. 297). However, evolutionary developmental biology has since shown that the latter process may in fact be critical in the generation of novel structures, and several studies (see Bateman and DiMichele 2002) showed that stress to the nucleus can cause gene duplications, the movement of transposable DNA sequences, and an increase in recombination rates (that is, the rate of the "systematic mutations") that can result in rapid and extensive phenotypic changes.

However, Goldschmidt's presentation of these ideas went against the grain of the genetic science of the 1940s. To Goldschmidt, the gene wasn't a locus or an allele, but a unit of development (Goldschmidt 1940, p. 197). For Goldschmidt, the regulatory processes of development relieved the need for thousands of modifier genes, and for this reason, he attempted "to convince evolutionists that evolution is not only a statistical genetical problem but also one of the developmental potentialities of the organism." (Goldschmidt et al. 1951). However, in the absence of molecular data, Goldschmidt's ideas were generally ignored (see Gould 1980).

Waddington: A Theory of Bodies, a Theory of Change

Goldschmidt's ideas were considered when two of the most far-reaching syntheses of evolution, genetics, and development were attempted by Ivan Ivanovich Schmalhausen and Conrad Hal Waddington (whose work is discussed in Chapter 10 and in Appendix D). Waddington was trained in genetics, experimental embryology, paleontology, and evolutionary biology and was able to appreciate the links between them. In his 1953 essay "Epigenetics and evolution," Waddington analyzed the shortcomings of the population genetic account of evolution. He noted (p. 187) that the genetic approach to evolution had culminated in the Modern Synthesis, but he also noticed that "it has been primarily those biologists with an embryological background who have continued to pose questions."

Waddington's essay claimed that the Modern Synthesis failed to work in at least three areas. First (as has been stressed throughout this book), much variation appears to be regulated by the environment, not the inherited genotype. Second, as Goldschmidt had noted, the large taxonomic divisions of animals encompass an extent of physical differences that is not compatible with local races branching off. That is, accumulations of small mutations in a local group would not be sufficient to separate, say, amphibians from fish or reptiles from amphibians. Waddington noted that Goldschmidt's own hypotheses about how such divisions might take place were so unconvincing to geneticists that they obscured the very real cogency of Goldschmidt's arguments for the existence of large phenotypic gaps that could not be bridged by the accumulation of small allelic changes. Third, Waddington noted the different rates of evolution seen in the paleontological record.

Waddington (1953, p.190) claimed that in conventional evolutionary studies, the animal is considered either as a genotype (and is studied by geneticists) or as a phenotype (and is studied by ecologists and taxonomists). What is needed, he said, is an evolutionary study of those processes that transforms genotype into phenotype—the epigenetics of development. Following Goldschmidt, Waddington (pp. 190–191) declared:

> Changes in genotypes only have ostensible effects in evolution if they bring with them alterations in the epigenetic processes by which phenotypes come into being; the kinds of change possible in the adult form of an animal are limited to the possible alterations in the epigenetic system by which it is produced.

Waddington proposed two levels of evolution. First, there would be an internal evolution whereby changes in gene expression *in the embryo* would create the possibility for new variation. Such changes would have to be met by other embryonic changes in order for the organism to survive to birth. Second, there would be the natural selection on the organism itself.

Thus, Waddington hoped to combine evolution as a theory of bodies with evolution as a theory of change. He also hoped to combine the nascent field of developmental genetics (which he was helping to create) with the established science of population genetics. This respect for both traditions was characteristic of both Waddington and Schmalhausen and provided a paradigm for today's evolutionary developmental biology.

Evolution without Development

While the theoretical issues raised by Goldschmidt, Waddington, and Schmalhausen were being debated, the founders of the Modern Synthesis were reacting strongly against developmental views. Most of them argued that a theory of evolutionary change did not need to be underwritten by a theory of body construction (and, unfortunately, at that time the science of embryology was in no condition to provide such a theory). Moreover, to the architects of the Modern Synthesis, developmental biology was not only asking the wrong questions, but it was structurally incapable of asking the right ones. Morgan (1932) had rewritten the *history* of evolutionary biology in such a way that embryology was excluded, and then Ernst Mayr rewrote the *philosophy* of evolutionary biology so as to exclude embryology from having anything at all to say to evolutionary studies (see Amundson 2005; Levit and Meister 2006).

In her analyses of the origins of the Modern Synthesis, Polly Winsor (2003, 2006) points out that in order to keep evolutionary biology focused on intraspecific variation, Mayr invented a typological dichotomy that in fact did not exist. On one hand, Mayr postulated that a semi-mystical, scientifically discredited mode of thinking called "essentialism" was characteristic of embryologists and paleontologists. On the other hand, he felt there was a scientifically rigorous and philosophically valid way of approaching life called "population thinking," which is what geneticists do. Darwin, Mayr said, had shown that essentialism was wrong and founded the latter perspective.

But as Winsor points out, the type of essentialism Mayr evoked in the 1950s had no adherents in contemporary biology, and in fact biology won the battle against essentialism decades before Darwin. She further points out that this dichotomy was historically false and, despite its popularity, "is little more than a myth." Thus, even though Mayr did not believe that evolution was solely a matter of allele frequency changes, his philosophical analysis demanded that the only scientific focus of evolutionary biology be evolution within species.

Another fundamental assumption of the Modern Synthesis was that macroevolution (the origin of new species and higher taxa) could be explained by natural selection. Natural selection was thought to be so strong a force that the reason unrelated organisms looked alike was due to similar selection pressures, and not to genes inherited from a common ancestor.* In fact, Mayr (1966) and Dobzhansky (1955) explicitly rejected the possibility that widely shared characteristics were caused by homologous genes. As Mayr (1966) wrote:

> Much that has been learned about gene physiology makes it evident that the search for homologous genes is quite futile except in very close relatives (Dobzhansky 1955). If there is only one efficient solution for a certain func-

*As Amundson (2005) points out, the need to have interbreeding species prevented the discovery of homologous genes between species. He writes that Mayr's and Dobzhansky's belief in the strength of natural selection far overreached the data and the methodology. "Both Mayr and Dobzhansky were perfectly willing to predict that common characters among species were the result of adaptive convergence, not homologous genes, even though they both knew that the Mendelian blind spots prevented direct evidence for (or against) that prediction. They had faith in the power of natural selection to be able to produce any phenomenon that homologous genes could produce. ... It illustrates an aspect of scientific commitment that extends well beyond the data immediately at hand" (p. 217).

tional demand, very different gene complexes will come up with the same solution, no matter how different the pathway by which it is achieved.

Thus, Mayr believed that "eyes evolved independently at least 40 times in different groups of animals" (Mayr 1988, p. 72) and had converged on a common phenotype from many sources (Salvini-Plawen and Mayr 1977). As discussed in Chapter 9, however, evolutionary developmental biology has a much more parsimonious explanation for eye development.

Even today, the field of quantitative genetics largely derides developmental biology as unnecessary to predict the phenotypic outcome of the interactions among several genes and environmental circumstances. Lynch (2007) and de Jong (2005) have written that quantitative population genetics has everything needed to explain phenotypic evolution and that no new synthesis is needed. They claim that the heritability and selection coefficient accurately predict the response of a single trait to selection, and matrices of variance and covariance predict the response of any suite of traits (or measurements of traits) to selection. Since quantitative genetics is a theory about phenotypic means and variances and not about the creation of new phenotypes, it is unnecessary to know anything about how the phenotype arises, nor about how genes produce the phenotype. But as Pigliucci (2007; Pigliucci et al. 2006) has noted, Lynch and de Jong miss the point. "What the modern synthesis has not given us is a theory of form, and applying population genetics to genomics—as valuable an exercise as that is in its own right—isn't going to give us one either" (Pigliucci 2007).

Not only is the non-developmental perspective an incomplete view of evolution, it becomes an inaccurate one when the variances are not linear. Brandon and Nijhout (2007; Nijhout 2003, 2007) have shown that G × E (*g*enotype with *e*nvironment) interactions are not linear and are dependent on developmental history (i.e., the temporal and spatial contexts in which they occur). Quantitative genetics (because it is statistical) is a linear theory that can work with nonlinear relationships between the phenotypes of parent and offspring only if there is an equation that linearizes the relationship. In certain developmental systems there does not seem to be any linear relationship; rather, non-linearity is an important component of the phenotype's development. Moreover, Brandon and Nijhout see quantitative genetics as largely descriptive (and not explanatory) because it essentially says that the mean trait value in next generation will be the same as that of the current generation, modified only by its selection differential (roughly: difference of parental mean from population mean) and its heritability.

The exclusion of embryology from the Modern Synthesis may have been a necessary step in the formation of an evolutionary theory. Until the 1970s, embryology did not have much to say about the role of genes in development. Meanwhile, the fusion of genetics and population biology allowed the formation of a theory of evolutionary change that could explain and predict the maintenance of diversity within populations. However, in excluding developmental biology, the Modern Synthesis made assumptions that were unwarranted and have been shown to be false. The data from evolutionary developmental biology show that phenotype is not the mere unrolling of genotype, that similarities in extremely divergent organisms can be explained by homologous genes, and that embryology need not be essentialist. Moreover, techniques invented since the 1950s now allow us to study evolution outside of interbreeding populations, thus removing a major constraint of the Modern Synthesis.

As Jean Rostand (1962, p. 94) so presciently noted, "It is always a windfall for truth when well-established facts collide with a well-constructed theory." Integrating new facts from developmental biology into evolutionary theory is the ongoing work of evolutionary developmental biology.

References

Amundson, R. 2000. Embryology and evolution 1920–1960: Worlds apart? *History and Philosophy of the Life Sciences* 22: 35–352.

Amundson, R. 2001. Adaptation and development: On the lack of common ground. In S. Orzack and E. Sober (eds.), *Adaptation and Optimality*. Cambridge University Press, Cambridge, pp. 303–334.

Amundson, R. 2005. *The Changing Role of the Embryo in Evolutionary Thought: Structure and Synthesis*. Cambridge University Press, Cambridge.

Bateman, R. M. and W. A. DiMichele. 2002. Generating and filtering major phenotypic novelties: neoGoldschmidtian saltation revisited. In Q.C.B. Cronk, R. M. Bateman and J. A. Hawkins (eds.), *Developmental Genetics and Plant Evolution*. Taylor and Francis, London, pp. 109–159.

Brandon, R. N. and H. F. Nijhout. 2007. The empirical non-equivalence of genic and genotypic models of selection: A decisive refutation of genic selectionism and pluralistic genic selectionism, *Philos. Sci.* 73: 277–297.

de Jong, G. 2005. Evolution of phenotypic plasticity: Patterns of plasticity and the emergence of ecotypes. *New Phytol.* 166: 101–118.

De Beer, G. R. 1954. *Embryos and Ancestors*, Rev. Ed. Oxford University Press, Oxford.

Dobzhansky, Th. 1955. *Evolution, Genetics, and Man*. Wiley, New York.

Goldschmidt, R. B. 1940. *The Material Basis of Evolution*. Yale University Press, New Haven.

Goldschmidt, R., A. Hannah and I. Piternick. 1951. The podoptera effect in *Drosophila melanogaster. Univ. Calif. Publ. Zool.* 55: 67–294.

Gould, S. J. 1980. The return of hopeful monsters. *Natural History* 86 (June/July): 22–30. Reprinted in *The Panda's Thumb*. 1992. Norton, New York.

Hamburger, V. 1980) Embryology and the Modern Synthesis in evolutionary theory. In E. Mayr and W. Provine (eds.), *The Evolutionary Synthesis: Perspectives on the Unification of Biology*. Cambridge University Press, New York, pp. 97–112.

Harrison, R. G. 1937. Embryology and its relations. *Science* 85: 369–374.

Levit, G. S. and K. Meister. 2006. The history of essentialism vs. Ernst Mayr's "Essentialism story": A case study of German idealistic morphology. *Theory Biosci.* 124: 281–307.

Lillie, F. R. 1927. The gene and the ontogenetic process. *Science* 64: 361–368.

Lynch. M . 2007. *The Origins of Genomic Architecture*. Sinauer Associates, Sunderland, MA.

Morgan, T. H. 1932. *The Scientific Basis of Evolution*. Norton, New York.

Mayr, E. 1966. *Animal Species and Evolution*. Harvard University Press, Cambridge, MA.

Mayr, E. 1980. Prologue: Some thoughts on the history of the evolutionary synthesis. In E. Mayr and W. B. Provine (eds.), *The Evolutionary Synthesis: Perspectives on the Unification of Biology*. Harvard University Press, Cambridge, MA, pp. 9–10.

Mayr, E. 1988. *Toward a New Philosophy of Biology*. Harvard University Press, Cambridge, MA.

Nijhout, H. F. 2003. On the association between genes and complex traits. *J. Invest. Derm. Symp.* 8: 162–163.

Nijhout, H. F. 2007. Complex traits: Genetics, development, and evolution. In R. Sansom and R. Brandon (eds.), *Integrating Evolution and Development: From Theory to Practice*. MIT Press, Cambridge, pp. 93–112.

Pigliucci, M. 2007. Postgenomic musings. *Science* 317: 1172–1173.

Pigliucci, M., C. J. Murren and C. D. Schlichting. 2006. Phenotypic plasticity and evolution by genetic assimilation. *J. Exp. Biol.* 209: 2362–2367.

Rostand J. 1962. *The Substance of Man*. Doubleday, Garden City, NJ.

Salvini-Plawen and E. Mayr. 1977. On the evolution of photoreceptors and eyes. *Evol. Biol.* 10: 207–263.

Waddington, C. H. 1953. Epigenetics and evolution. In R. Brown and J. F. Danielli (eds.), *Evolution* (SEB Symposium VII). Cambridge University Press. Cambridge, pp. 186–199.

Wallace, B. 1986. Can embryologists contribute to an understanding of evolutionary mechanisms? In W. Bechtel (ed.), *Integrating Scientific Disciplines*. Martinus Nijhoff, Dordrecht, pp. 149–163.

Winsor, M. P. 2003. Non-essentialist methods of pre–Darwinian taxonomy. *Biol. Philos.* 18: 387–400.

Winsor, M. P. 2006. The creation of the essentialism story: An exercise in metahistory. *Hist. Philos. Life Sci.* 28: 149–174.

Appendix D

Epigenetic Inheritance Systems

The Inheritance of Environmentally Induced Traits

Ecological evolutionary developmental biology, or "eco-evo-devo," has the data to bring two controversial alternative inheritance systems back into the discussion of evolutionary biology. The first idea concerns the inheritance of environmentally acquired traits, an ancient idea usually associated with Jean-Baptiste Lamarck (1744–1829), but which was also used by Charles Darwin and many other Victorian naturalists. The second controversial idea usually goes by the name "genetic assimilation," and it concerns the genetic fixation of an adaptive, plastic response into the genome. In this hypothesis, a response that was once part of a phenotypically plastic repertoire is now part of the normative genetic "program."

If ecological evolutionary developmental biology is going to shed this far-too-cumbersome moniker and become mainstream evolutionary biology, it has to deal with these issues and show why they were considered wrong and why new data suggest that these theories need to be reconsidered as containing important insights into evolution.

The Ghost of Lamarck

Epigenetic inheritance systems recall the specter of a banished ghost—Lamarckian inheritance. The year 2009 is not only the bicentenary of Darwin's birth and the centenary of the Woltereck and the Johannsen papers described in Chapter 1, it is also the bicentenary of Lamarck's *Philosophie Zoologique*.

Lamarckian inheritance was based on physiology, behavior, and phenotypic plasticity: if you used your muscles, they grew bigger. Moreover, such muscular changes would be passed on to subsequent generations, so that the offspring of runners would have stronger leg muscles. In the absence of any notion of genes, this physiological model seemed appropriate. Lamarck's two "laws" of evolution were:

First law: In every animal that has not passed the limit of its development, more frequent and continuous use of any organ gradually strengthens, develops, and enlarges that organ and gives it a power proportional to the length of time it has been so used; whereas the permanent disuse of any organ weakens and deteriorates it by imperceptible increments, progressively diminishing its functional capacity until it finally disappears.

Second law: All the acquisitions or losses wrought by nature on individuals through the influence of their environment and the predominant use or permanent disuse of any organ, are preserved by reproduction, provided that the acquired modifications are common to both sexes, or at least to the individual that produces the young.

Lamarck concludes:

> Nature has produced all the species of animals in succession, beginning with the most imperfect or simplest, and ending her work with the most perfect, so as to create a gradually increasing complexity in their organisation; these animals have spread at large throughout all the habitable regions of the globe, and every species has derived from its environment the habits that we find in it and the structural modifications which observation shows us.

Here, Lamarck proposes plasticity as a model for inheritance, in which the organism, influenced by its environment, alters its body (and those of its descendants) through use and disuse.

Darwin also believed that it might be possible for acquired characteristics to be inherited. He speculated that the body's tissues produced "gemmules" that circulated through the body and entered into the germ cells (Darwin, 1883). Thus, if a tissue were bigger due to more usage in a more demanding environment, it could produce more gemmules, causing the phenotype to be transmitted into germ cells. The amount and the type of gemmule produced would determine its inheritance.

This notion of diffusible gemmules was disproved by Darwin's cousin, Francis Galton. Galton assumed that Darwin's postulated gemmules had to be transferred through the blood. So he transfused blood between gray rabbits and lop-eared rabbits until he felt the lop-eared rabbits were totally transfused with gray rabbit blood and vice versa. However, each type of rabbit continued to breed true to its type—no influence of blood-borne gemmules was ever observed. A little later, August Weismann proposed that only the germline counted in heredity, and that the germline was separate from the somatic lineages of cells that formed the body. Therefore, anything that affected the individual could not influence heredity if the germline was not affected. Weismann cut off the tails of mice for nineteen generations and showed that a tailless race did not develop. Commenting on these experiments, Thomas Huxley (1884) pointed out that circumcision among Jews and Muslims had not produced men with no foreskins.

However, numerous examples of inheritance of acquired characteristics *had* been reported. By rearing butterfly pupae at different temperatures, Standfuss and Fischer (quoted in Thomson 1908; p. 215) obtained butterflies that resembled related species, and the temperature-induced phenotype was transmitted to their progeny.

Even into the twentieth century, reports of inheritance of acquired characteristics occasionally were published (Lewontin and Levins 1976). For instance, geneticist Viktor Jollos (1934) reported that heat resistance induced by heat shock could be transmitted to offspring. As mentioned in Appendix A, numerous examples of the inheritance of acquired characteristics were proposed by the Viennese biologists of the Prater Vivarium. In 1920, Austrian biologists Paul Kammerer and Eugen Steinbach reported morphological changes in offspring and grandoffspring of rats that were exposed to high temperatures (Gliboff 2005; Logan 2007).

As genetic studies found more and more examples of Mendelian inheritance, examples of inherited acquired traits became suspect. Indeed, Lindegren (1949), surveying inheritance in the fungus *Neurospora*, wrote that about two out of every three newly discovered variants did not show Mendelian segregation. Therefore, he said, such variants were usually discarded. A similar assessment was made by Benkemoun and Saupe (2006), who suggested that these "anomalies" to Mendelian segregation might be caused by epigenetic inheritance, and that "we should have a closer look before putting them in the autoclave." So it is possible that human bias as to what can be best studied has prevented us from seeing further examples of epigenetic inheritance (see Jablonka and Raz 2009).

Epialleles and Germline Transmission of Alternative Chromatin Structures

As shown by examples from this book, epigenetic variation and inheritance is a reality. The basis of the barrier to the acquisition of acquired characteristics has been that changes to the body did not affect the genes in the germ cells. We now know this to be false for some traits, even if true for most. Environmentally induced DNA methylation and other chromatin alterations can be passed through the germline. Therefore, environmentally induced phenotypes can be passed from generation to generation by chromatin modifications. The table on the following two pages lists some of the documented epigenetic inheritance systems in animals.

New mechanisms of epigenetic inheritance are still being discovered. For example, the mammalian germline contains small, non-coding RNAs (called *piwi*-interacting RNAs) that can silence DNA regions. The inheritance of a variant of the cell proliferation gene, *Kit*, in mice is mediated epigenetically through one of these RNAs (Rassoulzadegan et al. 2006). So, while the genetic component of our evolution is mostly Mendelian, there is an epigenetic component that allows for the inheritance of certain acquired traits. This may be very important in those lineages where there is not a strict and early segregation of the germline (Buss 1987). Future research will need to ascertain how these genetic and epigenetic systems interact and how natural selection mediates which one predominates in a particular circumstance.

Plasticity-Driven Evolution

With the rediscovery of Mendel's laws, Johannsen (1909) made the distinction between genotype and phenotype, claiming that phenotype was formed through the combination of genotype and environmental conditions. Also in 1909,

SOME EXAMPLES OF TRANSGENERATIONAL EPIGENETIC INHERITANCE IN ANIMALS

Organism	Phenotype	Gene involved
Caenorhabditis elegans (Nematode)	Small and dumpy appearance	RNAi of *ceh13*
	Silencing of green fluorescent protein (GFP)	RNAi of GFP transgene
	Various effects, not reported	RNAi of 13 genes
Daphnia pulex (Water flea)	Expression of *G6PD S* and *N* variants	*G6PD* or its regulator
Drosophila melanogaster (Fruit fly)	Modifying ability of Y chromosome	Imprintor gene interaction
	Ectopic outgrowth in eyes	*Kr* ($Kr^{If\text{-}1}$ allele), (*TrxG* mutation)
	Eye color	Transgenic *Fab7* flanking *lacZ* and mini-*white* reporter transgenes
	(a) Eye color (due to derepression of mini-*white* reporter gene cloned downstream to transgenic *Fab7*) (b) Suppression of wing deformations	(a) Activation state of endogenous and transgenic *Fab7* elements containing CMM (b) Derepression of *sd*
	Susceptibility to tumors	Probably several loci, including heritable epigenetic variation in the *ftz* promoter
Ephestia kuehniiella (Moth)	Reversion of shortened antennae and associated mating disadvantage	Suppressor of *sa* (sa^{WT})
Schistocerca gregaria (Locust)	Gregarious behavior; body color	Not known
Myzus persicae (Peach potato aphid)	Loss of insecticide resistance	Probably amplified resistance genes
Vulpes vulpes (Fox)	Piebald spotting	Activation state of *Star* gene

Transgenerational persistance	Cause	Mechanism
Over 40 generations	Feeding with bacteria expressing dsRNA targeting *ceh13*	Chromatin remodeling; RNAi-mediated
At least 40 generations	Feeding with bacteria expressing dsRNA targeting GFP	Chromatin remodeling; RNAi-mediated
At least 10 generations	Feeding with bacteria expressing dsRNA targeting the 13 genes	Chromatin remodeling; RNAi-mediated
Spontaneous reversion rate between the two forms 1 in 10 and 1 in 2	Spontaneous and glucose induced. Presence of *S* form related to stressful conditions	Not known
11 generations	Transient effect of imprintor gene	Chromatin marking
At least 13 generations	Geldanamycin treatment given to $Kr^{If\text{-}1}$ strain, or transient presence of *TrxG* mutation vtd^3	Chromatin marking involved
At least 4 generations	Transient presence of GAL4 protein	Probably chromatin inheritance
(a) At least 4 generations (more than 4 years) (b) Not specified (stability lower and hard to detect)	(a) High temperature (b) High temperature	Probably chromatin inheritance
Increased tumorigenicity, 2 generations; modified *ftz* methylation, at least 1 generation	Crossing with hop^{Tum1} and Kr^1 mutants	DNA methylation; chromatin inheritance
Up to 5 generations; incompletely inherited from mother but almost fully inherited from father	Exposure of larva and pupa to lithium ions; alternate electrical field; or 25°C at late 5th instar larval and pupal phases	Probably chromatin inheritance
Several generations	Material in egg foam secreted by females	Not known
Stable inheritance of lost resistance in clones which have amplified DNA	Induced by DNA amplification	DNA methylation involved
Star (semidominant allele) activated in ~ 1% of domesticated animals; inherited for more than 2 generations	Spontaneous in foxes raised on fur farms; hormonal stress suggested	Possibly heritable chromatin modification

(*continued on next page*)

SOME EXAMPLES OF TRANSGENERATIONAL EPIGENETIC INHERITANCE IN ANIMALS (continue

Organism	Phenotype	Gene involved
Mus musculus (Mouse)	Probability of developing yellow coat color, obesity, susceptibility to diabetes and cancer	A^{vy} (see Chapter 2)
	Reduced body weight, reduced level of proteins involved in sexual recognition and possibly higher mortality between birth and weaning	Not specified, but connected to major urinary protein (*Mup*) and olfactory marker protein (*Omp*) genes
	Probability of kinked tail shape	*Axin-fused*
	White-spotted tail and feet	*Kit*
	Repression of recombination of the *LoxP* element and concomitant methylation	Transgenic *LoxP* and surrounding chromosomal sequences
	Genome stability	Many
	Glucose intolerance, fat storage	*PPARα* (see Chapter 7); *HGR*
	Tendency to develop tumors	Elevated expression of gene coding for LF (an estrogen-responsive protein) and *c-fos*
Rattus norvegicus (Rat)	Modified serotonin content in immune cells	Not specified
	Increased expression of genes coding for metabolic factors	Promoters of *PPARα* and *GR* in liver; increased expression of other RNAs
	Decreased spermatogenic capacity; elevated incidence of tumor, prostate and kidney diseases, serum cholesterol levels, immune system abnormalities; premature aging and male mating disadvantage	Methylation state of 15 different DNA sequences. Reduced *expression of ankyrin 28, Ncstn, Rab12, Lrrn6a* and *NCAM1* found in vinclozolin group as well as increased expression of *Fadd, Pbm1b, snRP1c* and *Waspip*
	Altered glucose homeostasis	Not specified
Homo sapiens (Human)	Cardiovascular mortality and diabetes susceptibility	Imprinted tandem repeat upstream of *INS-IGF2-H19* region
	Predisposition for tumor formation and hereditary nonpolyposis colorectal cancer	*MSH2*
	Angelman and Prader-Willi syndromes	15q11-q13

Source: E. Jablonka and G. Raz. 2009. Transgenerational epigenetic inheritance: Prevalence, mechanisms, and implications for the study of heredity and evolution. *Quart. Rev. Biol.* In press. See original paper for references as well as for transgenerational epigenetic inheritance in fungi, plants, and bacteria. Quoted with the permission of E. Jablonka.

Transgenerational persistance	Cause	Mechanism
At least 2 generations of *agouti* epigenotype	Spontaneous, but affected by diet	Chromatin marking, including DNA methylation
Preliminary results suggest transmission of the traits to the F_2 generation	Induced by transfer of mouse pronuclei at the one-cell stage to eggs of a different genotype; traits are transmitted to most offspring through the male germline	DNA methylation assumed to be involved
Spontaneous rate of inactivation 6%; rate of reactivation 1%	Spontaneous; influenced by diet. Injection of hydrocortisone during spermiogenesis reduces penetrance	Chromatin marking including DNA methylation
2 generations of outbred crossing; 6 generations of inbred crossing between paramutants	Transient presence of Kit^{tm1Alf} mutation	RNA inheritance; RNAi involved
Methylated state maintained for at least 3 generations	Transient presence of *Sycp1-Cre*; exposure of wild type to recombinase activity	DNA methylation
At least 3 generations	Irradiation	Chromatin methylation
2 generations; some effect in the third generation	Diet, endocrine disruptors	DNA methylation
Apparent in the F_1 and F_2 generations	Induced by diethylstilbestrol during pregnancy	DNA methylation probably involved
2 generations	Intramuscular administration of β-endorphin during 19th day of pregnancy	Not specified
At least 2 generations	Protein-restricted diet during pregnancy	DNA methylation
At least 4 generations; transmission through the male germline	Vinclozolin or methoxychlor treatment during gestation	DNA methylation
Parental and 3 generations	Low-protein diet in mother, from day 1 of pregnancy through lactation	Not specified; DNA methylation probably involved in F_1 animals
At least 2 generations	Food availability during childhood growth period	Possibly methylation; transmitted through male germline
3 generations	Not known	DNA methylation
Inherited from paternal grandmother (no imprint erasure in the father).	Spontaneous	DNA methylation

Epigenetic Inheritance in Bacteria: The Selection of Epigenetic Variation

In organisms without germ cells, there is definitely proof in principle (if not in nature) that hereditable epigenetic diversity can create sufficient variation on which selection can act. Genetically identical bacteria can inherit patterns of antibiotic resistance based on gene expression differences rather than allelic differences. In 2004, Balaban and colleagues demonstrated that when genetically identical antibiotic-sensitive strains of *E. coli* are subjected to antibiotic stress, most will die. However, a very small percentage will survive. These "persistent" bacteria had not acquired any gene for antibiotic resistance, nor had any of their genes mutated to cause such resistance. In fact, when these cells were cultured, most all of their progeny were sensitive to the antibiotic. There was pre-existent heterogeneity in the population, not of alleles, but of gene expression for antibiotic resistance.

Adam and colleagues (2008) confirmed and extended those results, using three different antibiotics at different concentrations. They showed that within a single genetically identical strain of *E. coli* there were gene expression differences such that genes that could provide antibiotic resistance were overexpressed in some cells. Moreover, bacteria that survived a low dosage of ampicillin produced progeny that could survive higher doses. About 20% of the cells survived 1 μg/ml ampicillin; this percentage was too high to be accounted for by spontaneous mutation. When re-plated on the same dosage of ampicillin, 50% of these cells survived. Successive exposures to increasing amounts of ampicillin yielded *E. coli* that could survive at 10 μg/ml of the antibiotic. Their reversion frequency remained high (over 90%), showing that this resistance was not due to mutation, and no plasmids were found to be causing this effect. While genetically identical to the cells that died from ampicillin exposure, the persistent cells had higher levels of transcription from two genes (those encoding β-lactamase and glutamate decarboxylase), which block ampicillin's lethal effects. Similar epigenetic inheritance was seen with other antibiotics. Therefore, antibiotic drug resistance—one of the most medically important phenotypes—can be conferred and selected not only by the selection of pre-existing allelic mutations but also by the selection of pre-existing epigenetic variation.

Woltereck, working with parthenogenetically pure lines of *Daphnia*, introduced the concept of the reaction norm to illustrate the phenotypic differences that a given genotype could display in response to environmental cues. Five years later, the Swedish biologist H. Nilsson-Ehle (1914) placed this into the larger concept of phenotypic plasticity to define the observed variations exhibited by a single genotype due to either environmental or stochastic factors. This view that the genotype represented a reaction norm became the dominant view of the Russian school of evolutionary biology that coalesced around Chetverikov in the 1920s (Adams 1980a). Indeed, in 1926, the young Dobzhansky equated the reaction norm with the gene, claiming that the reaction norm was the actual Mendelian unit of inheritance (see Sarker and Fuller 2003).

But such plasticity was not a major part of the British and American schools of evolutionary biology. When T. H. Morgan (1932) claimed that genetics provided the only scientific way to study evolution, he meant the *Drosophila* genetics that his laboratory had founded, not the more plastic notions of heredity operating in Europe. When Morgan's genetics merged with the mathematical models of Fisher, Haldane, and Wright to form the Modern Synthesis, the phenotype was thought to be a direct readout of the genotype. Any phenotypic variation due to environmental factors was given a "genetic" explanation by the use of such terms as "penetrance" and "expressivity" (Vogt 1926; Sarkar 1999). Although Dobzhansky originally stressed the importance of the reaction norm concept in evolution, by 1937 he had abandoned this approach, defining evolution as a "change in the genetic composition of populations" and describing the reaction norm as a property of the population rather than of the individual (see Sarkar and Fuller 2002).

One of the first people to realize the importance of plasticity for evolution was Lancelot Hogben. Hogben (1933) looked at the reaction norms in flies having two genotypes. He found that when he graphed the number of eye facets versus temperature for these two strains, he did not obtain identical or parallel lines. From this, he concluded that there was an interdependence of genes and environment and that phenotype production was a complex relationship between these factors. Although this analysis was largely overlooked in favor of Haldane's linear models, this analysis of genome-environment (G × E) interactions was further developed in the 1970s and became important in newer models of gene-environment interactions (see, for example, Lewontin 1974).

Meanwhile, Russian biologists continued to integrate the notion of phenotypic plasticity into their accounts of evolution. In one of his last publications, Alexsei Nikolaeovich Severtsov (1935), the founder of the Russian school of evolutionary morphology, wrote:

> At the present time, we morphologists do not have the full theory of evolution. It seems to us that in the near future, ecologists, geneticists, and developmental biologists must move forward to create such a theory, using their own investigations, based on ours…

To Severtsov, a complete theory of evolution must causally explain the morphological changes seen in paleontology through the mechanisms of genetics, ecology, and embryology. He felt that genetics alone could not provide the mechanism because it did not involve the "how" of evolution (Adams 1980b). Only ecology and embryology could do that. To this end, Severtsov and others worked on evolutionary syntheses that included developmental and ecological considerations. Georgii Gause, best known for his competitive exclusion principle in ecology, wrote one of the first statements of the genetic fixation of a plastic phenotype. He claimed that adaptive phenotypes can emerge as physiological responses to the environment and that these modifications can become genotypically fixed by mutations if there is a fitness advantage for selection (Gause 1947).

This integration of embryology, development, and ecology became the project of the Institute of Evolutionary Morphology, headed by Severtsov's student Ivan Ivanovich Schmalhausen. For Schmalhausen (1949), such plasticity was critical for evolution: "every genotype is characterized by its own specific norm of reaction, which includes adaptive modifications of the organism to different environments." His book, *Factors of Evolution: The Theory of Stabilizing Selection*, provides numerous examples of polyphenisms that appear to have been genetically fixed. Independently of Schmalhausen's work, Conrad Waddington (1942, 1952) linked the fixation of developmental plasticity to genetic variation and tried to incorporate this idea into evolutionary biology. Schmalhausen called his idea "stabilizing selection";

Waddington named his notion "genetic assimilation." Waddington had to present his ideas against a Western background that was suspicious or ignorant of developmental plasticity, while Schmalhausen had to present his ideas against a Soviet background that was suspicious of genes* (see Gilbert 1994).

In the West, a reaction against Lysenkoism made geneticists very wary of the notion of phenotypes being produced by the environment (Lindegren 1966; Sapp 1987). As mentioned in Appendix A, attempts to look at non-genomic contributions to development were a casualty of the Cold War. Indeed, the founders of the Modern Synthesis were unimpressed by genetic assimilation and other such theories based on plasticity. George Gaylord Simpson (1953) dismissed genetic assimilation as a "relatively minor" contribution to evolution, while Ernst Mayr (1963) and Theodosius Dobzhansky (1970) interpreted genetic assimilation as a failed attempt to support a model of Lamarckian/Lysenkoist inheritance. Certainly, a title such as that of Waddington's 1953 paper "Genetic assimilation of an acquired character" would raise a number of red flags.

The tendency to marginalize or dismiss plasticity-driven evolutionary schemes seems to have been transmitted to contemporary evolutionary biologists. Orr (1999) dismisses the idea of genetic assimilation as "a baroque hypothesis," and de Jong (2005) dismisses genetic assimilation as a mechanism of evolution because of what she feels to be "the lack of convincing examples." The data presented in Chapter 10 should silence both these claims.

Other Epigenetic Inheritance Systems

Jablonka and Lamb (2005) have noted that there are many inheritance systems working simultaneously in the formation of phenotype. We have mentioned several in this volume, and one of these alternative systems of inheritance (discussed in Chapter 3) concerns the transmission of symbiotic systems from generation to generation. Given the preponderance of developmental symbioses, the notion of the "organism" becomes less like an autopoietic individual and more like an ecosystem. Consider, for instance, that the hindgut of an individual worker of the termite species *Nasutitermes ephratae* is about 1 μm and contains over 200 types of bacteria,

* Lysenkoism was critical to the negative reception of both hypotheses. Waddington was viewed as making concessions to the Marxists, while Schmalhausen was seen as making concessions to the capitalists. Schmalhausen was one of the very few biologists (along with D. A. Sabinin and A. N. Formozov) who openly disagreed with Lysenko in 1947. His Order of the Red Banner did not save Schmalhausen from Lysenko's vitriolic public attack; indeed Lysenko crafted much of his infamous 1948 speech at the All-Union Academy of Agricultural Science against Schmalhausen (Adams 1980b). Schmalhausen was stripped of all his appointments and assistants., after which he retreated to his dacha and independently studied vertebrate embryology. (Sabinin committed suicide.) As Lysenko lost power, Schmalhausen was gradually brought back into Russian biology. His last years were spent re-formulating his evolutionary theories into a cybernetic framework (see Wake 1986; Levit et al. 2006). In the West, it probably didn't help that many of the major proponents of plasticity-driven evolution were either Russian or had well known left-wing sympathies.

representing at least a dozen phyla. It is this collection of bacteria, not the termite, that digests wood. When termites of a related species, *N. takasagoensis*, are given antibiotics, they lose their cellulose-digesting enzymes (Tokuda and Watanabe 2007; Warnecke et al. 2007). Thus, what constitutes an "individual" is not just the genetical insect, but the insect-symbiont system and its hundreds of distinct genotypes.

Symbiosis has numerous implications for evolution, including the co-evolution and co-development of host and symbionts; the integration of this host-symbiont complex with the ecosystem (*Nasutitermes*, for instance, are ecologically critical for digesting dead plants); the production of variation; and the mutual advantages that can be enhanced by natural selection when two or more lineages are associated across many generations (Moran 2007). In many instances, symbionts provide their hosts with protection against natural enemies (Oliver et al. 2005; Scarborough et al. 2005) or against thermal stress (Russell and Moran 2006). In addition, the ability of one organism to complement functions of another can enable the loss of redundant genes from one of the partners, resulting in obligate symbiosis (as we saw in Chapter 3, certain species of insects cannot undergo normal morphogenesis without the symbiont species). Natural selection may be selecting "relationships" between organisms rather than the organisms (or their genes) themselves.

As we have seen, there are many ways to obtain hereditable variation and to expand it in a population. Populations can evolve through allelic substitutions, environmentally induced epialleles, *cis*-regulatory substitutions that result in heterochrony, heterotopy, and heterometry, and the genetic fixation of environmentally induced responses. Modularity and gene duplication have important consequences, and interactions between developmental units constrain and bias the direction of evolution. Developmental plasticity and symbioses can even provide new "targets" for natural selection.

It is likely that all of these mechanisms function during evolution; indeed, philosopher of biology Richard Burian (1988) has proposed that this plurality of mechanisms is itself a product of evolution:

> I conclude that the lack of a single dominant disciplinary matrix in evolutionary biology is a consequence of the nature of evolutionary phenomena, and particularly of the role of historical accidents in affecting the evolutionary success, failure, and transformation of lineages. For this reason, I submit, we would be foolish to expect the unification of evolutionary theory within a single paradigm—and, what's more, we should count the failure to achieve such unification as a Good Thing.

This does not mean that there are many theories of evolution that all have the same scientific support. Evolution by natural selection, the mainstay of the Darwinian synthesis, is not in question. But we have realized that animals have been able to evolve in many different ways. Most of this diversification involves the mechanisms used for the generation of variation, and that is where development plays its enormous roles. Natural selection works on the array of variants that development offers it. Evolution by natural selection is not going away. It's just getting a whole lot richer.

References

Adams, M. B. 1980a. Sergei Chetverikov, the Koltsov Institute, and the evolutionary synthesis. In E. Mayr and W. B. Provine (eds.), *The Evolutionary Synthesis: Perspectives on the Unification of Biology*. Harvard University Press, Cambridge, MA, pp. 242–278.

Adams, M. B. 1980b. Severtsov and Schmalhausen: Russian morphology and the evolutionary synthesis, In E. Mayr and W. B. Provine (eds.), *The Evolutionary Synthesis: Perspectives on the Unification of Biology*. Harvard University Press, Cambridge, MA, pp. 193–225.

Adams, M., B. Murali, N. O. Glenn and S. S. Potter. 2008. Epigenetic inheritance based evolution of antibiotic resistance in bacteria. *BMC Evol. Biol.* 8: 52. DOI: 10.1186/1471-2148-8-52.

Balaban, N. Q., J. Merrin, R. Chait, L. Kowalik and S. Leibler. 2004. Bacterial persistence as a phenotypic switch. *Science* 305: 1622–1625.

Benkemoun, L. and S. J. Saupe. 2006. Prion proteins as genetic material in fungi. *Fungal Genet. Biol.* 43: 789–803.

Burian, R. M. 1988. Challenges to the Evolutionary Synthesis. *Evol. Biol.* 23: 247–269.

Buss, L. W. 1987. *The Evolution of Individuality*. Princeton University Press, Princeton, NJ.

Darwin, C. 1883. *The Variation of Animals and Plants under Domestication*, 2nd ed., Vol. 2. Appleton, New York.

de Jong, G. 2005. Evolution of phenotypic plasticity: Patterns of plasticity and the emergence of ecotypes. *New Phytol.* 166: 101–118.

Dobzhansky, Th. 1970. *Genetics of the Evolutionary Process*. Columbia University Press, New York (pp. 210-211).

Gause, G. F. 1947. Problems of evolution. *Transact. Ct. Acad. Sci.* 37: 17–68.

Gilbert, S. F. 1994. Dobzhansky, Waddington and Schmalhausen: Embryology and the Modern Synthesis. In M. B. Adams (ed.), *The Evolution of Theodosius Dobzhansky: Essays on His Life and Thought in Russia and America*. Princeton University Press, Princeton, NJ, pp. 143–154.

Gliboff, S. 2005. "Protoplasm … is soft wax in our hands": Paul Kammerer and the art of biological transformation. *Endeavor* 29: 162–167.

Haraway, D. J. 1979. The biological enterprise: Sex, mind, and profit from human engineering to sociobiology. *Rad. Hist. Rev.* 20: 206–237.

Hogben, L. 1933. *Nature and Nurture*. W. W. Norton, New York.

Huxley, T. H. 1884. Quoted in *The Medical Record*, May 31, 1884, p. 628.

Jablonka, E. and M. J. Lamb. 1995. *Epigenetic Inheritance and Evolution: The Lamarckian Dimension*. Oxford University Press, Oxford.

Jablonka, E. and M. J. Lamb. 2005. *Evolution in Four Dimensions*. MIT Press, Cambridge, MA.

Jablonka, E. and G. Raz. 2009. Transgenerational epigenetic inheritance: Prevalence, mechanisms, and implications for the study of heredity. *Q. Rev. Biol.* In press.

Johannsen, W. 1909. *Elemente der exacten Erblichkeitslehre*. Gustav Fischer, Jena.

Jollos, V. 1934. Inherited changes produced by heat-treatment *in Drosophila melanogaster*. *Genetica* 16: 476–494.

Lamarck, J. B. 1809. *Philosophie zoologique, ou exposition des considérations relatives à l'histoire naturelle des animaux*. Transl. H. Eliot. Macmillan 1914 (p. 113; 126). Online at http://www.ucl.ac.uk/taxome/jim/Mim/lamarck6.html.

Levit, G. S., U. Hossfeld and L. Olsson. 2006. From the "Modern Synthesis" to cybernetics: Ivan Ivanovich Schmalhausen and his research program for a synthesis of evolutionary and developmental biology. *J. Exp. Zool.* 306B: 89–106.

Lewontin, R. C. 1974. The analysis of variance and the analysis of causes. *Am. J. Human Genet.* 26: 400–411.

Lewontin, R. C. and R. Levins. 1976. The problem of Lysenkoism. In H. Rose and S. Rose (eds.), *The Radicalisation of Science*. Macmillan, London, pp. 32–65.

Lindegren, C. C. 1949. *The Yeast Cell, its Genetics and Cytology*. Educational Publishers, St. Louis.

Lindegren, C. C. 1966. *The Cold War in Biology*. Planarian Press, Ann Arbor, MI.

Lysenko, T. 1948. Quoted in Lindegren 1966, op. cit., p. 57.

Logan, C. A. 2007 Overheated rats, race, and the double gland: Paul Kammerer, endocrinology, and the problem of somatic induction. *J. Hist. Biolo.* 40: 683–725.

Mayr, E. 1963. *Animal Species and Evolution*. Harvard University Press, Cambridge, MA.

Moran, N. A. 2007. Symbiosis as an adaptive process and source of phenotypic complexity. In J. C. Avise and F. J. Ayala (eds.), *In the Light of Evolution*, Volume 1, *Adaptation and Complex Design*. National Academies Press, Washington, DC, pp. 165–182.

Morgan, T. H. 1932. *The Scientific Basis of Evolution*. Norton, New York.

Nilsson-Ehle, H. 1914. Vilka erfarenheter hava hittills vunnits rörande möjligheten av växters acklimatisering? *Kunglig Landtbruksakadamiens. Handlingar och Tidskrift* 53: 537 –572.

Oliver, K. M., N. A. Moran and M. S. Hunter. 2005. Variation in resistance to parasitism in aphids is due to symbionts and not hist genotype. *Proc. Natl. Acad. Sci. USA* 102: 12795–12800.

Orr, H. A.1999. An evolutionary dead end? *Science* 285: 343–344.

Rassoulzadegan, M., V. Grandjean, P. Gounon, S. Vincent, I. Gillot and F. Cuzin. 2006. RNA-mediated non-Mendelian inheritance of an epigenetic change in the mouse. *Nature* 441: 469–474.

Russell J. A. and N. A. Moran. 2006. Costs and benefits of symbiont infection in aphids: Variation among symbionts and across temperatures. *Proc. Biol. Sci.* 273: 603–610.

Sapp, J. 1987. *Beyond the Gene*. Oxford University Press, New York.

Sarkar, S. 1999. From the *Reaktionsnorm* to the adaptive norm: The norm of reaction 1909–1960. *Biol. Philos.* 14: 235–252.

Sarkar, S. and T. Fuller. 2002. Generalized norms of reaction for ecological developmental biology. *Evol. Dev.* 5: 106–115.

Scarborough, C. L., J. Ferrari and H. C. Godfray. 2005. Aphid protected from pathogen by endosymbiont. *Science* 310: 781.

Schmalhausen, I. I. 1949. *Factors of Evolution: The Theory of Stabilizing Selection*. Blakiston, Philadelphia.

Severtsov, A. N. 1935. *Modes of Phyloembryogenesis.* Quoted in Adams 1980b, p. 217.

Simpson, G. G. 1953. The Baldwin Effect. *Evolution* 7: 110–117.

Thomson, J. A. 1908. *Heredity*. Putnam, New York.

Tokuda, G. and H. Watanabe. 2007. Hidden cellulases in termites: Revision of an old hypothesis. *Biol. Lett.* 3: 336–339.

Vogt, O. 1926. Psychiatrisch wichtige Tatsachen der zoologisch-botanischen Systematik. *J. Psychol. Neurologie* 101: 805–832.

Waddington, C. H. 1942. The canalization of development and the inheritance of acquired characters. *Nature* 150: 563.

Waddington, C. H. 1952 Selection of the genetic basis for an acquired character. *Nature* 169: 278.

Waddington, C. H. 1953. Genetic assimilation of an acquired character. *Evolution* 7: 118–126.

Wake, D. B. 1986. Foreword to *Factors of Evolution. The Theory of Stabilizing Selection* by I. I. Schmalhausen. University of Chicago Press, Chicago, pp. v–xii.

Warnecke, F. and 39 others. 2007. Metagenomic and functional analysis of hindgut microbiota of a wood-feeding higher termite. *Nature* 450: 560–565.

Wheelis, M. 2007. Darwin: Not the first to sketch a tree. *Science* 315: 597.

Wolfe, A. J. 2002. *Speaking for Nature and Nation: Biologists as Public Intellectuals in Cold War Culture.* Ph.D. Dissertation, University of Pennsylvania.

Woltereck, R. 1909. Weitere experimentelle Untersuchungen über Artveränderung, speziell über das Wesen quantitativer Artunderscheide bei Daphniden. *Versuch. Deutsch. Zool. Gest.* 1909: 110–172.

Opening Plate Credits

PART 1

Male dung beetle (*Onthophagus mouhoti*). See pp. 21–23. From *BioScience* 55 (2005), cover photo. Courtesy of A. P. Moczek.

Chapter 1

Photograph: Summer and winter morphs of a female checkered white butterfly (*Pontia protodice*). See p. 16. Courtesy of T. Valente.

H. Frederik Nijhout. 1999. Control mechanisms of polyphenic development in insects. *BioScience* 49: 181–192.

Chapter 2

Photograph: Genetically identical *viable Agouti* mice. See pp. 43–45. Courtesy of R. L. Jirtle.

Alfred North Whitehead. 1919. *The Concept of Nature.* Reprinted 1957 by the University of Michigan Press, Ann Arbor, p. 163.

Greg Gibson. 2008. The environmental contribution to gene expression profiles. *Nature Rev. Genet.* 9: 575–581.

Chapter 3

Photograph: Bobtail squid (*Euprymna scolopes*). See pp. 89–91. Courtesy of Margaret McFall-Ngai.

Lewis Thomas. 1974. *The Lives of a Cell: Notes of a Biology Watcher.* Viking Press, New York, p. 126.

J. Xu and J. I. Gordon. 2003. Honor thy symbionts. *Proc. Natl. Acad. Sci. USA* 100: 10452–10459.

Chapter 4

Photograph: "Orphan embryo" of a European crested newt (*Triturus cristatus*). See pp. 119-121. Copyright Blinkwinkel/Alamy.

John Dryden. 1681. *Absalom and Achitophel*, Part I

PART 2

A wild-caught northern leopard frog (*Rana pipiens*) with extra hindlimbs. See pp. 189–191. Courtesy of the U.S. Geological Survey.

Chapter 5

Photograph: Cyclopic lamb. See pp. 185–186. Courtesy of L. James, USDA ARS Poisonous Plant Research Lab.

Rachel Carson. 1962. *Silent Spring.* Houghton Mifflin, New York.

Chapter 6

Photograph: Underdeveloped and sperm-deficient seminiferous tubules in a male rat, attributed to exposure to the fungicide vinclozolin. See pp. 217–218. Courtesy of Michael Skinner.

Jeremiah 2:7 (King James Version). Quoted in Michael Lannoo (2008), *Malformed Frogs: The Collapse of Aquatic Ecosystems*, University of California Press.

Paul McCartney and John Lennon. 1965. Lyric from *Yesterday*.

Chapter 7

Photograph: Chromosome 1 from 50-year-old identical twins showing differential methylation patterns. See pp. 270–271. Courtesy of M. Esteller.

T. S. Eliot. 1935. *Burnt Norton*. http://www.tristan.icom43.net/quartets/norton.html.

Jay Leno, quoted in S. von Haehling, W. Doehner and S. Anker. 2006. Obesity and the heart: A weighty issue. *J. Amer. Coll. Cardiol.* 47: 2274–2276.

PART 3

Adult tunicates (*Rhopalaea* spp.), Sulawesi, Indonesis. See p. 300. Copyright Norbert Wu/Minden Pictures.

Chapter 8

Photograph: Genetically polymorphic ladybugs (*Adalia bipunctata*). See pp. 297–298. Copyright Papilio/Alamy.

Jean Rostand. 1962. *The Substance of Man*. Doubleday, New York, p. 21.

J. B. S. Haldane. 1953. Foreword to R. Brown and J. F. Danielli (eds.), *Evolution: Society of Experimental Biology Symposium 7*. Cambridge University Press, Cambridge, pp. ix–xix.

Chapter 9

Photograph: Ectopic expression of the *Drosophila Pax6* gene results in an extra pair of eyes. See pp. 326–327. Courtesy of W. Gehring.

David Quammen. 2007. Quoted in "Wordsmiths of Southwestern Montana." *Outside Bozeman* (Winter 2006/2007), p. 21.

Julian Huxley. 1942. *Evolution: The Modern Synthesis*. Allen and Unwin, London, p. 8.

Chapter 10

Photograph: Temperature-mediated color morphs of tobacco hornworm (*Manduca sexta*) larvae. See pp. 384–385. Courtesy of Fred Nijhout.

Charles Darwin. 1876. Letter to Moritz Wagner. Available at http://www.fullbooks.com/The-Life-and-Letters-of-Charles-Darwinx29407.html

Leigh Van Valen. 1973. Festschrift. *Science* 180: 488.

Coda

Fujiwara Yoshikiyo. 1187. *Senzai Wakashu*, Vol. 16. Translated by M. Hoshi.

Index